F. Ehrendorfer (ed.)

Woody plants — evolution and distribution since the Tertiary

Proceedings of a symposium organized by Deutsche Akademie der Naturforscher LEOPOLDINA *in Halle/Saale, German Democratic Republic, October 9–11, 1986*

Springer-Verlag Wien New York

Prof. Dr. FRIEDRICH EHRENDORFER
Institut für Botanik der Universität Wien,
Wien, Austria

Printed in Austria by A. Holzhausens Nfg., A-1070 Wien

With 146 Figures

Special Edition of "Plant Systematics and Evolution", Vol. 162, 1989

ISBN 3-211-82124-4 Springer-Verlag Wien-New York
ISBN 0-387-82124-4 Springer-Verlag New York-Wien

Contents

Ehrendorfer, F.: Introduction 1

Basic principles and examples

Krutzsch, W.: Paleogeography and historical phytogeography (paleochorology) in the Neophyticum 5

History of floras and vegetation types

Christophel, D. C.: Evolution of the Australian flora through the Tertiary . . . 63

Mai, D. H.: Development and regional differentiation of the European vegetation during the Tertiary 79

Palamarev, E.: Paleobotanical evidences of the Tertiary history and origin of the Mediterranean sclerophyll dendroflora 93

Van der Hammen, T.: History of the montane forests of the northern Andes 109

Straka, H., Ohngemach, D.: Late Quaternary vegetation history of the Mexican highland 115

Phylogeny of woody plants

Klaus †, W.: Mediterranean pines and their history 133

Crane, P. R.: Paleobotanical evidence on the early radiation of nonmagnoliid dicotyledons 165

Endress, P. K.: Aspects of evolutionary differentiation of the *Hamamelidaceae* and the Lower *Hamamelididae* 193

Kvaček, Z., Walther, H.: Paleobotanical studies in *Fagaceae* of the European Tertiary 213

Manchester, S. R.: Early history of the *Juglandaceae* 231

Gregor, H. J.: Aspects of the fossil record and phylogeny of the family *Rutaceae* (*Zanthoxyleae, Toddalioideae*) 251

Köhler, E., Brückner, P.: The genus *Buxus* (*Buxaceae*): aspects of its differentiation in space and time 267

Ecogeographical analysis of extant forest floras

Kubitzki, K.: The ecogeographical differentiation of Amazonian inundation forests 285

Browicz, K.: Chorology of the Euxinian and Hyrcanian element in the woody flora of Asia 305

Meusel, H., Jäger, E. J.: Ecogeographical differentiation of the Submediterranean deciduous forest flora 315

Pl. Syst. Evol. 162, 1 – 3

Plant Systematics and Evolution

Introduction

The last decades have witnessed great advances in paleogeography (reconstructing the former distribution and topography of land and sea) and paleoclimatology (inquiring into the climatic changes of the past). Paleobotany has enormously expanded the documentation of fossil plant groups, floras and vegetation types, supporting its conclusions by technically much improved analyses of microfossils (pollen) and anatomical details. An increasing quantity and quality of all these informations from the geosciences is available when we follow the history of the biosphere up to the present. Simultaneously, research from the biosciences on the morphology, ecology, distribution, systematics and evolution of extant vascular plants, and on the ecogeographical differentiation of the vegetation cover of our planet, has made enormous progress. Thus, a synthetic geo- and bioscientific approach becomes more and more feasible and urgent for further advances in the many problems of common concern. Nevertheless, considerable obstacles for an exchange of relevant informations and subsequent cooperative efforts result from the fact, that paleo- and neobotanists usually work in separate institutions, use methods of their own, and publish in different journals.

This situation has stimulated the organization of a symposium by the "Deutsche Akademie der Naturforscher LEOPOLDINA", attractive to both, paleo- and neobotanists. Under the title "Evolution und Arealbildung temperater und tropischer Holzpflanzen vom Tertiär bis zur Gegenwart" (Evolution and distribution of temperate and tropical woody plants from the Tertiary to the present), this symposium was held from October 9 – 11, 1986 in Halle/S., G.D.R. The president of the Academy, H. BETHGE opened the meeting, greeted the participants which had come from all over the world, and stressed the traditional efforts of the Academy LEOPOLDINA in supporting the fields of contact and international cooperation between different scientific disciplines. After a brief introduction by F. EHRENDORFER the scientific sessions started with contributions grouped under four headings:

I. Basic principles and examples ("Allgemeines zur Problematik der tertiären Paläogeographie, Klima-, Floren- und Vegetationsgeschichte")

W. KRUTZSCH (Berlin): Paleogeography and historical phytogeography (paleochorology) in the Neophyticum ("Paläogeographie und historische Pflanzengeographie im Neophytikum").

II. History of flora and vegetation types ("Klima-, Floren- und Vegetationsgeschichte seit dem Tertiär")

D. C. CHRISTOPHEL (Adelaide): Evolution of the Australian flora through the Teriary[1].

[1] Not personally presented during the symposium.

D. H. MAI (Berlin): Development and regional differentiation of the European vegetation during the Tertiary ("Vegetationsentwicklung und regionale Differenzierung im Tertiär Europas").

E. PALAMAREV (Sofia): Paleobotanical evidence for the Tertiary history and origin of the Mediterranean sclerophyll dendroflora ("Paläobotanische Hinweise auf die Entstehung mediterraner Hartlaubelemente im Tertiär").

F. SCHAARSCHMIDT (Frankfurt/M.): The flora of Messel – an important component of the Eocene vegetation in C. Europe ("Die Flora von Messel – ein wichtiges Glied der eozänen Vegetation Mitteleuropas")[2].

D. DILCHER (Bloomington, Ind.): The Eocene flora from SE. North America and its affinities[2].

T. VAN DER HAMMEN (Amsterdam): History of the montane forests of the northern Andes.

H. STRAKA & D. OHNGEMACH (Kiel): Late Quaternary vegetation history of the Mexican Highland ("Spätquartäre Vegetationsgeschichte im Hochland von Mexiko").

III. Phylogeny of woody plant groups ("Die Fossilgeschichte und Differenzierung einzelner Holzpflanzengruppen vom Tertiär bis zur Gegenwart")

W. KLAUS † (Vienna): Mediterranean pines and their history ("Zur Geschichte der Mediterran-Kiefern")[1].

P. R. CRANE (Chicago): Paleobotanical evidence on the early radiation of non-magnoliid dicotyledons (extended version of the original contribution "*Trochodendron, Cercidiphyllum,* and their allies").

P. K. ENDRESS (Zürich): Aspects of evolutionary differentiation of the *Hamamelidaceae* and the Lower *Hamamelididae* ("Zur systematischen Differenzierung und Evolution der *Hamamelidaceae* und verwandter Familien").

H. WALTHER & Z. KVAČEK (Dresden, Prague): Paleobotanical studies in *Fagaceae* of the European Tertiary ("Paläobotanische Untersuchungen zur Entwicklung der *Fagaceae* in Europa seit dem Tertiär").

E. J. ROMERO (Buenos Aires): Our present understanding of *Nothofagus,* its history and relationships[3].

S. R. MANCHESTER (Bloomington, Ind.): Early history of the *Juglandaceae* ("The fossil history and evolution of the *Juglandaceae*").

W. MORAWETZ (Vienna): Differentiation patterns of some extant *Magnoliidae* groups as a tool for historical interpretations[2].

H. J. GREGOR (Munich): Aspects of the fossil record and phylogeny of the family *Rutaceae* (*Zanthoxyleae, Toddalioideae*) ("Beiträge zur Fossilgeschichte und Phylogenie der *Rutaceae*").

E. KÖHLER & P. BRÜCKNER (Berlin): The genus *Buxus* (*Buxaceae*): aspects of its differentiation in space and time ("Die Differenzierung und räumliche Entfaltung der Gattung *Buxus, Buxaceae*").

IV. Ecogeographical analysis of extant forest floras ("Vergleichende chorologisch-ökologische Florenanalysen")

K. KUBITZKI (Hamburg): The ecogeographical differentiation of Amazonian inundation forests ("Die ökogeographische Stellung der amazonischen Überschwemmungsflora – ein Beitrag zur Pflanzengeographie Amazoniens").

K. BROWICZ (Kórnik, Poland): Chorology of the Euxinian and Hyrcanian element in the woody flora of Asia[1].

H. MEUSEL & E. JÄGER (Halle/S.): Ecogeographical differentiation of the Submediterranean deciduous forest flora ("Submediterrane Gehölzsippen in ihrer ökogeographischen Differenzierung").

[1] Not personally presented during the symposium.
[2] Not included in this volume.
[3] To be presented in one of the next issues of this journal.

The scientific sessions were concluded by an extensive general discussion and framed by an attractive social and cultural program, offered by the Academy LEOPOLDINA. All this contributed to a very pleasant atmosphere during the week in Halle/S. and greatly stimulated personal contacts as well as an exchange of ideas and informations.

Thanks are due to the president, H. BETHGE, and the vicepresident, B. PARTHIER, of the Academy LEOPOLDINA, and to all their collaborators for the perfect organization of the symposium and for their wonderful hospitality in Halle/S. For great help in preparing the scientific program and in establishing the necessary international contacts I want to express my gratitude to the co-organizer H. MEUSEL (Halle/S.) as well as to E. KÖHLER (Berlin), D. H. MAI (Berlin), R. SCHUBERT (Halle/S.), H. WALTER (Dresden), and W. MORAWETZ (Vienna).

The main results of our LEOPOLDINA Symposium are now presented in this special volume of "Plant Systematics and Evolution". That this has become possible in spite of considerable difficulties is due to the joint efforts of authors, translators, the editorial staff, and the publishers; this is gratefully acknowledged. We all hope that these results will stimulate the international scientific community to more attention and cooperation in respect to the somewhat neglected problems concerning the historical development of our present plant world.

F. Ehrendorfer

Pl. Syst. Evol. 162, 5–61

—Plant—
Systematics
and
Evolution

Paleogeography and historical phytogeography (paleochorology) in the Neophyticum*

WILFRIED KRUTZSCH

Received February 5, 1987; in revised form June 25, 1988

Key words: Angiosperms, *Nypa, Ctenolophon, Bombacaceae, Olacaceae, Symplocos, Alangium, Nepenthes, Pachysandra* and *Sarcococca, Restionaceae, Chloranthaceae, Austrobaileyaceae.* – Paleogeography, paleochorology, chorology, Neophyticum.

Abstract: After presenting the methodological bases of historical phytogeography (paleochorology; Fig. 1), a new paleographic map series is introduced (Fig. 2). They serve for ten examples of paleochorological analyses, in which the major phases of the development of a group in space and time are documented. In addition to purely tropical taxa (*Nypa, Ctenolophon*), others which originated in the warm and humid N. Tethyan belt (*Bombacaceae* p.p., *Olacaceae* p.p., *Symplocos, Alangium, Nepenthes*) or which presently occur as eastern N. American/E. Asian disjuncts (*Sarcococca* and *Pachysandra*) are discussed (Figs. 3 – 14). Finally, analyses of two very old angiosperm groups with complicated paleochorological patterns (*Restionaceae* and *Ascarina, Chloranthaceae* p.p.; Figs. 15 – 16) are presented.

Efforts to clarify the evolution of plant groups in space and time have a long tradition, but only during the last 50 or 60 years have relevant methods been so improved as to allow deeper insight into the problem. Even the documentation of recent distribution patterns in Higher Plants has reached a satisfactory level only in the temperate N. hemisphere, whereas much is still to be done in the tropical regions. Paleobotanical information depends on the available fossil material, i.e., its quantity, level of analysis, reliability of assignment, and particularly its representation in space and time. Even the current methods of a comparative interpretation of historical and recent data are research products of the last decades. Exemplary in this respect are the efforts for a world-wide synthetic approach by AXELROD and his collaborators. In the earlier studies by VAN STEENIS, MELVILLE, SMITH and many others, the present distribution areas of plant groups formed the basis and all fossil data were related to it (landbridge and drift theories, etc.). In contrast to this, a primarily historical approach had already begun with the theory of continental drift as suggested by A. WEGENER. Today, it is our task to continue with this approach and develop the theories about the differentiation of floras and individual plant groups. Furthermore, remarkable progress has been made during the last 30 to 40 years in understanding the geological and paleogeographical

* Part I, to be followed by part II in this journal.

development of the world's surface; this knowledge should also be fully used for our purpose. Therefore, a methodological chapter concerning the basis of paleochorology and paleogeography has been placed in front of this contribution.

Methodological bases of paleochorology

All phytogeographical studies are preceeded by a series of analytical steps, only the last of which actually lead to what can be called historical phytogeography or paleochorology (cf. Fig. 1).

1) Description of the present condition of extant taxa (families, tribes, genera, subgeneric taxa, species, etc.) with respect to their definition, affinities, and distribution. An already classic example is the chorological atlas for Europe by H. Meusel & al. (1965). Such data form the basis of all further discussions and interpretations. This also applies to paleobotany, for which much less material is available.

2) Comparative structural analysis of taxa and distribution areas with respect to their evolution and phylogeny. Interpretation of centres of diversity, cladistic analysis of taxa, origin of distribution boundaries, comparative analysis of area types. A search for the origins of these phenomena in the past begins. These time indications are not definite and remain relative, except for certain recent ice-age taxa. This is the first step towards historical phytogeography.

3) Occasional use of fossil evidence at the following stage of analysis often is uncritical and at random (usually only in cases where fossils corroborate the theory in question). Usually, the fossils are not taken seriously and serve as "alibis". This is partially due to many older "imaginative" determinations and placements of fossils, and to the special paleobotanical nomenclature for different organs. It is also due to the lacking or incomplete exchange of information between botanists and paleobotanists. The latter are, in many countries, organized as geo-scientists, the former as bio-scientists. The present Leopoldina Meeting is an important step forward to improve this situation.

4) Fossils as evidence for the age of taxa are of general interest to the botanists concerned with phylogenetics. This is illustrated, i.a. by the synthetic studies of J. Muller (1970, 1981), in which the first occurrence of taxa has been summarized in tabular form. Muller also was among the first to indicate geographical pathways for several fossil groups (e.g., *Clavatipollenites,* p. 47 – 50).

5) Use of fossils to explain present geographical disjunctions as a method of analysis has been frequently employed in the last two decades in botany and paleobotany by entering the fossil evidence for taxa in extant distribution maps. Examples are *Mastixioideae, Symplocaceae, Magnoliaceae,* and many others (cf., contributions by Mai, Axelrod, etc.). European paleobotanists often have studied taxa which today exhibit the "ASA GRAY disjunction" (i.e., eastern N. America and SE. Asia, with fossil evidence in the Tertiary of Europe).

6) Fossils of certain taxon may be documented on separate maps according to their historical position. So far, one has mostly used modern geographical maps for this, only occasionally including paleogeographic references (e.g., on land and sea in former geological periods). An example thereof are the works by Tralau in the late 60's on *Ginkgo* and its Arctotertiary forms, and the comprehensive maps by Vachrameev and other Soviet authors in the 60's and 70's.

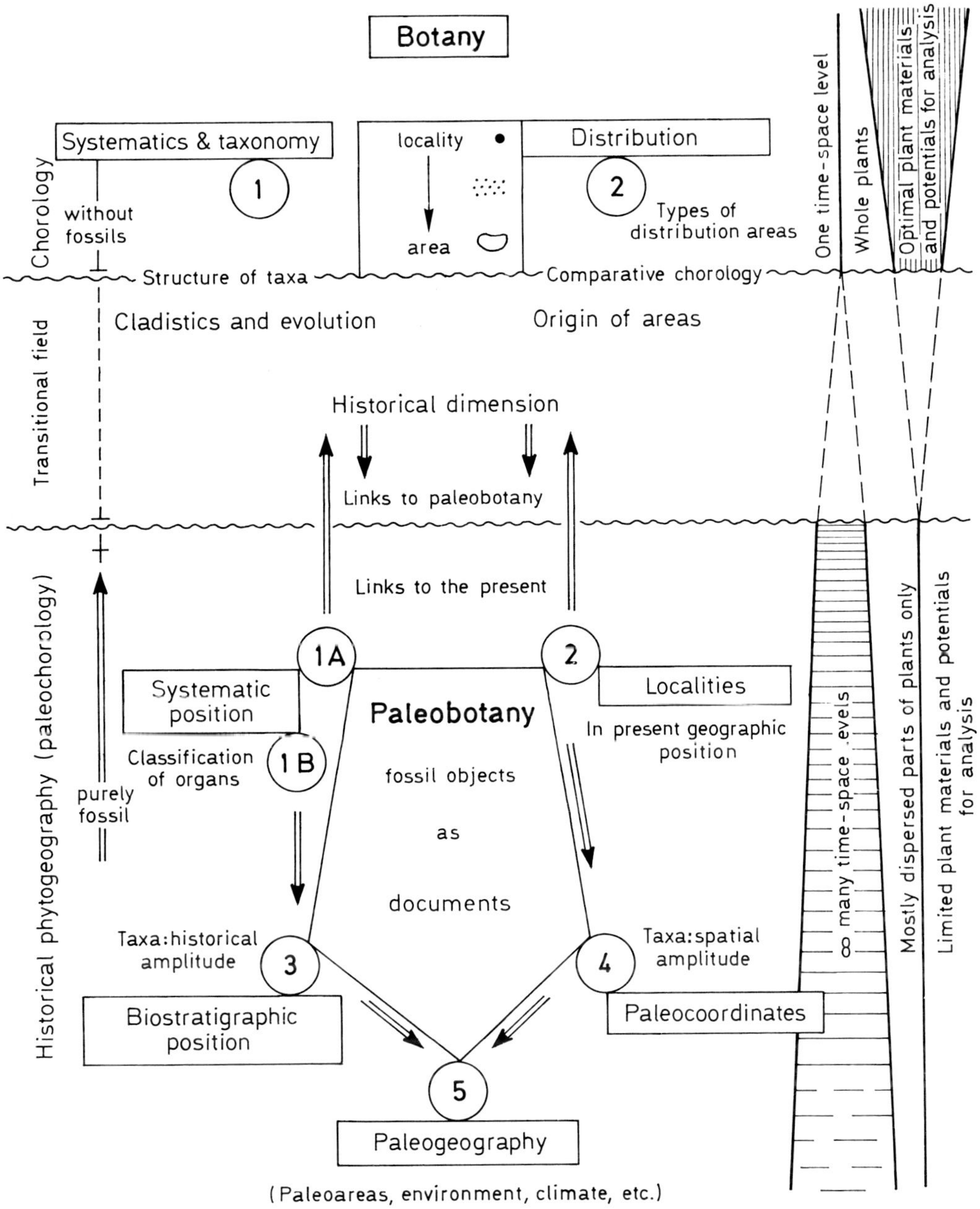

Fig. 1. Relationships between the various elements of paleochorology

7) Single taxa may be represented on a series of separate maps according to their historical development from earliest occurrences to the present. These maps should contain available paleogeographic and also paleoecological evidence. This requires a large amount of fossil material from different time periods and areas, a goal at present attainable only for a small number of taxa. Furthermore, satisfactory paleogeographic maps have only recently become available (Krutzsch, in press b; and Fig. 2).

As is shown in Fig. 1, historical phytogeography rests on five foundations, in contrast to present-day chorology which is based on taxonomy and distribution alone. The three additional aspects of importance for problems of historical phytogeography are: historical position, paleocoordinates, and paleogeography.

For an extant analysis infinite data and samples are available and the methods of investigation are well-developed. Fossil finds, however, are relatively few and from many different time periods. Their analysis is hindered by their derivation from many different organs and by restricted methods of investigation. The whole plant and its environment (climate, soil, etc.) form the basis for all analyses. For fossils all this evidence is quite difficult and often only indirectly to obtain. Nevertheless, fossils are the most important documents for real evolution.

Paleobotany thus complements neobotany (e.g., its cladistic theories) in the historical dimension, and only both together can hope to solve phylogenetical problems (MEUSEL 1978). Paleobotany, therefore, must closely link with the extant analyses of taxonomy and historical phytogeography which often shed light on paleobotanical findings.

There are also fossil plant groups which have not survived to the present day or for which no extant representatives are known. Their morphology and occurrences allow conclusions as to the accompanying flora and relevant patterns of temporal/spatial distribution. If, on the contrary, little or no fossil evidence is available, phylogenetic analyses are still possible with the help of taxonomic analyses, modern chorological data, and reference to available paleobotanical information on vegetation development (e.g., during the Upper Cretaceous). An example is the polymorphic genus *Rhododendron* (KRUTZSCH, in press a).

Paleogeographic maps

Up to now, no reliable map series has been available for paleogeographic survey analyses. Such map series are necessary for proper phytogeographic investigations and must take into account recent advances in the understanding of plate tectonics, geophysics, paleomagnetism, marine geology, etc.

Modern plate tectonics theory has only the idea of mobility in common with the old Wegenerian theory of continental drift, to which many biologist have remained loyal during the "crisis" years in the 50's and 60's. Yet, some of the continental positions do not differ widely between the new and old theories, especially with respect to Gondwana.

To correctly understand the temporal/spatial development of cenophytic vegetation, also with respect to climate, the period from the beginning break-up of the Pangaea to the present must be considered. This spans about 200 million years, the latter two-thirds of which correspond to the Cenophyticum. During the last three years, the author has attempted to illustrate the continent/ocean and land/sea distribution in a series of 15 maps (Fig. 2). The sequence starts with time spans of 20 million years and continues after the Eocene with increasingly shorter spans of 10, 5, and 2 million years. Major glaciated areas are described for the Pleistocene period. Details are explained elsewhere (KRUTZSCH, in press b).

The positional relationships between the continents and oceans rely heavily on the map series by SMITH & al. (1981); some aspects (e.g., those concerning the Beringia area) have been taken from BARRON & al. (1981). Constructions by ZIEGLER (1982) and RÖGL & STEININGER (1983), as well as older data and records were used (TERMIER, THENIUS, SCHWARZBACH, KRUTZSCH & LOTSCH, SCHUCHERT, WEYL, and others).

- ● locality of representative and documented occurrence (fossil or recent)
- M macrofossils (mostly leaves)
- ? questionable occurrence
- + extinct fossil occurrence
- — lacking in corresponding rich fossil floras
- glaciated areas
- pack-ice
- drift-ice
- taiga, loess steppe, tree-line
- documented fossil and/or recent distribution area: numbers refer to pollen type or species numbers
- potential fossil distribution area
- centres of diversity
- distribution boundaries: ascertained/assumed

- direction of expansion or retreat: ascertained
- direction of expansion or retreat: assumed
- direction of plate drift
- VVV volcanic island chains (arcs)
- .... isolation by sea
- mountain barriers (xeric)
- xeric and very xeric areas
- semi-dry palm savanna areas
- N. and S. limits of evergreen laurophyll zone

Sa	Santonian	
LEo	Lower	Eocene
MEo	Middle	Eocene
UEo	Upper	Eocene
Ol	Oligocene	
Mi	Miocene	
Pl	Pliocene	
Qu	Quaternary	
Pt	Pleistocene	
Ho	Holocene	

Key to paleographical and paleochorological maps (Figs 2a–o, 4, 6–9, 11, 12, 14–16)

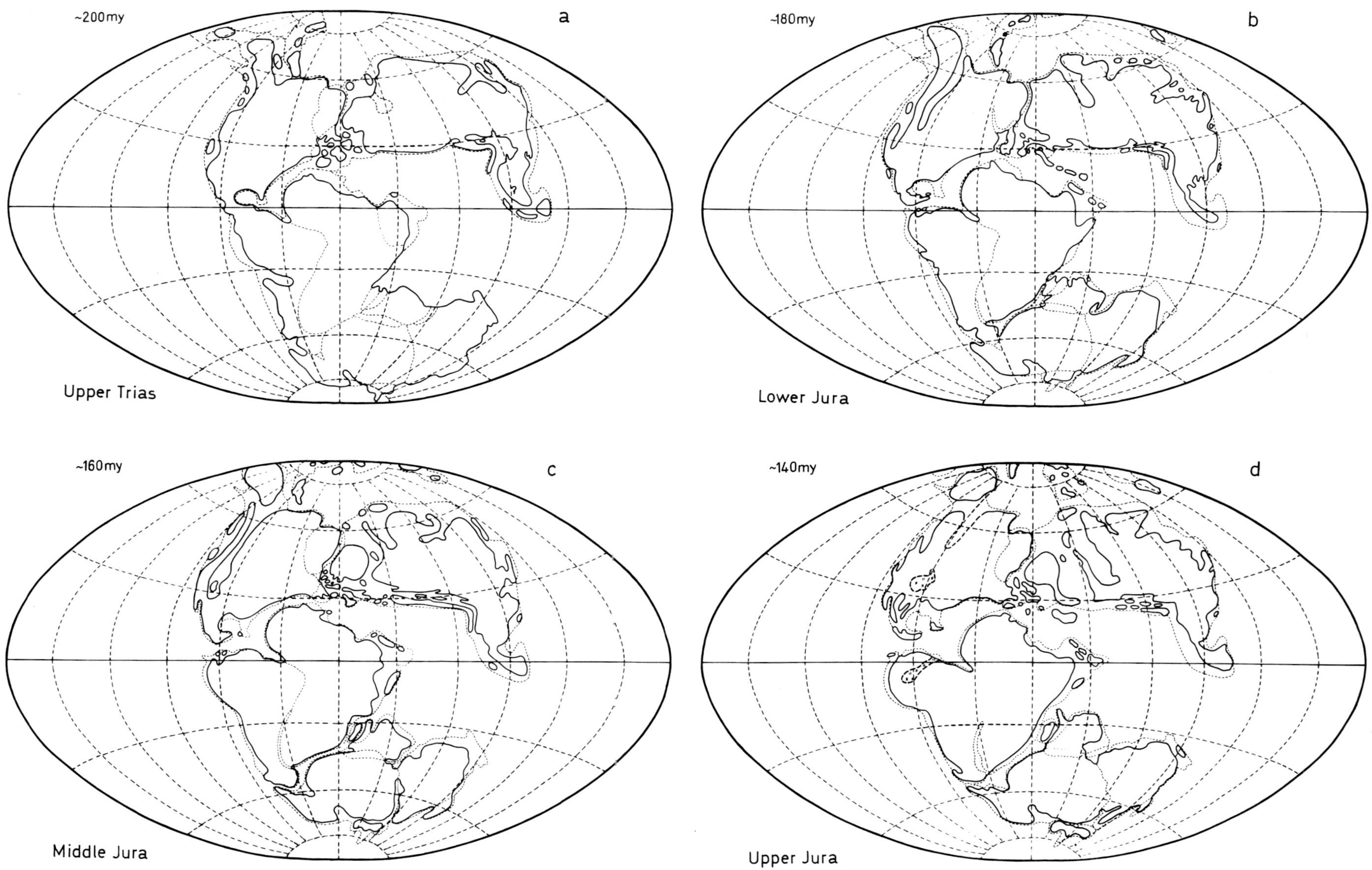
~200my
a
Upper Trias
~180my
b
Lower Jura
~160my
c
Middle Jura
~140my
d
Upper Jura

Fig. 2

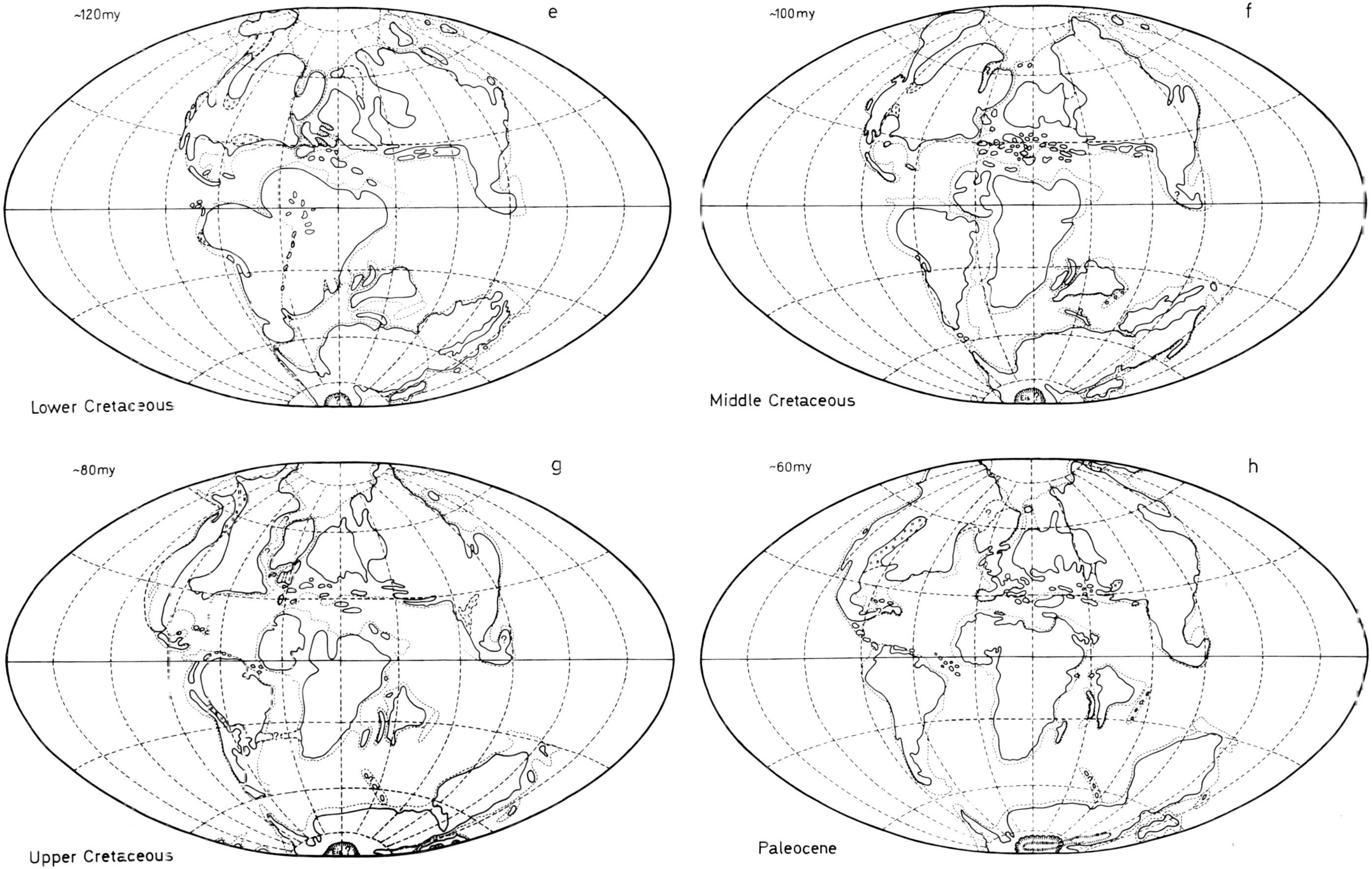
~120my
e
Lower Cretaceous
~100my
f
Middle Cretaceous
~80my
g
Upper Cretaceous
~60my
h
Paleocene

Fig. 2

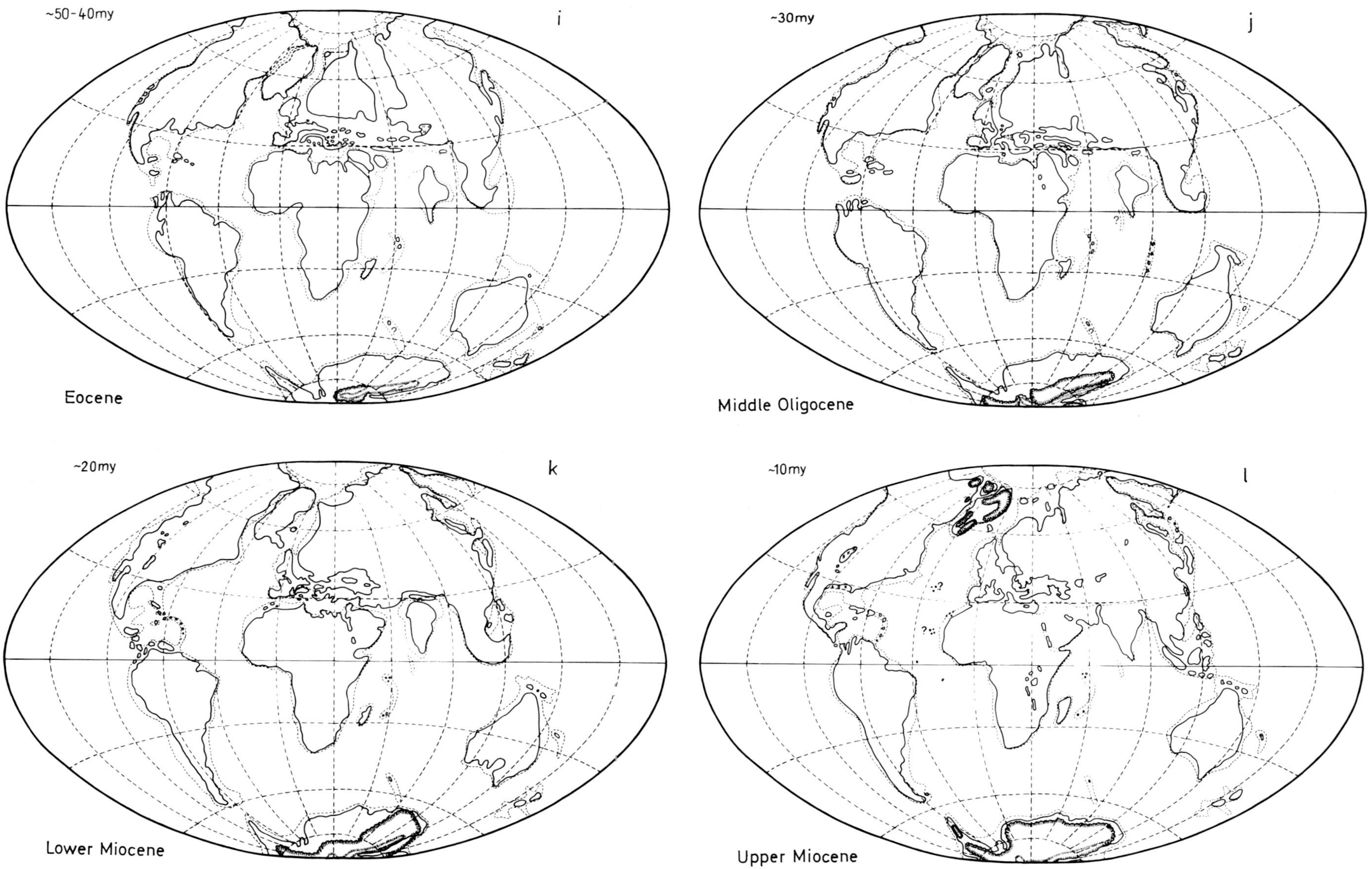
~50-40my
i
Eocene
~30my
j
Middle Oligocene
~20my
k
Lower Miocene
~10my
l
Upper Miocene

Fig. 2

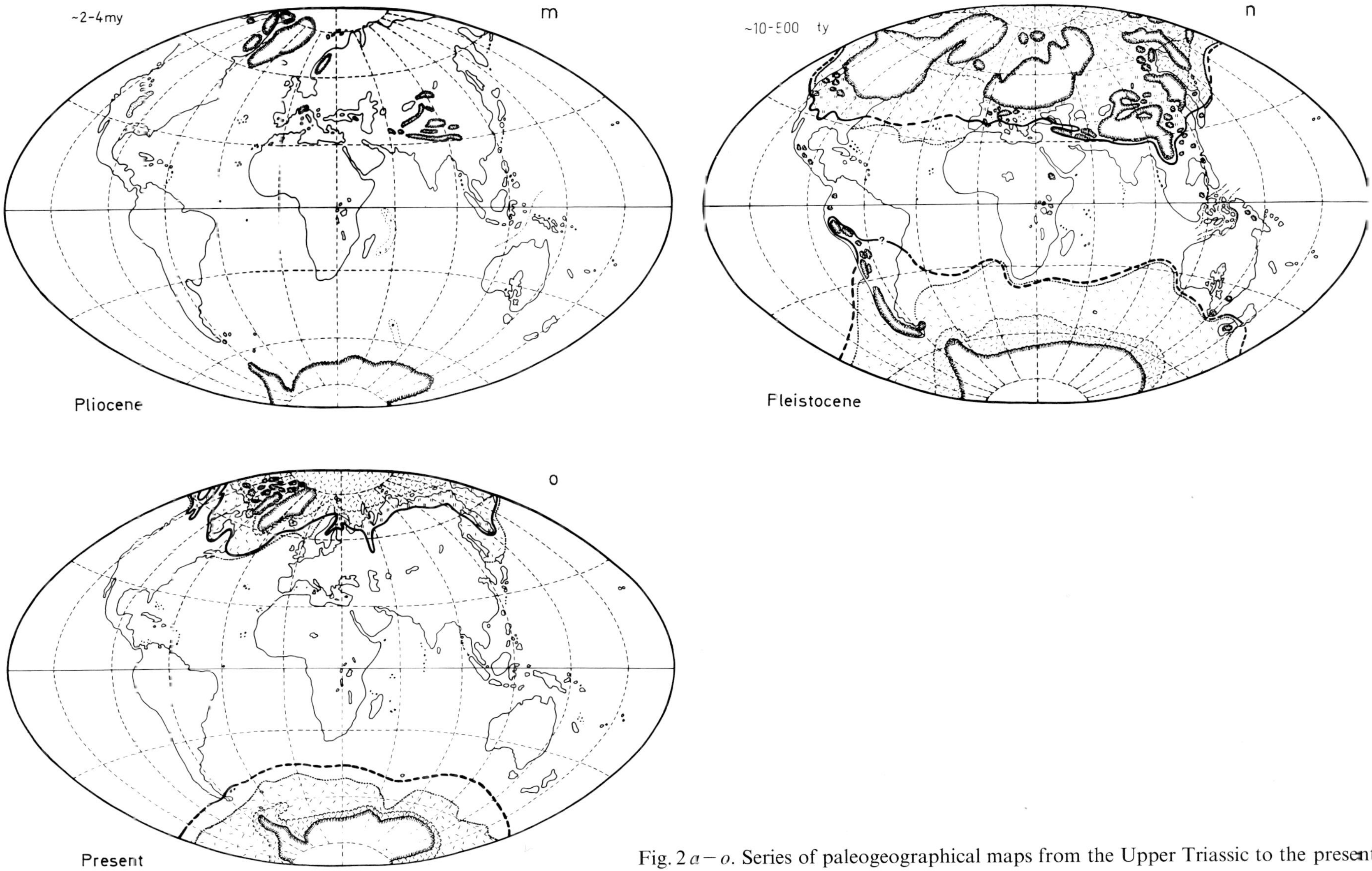

Fig. 2 *a* – *o*. Series of paleogeographical maps from the Upper Triassic to the present

It is difficult to depict the shelf areas and the constantly changing coast lines, continuously undergoing local transgressions and regressions. It is impossible to depict them in time periods of 20 million years. These maps, therefore, are only a temporary solution until other maps for shorter time periods have been worked out. The weak points and the limits of such maps are most obvious to the author; nevertheless, they had to be constructed for the purpose of historical phytogeographical analyses.

It was necessary to have one map for each time period, and to ensure minimal distortions. This was not possible with either Mercator projections or separate representations of the polar and equatorial areas. Small distortions in the N/S or E/W directions were unavoidable, since the maps were constructed without a computer. This should, however, have no effect on the map's usefulness. It is important to have at least provisional methods which allow the arrangement of complex data into map form and facilitate interdisciplinary comparisons of temporal/spatial analyses. Data may be of a varying nature, e.g., paleobotanical, paleozoological (continental and marine groups), or geological (climate zones, moor and coal extension, saline areas, weathering formations, mountain ranges, vulcanic areas, etc.). In a few years, one should be able to determine where changes and improvements in the maps are necessary, e.g., developing new concepts for certain areas such as SE. Asia, parts of S. America, and Antarctica.

For a paleochorological representation of angiosperms, it is necessary to refer to the Upper, and sometimes to the Middle and Lower Cretaceous. For conifers and other gymnosperms one has to rely upon data from the Mesophyticum, because of their earlier origins. The same is true for many fern groups and Lower Plants.

Examples of paleochorological analyses of single taxa

Approximately 30 examples have been prepared or are in preparation, ten of which will be presented here. They all relate to already well-studied developmental models and "typical" differentiation processes. All available data have been taken into account; however, certain gaps in the evidence are unavoidable, due to ongoing research. This information flow will necessitate future revisions, especially in details. Nevertheless, the maps are not hypothetical constructions as was once usual in the absence of fossil evidence: they are based exclusively on historically classified fossils and corresponding paleogeographic features.

The following examples rely partially on pollen evidence and partially on a combination of evidence from several organ groups (pollen, fruits, and seeds; pollen, fruits, seeds, and leaves; etc.). Examples based solely on leaf or fruit remnants are not presented here, as they should be analyzed by specialists in the future (e.g., in such important families as the *Mastixioideae, Lauraceae, Theaceae, Monimiaceae, Aceraceae,* etc.). On the other hand, work on *Fagaceae, Myricaceae, Platanaceae, Proteaceae,* etc., has not been completed and will be dealt with in the second part of this contribution.

Each example is illustrated by different numbers of maps, depending upon the available data, the distribution pattern, and the rate of change. The causes of the different rates in the change of the paleoareas is not clear, excepting the major paleogeographic and climatological changes. Emphasis should be placed on mountain barriers, dry or wet areas, land bridges and island chains as linking elements, and sea straits as dividing elements, etc. Concrete evidence, where available, has been presented and discussed in the examples.

The taxonomic categories covered by the different examples are quite varied, ranging from entire families to single genera, depending on the fossil evidence.

Even with monotypical relics, one encounters problems (ANDREANSKY 1951). In some cases, generic groups are summarized under their fossil names (based on certain organs only), when they cannot be classified separately.

The records are accompanied only in part by original citations. Otherwise, references concern important synthetic works (e.g., J. MULLER 1981, HERNGREEN & CHLONOVA 1981, and the Indian-French contribution of THANIKAIMONI & al. (1984). In many cases it was unimportant under how many, or what fossil names or "artificial" terms a group was classified, as long as it was clear to which taxon it belonged. Proper caution must of course be taken to avoid false classifications.

Nypa (*Arecaceae*)

Today, *Nypa* is a monotypic genus. The fossil record mainly consists of fruits (known since the late 18th century) and pollen findings (from the last 30 years). For surveys, the reader is referred to TRALAU (1964), MULLER (1964, 1970, 1974, 1979, 1981), MOORE (1973), THANIKAIMONI & al. (1984), KAR (1985), HERNGREEN & CHLONOVA (1981), and GERMERAAD & al. (1968).

For paleoclimatology it is of importance that *Nypa*, a member of the mangrove vegetation, grows in a ± frost-free climate only. Today, *Nypa* is ecologically narrowly restricted to the tropical coast of SE. Asia. So far, fossil findings have been combined into single maps (e.g., KAR 1985: Fig. 32).

Fossil fruits of *Nypa* are presently classified as a single species, but pollen normally can be divided into two types (*N. echinatus* and *N. baculatus:* MULLER 1968), and more detailed analyses allow further differentiation. In the Upper Cretaceous flora of S. Arabia, four pollen species have been identified on the basis of spinosity types; one of them is identical with the extant *N. fruticans* (Fig. 3). Further pollen types from the Upper Cretaceous and the Lower Tertiary can be grouped close to the *Nypa* type (e.g., *Grimsdalea* GERMERAAD & al. 1968); undoubtedly, they also belong to the palms, but it is not known whether their ecology corresponded to that of *Nypa*. Thus, it is possible that in these early periods of mangrove formation, other palms grew in this vegetation type and were eliminated later by other angiosperms. Many typical members of the present-day mangrove formation, e.g., *Sonneratiaceae, Rhizophoraceae, Brownlownia, Avicennia, Acanthus,* etc. first appeared in this formation during various periods of the Tertiary (from the Eocene to the Miocene) and have gradually given it its present composition. Not all of these typical mangrove plants have yet been found as fossils. Furthermore, it is relevant that the extant mangroves exhibit a remarkable floristic and regional differentiation (cf. numbers in Fig. 4c from WALTER 1962: Table 40). It appears very unlikely that *Nypa* was not originally a member of the mangrove formation and adapted only later to this special habitat (MULLER 1979, KAR 1985).

It is possible to distinguish three or four main phases in the space and time differentiation of *Nypa* (cf. Fig. 4). The first phase dates back to the upper Cre-

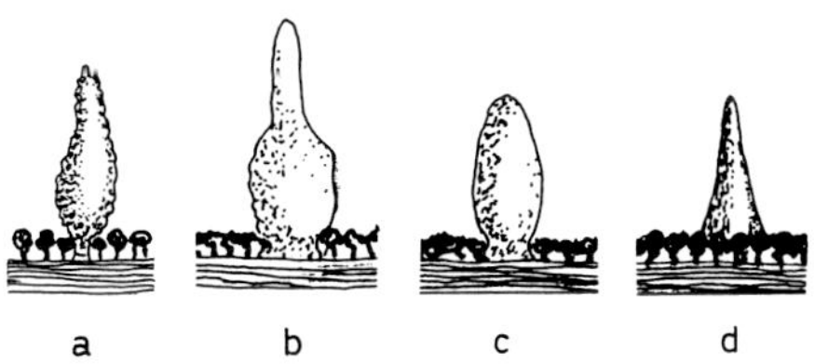

Fig. 3. Exine details of diverse *Nypa* (*Arecaceae*) pollen types from the Middle to Upper Cretaceous of S. Yemen. *a* Spec. 1, *b* spec. 2, *c* spec. 3, *d* spec. 4 (= *N. fruticans*); × > 4000

taceous and is documented by numerous fossil finds from the C. African and S. American Atlantic coasts. Further findings are from northwestern S. America (Peru, Venezuela, Columbia) and from the coast of the S. Tethys (e.g., Egypt: CHANDLER 1954, SCHRANK 1984: S. Arabia, and the Indian Maastrichtian from Bengal and Cauvery: VENKATACHALA & SHARMA 1974). Thus, the oldest known occurrence of *Nypa* is along the coasts of NW., N. and E. Gondwanaland. Because all these finds come from coastal areas, there is no doubt about the mangrove nature of these early *Nypa* populations.

In the N. Tethys coastal area, which breaks up into islands in the south, Upper Cretaceous reef formations (ZIEGLER 1982) but no *Nypa* specimens have been found. Judging from the reefs, mangrove formations would have been paleoclimatologically possible. *Nypa* has been found in SE. Asia in the transition from Cretaceous to Tertiary (Borneo: MULLER 1968; but cf. HERNGREEN & CHLONOVA 1981), whereas its existence in the Santon of Japan is rather unlikely, considering the object itself and the paleoclimatological situation at that time (MIKI 1972: cf. also MULLER 1981: 12).

Most finds relevant to the second phase (Eocene) were no longer made in the coastal areas of the C. S. Atlantic and the N. Gondwanaland coast (MOHR 1986: S. Atlas, GREGOR & HAGN 1982), but rather in the Tethys areas including the N. Tethyan coastal zone (N. Italy) and the epicontinental shelf coasts in S. Laurasia (Londonian, Parisian, Belgian, N. German basins, Helmstedt and Halle areas, Poland, Hungary, Ukraine). Thus, a shift towards the N. of the mangrove belt area (with *Nypa*) took place, which corresponds to the warmth optimum during the Tertiary.

In southern N. America, only dubious fragments have been found and the finds in Texas have been criticized several times (TRALAU 1964), although the climatic conditions (W. end of the N. Tethys) give no cause for doubt.

The other directions of expansion are towards the E. and along the W. Pacific edge, towards the S. (Australia: KEMP 1981) and N. (Japan). In India, *Nypa* occurs from the Miocene upwards (KULKARNI & PHADTARE 1981); in Kutch (NW. India) it occurs only as far back as the Oligocene (KAR 1985).

In the third phase (Miocene), *Nypa* disappears almost completely from the species-poorer mangroves of the W. hemisphere. An exception is the entry by KEDVES (1986: 124; in reference to BIAGGI & al. 1976) for the Oligocene/Miocene of Chiapas, Mexico. It has been found outside of its present area only in NE. India (Bengal) and in S. Arabia. In its present distribution area, the mangrove contains the most species (15 or more representatives; WALTER 1962). In the Pliocene or Pleistocene, *Nypa* probably reached its present area in the E., from New Guinea to the Salomon Islands.

Summary. The mangrove palm *Nypa* is a very old representative of the tropical coastal vegetation. Its main dispersal occurred along the edges of the middle S. Atlantic during the Cretaceous and the coastal areas of the Tethys to the E. and W. during the early Tertiary. Its area narrowed towards the E. and almost reached its present status in the Miocene. After the end of the upper Cretaceous, it probably spread constantly towards the E. in tropical regions, whereas its area in the N. (Japan) and S. (Australia) lasted only a short time during the Eocene climatic optimum. A general displacement of the area from the western to the eastern tropics is evident.

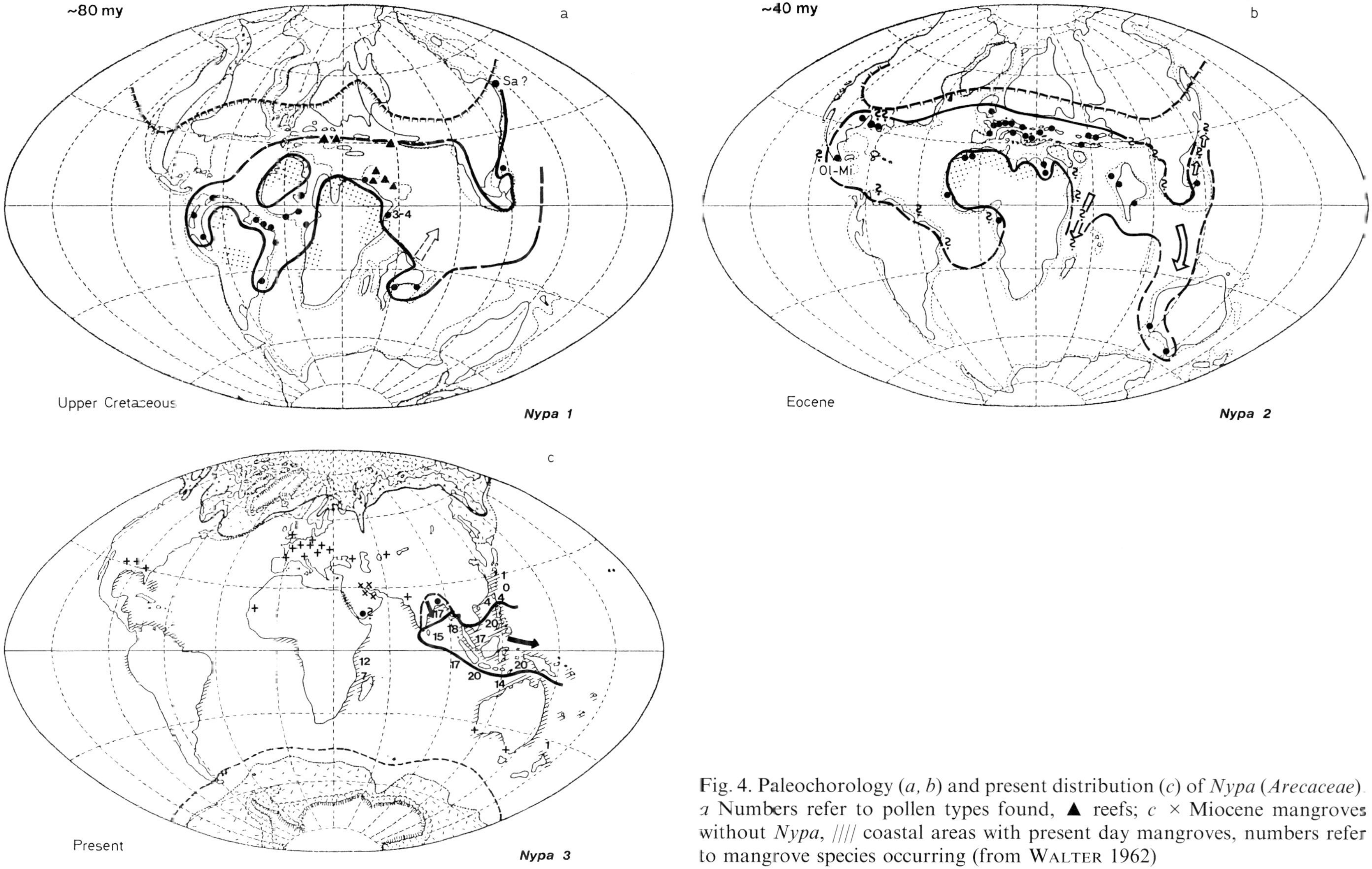

Fig. 4. Paleochorology (*a*, *b*) and present distribution (*c*) of *Nypa* (*Arecaceae*). *a* Numbers refer to pollen types found, ▲ reefs; *c* × Miocene mangroves without *Nypa*, //// coastal areas with present day mangroves, numbers refer to mangrove species occurring (from WALTER 1962)

Ctenolophonaceae

Ctenolophonaceae, in former sense a subfamily of *Linaceae,* contains only the genus *Ctenolophon,* presently with five species, one (*C. englerianus*) in C. and W. Africa, the other four (including *C. philippinensis* and *C. parvifolius*) in SE. Asia. All are trees in fresh water swamps and humid forests (AUBREVILLE 1969, BADRE 1973). The only fossils are characteristic polycolporate pollen grains; extant pollen conforms to two main morphotypes, the *C. parvifolius* and the *C. englerianus* types (SAAD 1962; MULLER 1968, 1981; THANIKAIMONI & al. 1984; etc.). The former is finely structured, the latter has bulging exine thickenings or foldings in the sectors between the colpi, which meet at the polar regions and create diverse patterns. The fossils can also be distinguished and defined on the basis of size, exine sculpture and thickness, annulus thickening around the little-studied germination points, etc.

No comparative morphological analyses are yet available which fully take into account the differences at the species level of the fossil and extant material. A good beginning has been made by THANIKAIMONI & al. (1984). Approximately twenty fossil pollen types, mostly from the Indian Miocene, have been described and classified in a variety of "pollen genera" with more than 15 species described. Because such ctenolophonoid pollen forms appear nowhere other than in this family, classification into the single genus of the family should be sufficient. A more thorough and comparative morphological analysis of the fossil specimens should further characterize the taxa, ascertain possible synonyms, etc. The varied *C. costatus* group deserves special attention and definition.

The classification of the extant family into two main pollen morphotypes is mostly supported by the still incomplete fossil evidence. Because the extant species are evolutionary dead-ends, it is impossible to trace their development without relying upon the fossil evidence (e.g., MULLER 1981).

As far as is known (Fig. 5), the following evolutionary trends are recognizable: (1) Apertures (colpi) increase from the basic number 6 to 7, 8, 9, and even 10. The original symmetry is thereby lost. (2) The polycolporate, non-elaborate *C. parvifolius* type appears to be the basic morphotype; from it the elaborate forms of the *C. englerianus* type could have arisen through folding and thickening of the exine. However, according to present knowledge, the simple bulged type appears earlier (Maastrichian). Yet, a little later in the Paleocene, both developmental lines are found together. (3) Increase in wall thickening and size (?) of the *C. parvifolius* type, which also differentiates with respect to structure and sculpture. (4) The main evolution concerns the *C. englerianus* type; it has progressed into different, sometimes opposing directions, shown in Fig. 5. This should facilitate our future understanding of the problem and encourage classification of the *C. costatus* group. Structure and thickness of the bulges and their distribution among pollen are of utmost importance, and the two fields may vary in their differentiation.

The trends recognized by KAR (1985) now appear to be even more differentiated and variable. More thorough analyses of the individual types, with special attention to apertures, are necessary. The author (e.g., in KRUTZSCH & VAN HOORNE 1977) has long carried out and called for such investigations, which cannot be made with SEM alone. A good start in this direction has already been made by HOEKEN-KLINKENBERG (1964). Finally, it should be noted that the extant pollen forms classified as *C. englerianus* by GERMERAAD & al. (1968: Plate XV, Figs. 1 – 2) and in THANIKAIMONI & al. (1984: Plates 9 and 10) do not coincide with respect to their morphology. In one case an almost closed bulged ring on

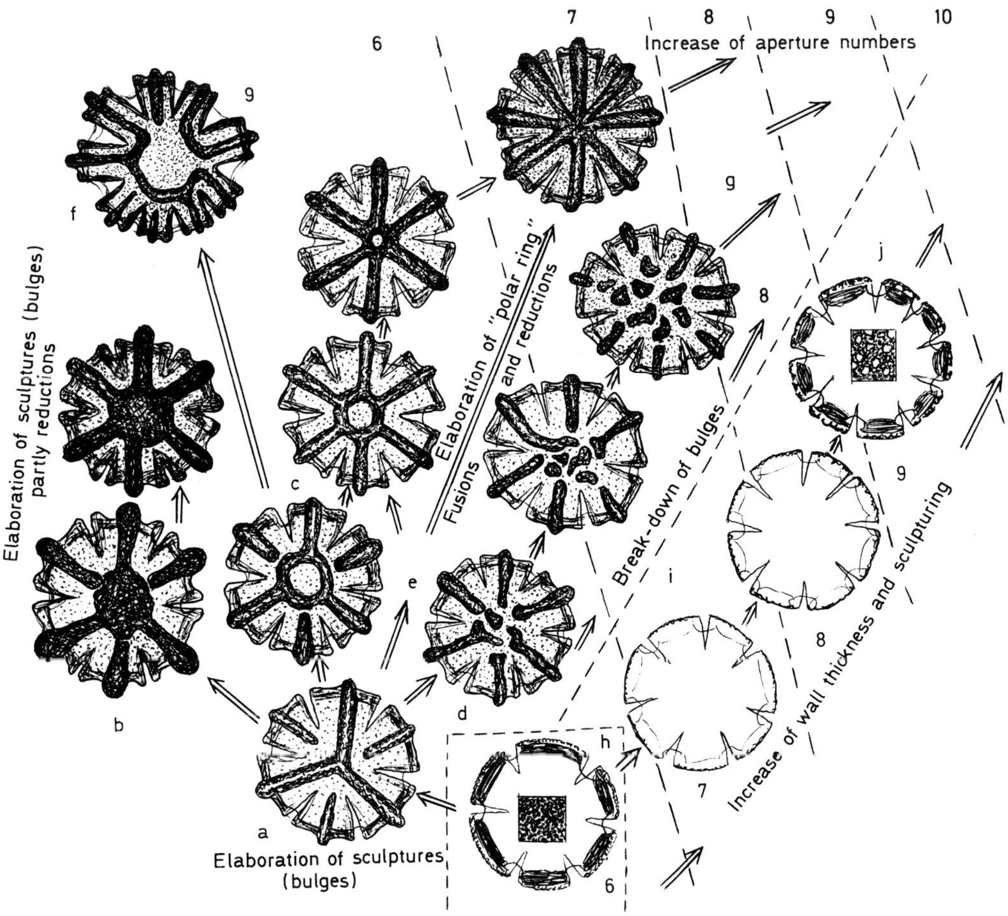

Fig. 5. *Ctenolophon*. Trends in the differentiation of fossil and extant pollen types. *a – c* Yemen spp. 1 – 3 (Upper Cretaceous); *d C. costatus*, holotype; *e C. lisamae*, small type; *f, g C. englerianus*, two different types (extant); *h C. williamsii; i C. philippinensis*; *j C. parvifolius*; scales rary, × 500 – 800

the polar fields is pictured; in the other, this ring is completely dissolved into a corrugate-hamulate knobbed pattern. Obviously, pollen in different stages of evolutionary advancement coexist in this extant African species. The reason for this heterogeneity and any possible taxonomic implications should be investigated.

Paleochorology. The following observations apply to the genus as a whole. Its major paleogeographic shifts have been documented in the literature and new finds in E. Africa and S. Arabia have established the connection between C. Africa and India. The complete development is more easily illustrated by a series of five maps (Fig. 6).

The oldest known fossils of *Ctenolophon* date back to the Maastrichtian from C. Africa (Nigeria, MULLER 1981: 73) and belong to the *C. englerianus* group. In S. Arabia three different species of the group occur (pers. obs.; cf. Fig. 5 a – c) at the boundary between the Tertiary and the Cretaceous (more exact dating is not yet possible). The genus spread to the W. during the Paleocene and appears in northwestern S. America (Venezuela, Colombia, GERMERAAD & al. 1968; Guayana:

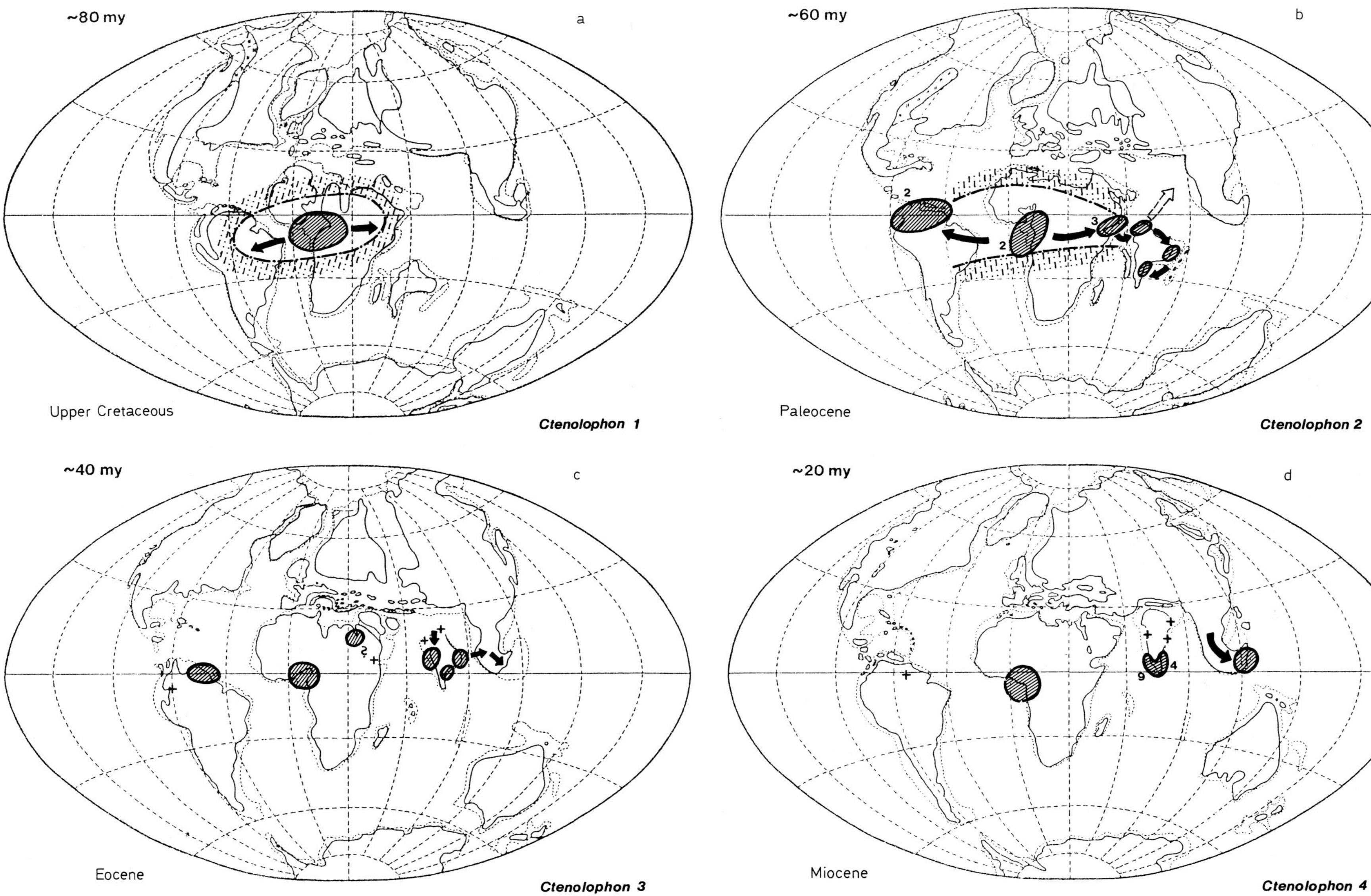
~80 my
a
Upper Cretaceous
Ctenolophon 1
~60 my
b
Paleocene
Ctenolophon 2
~40 my
c
Eocene
Ctenolophon 3
~20 my
d
Miocene
Ctenolophon 4

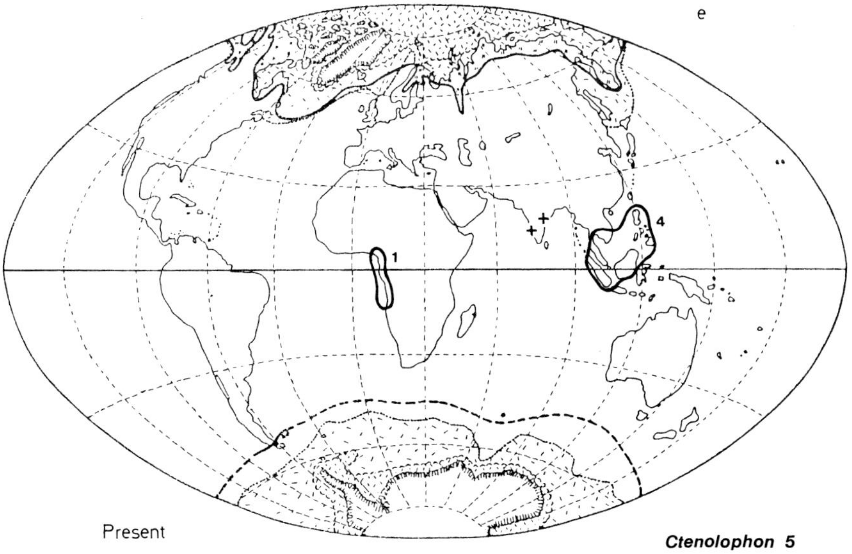

Fig. 6. Paleochorology (*a–d*) and present distribution (*e*) of *Ctenolophon* (*Linaceae* or *Ctenolophonaceae*, resp.); —·—·— limits of rain forest towards semi-dry palm savanna

WIJMSTRA 1971) but disappears again in the early Eocene. In S. America and S. Arabia, only representatives of the *C. englerianus, C. lisamae,* and *C. costatus* groups have been found. However, the *C. parvifolius* type occurs in C. Africa in the early and middle Paleocene and early Eocene (ADEGOKE & al. 1978, MULLER 1981, and others). Indian finds from the early Tertiary prove that *C. parvifolius* must have spread to the E. (THANIKAIMONI & al. 1984: 81). In Africa, it occurs only in the early Tertiary to early Eocene, whereas the *C. englerianus* group can be traced – with sometimes incomplete fossil evidence – up to the present.

Ctenolophon spread in the early Paleocene from the N. African/S. Arabian area to India as this island-continent's NW. tip drifted close by the NE. African/Arabian coast. From Kutch, where *Ctenolophon* first occurred on the Indian subcontinent in the Paleocene (KAR 1985), it rapidly spread to the NE. in the direction of Bengal and Assam. At the end of the Paleocene and in the early Eocene, it turned toward the S. to settle the coastal plains along the E., S., and SW. Indian coasts. The genus' centre constantly remained in the equatorial zone. Extinction occurred first in the NW., then in Bengal and Assam by the end of the Oligocene or already in the Eocene, as India drifted further N. It reached S. India in the Miocene and attains its greatest diversity of form (4 to 9 known pollen types). ERDTMAN (1955) first described and characterized the fossil pollen from this region.

Before *Ctenolophon* disappeared in N. and NE. India, it was already established on the neighboring SE. Asian continent during the Oligocene-Miocene transition (MULLER 1968, MORLEY 1977). It spread in a southeasterly direction to Borneo during the Miocene. Both main groups were involved in the leap from the "Noah's Ark" of India to SE. Asia. Today they are represented by four species in Malaysia, Sumatra, Borneo, the Philippines, S. Vietnam, Campuchea, and Thailand. *Ctenolophon* probably disappeared between the Miocene and Pliocene in S. India. SMITH (1973: 59) had postulated a contrary direction of spread, from E. to W.

Summary. This tropical lowland genus is small in size, clearly circumscribed, and palynologically well-documented. It is characterized by a short-lived expansion from the C. African tropics to the northwestern S. American neotropics in the early Tertiary, and a more complex and longer lasting eastward expansion during the Tertiary from E. Africa to SE. Asia, connected with the drifting of India. The present disjunction between C. W. Africa and SE. Asia is easily explained by the genus' fossil record. Further comparative morphological structural analyses will shed light on the details of the chorological developmental phases of this genus.

Bombacaceae (pro parte)

The family includes about 45 genera and more than 200 species. The fossil record consists primarily of pollen and some macrofossils (leaves). The *Bombax* type pollen forms are the most important for paleobotany and are included in the pollen genus *Bombacacidites* (Couper 1960). Five other basic types can be distinguished (Fuchs 1967), but are poorly documented by fossil data. The *Bombacacidites* pollen type corresponds mostly to genera of the tribe *Adansonieae* (*Bombaceae*) and occurs only rarely in other tribes (Fuchs 1967). Older studies on extant *Bombacaceae* pollen can be found in Erdtman (1952), Chaudefaud (1955), Nair (1962), Robyns (1963 a), and Tsukada (1964). The important new paper on extant *Bombacaceae* in "World Pollen and Spore Flora 14", has reached the author only after this manuscript was finished (Nilsson & Robyns 1986). The last 30 years have resulted in numerous *Bombacacidites* finds in many parts of the world, as they are easily recognizable in slides. A paleochorological evaluation of that part of *Bombacaceae* thus appears feasible.

Macrofossil data (leaves) complement the pollen finds, particularly with respect to the early Tertiary (Leopold & MacGinitie 1972, MacGinitie 1969, Axelrod 1979, Axelrod & Raven 1985, and others). The boundaries between two neighbouring families, the *Sterculiaceae* and *Tiliaceae,* are still difficult to determine with certainty in the case of leaves and pollen (Kirchheimer 1957, Krutzsch 1970 b). Because the two fossil organ groups are so different, the paleochorological analysis must be based on both.

Despite these limitations, there arises an interesting new paleochorological picture, which differs markedly from earlier theoretical attempts (especially that of Croizat 1952). *Bombacacidites* no longer represents a primarily tropical African group, but rather an element which first occurs in the N. Tethyan flora and only then migrates to the equatorial tropics.

Four to five main paleochorological phases can be reconstructed on maps (Fig. 7). The oldest finds are those from the lower Campanian and Maastrichtian of New Jersey (Fig. 7 a) as published by Wolfe (1975, 1976). His figures show three well-differentiated bombacaceoid type pollen forms. However, the forms from the Maastrichtian and Paleocene of Alaska, classified as "*Bombacacidites nacimentoensis*" by Wilson (1978), do not belong here.

In the second phase, at the end of the Cretaceous and beginning of the Tertiary (Fig. 7 b), a rapid expansion set in, which is documented by fossil finds from the N. Tethyan belt, reaching from California (Drugg 1967) to New Mexico (Anderson 1960) and Texas (Elsik 1968, and others), and finally to Europe (Krutzsch 1957, 1970). Simultaneously, *Bombacacidites* forms migrated from N. America via

C. America, over a chain of vulcanic islands to S. America (Paleocene, Colombia and Venezuela), and from there to C. Africa (Paleocene/Eocene, Nigeria: WOLFE 1975).

In the following phases (Fig. 7 c, d), the southward expansion continues from S. America via the Antarctic to Australia (starting in the lower Eocene) and New Zealand (Oligocene to Miocene). For details concerning this expansion, the reader is referred to AXELROD & RAVEN (1978: 89), MULLER (1981: 47), and KEMP (1981).

The earlier, speculative migratory patterns as envisaged by AXELROD & RAVEN (1974: 579, 623) and CROIZAT (1952; cf., also KAR 1985), have been superseded by the newest fossil data. CROIZAT assumed the center of the family's E.-W. expansion in C. Africa; AXELROD & RAVEN postulated a migration in the lower Eocene from S. America towards the N. But the fossil data indicate the opposite migratory direction.

In the third main phase, during the Eocene (Fig. 7 c), *Bombacacidites* was distributed (similar to the Paleocene) in the N. American N. Tethyan belt (macrofossils from the Rocky Mountains region: LEOPOLD & MACGINITIE 1972) from the W. to the S. and SE. (TSCHUDY & LOENEN 1970, AXELROD & RAVEN 1985) and to Europe (KRUTZSCH 1970 b, KEDVES 1974, and others). The genus was represented by a variety of species, but disappeared in C. Europe with the onset of cooler climates after the Middle Eocene. It has been demonstrated in the Eocene in the S. Atlas Mountains of Morocco (MOHR 1986, OLLIVIER-PIERRE 1982). It occurs in the middle Eocene in Cuba (ARECES-MALLEA 1985) and later in the Oligocene in Puerto Rico (GRAHAM 1972, GRAHAM & JARZEN 1969). There is evidence from the Venezuelan (NOREM 1955, cited by ARECES-MALLEA 1985) and Colombian (GONZÁLES-GUZMAN 1967) Eocene; from Guyana (LEIDELMEYER 1966); from the Eocene of the Amazonian delta (BOER & al. 1965); and from Brazil (THIERGART 1940). It occurs frequently upwards into the Pliocene in tropical W. Africa (BOLTENHAGEN & al. 1985, AXELROD & RAVEN 1978, MENZEL 1920, and others). In the African Paleogene there was probably a gradual expansion to the E., as *Bombacacidites* appears with several pollen types in the Ethiopian Oligocene/Miocene (unpubl. obs.) and the Pliocene/Pleistocene of Burundi (SAH 1967). Furthermore, Madagascar was probably reached during the Tertiary in this way.

In the fourth main phase (Oligocene to mid-Miocene, Fig. 7 d), *Bombacidites* retreats from the N. Tethyan belt towards the E. and SE. during the middle Tertiary. It reaches N. India in the Oligocene (KAR 1985) and spreads rapidly in SE. Asia, forming a new, secondary evolutionary centre, which still exists. Otherwise, the group disappears completely in N. America and in almost all of Europe, particularly in C. Europe. Only in Portugal does it extend to the Oligocene/Miocene transition (PAIS 1981), to the Miocene in Spain (BESSEDIK 1984) and to the Early Pliocene in the NW. Mediterranean (SUC 1980, and others). This disappearance in the N. regions is due to the considerable temperature drop and the simultaneous southward shift of the paleolatitudes as of the Oligocene.

A migration from Africa to SE. Asia of the *Bombacaceae*, as postulated by AXELROD & RAVEN (1974), probably never has occurred. The *Bombacaceae* taxa of SE. Asia and the W. Pacific did not migrate via Africa; rather, they retreated to the SE. with the N. Tethyan flora during the last 20 to 30 million years (Fig. 7 e). This corresponds to the taxonomic structure of the family (AXELROD 1970: Fig. 4). Only the *Adansonieae*, probably the most primitive of all six tribes, is represented

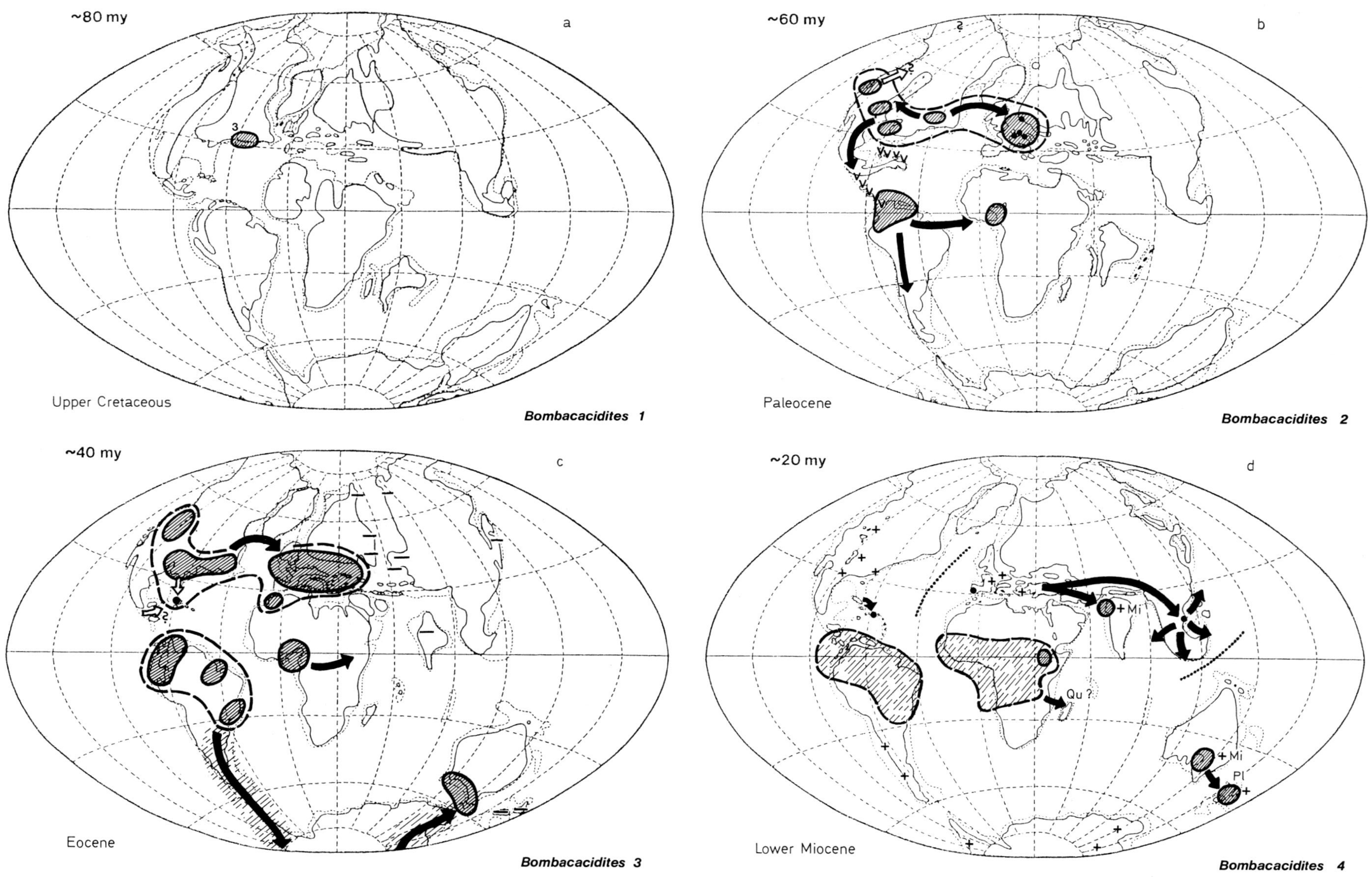
~80 my
a
3
Upper Cretaceous
Bombacacidites 1
~60 my
b
Paleocene
Bombacacidites 2
~40 my
c
Eocene
Bombacacidites 3
~20 my
d
Mi
Qu ?
Mi
Pl
Lower Miocene
Bombacacidites 4

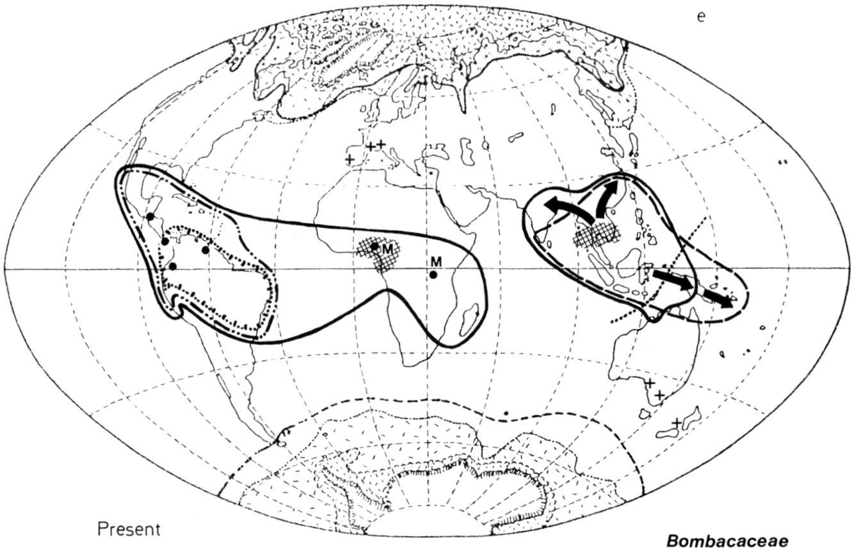

Fig. 7. Paleochorology of *Bombacidites* ($a-d$) and present distribution of *Bombacaceae* (e); in e "*Adansonieae*" (———, 7 genera), *Duronieae* (– – – –, 8 genera), *Hampieae, Ceibeae*, and *Matisieae* (–··–··–, each with 3 genera), and *Catostemmateae* (|..|..|..|, 2 genera); dots are Pl and Qu finds

in all three present-day main areas of the family, and in Africa, it is the only tribe altogether from the Paleocene onwards. No new evolutionary lines on the subtribal level have occurred there, but there have been adaptations to very specific environmental conditions, as exemplified by the baobabs and their flower biology (chiropterophily). In the family's W. secondary developmental centre in S. America, which experienced secondary spread to C. America and the Caribbean during the Pliocene, in addition to the *Adansonieae*, there are four other endemic tribes (*Hampeae, Matisieae, Ceibaeae, Catostemateae*). In the E. secondary developmental centre in SE. Asia, the sixth tribe *Duronieae* (with eight genera), is endemic. The *Adansonieae* are represented there not only in relic areas in NE. Australia (AXELROD 1970; Fig. 4), but are also more widely distributed in Indomalaysia. The *Bombacaceae* have spread to New Guinea, the New Hebrides, and N. Australia during the Pliocene and Pleistocene, thereby crossing the WALLACE-line; another secondary expansion was towards the NW. to India. Thus, one can differentiate between two separate settlement periods of *Bombacaceae* in Australia: an old settlement in the S., which began in S. America in the Paleocene via the Antarctic and lasted up to the Oligocene; and a very young settlement, which began in the NW. during the last $1-5$ my and is still in progress.

It is still uncertain where the origins of the five *Bombacaceae* tribes (other than the *Adansonieae*) are to be found. They may lie in the existing centres or in the common predecessors before migration to these centres. This should be the subject of future studies.

Summary. The *Bombacaceae* and particularly their most primitive tribe, the *Adansonieae*, are represented in the fossil record mainly by microfossils (classified as belonging to the pollen genus *Bombacacidites*) and some macrofossils. They first

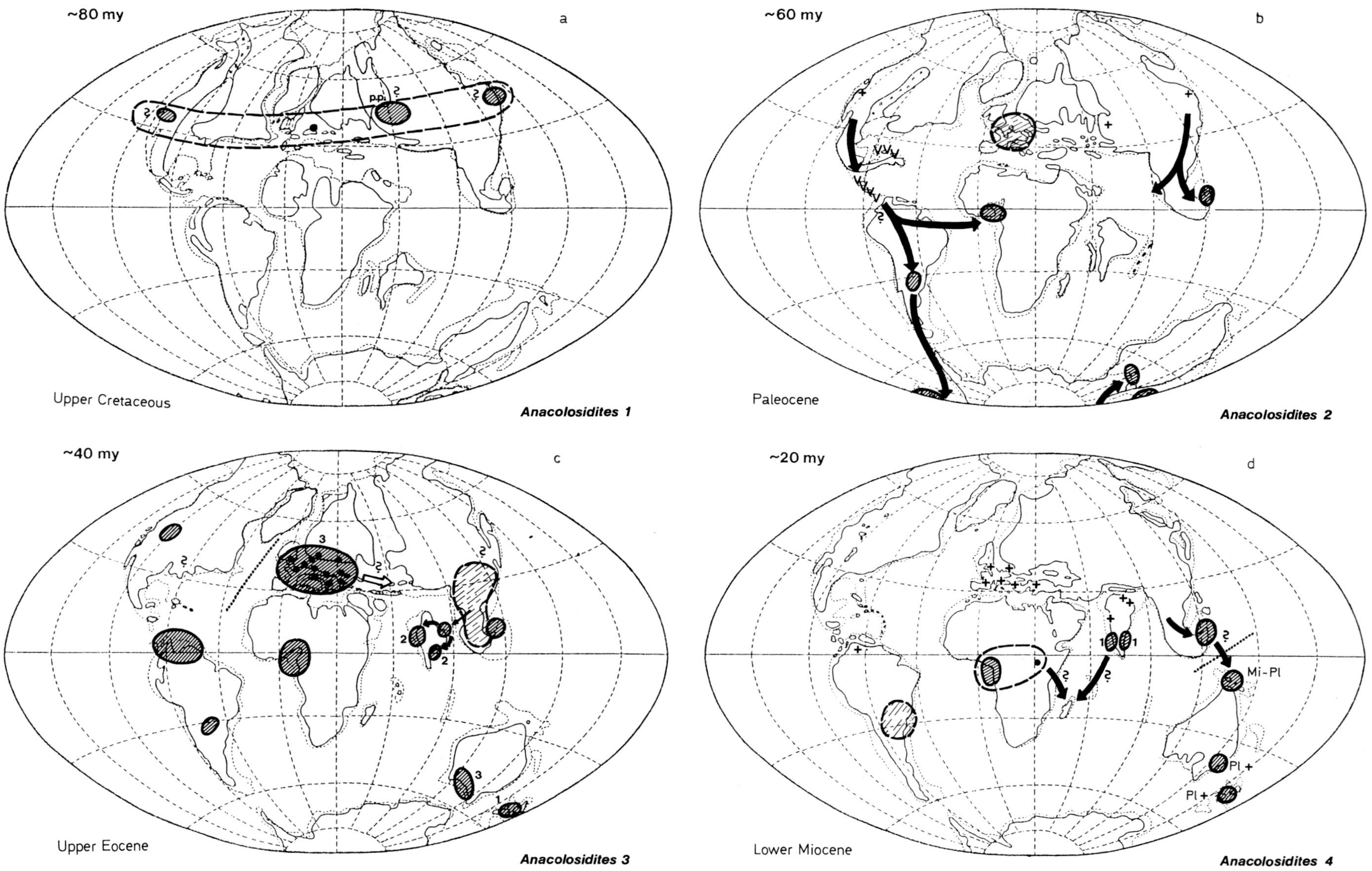
~80 my
a
p.p.
Upper Cretaceous
Anacolosidites 1
~60 my
b
Paleocene
Anacolosidites 2
~40 my
c
3
2
2
3
1
Upper Eocene
Anacolosidites 3
~20 my
d
1
1
Mi–Pl
Pl +
Pl +
Lower Miocene
Anacolosidites 4

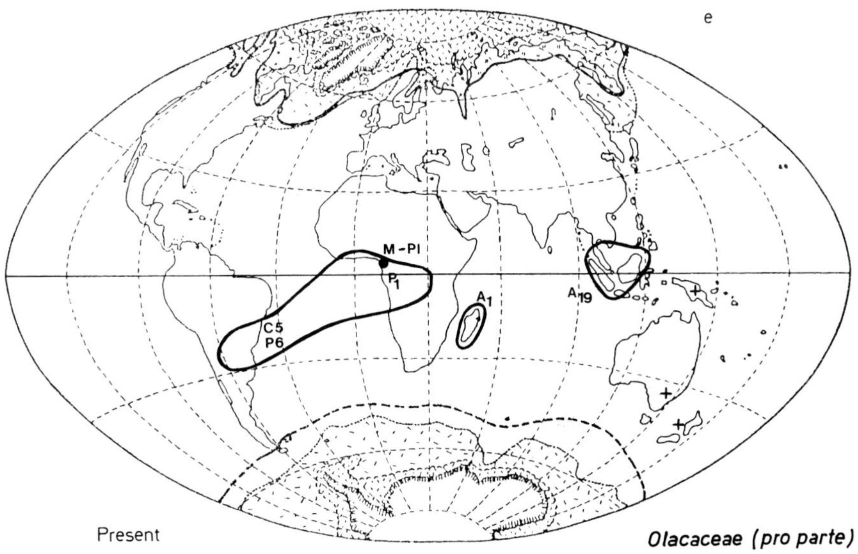

Fig. 8. Paleochorology (*a – d*) of *Anacolosidites* and present distribution (*e*) of *Olacaceae* (pro parte). *Anacolosa* (A), *Ptychopetalum* (P), and *Cathedra* (C), with numbers of species and a Pliocene macrofossil find

occurred in the N. Tethyan flora and reached the S. American tropics via C. America over volcanic arcs in a second expansion phase during the Cretaceous/Tertiary transition. They moved from S. America to C. Africa in the Paleocene. In a relatively short southward migratory phase they spread from S. America over the W. Antarctic to S. Australia and New Zealand, where they soon disappeared. They exist up to the present in C. Africa and S. America, where they have formed a centre of diversity extending to the Caribbean and C. America during the Pliocene/Pleistocene. With the reduction and final extinction of the tropical flora along the N. Tethys during the Upper Paleogene, the *Bombacaceae* retreat to N. India and reach SE. Asia during the Miocene. Here, they form an E. evolutionary centre and expand in the Pliocene/Pleistocene, to New Guinea and N. Australia. It cannot be documented with fossil material that the family's original centre was located in Africa (CROIZAT 1952), nor that a migratory route from Africa to SE. Asia via India existed.

Olacaceae (pro parte)

The family *Olacaceae* contains 27 genera with c. 230 species, of which only three are of interest here: *Anacolosa* (19 species in SE. Asia, 1 species in C. Africa, 1 species in Madagascar); *Ptychopetalum* (1 species in W. Africa, 5 species in S. America); *Cathedra* (5 species, only in Brazil). All three genera are presently limited to the tropical zones.

The only fossils are those described in the pollen genus *Anacolosidites* (COOKSON & PIKE 1954), which correspond to the three extant genera. Up to now, about ten different pollen types have been described (for recent reviews, see KRUTZSCH 1970, MULLER 1981, THANIKAIMONI & al. 1984). In 1934, R. POTONIE discovered the first *Olacaceae* species, *A. efflatus,* then not recognized as such, in the Middle Eocene of the Geiseltal.

With respect to paleochorology, four to five main phases can be distinguished. The oldest finds (Fig. 8 a) come from the warm and humid N. Tethyan vegetation belt at the end of the Upper Cretaceous period (Maastrichtian). Questionable evidence is available from California (CHMURA 1973), Siberia (CHLONOVA 1974), and Japan (MIKI 1977); positive evidence, from C. Europe (KRUTZSCH & LENK 1969).

The species *A. insignis* from the Campanian of N. America, described by NORRIS & al. (1975), does not belong to the *Anacolosidites* form group, but rather to a fossil group (SAMOILOVIC 1961) from high latitudes of Siberia. It is intermediate in its pollen morphology, between *Olacaceae, Santalaceae,* and *Loranthaceae*. The same is true for *A. striatus* (e.g., WIGGINS 1982; Campanian, Alaska). Other pollen types, classified (mostly in the Soviet literature) as *Anacolosidites* (e.g., *A. supplingensis,* ZAKLINSKAJA 1963), do not belong to the *Olacaceae*, but rather are members of the *Normapolles* group and the pollen genus *Interpollis*. The author has also temporarily classified them as *Anacolosidites* (KRUTZSCH 1959–1961).

In a second phase, during the Paleocene (Fig. 8 b), *Anacolosidites* migrated to the south. They appeared in S. America, C. Africa and Borneo (ARCHANGELSKI 1973, GERMERAAD & al. 1968) and reached Australia (HARRIS 1965) and New Zealand (COUPER 1960, MILDENHALL 1980) via Antarctica.

In a third phase (Fig. 8 c) they can be found in northwestern and southern S. America and in central W. Africa. They extended from the N. American Gulf coast to W. Europe and C. Asia in the N. Tethyan vegetation belt, where there is no evidence yet for their existence in the Paleocene. In India, they first appear in the Eocene (THANIKAIMONI & al. 1984, MULLER 1981, and others) and are represented with several species in Assam, Bengal, Bombay, and Cauvery. It is uncertain how *Anacolosidites* reached India, probably from the N. or NE.

In the fourth phase, during the early Neogene (Fig. 8 d), *Anacolosidites* has already disappeared from Europe. There is no more evidence for an occurrence in northwestern S. America, but the group exists in Brazil and C. Africa up to the present. In N. India, it disappeared in the lower Miocene, but survived in Kerala and Cauvery (SE. and SW. India) into the upper Miocene. There are no fossils yet from the Oligocene and Miocene of Australia and New Zealand. It is therefore uncertain whether the occurrences in the New Guinean Upper Miocene and N. Australian Pliocene (COOKSON & PIKE 1954) are due to migrations from former S. Australian areas, or from the NW., e.g., from Laurasia over the WALLACE-line (as shown in Fig. 8 d).

There is too little known about the pollen genus *Olaxipollis* W. KR., corresponding to the *Olax* type within the *Olacaceae* group, to undertake a paleochorological analysis. However, the situation apparently resembles that of *Anacolosidites:* its stratigraphic spread indicates that *Olaxipollis* was a typical members of the N. Tethyan floral zone in Europe during the warmer phases of the Lower Eocene to Miocene. In contrast, *Olax* today occurs primarily in W., C., and also E. Africa. It extends to Madagascar, Sri Lanka, India and SE. Asia (from the W.- Himalayas to W. Indonesia), but is completely missing in the New World.

Summary. *Anacolosidites*, like the preceeding example *Bombacacidites*, is not primarily a member of the tropical flora. The oldest fossils date from the end of the Cretaceous of the N. Tethyan belt, extending from California in the W. to Europe, and Birma in the E. Only in the earliest Tertiary did *Anacolosidites* migrate to the equatorial tropics, to S. America, C. Africa, SE. Asia and Borneo. From S. America, it reached S. Australia and New Zealand via Antarctica. It has been found in the Eocene of the N. Tethyan zone from Europe to C. Asia. In India, it first appears in the Eocene, but persists in several species up to the Upper Miocene. It is lacking in Europe from the Oligocene onwards, as well as in the Oligocene and Miocene of Australia and New Zealand. Therefore, the New Guinean upper Miocene and Australian Pliocene finds might be due to an early migration from the NW. over the WALLACE-line.

Symplocos (Symplocaceae)

Symplocos is the only genus of the family *Symplocaceae* with 300 to 400 extant species. The fossils consist of fruit and pollen. Some leaf fossils from the Tertiary which have been classified as *Symplocos* are doubtful (RÜFFLE, pers. comm.; cf., also BARTHEL & RÜFFLE 1976).

The pollen classified in the Soviet literature as *Symplocacites* MČEDLISHVILI (1961) do not belong to *Symplocos*; their botanical position is unknown. Furthermore, many other pollen forms named "*Symplocos*", do not belong here (cf., WILSON 1978: Maastrichtian, Alaska; PENNEY 1969: Upper Cretaceous and old Tertiary of Australia and New Zealand; VENKATACHALA & KAR 1969, KAR 1985, SAH & KAR 1970: different species from the Indian Tertiary; OLTZ 1969: Paleocene in Montana; DRUGG 1967, JARZEN & NORRIS 1975, NORRIS & al. 1975, ROUSE & al. 1971, and many others).

On the other hand, true *Symplocos* pollen are often "hidden" behind names which do not reveal their actual botanical identity and so hinder any search for the earliest *Symplocos* representatives. A critical treatise of the fossil *Symplocos* pollen can be found in MULLER (1981, 1984). KIRCHHEIMER (1944, 1950, 1957, etc.) and MAI (1970) have published critical reviews on the fruits.

About 25 fossil fruit species occur in the C. European Tertiary and there are about as many fossil pollen species, but they have not yet been exactly described. Thus, a paleochorological analysis would have a good basis, particularly since the fossils are from many different locations and occur throughout the Tertiary. Unfortunately, most of these data are still unpublished.

The present distribution of the *Symplocaceae* in southern N., C., and S. America and in Australasia represents a southern variant of the ASA GRAY disjunction Its origin can be traced through five to six paleochorological phases.

As far as can be interpreted from the literature, only one species from the many other Cretaceous finds classified as *Symplocos*, can actually be placed in this genus, i.e., *Triporopollenites andersonii* DRUGG (CHMURA 1973: p. 20, Plate 32, Fig. 12): it is the genus' oldest representative (Fig. 9 a). DRUGG (1967) compared this species with the extant *S. pringlei*, but there still remain some doubts regarding its classification as *Symplocos*. The other pollen form (*T. scabroporus* NEWMAN) which was classified as *Symplocos* by CHMURA (1973), certainly does not belong there.

In a second phase (Fig. 9 b), *Symplocos* can be found, although rarely, in the S. U.S.A. (TSCHUDY 1973). In a third phase (Lower Eocene), representatives of

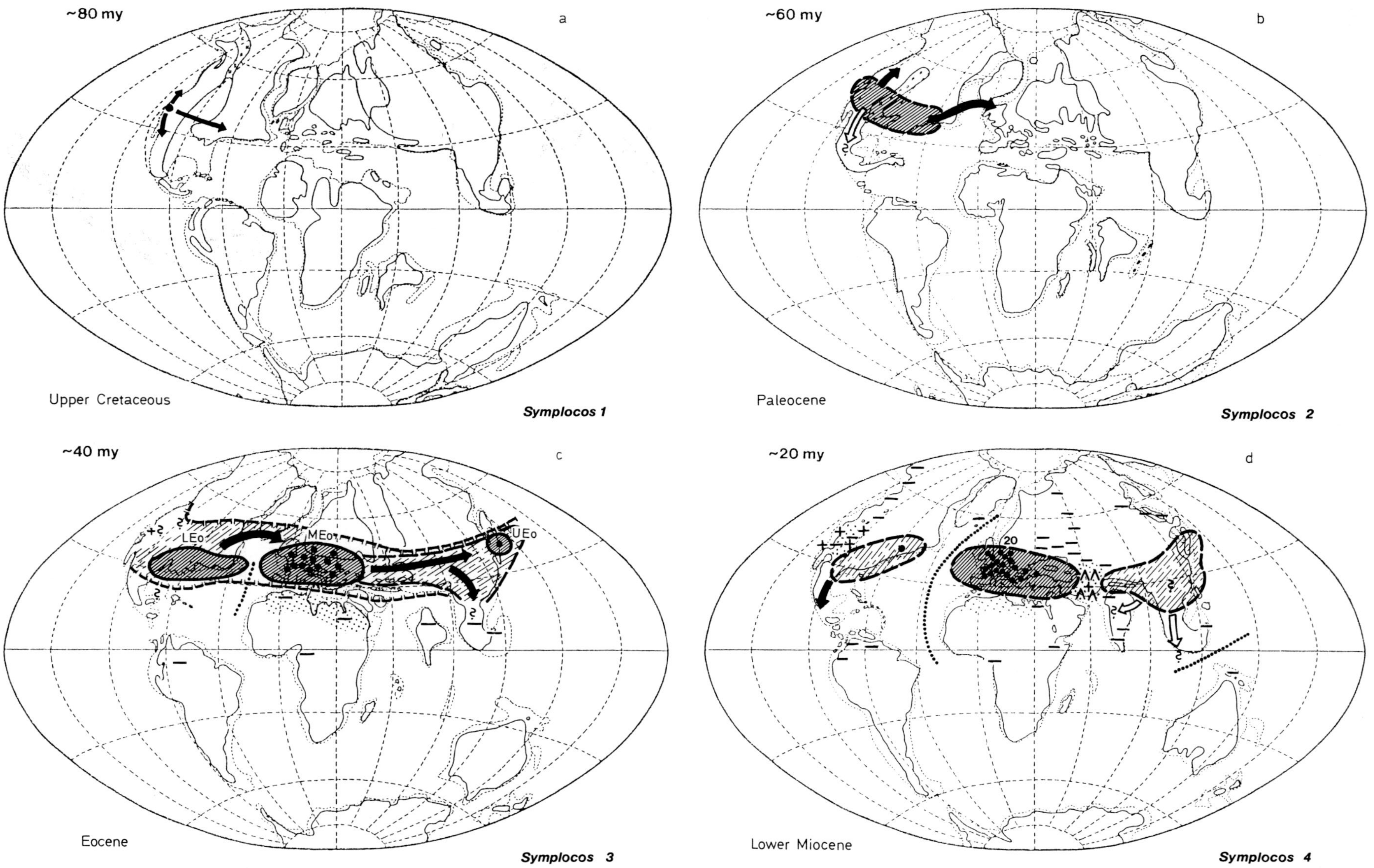

~80 my
a
Upper Cretaceous
Symplocos 1
~60 my
b
Paleocene
Symplocos 2
~40 my
c
LEo
MEo
UEo
Eocene
Symplocos 3
~20 my
d
20
Lower Miocene
Symplocos 4

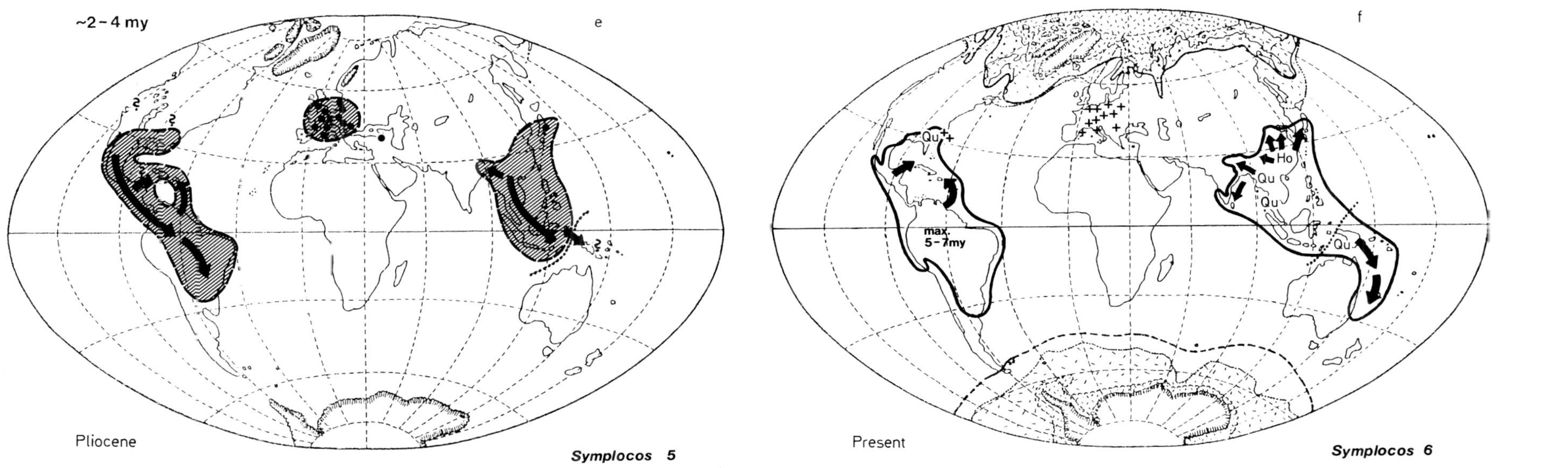

Fig. 9. Paleochorology (*a*−*e*) and present distribution (*f*) of *Symplocos* (*Symplocaceae*); in *f* Quaternary (Qu) and Holocene (Ho) expansions, and maximum immigration age

the *Symplocaceae* occur more frequently in the SE. U.S.A. (FAIRCHILD & ELSIK 1969, FREDERIKSEN 1980; and others), and extend eastward to W. and C. Europe (Fig. 9 c). They are represented here, especially in the G.D.R., by ten species, whose individual and species-related frequency depends upon the facies and ecological milieu. This easterly expansion continued in the Eocene towards E. Asia and reached Japan in the upper Eocene (TANAI 1972). *Symplocos* thus spread over the entire warm-humid N. Tethyan flora belt. Up to now, there are still no unequivocal old Tertiary *Symplocos* finds from Africa, India, Australia, or southern SE. Asia.

In a fourth phase (Oligocene/Miocene, Fig. 9 d), this old Tertiary belt area broke down into three disjunct parts: a very species-poor N. American area, a very species-rich European-Asian area, and a Chinese-Japanese area.

Since the upper Eocene, the constantly expanding N. Atlantic has separated the N. American warmer old Tertiary flora from the European. From Iran to Afghanistan, a mountain and drought barrier has developed between Europe and E. and S.E. Asia since the Upper Oligocene. Only now, a first migration of *Symplocos* from the NE. towards India and further expansion from SE. Asia to Indonesia may have occurred. From C. America, *Symplocos* may have reached the Caribbean and northern S. America by island jumps, at this time or during the Pliocene (a question which could be answered by future fossil finds).

In a fifth phase (Pliocene, Fig. 9 e), *Symplocos* apparently became extinct in the N. and NW. and retreated towards southern N. America. After the appearance of the permanent Central American land bridge, at the latest (i.e., 5 – 7 my ago), the genus expanded rapidly towards S. America where it developed its present western secondary centre of diversity in the mountain forests. In Europe, *Symplocos* existed with three or four species in Europe until the beginning of the Pleistocene. It had completely settled Indonesia, and reached the WALLACE-line by the end of the Pliocene. The genus crossed this line in the Pleistocene and reached New Guinea and its present distribution boundaries in NE. Australia. Probably S. India and Sri Lanka were settled from the E. or NE. in the Pleistocene.

AXELROD & RAVEN (1974: 571) view *Symplocos* as an old Laurasian element which spread from SE. Asia to Australia in the Miocene or later. They suppose that it reached S. America from the N. or via Africa. The latter possibility is unlikely, since *Symplocos* evidently never occurred in Africa. An analysis of fossil species and extant taxonomic relationships should reveal further interesting details about the history of *Symplocos*.

Summary. The paleochorology of the genus *Symplocos*, based equally on pollen and carpological finds, is characterized by its successive settlement of the warm-humid N. Tethyan flora belt. This settlement began in California in the late upper Cretaceous, and constantly expanded eastwards until it reached Japan at the end of the Eocene. The belt areal then disintegrated into a N. American, an European, and an E.-SE. Asian disjunction. The European representatives existed until the beginning of the Pleistocene and then died out. In the Pliocene, the Caribbean and, at the latest after the emergence of the C. American permanent land bridge, large areas of S. America were settled from N. America, resulting in a western secondary centre of diversity. In the Miocene, Indomalaysia was fully settled up to the Philippines and the WALLACE-line. The line was crossed towards the E. and SE. and the W. Pacific was reached during the Pleistocene. Another secondary, eastern and very young centre of diversity arose during this process.

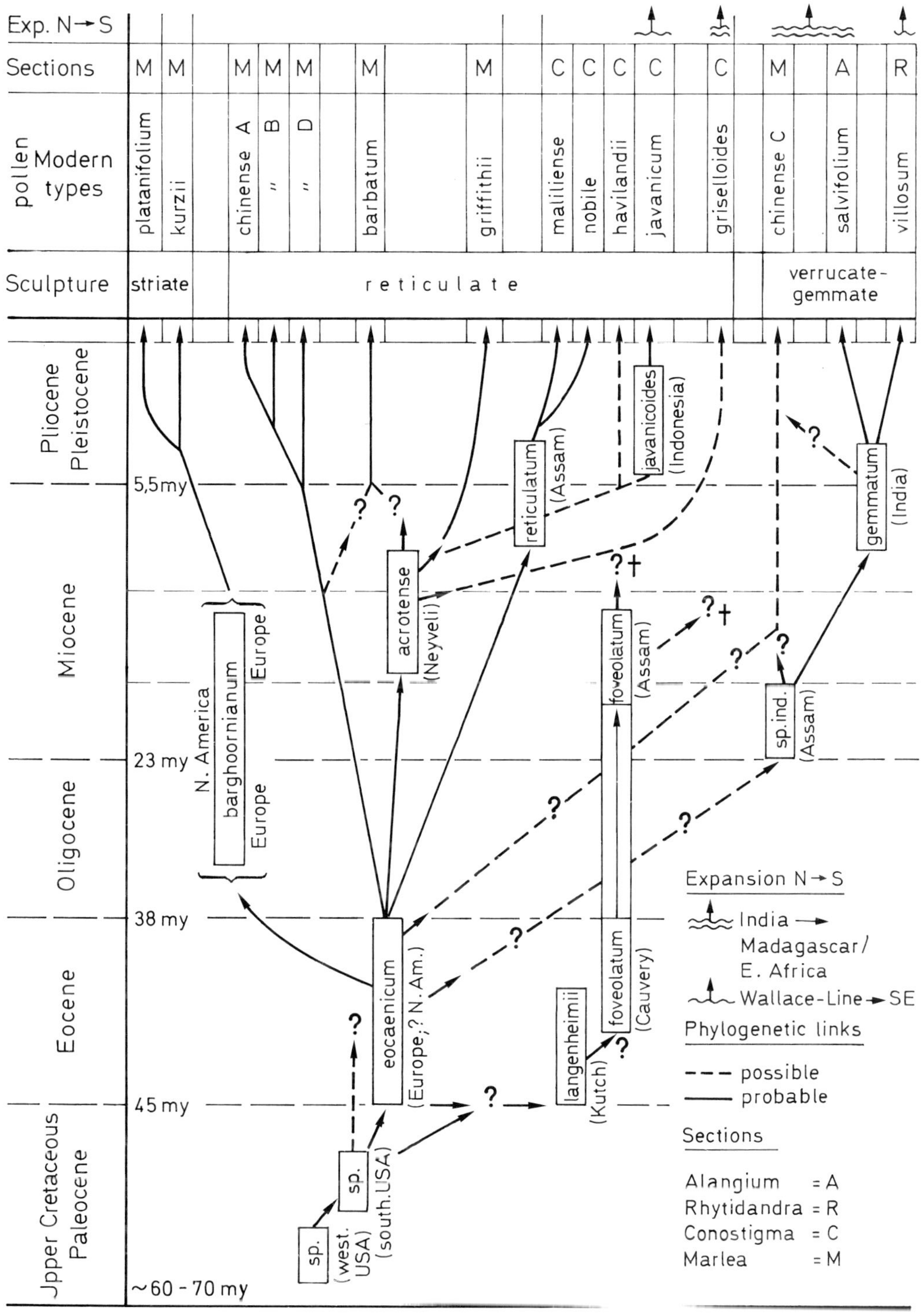

Fig. 10. Illustration of relationships among fossil and extant pollen types of *Alangium* (*Alangiaceae*) in time and space (modified from REITSMA 1970)

Alangium (*Alangiaceae*)

The family *Alangiaceae* (Figs. 10 and 11) is monotypic and *Alangium* has 19 species (classification according to Bloembergen 1939).

Its pollen was analyzed by Reitsma (1970), including the chorology of the extant pollen type groups. Fossils exist as pollen, leaves, fruits, and wood fragments.

Although Muller (1981: 77) based his analysis of the genus primarily on three pollen types, Reitsma (1970) showed that many more differentiation criteria exist. This requires a classification into a larger number of pollen types. Also, among the fossil pollen, there are more than three types which often can be related to the extant groups. Fig. 10 illustrates the temporal/spatial relationships between these pollen types. Macrofossils are too rare (Mai 1970) to allow more than documentation of the genus' record.

Paleochorology. A pollen form which is thought to be the oldest known evidence of *Alangium* (Fig. 11 a) was recently found in the Maastrichtian of Wyoming and Edmonton by Farabee & Canright (1986: 55, Plate 19, Figs. 11 – 12, not 10) and published as *Margocolporites lihokus*. An *Alangium* find from the Paleocene of the Gulf coastal area was erroneously classified as *Nyssa* spec. (Fairchild & Elsik 1969; Plate 38, Fig. 49). It is closely related to *A. eocaenicus*.

More evidence comes from the Eocene, primarily from the N. Tethyan belt (Fig. 11 b). In N. America there are mainly macrofossils from the Eocene of Wyoming (Leopold & MacGinitie 1972) and S. Alaska (Wolfe 1972). From the Lower to Upper Eocene of W., C., and SE. Europe, more than 100 pollen fossil localities have been documented (*A. eocaenicum*, Krutzsch 1962, and other authors), together with a series of different macrofossils from W. and C. Europe (Chandler 1961, 1964; Mai 1970), Japan (Tanai 1972). An entry from Kulkova (1971) from the NE. Siberian Eocene (Jana-Indigirka basin) is probably not evidence of *Alangium*, although it is similar to pollen of this family.

In India, macrofossils of two different types (Thanikaimoni & al. 1984, Kar 1985) were found: *A. langenheimii* from Kutch (NW. India, Lower and Middle Eocene), very similar to the N. Tethyan *A. eocaenicum*; and *A. foveolatum* from Cauvery (SW. India, Eocene to Miocene). These finds were probably deposited during the Eocene when the Indian continent was drifting through the tropical region.

The species *A. regularie* is documented from the Eocene of Indonesia (Morley 1982). *Alangium* at that time had already occupied SE. Asia (including India), with types related to the extant sect. *Marlea*.

The genus has been documented in C. Europe from the Oligocene to the Miocene (Fig. 11 c) by *A. barghoornianum* (Krutzsch 1961, 1970; Nagy 1969, 1972; Pacltová 1963, 1966; and others). The same species, which is related to the extant *A. kurzii* type (Traverse 1955), has also been found in eastern N. America, Europe, and recently as a relic in the Lower Pliocene of the Colchis (Stuchlik & Shatilova 1987). Fruit fragments are documented for Europe and N. America (Kirchheimer 1937, Mai 1970, Eyde 1968, Eyde & al. 1969).

Fig. 11. Paleochorology (*a – c*) and present distribution (*d*) of *Alangium* (*Alangiaceae*). Circles with crosses in *d* stand for L. Pliocene fossil finds

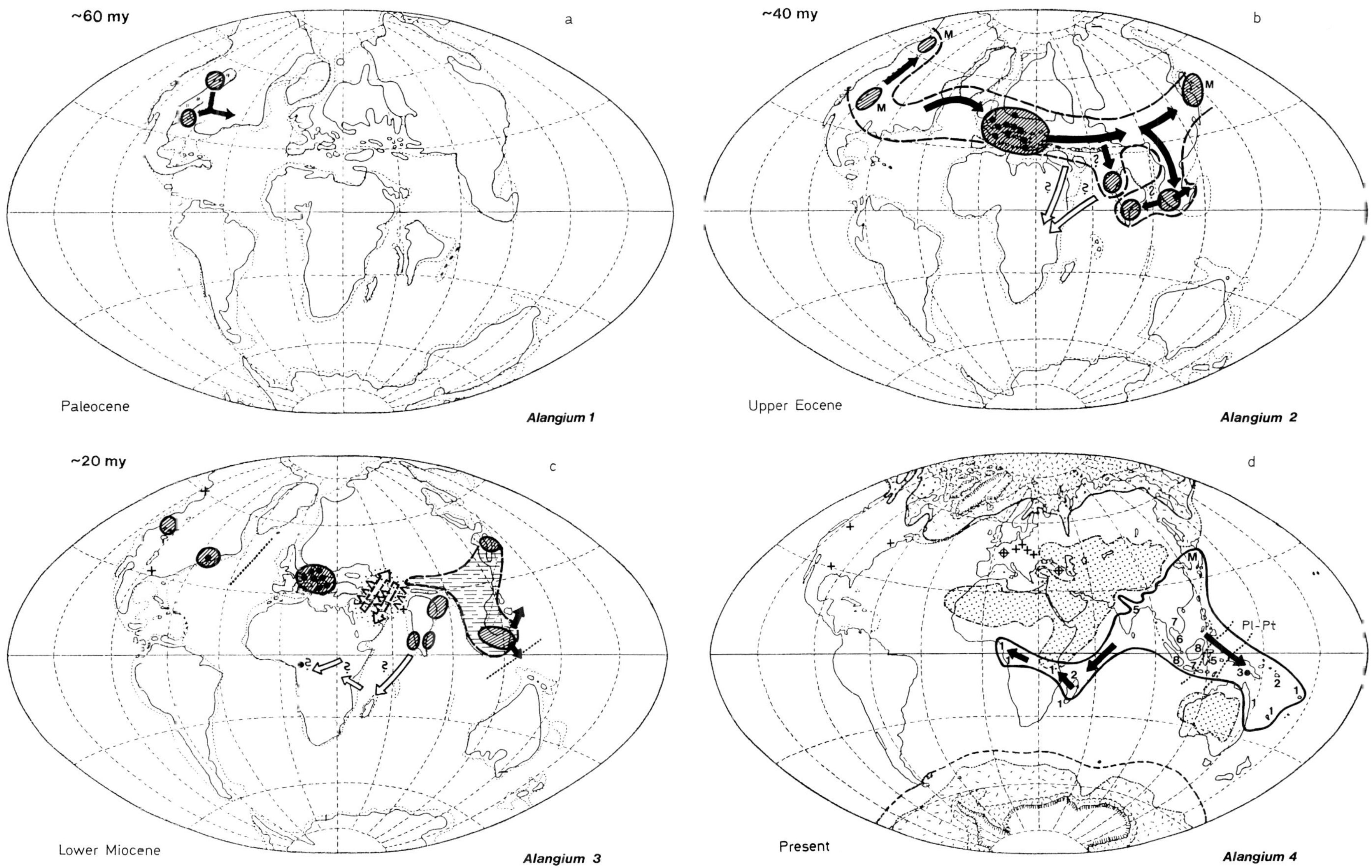
~60 my
a
Paleocene
Alangium 1
~40 my
b
M
Upper Eocene
Alangium 2
~20 my
c
Lower Miocene
Alangium 3
d
Pl-Pt
Present
Alangium 4

Alangium reached a peak in the Indian Miocene with four pollen species (*A. foveolatum, A. reticulatum, A. acrotense* and *A.* spec., in evidence from Cauvery, Assam, and Neyveli) and wood fragments (from Pondicherry, Madras; AWASHTI 1968). The genus is documented by pollen from Indomalaysia (MORLEY 1982 as well as Japan SATO 1963), where fruit fossils were also found (MIKI 1956).

In the Upper Miocene, *Alangium* occurred in Borneo (MULLER 1975) with pollen of the *A. javanicoides* type, i.e., from the advanced, now purely tropical sect. *Conostigma* (Fig. 10). According to the pollen morphological typing of REITSMA (1970), this is a group originating from an *A. griffithii* precursor, to which belong the extant species *A. maliliense, A. nobile, A. havilandii,* and *A. japanicum* (together referred to by MULLER 1981 as *A. havilandii* type s. l. only). The latter is one of the three groups which crossed the WALLACE-line in the Pliocene and reached New Guinea and the New Hybrides (Upper Pliocene: COOKSON 1957). Only *A. villosum* (with several subspecies and varieties) penetrated into NE. Australia, forming the border line of the genus in the SW. Pacific.

Almost all of the *Alangium* areas outside of its present boundaries had disappeared by the Pliocene. Whereas *A. kirchheimerii* SZAFER from the lower Pliocene in Kroscienko (Poland) was not confirmed by MAI (1970), the genus has been documented in the Lower Pliocene of Alsace (GEISSERT & GREGOR 1981). During the latest Tertiary in Japan the extant E. Asian species are already present (REITSMA 1970; 279).

Summary. *Alangium* was initially a N. Tethyan element which spread early to Indonesia and India (at that time an island-continent) at the beginning of the Eocene. In the Middle Tertiary, the main area of the genus was in SE. Asia and India, expanding to the SE., but the disjunctions in Europe and N. America became extinct until the end of the Miocene. How and when *Alangium* reached Africa and Madagascar with two to three species, remains uncertain. Most probable is a migration during the early Tertiary from S. India via the Seychelles and Madagascar.

Nepenthes (*Nepenthaceae*)

This family is monotypic (Fig. 12) and is represented by about 90 extant species from SE. Asia to New Guinea, New Caledonia, and N. Australia in the E.; and to Sri Lanka, S. India, the Seychelles, and Madagascar in the W. and SW. These species are herbaceous ground plants, vines and epiphytes in humid forests and open swamp areas, e.g., in the Indonesian mountains (DANERT 1971, STEIN 1978). They avoid dry regions. Only one extant pollen type is known: nonaperturate, loosely and finely spinous tetrads (RAO & ONG 1972: THANIKAIMONI & VASANTHY 1972, KRUTZSCH 1970c, 1985). The only fossils are pollen.

Paleochorology. The fossil pollen finds from the European Eocene (France to SE. Europe and the Caucasus area) are of special importance (Fig. 12b). Thus, in its earliest known phase, *Nepenthes* was an element of the N. Tethyan humid evergreen flora. It must have spread to S. and SE. Asia before the formation of

---

Fig. 12. Paleogeography of "*Droseridites*" *senonicus* (*a*) and *Nepenthes* (*Nepenthaceae*) (*b*–*d*), up to the present. *a* In respect to the dubious placement of this pollen species, cf. the text, *c* small arrows signal expansion of xeric areas

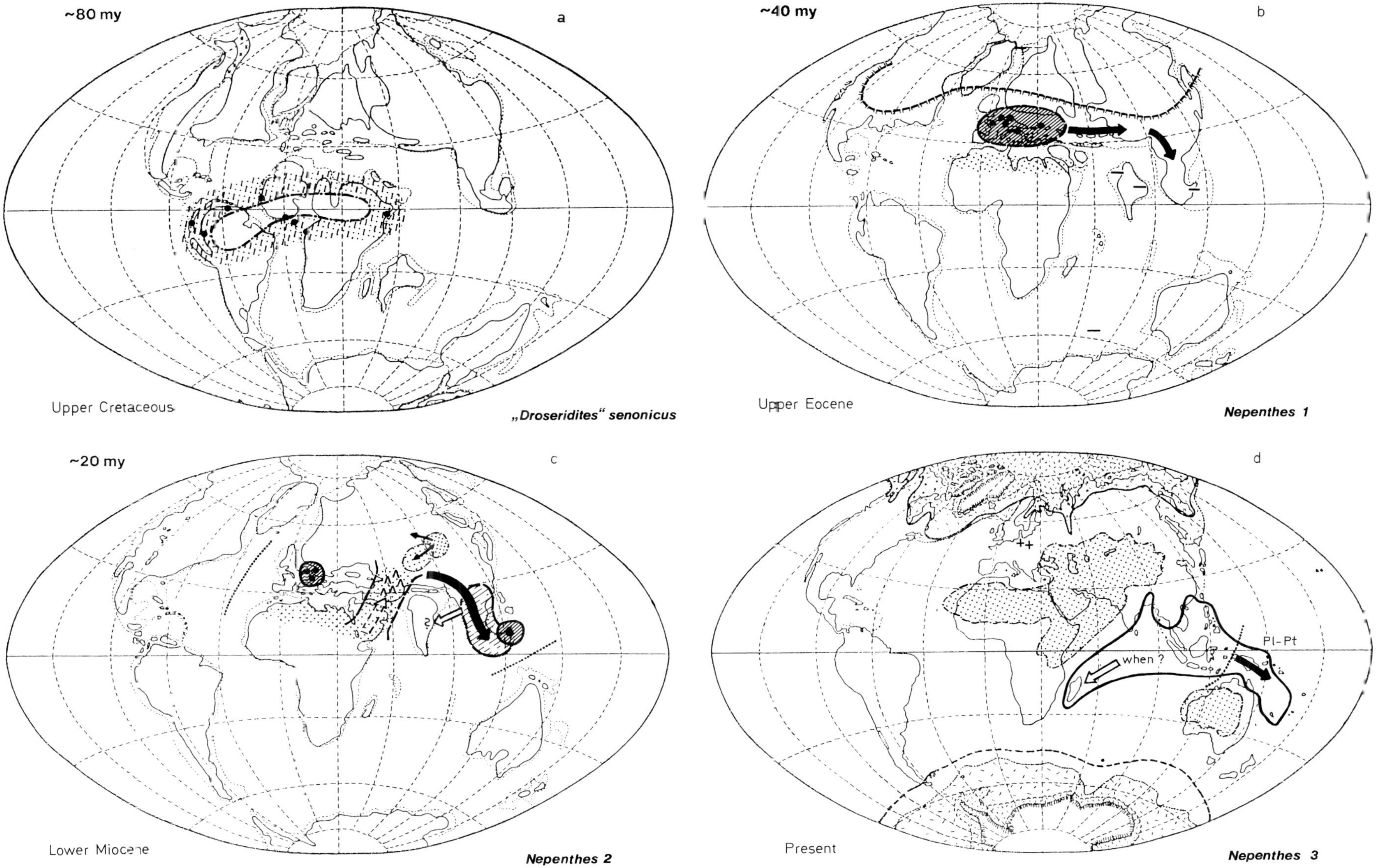
~80 my
a
Upper Cretaceous
„Droseridites“ senonicus
~40 my
b
Upper Eocene
Nepenthes 1
~20 my
c
Lower Miocene
Nepenthes 2
d
when ?
Pl-Pt
Present
Nepenthes 3

the mountain- and drought-barriers (in Iran/Afghanistan) at the end of the Paleogene to the early Neogene (Fig. 12 c). Borneo was reached in the Miocene (MULLER 1974). S. and SE. Asia became not only a refugial area, but also a new, secondary centre of diversity and development for the genus. In C. Europe, *Nepenthes* survived, the Oligocene and occurred in the Lower Miocene as a disjunct relic in a few warm phases and ecological niches (e.g., in the Palustricum) before it died out there in the Middle Miocene.

In the Pliocene/Pleistocene (Fig. 12 d), *Nepenthes* crossed the WALLACE-line in E. Indonesia and reached its present E. boundary in the Pleistocene. It probably migrated from Burma to India, but it is uncertain when this occurred and how it reached its present southern limits (in S. India, Sri Lanka, the Seychelles, and Madagascar).

This paleochorological sketch probably can be developed in more detail with the help of further fossils, particularly from outside of Europe (e.g., from N. America).

An important note concerns the forms, known in the literature as *Droseridites senonicus* (JARDINE & MAGLOIRE 1965, HERNGREEN & CHLONOVA 1981; and others). They are similar to *Nepenthes* pollen, formerly also placed in the pollen genus *Drosiderites* (KRUTZSCH 1985). However, *Drosiderites* appears much earlier in other regions of the earth and in less humid climates (Fig. 12 a). It is limited to the equatorial Cretaceous, the so-called palm-savannah belt of S. America and Africa. Recent personal observations in the S. Arabian region give proof of the existence of *D. senonicus* in the Middle to late Upper Cretaceous. The botanical placement of *D. senonicus* is not yet possible; perhaps these forms belong to palms, since some of the pollen grains open in a monosulcate manner. Their wall is usually thicker than those of the European *Nepenthes* types of the old Tertiary. On one side there is a sulcus-like thinning of the wall, but no folding. Similar loosely spinous, but clearly monosulcate pollen forms evidently are from palms and occur frequently in these samples. SEM analyses are needed to interpret *Droseridites*, which, despite its name, certainly does not belong to the *Droseraceae*!

The following literature data are either uncertain or have nothing to do with the *Droseridites* in the sense of its type species *D. spinosus* (COOKSON 1947: Tertiary Kerguelen): (a) *Droseridites* cf. *spinosus* (NAGY 1985: Plate CV, Figs. 8 – 10; Hungarian Miocene). (b) *Spinulotetradites juxtus* (KAR 1985; Plate 16, Figs. 10 – 12; lower Eocene from Kutch, India), a tricolporate, short-spinous type. (c) *Droseridites parvus* (DUTTA & SAH 1970: Plate 7, Figs. 6 – 8; lower Eocene from Assam, Cherry-fm), with a wart-like sculpturing, is presently impossible to interpret; it is said to also occur in the Paleocene (Tura-fm) of the *Retialetes emendatus* zone in Assam (SAH & SINGH 1974). (d) Regarding the type species of *Droseridites* COOKSON and the fossil from the old Tertiary of Spitzbergen (MANUM 1962) see KRUTZSCH (1985). Neither *Droseraceae* nor *Nepenthaceae* can be linked at present with these four to five types.

Summary. As noted earlier (KRUTZSCH 1985), *Nepenthes* was not an original member of the SE. Asian flora, but rather a representative of the N. Tethyan evergreen vegetation zone (cf. AXELROD & RAVEN 1974: 581). Its original area disintegrated in the Middle Tertiary, during which a C. European-SE. Asian disjunction arose. In the Middle Miocene, the European relics died out under the pressure of more temperate elements. A new secondary centre of diversity developed in SE. Asia, which expanded in the Pliocene/Pleistocene to NE. Australia and New Caledonia and, in an unknown way and period of time, to Madagascar.

Pachysandra and *Sarcococca* (*Buxaceae*)

Both genera (Figs. 13, 14) differ in many characteristics. However, they do possess a common, crotonoid-periporate pollen type, which in the last 25 years has been frequently found as fossils and is classified as the pollen genus *Erdtmanipollis* KRUTZSCH. It is not possible to palynologically distinguish between the fossils of both extant genera. In addition, there are also a few carpological remains from the European Tertiary.

In contrast to *Buxus, Pachysandra* and *Sarcococca* are morphologically advanced taxa. However, their fossil record antedates by 40 to 50 million years that of *Buxus* and diverse buxoid types, making them approximately twice as old (MULLER 1984).

Chorologically, both genera conform to typical E. Asian-N. American disjunctions, with *Sarcococca* in the tropical-subtropical and *Pachysandra* more in the temperate zone. They are represented in E. Asia by 3 and 8 species, respectively, and in N. America by 1 species each. This typical ASA GRAY disjunction did not arise via Beringia (LEOPOLD & MACGINITIE 1972), but by disintegration of a primary European area. 3 to 4 paleochorological main phases can be distinguished, the first (Fig. 14 a) dating from middle to the late Upper Cretaceous (C. European Santonian to Campanian, in the centre of the *Normapolles* region of the warm N. Tethyan belt.

Besides *Erdtmanipollis*, other similar fossil pollen genera from C. Europe are known from this early phase. They have been classified (KRUTZSCH 1966 etc.) as old extinct

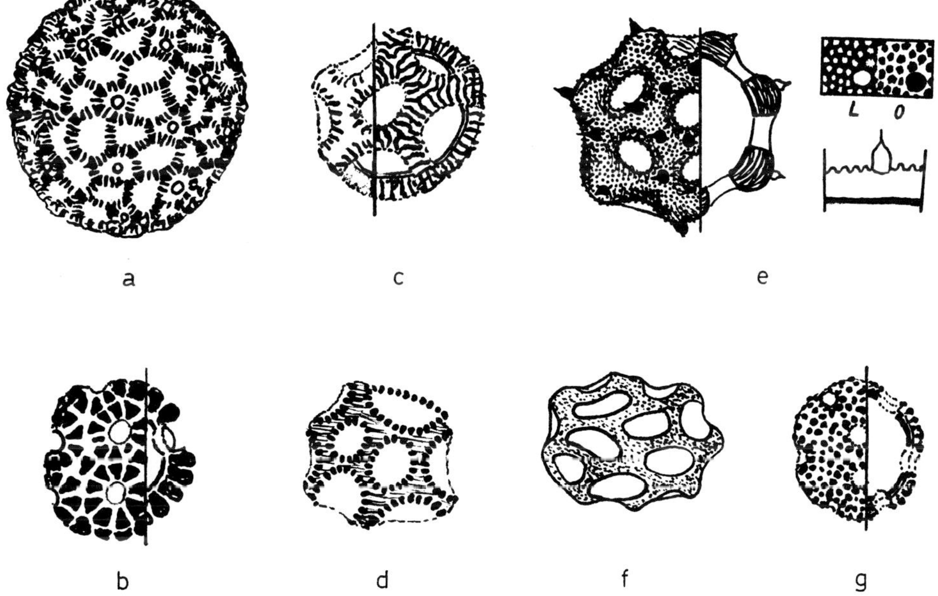

Fig. 13. Ancient pollen genera related to *Buxaceae* from C. Europe (with stratigraphic data). *a Erdtmanipollis* (div. species, Campanian-Pliocene), *b Grootipollis* (2 species, Campanian-Maestrichtian), *c Cristaepollis* (4 species, Campanian-Maestrichtian/Danian), *d Monocirculipollis* (1 species, Maestrichtian/Danian – earliest Eocene), *e* and *f Reductipollis* (2 species, Maestrichtian/Danian; reminiscent of *Buttatina*), *g Graniperipollis* (2 species, Coniacian/Santonian and Maestrichtian/Danian); × 1000

Buxaceae types (*Cristaepollis, Monocirculipollis, Reductipollis, Graniperipollis* and *Grootipollis*), despite the lack of evidence other than pollen morphology (Fig. 13). They occur in the Coniacian to Santonian of S. Bohemia (C.S.S.R.), in the Aachen Campanian, and in N. Germany from the Maastrichtian to the Paleocene, with occasional examples in the early Eocene (KRUTZSCH 1970c). As they are rare in otherwise rich palynofloras, it is possible that they may yet be found in other areas.

By the end of the upper Cretaceous and Paleocene, representatives of *Erdtmanipollis* had spread to NW. America. The claim by LEOPOLD & MACGINITIE (1972: 185, Fig. 14) that this group arose in N. America, is due to an ignorance of the European fossil material of the last decades. In addition, *Erdtmanipollis* has been found in California (CHMURA 1973), Montana (OLTZ 1969), Wyoming (STANLEY 1965), Alberta, Canada (SNEAD 1969), and the Yukon area (LEOPOLD & MACGINITIE 1972, SRIVASTAVA 1972, FARABEE & CANRIGHT 1986, and others). The question whether these northerly finds are fully autochtonous, or reached these regions later, at least partly by tectonic shifts, remains unanswered.

In the Eocene (Fig. 14b), *Erdtmanipollis* no longer occurred in northern latitudes of N. America, but had retreated to the Rocky Mountain areas, California, and the Gulf coast of Alabama TSCHUDY (1973). In Europe, the group is represented by 2 to 3 pollen species in the Lower and Middle Eocene and by Upper Eocene macrofragments of *Pachysandra* from the Geiseltal (MAI & WALTHER 1985). In W. Siberia, there are finds from the Upper Paleocene to Upper Eocene from the lower Ob area (MČEDLISHVILI & SAMOILOVIČ 1961), but in E. and SE. Asia, further evidence is absent.

By the Miocene (Fig. 14c), the ± continuous area of *Erdtmanipollis* had fallen apart into three disjunctions: a N. American, an European, and a (new) E. Asian-Japanese. For N. America, see LI (1951), GRAY & SOHMA (1964), LEOPOLD & MACGINITIE (1972), and others. The European finds are concentrated in the Upper Oligocene and Lower Miocene (KRUTZSCH 1962, 1966, 1970); in addition, there are macrofossils of *Sarcococca* (MAI 1970).

By the end of the Miocene, *Erdtmanipollis* disappeared in the western states of the U.S.A., perhaps due to the expanding drought. Both *Pachysandra* and *Sarcococca* survived the Pleistocene with only one species each in the SE. U.S.A. and Guatemala. In Europe, *Erdtmanipollis* died out during the Upper Pliocene (KRUTZSCH 1966). In E. Asia, nothing concrete is known about the recent *Buxaceae* history. Extant occurrences can only be interpreted as refugial areas from the former N. Tethyan zone. Expansions of *Sarcococca* into N. India, Java, and the Philippines, may be quite young (Pleistocene).

Summary. *Erdtmanipollis* has a relatively simple chorological history which began in C. Europe. It first expanded in the warm-humid N. Tethyan floristic belt westward to Pacific N. America, and for a short period northward to Alaska. Then it expanded from C. Europe eastward to Asia with occasional northward movements along the W. Siberian sea during the older Tertiary. In the middle to late Upper

---

Fig. 14. Paleochorology of *Erdtmanipollis* (*a*–*c*) and present distribution (*d*) of *Pachysandra* and *Sarcococca* (*Buxaceae*). *a* Total area of *Erdtmanipollis* and related ancient pollen genera (△, cf. Fig. 13) during the Upper Cretaceous (– – –), *b* and *c* with macrofossils of *Pachysandra* (P–M) and *Sarcococca* (S–M)

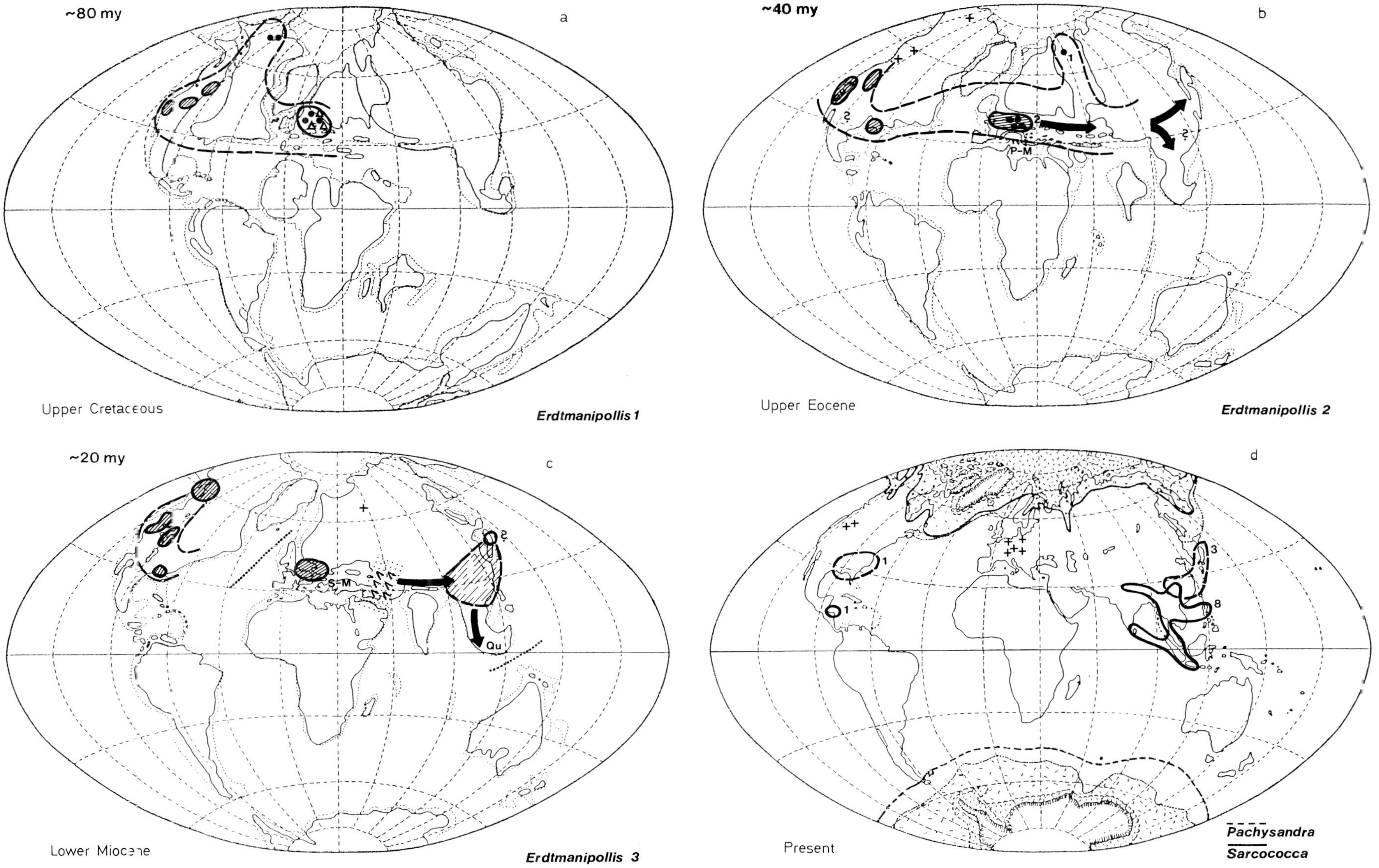

~80 my
a
Upper Cretaceous
Erdtmanipollis 1
~40 my
b
P-M
Upper Eocene
Erdtmanipollis 2
~20 my
c
S-M
Qu
Lower Miocene
Erdtmanipollis 3
d
Present
Pachysandra
Sarcococca

Cretaceous of C. Europe, when *Erdtmanipollis* first appeared, related fossil pollen forms have been found. They can be regarded as early, soon extinct *Buxaceae*, which indicates an European origin for the family.

Restionaceae

Today, this family (Fig. 15) includes 25–30 genera and 300–400 species, almost all from the S. hemisphere, with centres of frequency in S. Africa and W. and E. Australia. For taxonomic reviews, see MELCHIOR (1964) and CUTLER (1972); for ecological and climatic requirements, HOCHULI (1979: 126) and CAMPELL (1964); for pollen morphology, CHANDA (1966) and LINDER & FERGUSON (1985).

The only fossils are pollen finds, united in the pollen genus *Milfordia* (according to ERDTMAN 1960). Today there are 5 to 6 known pollen species (or types), which can be classified according to their apertures into two main groups, a wide- or large-pored (ulcerate) and an annulate or small-pored (gramineoid) type, linked by intermediates. The needle-point perforation of the foveolate-verrucate exine sculpture is typical for most of the *Restionaceae* (and a few *Centrolepidaceae*); only here do such structures occur.

Some authors classify the two main types into different pollen genera: *Milfordia* and *Restioniidites* ELSIK. The former group converges into the extant *Hypolaena laterifolia* morphotype, the latter into the *Restio subverticillatus* morphotype. For more information, see MULLER (1981, 1984), LINDER & FERGUSON (1985), and others. Critical summaries of the fossil evidence of the last 30 years can be found in KRUTZSCH (1970 a, older synonyms), KEDVES (1965), HOCHULI (1979), and MULLER (1981, 1984). HOCHULI also comments on the relationships of the macrofossils of *Rhizocaulon* SAPORTA, a problem not yet solved.

Paleochorology. Although HOCHULI (1979) has already written much on this subject, entering the data on a series of paleogeographic maps allows a clearer understanding of the subject and illuminates the role played by the *Restionaceae* in the N. hemisphere. The present maps also contain additions and some unpublished evidence.

The temporal and spatial development of the pollen genus *Milfordia* can be divided into 6 main phases (Fig. 15 a–f), covering a long and highly differentiated history, beginning in the Middle Cretaceous (Fig. 15 a). Although MULLER (1984) expressed doubts concerning the hitherto oldest find (in the Alb of the deep-sea drilling DSDP, leg. 53, site 418 B, Wn. middle Atlantic; HOCHULI & KELTS 1980, HOCHULI 1979: 125), there seems to be no reason to classify this object otherwise, except perhaps as a separate species of *Milfordia*. It is possible that the object's present position, c. 200 km ESE of Puerto Rico, was once much nearer to Africa than to the Caribbean area and moved with the ocean floor to its present location millions of years ago (Fig. 15 a).

From the second phase (Upper Cretaceous, Fig. 15 b), the only evidence is from Africa, which suggests that Africa was this group's cradle. Evidence comes from Senegal (*M. jardinei*), C. Africa (questionable), Nigeria (LAWAL & MOULLADE 1986), Kamerun (BOLTENHAGEN & al. 1985: 179), and the S. Arabian region (pers. obs., middle and late Upper Cretaceous, at least 2 different species). There are possible macrofossils of *Rhizocaulon* from the Tethys area of S. Europe, but they have not been proven to be *Restionaceae* yet. All of the Upper Cretaceous *Restionaceae* finds mentioned probably belonged to the palm-savannah belt, which surrounded the developing and expanding tropical-equatorial rain forest region during the Upper Cretaceous (KRUTZSCH, in prep.). No *Restionaceae*-like finds as old as these have been made in the fossil microfloras of Europe or the S. hemisphere (southern S. America, S. Africa, etc.).

In the third phase (transition from Cretaceous to Tertiary/Paleocene, Fig. 15 c), the first demonstrable large expansion of *Milfordia* took place, first in Europe (S. Spain, KEVDVES, in prep; later elsewhere) and S. America (S. Argentinia: ARCHANGELSKI 1973). It is unclear which land or island-hopping route the forms followed to Europe. Two routes are possible: From Senegal over NW. Africa (fossils: MOHR 1986) to SW. Europe and/or from the S. Arabian area over island chains to the Near East.

In the fourth phase (up to the Eocene, Fig. 15 d), the expansion continued rapidly. The oldest evidence comes from the N. Tethys in N. America, beginning with the middle Wilcoxian (FAIRCHILD & ELSIK 1969, and others). In the E., the Turgai area (ČIGURJAJEVA 1956) was reached, but the expansion towards the E. certainly reached further in the evergreen laurophyll zone. At the end of the Eocene and/or beginning of the Oligocene, S. Africa (Kapland: THIERGART & al. 1963) and S. Australia were reached, starting from southern S. America via W. Antarctica (HEKEL 1972, MARTIN 1978, and others). There is no Eocene evidence from the wet tropical African zone, probably because *Restionaceae* are pronounced heliophytes. In the European Eocene, *Milfordia* occcurred in abundance in palustrine and coastal paralian areas, where its pollen can make up to 20% of the total. In C. Europe alone, there are more than 100 such sites (mostly unpublished data).

In the fifth phase (Oligocene, Fig. 15 e), *Milfordia*'s well-documented spread continued from W. Europe to C. Asia. In W. Siberia, it reached the northernmost point in its history (Tasow Peninsula, Middle Oligocene: LJUBOMIROVA 1960). There is no evidence from the Oligocene of southern and southeastern N. America; the date of extinction there is uncertain and no evidence from the N. American Tertiary has been available to me. Fossil appearance in Egypt and the Sudan (KEDVES 1985), may be linked to Europe or NW. Africa.

Milfordia probably existed from the Eocene up to the present in southern S. America and in SE. to SW. Australia, although there is no concrete fossil evidence to date. Separate developments must have continued over millions of years in S. Africa and Australia, leading to the family's present diversity. The two present centres have no common species or genus, a fact corroborated by the differentiated pollen types in both areas (LINDER & FERGUSON 1985). In the Oligocene (COUPER 1960) or somewhat later (MILDENHALL 1980), the *Restionaceae* migrated from SE. Australia to New Zealand and also reached the Oligocene islands of the Ninety East Rift in the Indian Ocean (KEMP & HARRIS 1977). Its northward expansion in Australia proceeded mainly in the Miocene, during which it adapted to the semihumid and increasingly arid climatic conditions. New Guinea probably was settled from the S. towards the end of the Tertiary and during the Pleistocene.

The single isolated species, which occurs today on the SE. Asian mainland and does not belong to an Australian genus, may represent a last relic from the N. Tethyan distribution area of the family, pushed towards the SE. in the older Tertiary. The *Restionaceae* did not cross the WALLACE-line from the SE. nor come near it in the NW. and there is no evidence from the Indonesian region (excepting Irian,

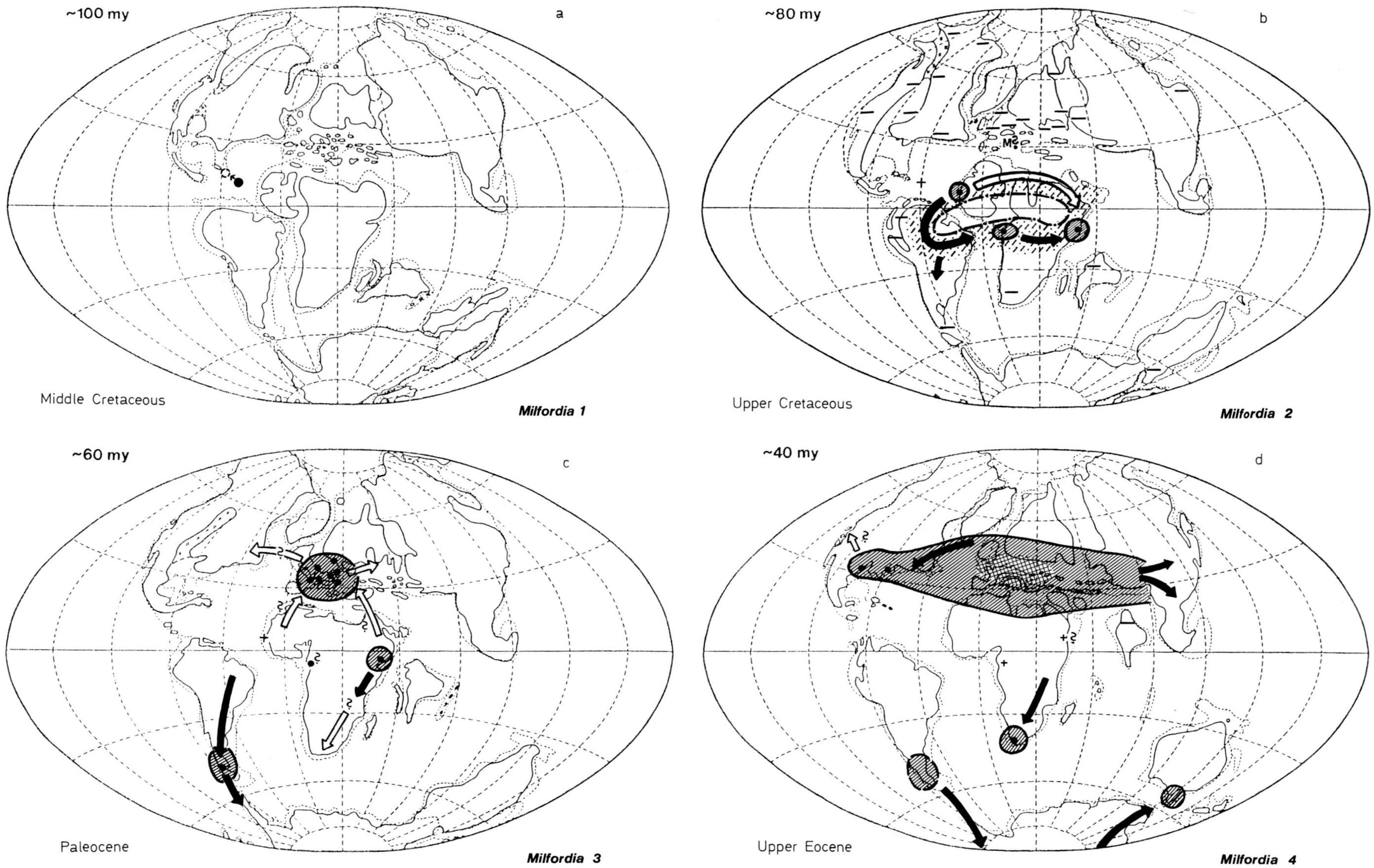

~100 my
a
Middle Cretaceous
Milfordia 1
~80 my
b
Upper Cretaceous
Milfordia 2
~60 my
c
Paleocene
Milfordia 3
~40 my
d
Upper Eocene
Milfordia 4

Fig. 15

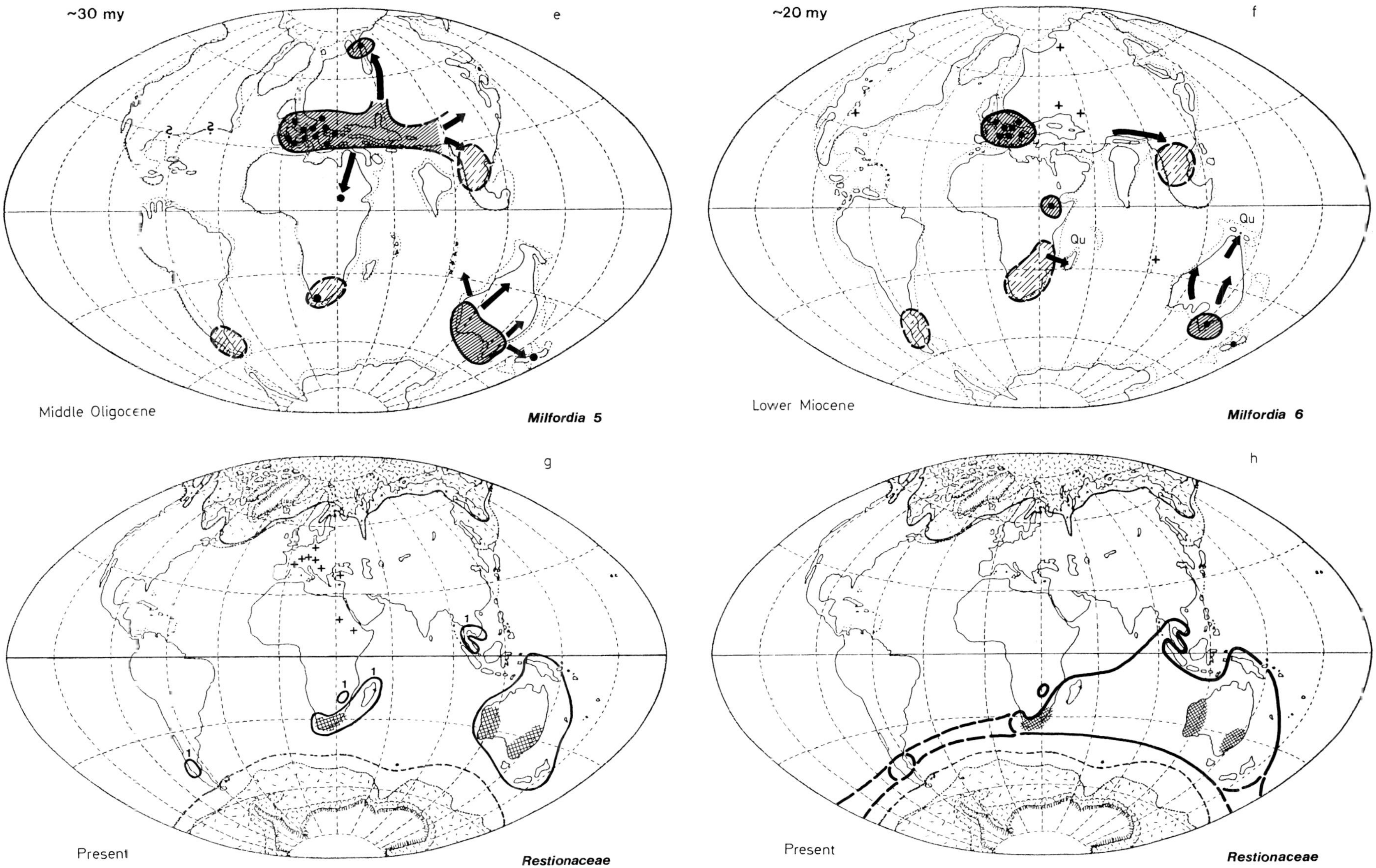
~30 my
e
Middle Oligocene
Milfordia 5
~20 my
f
Qu
Qu
Lower Miocene
Milfordia 6
g
1
1
1
1
Present
Restionaceae
h
Present
Restionaceae

Fig. 15

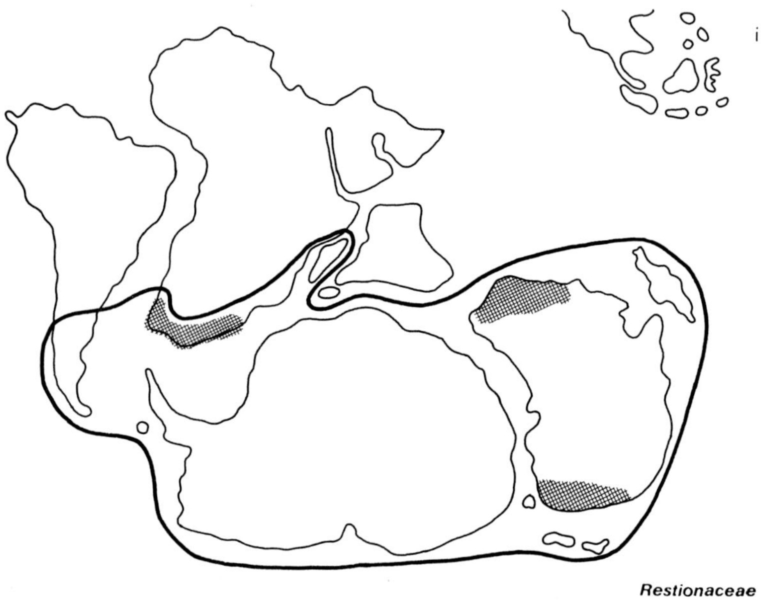

Fig. 15. Paleochorology of *Milfordia* (*a–f* pollen only) and present distribution (*g*) of *Restionaceae*). *a* Deep see drilling with first occurrence of *Milfordia*, and possible original position (arrow), *b* African pollen fossils of *Milfordia*, and main macrofossil find of *Rhizocaulon* (M?); *h, i* interpretation of present distribution according to extension of a S. Hemisphere belt area (*h*) or fractionation of a Gondwanaland area (*i*; after AXELROD & RAVEN 1974)

whereas the *Centrolepidaceae* are a somewhat different case (see map in MULLER 1984).

The N. Tethyan distribution belt disintegrated during the later Tertiary, with the last C. European relics disappearing at the beginning of the early Pliocene. Already during the Miocene, their occurrence was limited to ecologically favourable niches and phases. Their decrease in the N. Tethyan flora probably is due to successive displacement by the more vigorous and modern *Cyperaceae* and *Gramineae* (HOCHULI 1979).

The chorological history sketched here for the *Restionaceae* is in stark contrast to other authors' previous interpretations. They link the occurrence in southern S. America partly with S. Africa and partly with New Zealand (R. SCHUSTER 1974, SMITH 1973). The SE. Asiatic species is supposed by HOCHULI (1979) to have reached the N. hemisphere from the S. (Fig. 15 h), rather than to have survived as a relic. The plate tectonic oriented map with the distribution presented by AXELROD & RAVEN (1974) (Fig. 15 i) shows an almost ideal S. hemisphere distribution area, except for the family's unexplainable absence in India. However, this interpretation turns out to be non-historical and inadequate when the fossil data are included as evidence of the historical relationships.

The case of the *Restionaceae* exemplifies the limitations of chorological interpretations based on extant distribution data only, a situation which becomes more problematic with increasing age of the group (cf. HARTMANN 1986). First, it is necessary to recognize the general patterns in the temporal/spatial differentiation of cenophytic woody plants on the basis of paleochorological analyses. Only then will one be able to interpret complicated extant distribution patterns in conjunction with taxonomic or cladistic analyses and ± without fossil evidence (p. 43). However, such interpretations generally will be conclusive only in relatively young groups (e.g., from the Holarctic).

A mere comparison of the distribution patterns of the *Restionaceae* with those of the *Podocarpus, Araucariaceae, Proteaceae, Nothofagus, Dilleniaceae, Gunneraceae, Epacridaceae* (cf. Hochuli 1979: 119) as set forth by Axelrod & Raven (1972) is inadequate. Each group first has to undergo a paleochorological examination before a comparison can lead to a typification of extant distribution areas. This is particularly important for ancient groups such as *Podocarpaceae* and *Araucariaceae*, whose origins lie deep in the Mesozoic.

Summary. The present-day distribution area of the almost purely S. hemisphere *Restionaceae*, now split into 4 to 5 major disjunctions, does not, as has been supposed, represent the vestiges of a disintegrated Gondwanaland area. Instead, the occurrences in S. America, S. Africa, Australasia, and SE. Asia are the results of long migrations.

Beginning in the savanna and palm zone of Africa, the *Restionaceae* spread in different directions and with different speeds during the Middle and Upper Cretaceous. They reached S. Africa and S. America and from there, via Antarctica, S. Australia, subsequently New Zealand, and later New Guinea. At the beginning of the Tertiary, they spread northwards into the N. Tethyan warm-humid (occasionally semihumid) vegetation belt, forming a centre of diversity there. Although they have survived in S. America, S. Africa and Australia in disjunct relic areals, where some new centres of evolution and diversity have developed; they died out in the N. during the Neogene. Only one relic species still occurs in SE. Asia. Their disappearance probably began in southern N. America and ended in C. Europe in the Lower Pliocene. It is important to note their complete absence in India and Indonesia. They did not cross the Wallace-line from the SE. nor come near it in the NW. But they persisted in the "southern Tethys area", i.e., in N. and NE. Africa from the Eocene (Morocco) into the Oligocene (Egypt, Sudan), and in E. Africa (Ethiopia) into the Oliocene/Miocene (Ethiopia), perhaps by exclusion from the N. Tethys area.

Ascarina (*Chloranthaceae*) and *Austrobaileya* (*Austrobaileyaceae*)

Ascarina (with 11 species) is one of the 5 genera of *Chlorantaceae. Austrobaileyaceae* represents a monotypic family. The present distribution areas are W. Pacific, with *Ascarina* (resp. *Ascarinopsis*) also extending to Madagascar (Fig. 16g). The only fossil evidence is pollen. It belongs to the oldest, or one of the oldest known angiosperm types, first characterized by Couper (1958) as *Clavatipollenites* from the Barremian of Great Britain. The Mesozoic and S. hemispheric Tertiary *Clavatipollenites* data have been summarized in tabular form by Muller (1981). This represents a first step towards our understanding of the complex paleochorology of this primitive angiosperm group. When transferred to corresponding paleogeographic maps, this chorological development becomes more clearly visible.

In the meantime, Muller's analysis has been enlarged by further data (e.g., Burger 1981) and the excellent ultrastructural morphological analysis by Walker

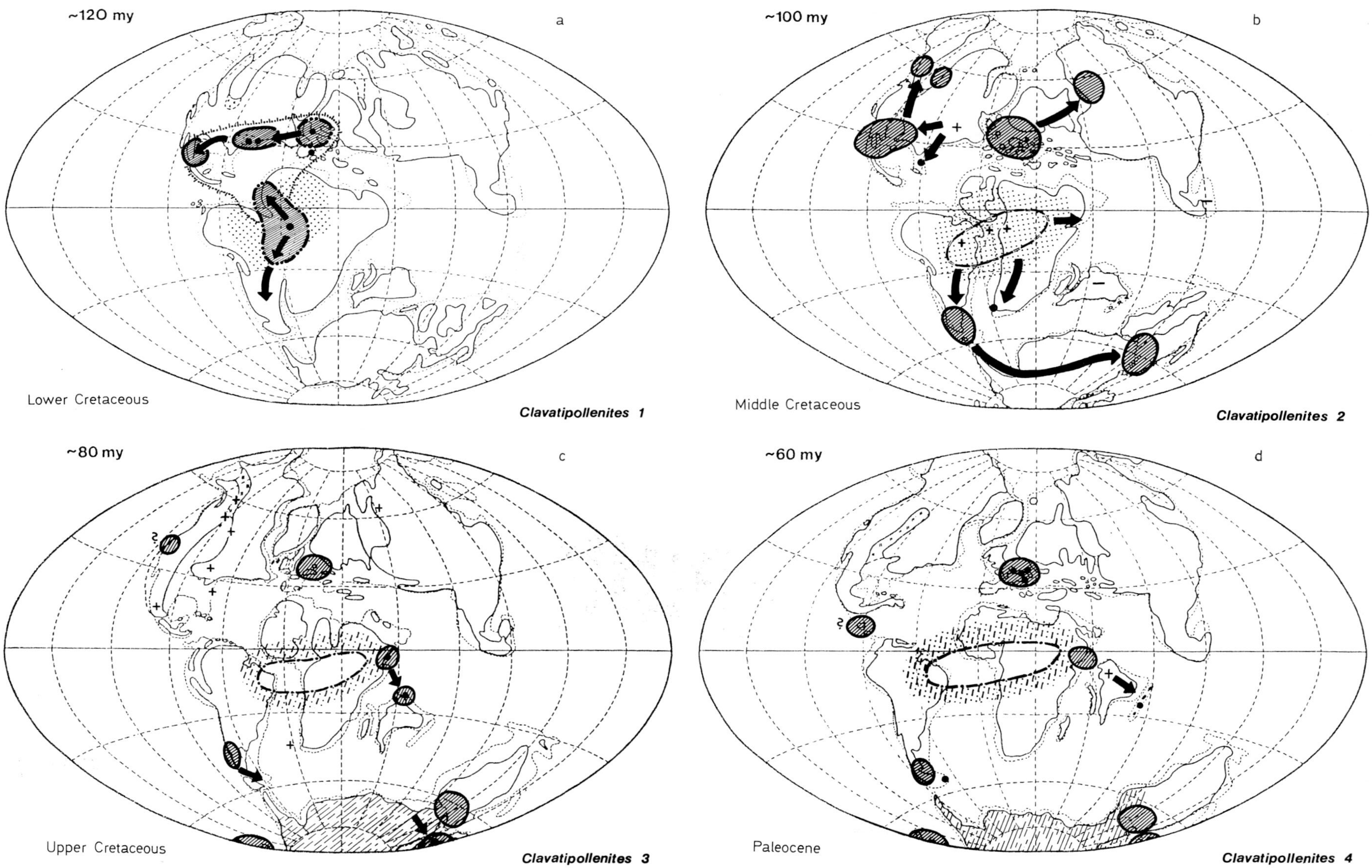
~120 my
a
Lower Cretaceous
Clavatipollenites 1
~100 my
b
Middle Cretaceous
Clavatipollenites 2
~80 my
c
Upper Cretaceous
Clavatipollenites 3
~60 my
d
Paleocene
Clavatipollenites 4

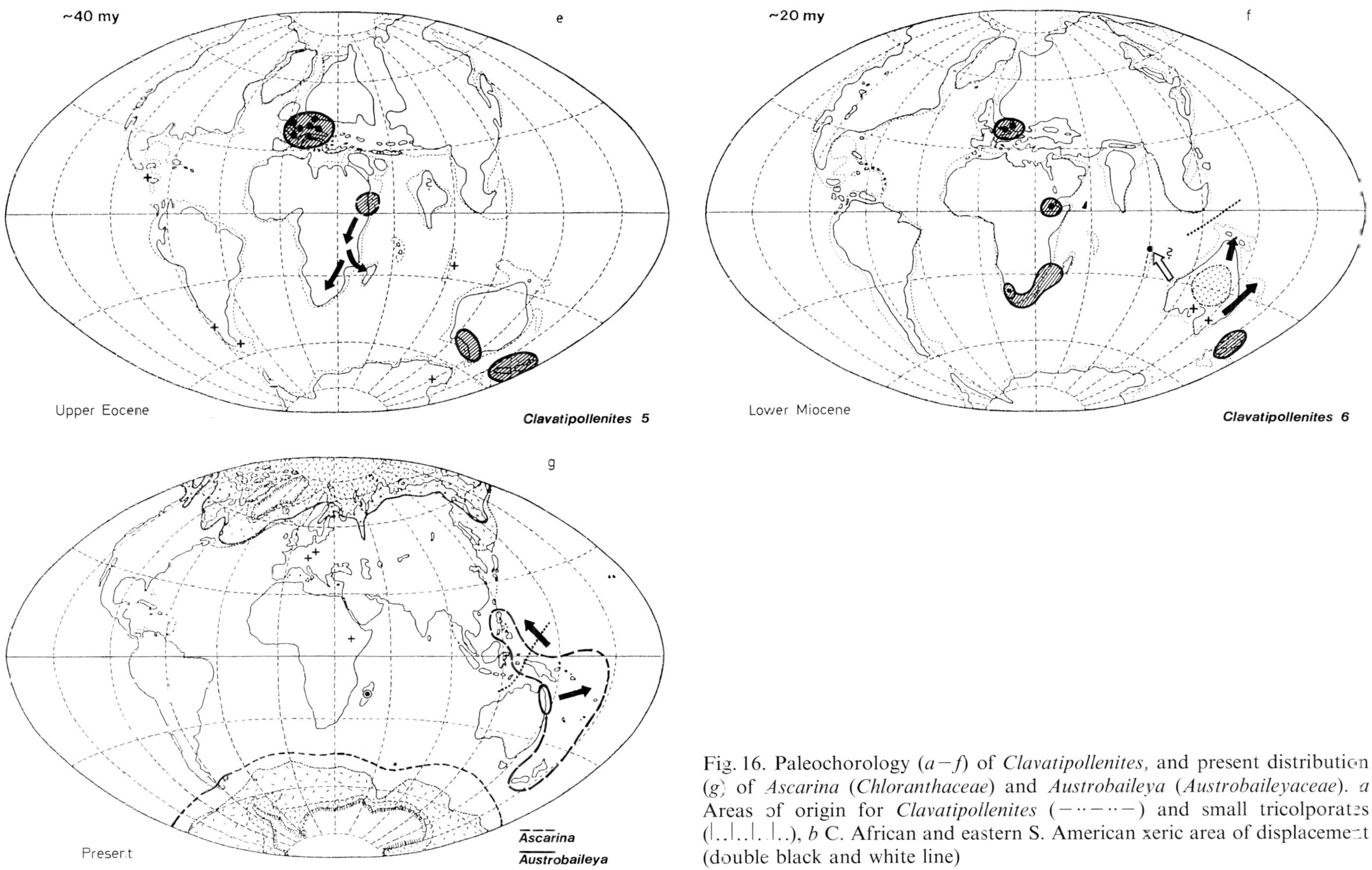

Fig. 16. Paleochorology (*a–f*) of *Clavatipollenites*, and present distribution (*g*) of *Ascarina* (*Chloranthaceae*) and *Austrobaileya* (*Austrobaileyaceae*). *a* Areas of origin for *Clavatipollenites* (—‥—‥—) and small tricolporates (|..|..|..|..), *b* C. African and eastern S. American xeric area of displacement (double black and white line)

& WALKER (1985) has allowed, to a certain degree, linking of *Clavatipollenites* to the extant genus *Ascarina*. Up to now, the *Austrobaileyaceae* also were being considered in this respect (ENDRESS & HONEGGER 1980), but this has been rejected by MULLER (1981). The genera are usually classified as *Piperales/Chloranthales* and *Laurales/Magnoliales*, respectively. But their general affinity is likely (MEEUSE 1975, RÜFFLE 1986/87).

The historical analysis by MULLER (1981) and BURGER (1981) has been supplemented by new finds in different parts of the world (Africa, Arabia, U.S.S.R.) and the reconsideration of a previously ignored European fossil. In 1952, 6 years before COUPER (1958), a pollen species was described (MÜRRIGER & PFLUG) from the Hessian upper Oligocene of the Mainz basin (pit "Emma" near Marxheim). It was named *Inaperturopollenites emmaensis* and interpreted as a nonsaccate conifer. This and similar forms are wide-spread, although occurring infrequently throughout the European Tertiary. The palynologists who worked on the Mesozoic and Tertiary during the 1950's to 1970's did not coordinate their pollen nomenclature. Thus, it did not occur to anyone that these two form groups, the *I. emmaensis* group of the European Tertiary and the *Clavatipollenites* of the late Mesozoic, could be very similar or even identical. Even the present author erected a separate genus *Emmapollis* for the former group and described a separate pollen species, *E. pseudoemmaensis* (KRUTZSCH 1970: Atlas VII, Pl. 16), as the Marxheim typus no longer exists and the old illustrations are vague. The SEM analysis of *Emmapollis* is under way, in collaboration with M. KEDVES, but it is undoubtedly closely related or even identical to *Clavatipollenites*. *Ascarina* evidence from the Parisian basin Eocene (GRUAS-CAVAGNETTO 1978) has been questioned by MULLER (1981: 12) and others because it is isolated. The *Emmapollis-Clavatipollenites* group can be traced back at least to the Middle Cretaceous in C. Europe (e.g., Cenoman, Königstein, Saxony: KRUTZSCH 1963) and extends up to the Miocene. It had a habitat preference similar to that of *Milfordia,* that is, it occurred in open moors and paralic regions. All this allows an enlargement of the paleochorological picture (as interpreted by MULLER) as a group extending into the N. Tethyan area of Eurasia, somewhat similar to the *Restionaceae,* but with a different timing.

Paleochorology. Seven phases can be reconstructed on maps (Fig. 16 a – g). An origin can be placed in the Lower Cretaceous of the W. African, NW. Gondwanaland-C. Atlantic area (similar to the origins of the earliest tricolporate angiosperm pollen: DOYLE 1977, and others). The primary phases of expansion could be more clearly illustrated by dividing them into a pre-Apt, Lower Apt, Upper Apt, Alb, and Cenoman phase (cf. KREMP 1978, BURGER 1981, and others). Up to the Middle Cretaceous, expansion continued into the NW. regions (Texas, Mexico, N. Caribbean) and from W. equatorial Africa towards S. America (Fig. 16 a).

In the second main phase (Middle Cretaceous, Fig. 16 b), NW. and C. Canada and starting in C. Europe, W. and NW. Asia are reached (*Clavatipollenites incisus* CHLONOVA 1984, Alb/Cenoman). The group appears in southern S. America in the Apt and from there spreads via Antarctica to S. Australia, where it is first found in the Alb (BURGER 1981). In C. Africa and adjacent S. America, however, the group dies out in the Middle Cretaceous as a result of dry climatic conditions.

In the third main phase, in the Upper Cretaceous (Fig. 16 c), *Clavatipollenites* is represented in E. Africa and the S. Arabic region, albeit infrequently. From here, it can be followed (together with *Ctenolophon*) to NW. India (Kutch: Mathur & al. 1970).

However, the latest finds from the Lower Eocene of Kutch (Kar 1985), which were classified as *C. cephalus*, are questionable from the pictures published. In New Zealand, members of the group spread during the Upper Cretaceous via Antarctica to the islands in the Campbell Plateau, where they still exist today. In Australia they probably migrated northwards along the E. coast during the early Tertiary, crossing the Wallace-line NW. of New Guinea towards the end of the Tertiary before stopping at their present boundaries in the Philippines. According to the *Ascarina* interpretation of *Clavatipollenites,* they have died out in Australia. If *Austrobaileya* is not exluded from the *Clavatipollenites* complex, they have survived in a small refuge area in NE. Australia. In the N. hemisphere outside of C. Europe, there are only doubtful Californian fragments (*Cl. evittii,* Chmura 1973) from the Upper Cretaceous.

In the fourth main phase, the Paleocene (Fig. 16 d), *Clavatipollenites* may have survived as relics on various C. American islands from the Cretaceous. But as of the Paleocene, that is, 45 my ago, they apparently completely disappeared from the Americas.

The old, wide (but not world-wide: Endress & Honegger 1980) Cretaceous distribution area of *Clavatipollenites* thus had already ceased to exist at the beginning of the Eocene. In the Eocene (Fig. 16 e), only sporadic evidence occurs outside of Australia, in E. Africa (Ethiopia) and C. Europe. In the Oligocene the distribution area again extends southwards (as first in the Middle Cretaceous). Furtheron, Madagascar is reached via the then at least partly dry Madagascar channel (Förster 1975). On the island members of the group still exist as relics (*Ascarinopsis,* resp. *Ascarina,* sect. *Madagascarina*; Jérémie 1980). An immigration from the east – as formerly assumed by Axelrod & Raven (1974) – is quite unreliable. In the Lower Miocene fossils (Fig. 16 f) are known from the western Cape Province (Coetzee & Muller 1984). Populations finally are eliminated by the advance of Mediterranean climate, others persist in E. Africa and C. Europe until they become extinct during Miocene/Pliocene.

The geographical reduction of *Clavatipollenites* which had begun in C. Africa during the Middle Cretaceous probably was due to increasing aridity (J. Muller 1981: 11). A similar reduction occurred during the late Tertiary in areas of C. Australia. As a whole, the group has preferred warm and/or humid habitats on coastal plains and estuaries, but also tropical-montane regions (Jérémie 1980).

Summary. The *Clavatipollenites* complex has a history similar to that of the *Restionaceae*. During the late Lower Cretaceous, the forms migrated from the E. middle Atlantic and N. Godwanaland region to Europe, where they spread rapidly and migrated eastwards to W. Siberia. In N. America they spread up to Canada, in Africa and S. America southward via Antartica to Australia. All of this happened within a time period of 10 to 15 my. However, during the Middle Cretaceous, they were displaced from the equatorial African/S. American regions by increasing aridity, surviving in relic areas of N. America and the Caribbean until the Paleocene and in Europe up to the Miocene. With the exception of these European and a few E. and S. African finds and the present relic occurrence in Madagascar, from the Eocene they are limited to the Australasian region, where they start in the New

Zealand Maastrichtian and still exist. From S. Australia, they spread along the W. Pacific coastal zones northward via New Guinea to the Philippines, a process which lasted until relatively recent times. In Australia they were displaced during the late Tertiary as a result of increasing aridity (if considering *Ascarina*) or survived in small relic areals in the NE. (if considering *Austrobaileya*). In both cases, these corresponding extant taxa mainly occupy SW. Pacific "relic areas" showing latest migrations to the Pacific area.

General conclusions

The ten analysis examples clearly show what progress can be made by using paleochorological methods in contrast to interpreting the existing "extant and fossil distribution maps". Of course, corresponding abundant and unequivocal fossil data must be available. It is, however, unimportant which organs are taken into consideration, whether data from palynology, carpology, xylotomy, leaf fossils, etc. are used. The older an analysis object is, the more complicated is its developmental and dispersal pattern. Holarctic taxa, however, have a relatively simple chorological history. When a taxon spreads from a centre (e.g., Africa) to different geographic areas (e.g., N. and S. hemispheres), its further developmental processes are independent, adapting it to the varying orographic, ecological, and climatic possibilities and conditions. New secondary developmental centres or refuge disjunctions may develop, or the taxon may die out. The paleochorological distribution patterns are quite heterogeneous, also with respect to the time periods.

Basically, the how and when for numerous taxa must be worked out before any typification of the historical processes can be achieved. Extant distribution patterns rarely give valid informations about the chorological development. The new method of serial maps can be diversely employed and extended, and should open a new avenue of cooperation between paleo- and neobotany.

I would like to thank my colleagues at the Museum of Natural History, Berlin, and those colleagues in the G.D.R. and Č.S.S.R. who lent me literature, supported me with critical discussions, or helped me with technical work. Special thanks are due to my colleague L. RÜFFLE, without whose stimulating comments and critical discussions this work would have been impossible.

References

ADEGOKE, O. S., CHENE, J. R. E. DE, AGUMANU, A. E., 1978: Palynology and age of the Kerri-Kerri Formation Nigeria. – Rev. Esp. Micropaleont. **10**: 267–283.

ANDERSON, R. Y., 1960: Cretaceous-Tertiary palynology, eastern side of the San Juan Basin, New Mexico. – N. Mex. Bur. Mines Min. Resources Mem. **6**: 1–58.

ANDREANSKY, G., 1951: Die entwicklungsgeschichtliche Bedeutung der monotypischen Pflanzengattungen und der monogenerischen Familien. – Biol Univ. Hung. **1**: 9–13.

ANONYMUS, 1961: Pollen und Sporen Westsibiriens. Jura – Paläozan. – Trudy WNIGRI **117**, Leningrad. (In Russian.)

ARCHANGELSKY, S., 1973: Palinologia del Paleoceno de Chubut. I. Descripciones Sistematicas. – Ameghaniana **10**: 339–399.

ARECES-MALLEA, A. E., 1985: New species of *Bombacacidites* COUPER emend. KRUTZSCH of the middle Eocene in Cuba. – Rev. Technologica ser. Geol. 1, **15**: 3–7.

AUBRÉVILLE, A., 1969: Essais sur la distribution et l'histoire des angiospermes tropicales dans le monde. – Adansonia **9**: 189–247.

AWASHTI, N., 1968: A new fossil wood belonging to the family *Alangiaceae* from the Tertiary of South India. — Palaeobotanist **17**: 322–325.

AXELROD, D. I., 1970: Mesozoic paleogeography and early angiosperm history. — Bot. Rev. **36**: 277–319.

— 1979: Desert vegetation, its age and origin. — In GOODIN, J. R., NORTHINGTON, D. K., (Eds.): Arid land plant resources, pp. 1–72. — Lubbock/Texas: Tex. Tech. Univ.

— 1983: Biogeography of oaks in the arcto-Tertiary province. — Ann. Missouri Bot. Gard. **70**: 629–657.

— RAVEN, P. H., 1972: Plate tectonics and Australasian palaeobiogeography. — Science **176**: 1379–1386.

— — 1974: Angiosperms biogeography and past continental movements. — Ann. Missouri Bot. Gard. **61**: 539–673.

— — 1978: Late Cretaceous and Tertiary vegetation history of Africa. — In WERGER, M. J. A., (Ed.): Biogeography and ecology of Southern Africa, pp. 77–130. — Den Haag: W. Junk.

— — 1985: Origins of the Cordilleran flora. — J. Biogeography **12**: 21–47.

BADRÉ, F., 1973: *Ctenolophonaceae*. — Flora Gabon **21**: 55–58.

BARRON, E. J., HARRISON, C. G. A., SLOAN II, J. L., HAY, W. W., 1981: Paleogeography, 180 million years ago to the present. — Eclogae Geol. Helv. **74**: 443–470.

BARTHEL, M., RÜFFLE, L., 1976: Ein Massenvorkommen von Symplocaceenblättern als Beispiel einer Variationsstatistik. — Abh. Zentr. Geol. Inst./Pal. Abh. **26**: 291–306.

BESSEDIK, M., 1984: The Early Aquitanian and Upper Langhian-Lower Serravallian environments in the Northwestern Mediterranean region. — Paléobiol. Cont. (Montpellier) **14 (2)**: 159–162.

BIAGGI, R. E., FISK, L. H., ENRIQUE, , M.-H., 1976: Plant microfossils from the La Quinta Formation (Oligo-Miocene) of Chiapas, Mexico. — 4. Int. Palyn. Conf., Lucknow/India, **1**: 14.

BOEMBERGEN, S., 1939: A revision of the genus *Alangium*. — Bull. Jard. Bot. Boitenzorg, Ser. 3, **16**: 139–235.

BOER, DE N. P., HAMMEN, TH., V. D., WIJMSTRA, A. A., 1965: Palynological study of an age of some borehole samples from the Amazonas delta area, NW-Brazil. — Geol. Mijnb. **44**: 254–258.

BOLTENHAGEN, E., DEJAX, J., SALARD-CHEBOLDAEFF, M., 1985: Evolution de la végétation tropicale Africaine du Cretacé a l'actuel d'apres les donnés de la Palynologie. — Bull. Sect. Sci., Min. de l'Education Nat. **8**: 165–194.

BURGER, D., 1981: Observations on the earliest angiosperm development with special reference to Australia. — 4. Int. Palyn. Conf. Lucknow (1976/77) **3**: 418–428.

CAMPELL, E. O., 1964: The Restiad peat bog at Motumaoha and Moanatuatua. — Transact. Roy. Soc. New Zealand **2**: 219–227.

CHANDA, S., 1966: On the pollen morphology of the *Centrolepidaceae, Restionaceae* and *Flagellariaceae*, with special reference to taxonomy. — Grana Palynol. **6**: 355–415.

CHANDLER, M. E. J., 1954: Some Upper Cretaceous and Eocene fruits from Egypt. — Brit. Mus. Bull., Nat. Hist., Geol. **2**: 149–187.

— 1961: The Lower Tertiary floras of Southern England 1: Paleocene floras, London clay flora (supplement). — London: British Museum Natural History.

— 1964: The Lower Tertiary flora of Southern England 4: A summary and survey of findings in the light of recent botanical observations. — London: British Museum Natural History.

CHAUDEFAUD, M., 1955: Remarques sur quelques pollens des plantes tropicales particulièrement interessantes des points de vue palynologique ou systématique. — Rev. Gén. Bot. **62**: 641–660.

CHLONOVA, C. A., 1974: Palynology of Cretaceous deposits of Siberia and Far East. — Trans. Inst. Geol. Geophys. Akad. Sci. U.S.S.R., Sib. Branch, **96**: 1—6. (In Russian.)
— 1984: Stratigraphical ranges and distribution of the morphological types of Cretaceous angiospermaceous pollen. — In: Problems of modern palynology, pap. Sovj. palynol. VI. int. palyn. conf. (Calgary, Canada, 1984), pp 13—17. — Novosibirsk. (In Russian.)
CHMURA, C. A., 1973: Upper Cretaceous (Campanian-Maestrichtian) angiosperm pollen from the Western San Joaquim Valley, California. — Palaeontographica B, **141**: 89—171.
ČIGURJAJEVA, A. A., 1956: Atlas der Mikrosporen aus tertiären Ablagerungen der UdSSR. — Charkow, Saratow. (In Russian.)
COETZEE, J. A., MULLER, J., 1984: The phytogeographic significance of some extinct Gondwana pollen types from the Tertiary of the Southwestern Cape (South Africa). — Ann. Missouri Bot. Gard. **71**: 1088—1099.
COOKSON, I. C., 1947: Plant microfossils from the Lignite of Kerguelen Archipelago. — B.A.N.Z.A.R.E. Reports, Ser. A II **(8)**: 127—142.
— 1957: On some Australian Tertiary spores and pollen grains that extend the geological and geographical distribution of living genera. — Proc. Roy. Soc. Victoria **69**: 41—54.
COOKSON, I. C., PIKE, K., 1954: Some dicotyledonous pollen types from Cenozoic deposits in the Australian region. — Austral. J. Bot. **2**: 197—219.
COUPER, R. A., 1958: British Mesozoic microspores and pollen grains. — A systematic and stratigraphic study. — Palaeontographica B **103**: 75—179.
— 1960: New Zealand Mesozoic and Cenozoic plant microfossils. — New Zealand Geol. Survey Palaeont. Bull. **32**: 1—87.
CROIZAT, L., 1952: Manual of phytogeography. — The Hague: Uitgeverij-Junk.
CUTLER, D. F., 1972: Vicarious species of *Restionaceae* in Africa, Australia and South America. — In VALENTINE, D. H., (Ed.): Taxonomy, phytogeography and evolution, pp. 73—83. — London: Academic Press.
DANERT, S., 1971: Ordnung *Sarraceniales, Nepenthaceae.* — In: Urania-Pflanzenreich. Höhere Pfl. 1. — Leipzig.
DOYLE, J. A., 1977: Patterns of evolution in early angiosperms. — In HALLAM, A., (Ed.): Patterns of evolution, pp. 501—546. — Amsterdam: Elsevier.
DRUGG, W. S., 1967: Palynology of the Upper Moreno Formation (Late Cretaceous—Paleocene) Escarpado Canyon, California. — Palaeontographica B **130**: 1—71.
DUTTA, S. K., SAH, S. C. D., 1970: Palynostratigraphy of the Tertiary sedimentary formations of Assam: 5. Stratigraphy and palynology of South Shillong Plateau. — Palaeontographica B **131**: 1—72.
ELSIK, C. W., 1968: Palynology of a Paleocene Rockdale Lignite, Milam County, Texas. 1/2. — Pollen & Spores **10**: 283—345, 519—664.
ENDRESS, P. K., HONEGGER, R., 1980: The pollen of the *Austrobaileyaceae* and its phylogenetic significance. — Grana **19**: 177—182.
ERDTMAN, G., 1952: Pollen morphology and plant taxonomy. Angiosperms. — An introduction to palynology 1. — Stockholm: Almquist & Wiksell.
— 1955: Pollen grains of cf. *Ctenolophon* from Tertiary deposits in India. — Bot. Not. **108**: 143—145.
— 1960: On three new genera from the Lower Headon Beds, Hampshire. — Bot. Not. **131**: 46—48.
EYDE, R. H., 1968: Flowers, fruits and phylogeny of *Alangiaceae.* — J. Arnold Arbor. **49**: 168—192.
— BARTLETT, A., BARGHOORN, E. S., 1969: Fossil record of *Alangium.* — Bull. Torrey Bot. Club **96**: 288—314.
FAIRCHILD, W. C., ELSIK, C. W., 1969: Characteristic palynomorphs of the Lower Tertiary in the Gulf coast. — Palaeontographica B **128**: 81—89.

FARABEE, M. J., CANRIGHT, J. E., 1986: Stratigraphic palynology of the Lower Part of the Lance Formation (Maestrichtian) of Wyoming. – Palaeontographica B **199**: 1–89.

FÖRSTER, R., 1975: Die geologische Entwicklung von Süd-Mozambique seit der Unterkreide und die Ammoniten-Fauna von Unterkreide und Cenoman. – Geolog. Jahrb. B **12**: 1–324.

FREDERIKSEN, N. O., 1980: Sporomorphs from the Jackson Group (Upper Eocene) and adjacent strata of Mississippi and Western Alabama. – U.S. Geol. Surv., Prof. Pap. **1084**: 1–55.

FUCHS, H. P., 1967: Pollen morphology of the family *Bombacaceae*. – Rev. Palaeobot. Palynology **3**: 119–132.

GEISSERT, F., GREGOR, J., 1981: Einige interessante und neue sommergrüne Pflanzenelemente (Fruktifikationen) aus dem Elsäßer Pliozän (Genera *Sabia* COLEBR., *Wikstroemia* Mirador formations (Lower and Middle Eocene, Tibú Area, Colombia). – Thesis, Landesver. Naturkunde Natursch., N. F. **12** (3/4): 233–239.

GERMERAAD, J. H., HOPPING, C. A., MULLER, J., 1968: Palynology of Tertiary sediments from tropical areas. – Rev. Palaeobot. Palynol. **6**: 189–348.

GONZÁLES-GUZMAN, A. E., 1967: A palynological study on the Upper Los Cuervos and Mirador formations (Lower and Middle Eocene, Tibú Area, Colombia). – Thesis, Leiden.

GRAHAM, A., 1972: Outline of the origin and historical recognition of floristic affinities between Asia and Eastern North America. – In GRAHAM, A., (Ed.): Floristics and paleofloristics of Asia and Eastern North America, pp. 1–18. – Amsterdam: Elsevier.

– JARZEN, D. M., 1969: Studies in neotropical paleobotany I: The Oligocene communities of Puerto Rico. – Ann. Missouri Bot. Gard. **56**: 308–357.

GRAY, J., SOHMA, K., 1964: Fossil *Pachysandra* from Western America with a comparative study of pollen in *Pachysandra* and *Sarcococca*. – Amer. J. Sci. **262**: 1159–1197.

GREGOR, H.-J., HAGN, H., 1982: Fossil fructifications from the Cretaceous-Palaeocene-boundary of SW-Egypt (Danien, Bir Abu Manqar). – Tertiary Res. **4**: 121–147.

GRUAS-CAVAGNETTO, C., 1978: Étude palynogique de l'Eocène du Bassin Anglo-Parisien. – Mem. Soc. Geol. France. N.S. 56, Mem. **131**: 1–64.

HARRIS, W. K., 1965: Basal tertiary microfloras from the Princetown area, Victoria, Australia. – Palaeontographica B **115**: 75–106.

HARTMANN, G., 1986: Biogeographie und Plattentektonik. Gondwana und die rezente Verteilung der Organismen. – Naturwissenschaften **73**: 471–480.

HEKEL, H., 1972: Pollen and spore assemblages from Queensland Tertiary sediments. – Publ. Geol. Surv. Queensland **355**, Palaeont. Pap. **30**: 1–33.

HERNGREEN, G. F. W., CHLONOVA, A. F., 1981: Cretaceous microfloral provinces. – Pollen & Spores **23**: 441–555.

HOCHULI, P. A., 1979: Ursprung und Verbreitung der Restionaceen. – Pflanzengeographische Aspekte von Fossilfunden aus Kreide und Tertiär. – Vierteljahresschr. Nat. Ges. Zürich **124**: 109–131.

– KELTS, K., 1980: Palynology of Middle Cretaceous black clay facies from DSDP sides 417 and 418 of the Western North Atlantic. – Init. Repts. Deep Sea Drill. Proj. **51–53**: 897–935.

HOEKEN-KLINKENBERG, P. M. J. VAN, 1964: A palynological investigation of some Upper-Cretaceous sediments in Nigeria. – Pollen & Spores **6**: 209–231.

JARDINE, S., MAGLOIRE, L., 1965: Palynologie et stratigraphie du Crétacé des bassins du Sénégal et de Côte d'Ivoire. – Mem. BRGM (Coll. Int. Micropal.) **32**: 187–222.

JARZEN, D. M., NORRIS, G., 1975: Evolutionary significance and botanical relationships of Cretaceous angiosperm pollen in the western Canadian interior. – Geosci. & Man **11**: 47–60.

JÉRÉMIE, J., 1980: Notes sur le genre *Ascarina* (*Chloranthaceae*) en Nouvelle-Caledonie et à Madagascar. – Adansonia, ser. 2, **20**: 273–285.

KAR, R. K., 1985: The fossil floras of Kutch. IV. Tertiary palynostratigraphy. – The Palaeobotanist **34**: 1–279.

KEDVES, M., 1965: A new *Restionaceae* pollen type from the Hungarian Lower Tertiary layers. – Advanc. Frontiers Plant Sci. **13**: 49–54.

– 1974: Étude palynologique d'un Lignite tertiaire de Blao, Vietnam, 2. – Acta Biol. Szeged **31**: 97–123.

– 1985: Études palynologiques sur les sediments prequarternaires de l'Egypte, Oligocene. – Rev. Esp. Micropal. **17**: 333–343.

– 1986: Introduction to the palynology of Pre-Quaternary deposits 2. – Akad. Kiado, Budapest; Studia biol. Hung. **20**: 144.

KEMP, E. M., 1981: Tertiary palaeogeography and the evolution of Australian climate. – In KEAST, A., (Ed.): Ecological biogeography of Australia, pp. 33–49. – The Hague, Boston, London: W. Junk.

– HARRIS, W. K., 1977: The palynology of early Tertiary sediments, Ninetyeast Ridge, Indian Ocean. – Palaeont. Ass., Spec. Papers Palaeontology **19**: 1–69.

KIRCHHEIMER, F., 1937: Beiträge zur Kenntnis der Teriärflora; Früchte und Samen aus dem deutschen Tertiär. – Palaeontographica B **82**: 71–141.

– 1944: Neue Erkenntnisse über die Beschaffenheit und den Lebensraum der Braunkohlenflora des ausgehenden Alttertiärs. – Biol. Zentralbl. **64**: 14–24.

– 1950: Die Symplocaceen der erdgeschichtlichen Vergangenheit. – Palaeontographica B **90**: 1–52.

– 1957: Die Laubgewächse der Braunkohlenzeit. – Halle/S.: Knapp.

KREMP, G. W. O., 1978: The earliest appearance worldwide of tricolpate pollen and the origin of angiosperms. – Abstracts Amer. Assoc. Stratigr. Palyn., **11**. Ann. Meet., Phoenix, Az.: 23.

KRUTZSCH, W., 1957: Pollen und Sporengruppen aus der Oberkreide und dem Tertiär Mitteleuropas und ihre stratigraphische Verteilung. – Z. angew. Geol. **3**: 519–549.

– 1959: Mikropaläontologische (sporenpaläontologische) Untersuchungen in der Braunkohle des Geiseltales. – Beih. Z. Geologie (Berlin) **21/22**: 1–425.

– 1961: Beitrag zur Sporenpaläontologie der präoberoligozänen kontinentalen und marinen Tertiärablagerungen Brandenburgs. – Ber. geol. Ges. DDR **5**: 290–343.

– 1962: Stratigraphisch bzw. botanisch wichtige neue Sporen- und Pollenformen aus dem deutschen Tertiär. – Geologie **11**: 265–308.

– 1963: Beitrag zur Kenntnis der Mikroflora der Niederschönauer Schichten. Eine kleine Mikroflora aus der Bohrung Königstein 1. – Ber. geol. Ges. DDR **8**: 224–240.

– 1966: Zur Kenntnis der präquartären periporaten Pollenformen. – Geologie **15**: 16–71.

– 1970a: Atlas der mittel- und jungtertiären dispersen Sporen- und Pollen- sowie der Mikroplanktonformen des nördlichen Mitteleuropas. 7. Monoporate, monocolpate, longicolpate, dicolpate und epliedroide (polyplicate) Pollenformen. – Jena: VEB G. Fischer.

– 1970b: Einige neue Pollenformen aus den Familien der Tiliaceen, Bombacaceen und Sterculiaceen aus dem mitteleuropäischen Alttertiär. – Jahrb. Geol. **3**: 275–307.

– 1970c: Die stratigraphisch verwertbaren Sporen- und Pollenformen des mitteleuropäischen Alttertiärs. – Jahrb. Geol. **3**: 309–379.

– 1985: Über *Nepenthes*-Pollen (alias „*Droseridites*" p. p.) im europäischen Tertiär. – Gleditschia **13**: 89–93.

– (in press, a): Gliederung, Verbreitung und Entstehung der Gattung *Rhododendron*. – Wiss. Zeit Schr. Humboldt-Univ. Berlin.

– (in press, b): Eine paläogeographische Kartenserie (Ob. Trias bis rezent) als Grundlage für die historische Pflanzengeographie (Paläochorologie) u. a. Synthesen. – Hercynia/ Halle.

KRUTSCH, W., (in press, c): Oberkretazische Mikrofloren aus dem südarabischen Raum. – Zeit Schr. Geol. Wiss. Berlin.

– (in prep., a): Paläogeographie und historische Pflanzengeographie 2. (Weitere paläochorologische Analysenbeispiele).

– (in prep., b): Paläogeographie und historische Pflanzengeographie 3. (Skizzenhafter Abriß der Vegetationsentwicklung im Neophytikum).

– LENK, G., 1969: Über 5 neue Pollenarten aus dem Maastricht der Bohrung Colbitz 10 (Calvoerder Scholle, DDR). – M. Ber. Deutsch. Akad. Wiss. **11**: 938–945.

– VANHOORNE, R., 1977: Die Pollenflora von Epinois und Loksbergen in Belgien. – Palaeontographica B **163**: 1–110.

KULKARNI, A. R., PHADTARE, N. R., 1981: Pollen of *Nypa* from the Lignitic beds of Ratnagiri District, Maharashtra. – Phytomorphol. **31**: 48–51.

KULKOVA, A., 1971: Eocene flora of Jana-Indigirka Lowland and its comparison with synchronous floras of other territories of Northern hemisphere. – Flora Kainozoic in Sibiria to data of palynology, for the 3. Int. Conf. Palynol., Nowosibirsk, Moskwa, pp. 7–21. (In Russian.)

LAVAL, O., MOULLADE, M., 1986: Biostratigraphie palynologique des sediments cretacés dans le bassin de la Haute Benoué (Nigeria NE). – Rev. Micropaleont. **29**: 61–83.

LEIDELMEYER, P., 1966: The Paleocene and Lower Eocene pollen flora of Guyana. – Geol. Meded. **36**: 49–70.

LEOPOLD, E. B., MACGINITIE, H. D., 1972: Development and affinities of Tertiary floras in the Rocky Mountains. – In GRAHAM, (Ed.): Floristics and paleofloristics of Asia and Eastern North America, pp. 147–200. – Amsterdam: Elsevier.

LI, H. L., 1951: Floristic relationships between eastern Asia and eastern North America. – Trans. Amer. Phil. Soc. **41**: 371–409.

LINDER, M. P., FERGUSON, I. K., 1985: On the pollen morphology and phylogeny of the *Restionales* and *Poales*. – Grana **24**: 65–76.

LJUBOMIROVA, K. A., 1960: Palynologische Unterlagen zur Stratigraphie des Paläogens der Tasowsker Halbinsel. – Trudy VNIGRI **158**: 66–75. (In Russian.)

MACGINITIE, H. D., 1969: The Eocene Green River Flora of northeastern Utah. – Univ. Calif. Publ. Geol. Sci. **83**: 1–140.

MAI, D. H., 1970: Subtropische Elemente im europäischen Tertiär 1. – Paläont. Abh. B **3**: 441–503.

– WALTHER, H., 1985: Die obereozänen Floren des Weißelster-Beckens und seiner Randgebiete. – Abh. Staatl. Mus. Mineral. Geol. Dresden **33**: 1–260.

MARTIN, H. A., 1978: Evolution of the Australian flora and vegetation through the Tertiary: evidence from pollen. – Alcheringa **2**: 181–202.

MATHUR, Y. K., SOODAN, K. S., MATHUR, K., BHATIA, M. L., JUYAL, N. P., PUNT, J., 1970: Microfossil evidences on the presence of Upper Cretaceous and Paleocene sediments in Kutch. – Bull. Oil Nat. Gas Commn **7**: 109–114.

MČEDLISHVILI, N. D., 1961: vgl. in ANONYMUS 1961.

– SAMOILOVIČ, S. R., 1961: vgl. in ANONYMUS 1961.

MEEUSE, A. D. J., 1974: Taxonomic affinities between *Piperales* and *Polycarpicae* and their implications in interpretative floral morphology. – In J. S. MURTY & al. (Eds.): Advances in plant morphology (S. Puri Commem.), pp. 3–27. – Meerut, Sarita Prakashan.

MELCHIOR, H., 1964: A. Englers Syllabus der Pflanzenfamilien. 12th edn. – Berlin: Bornträger.

MENZEL, P., 1920: Über Pflanzenreste aus Basalttuffen des Kamerungebietes. — Beiträge Geolog. Erforsch. deutsch. Schutzgebiete **18**: 17–32.

MEUSEL, H., 1978: Die Evolution der Pflanzen in pflanzengeographisch-ökologischer Sicht. – In BÖHME, H., HAGEMANN, R., LÖTHER, R., (Eds.): Beiträge zur Genetik und Abstammungslehre, pp. 521–555. – Berlin: VEB Volk & Wissen.

– JÄGER, E., WEINERT, E., 1965: Vergleichende Chorologie der zentraleuropäischen Flora. – Jena: G. Fischer.

MIKI, A., 1972: Palynological study of the Kiji group in Northeastern Honshu, Japan. – J. Fac. Sci. Hokkaido Univ. Ser. 4, Geol. Min. **12**: 513–551.

– 1977: Late Cretaceous pollen and spore floras of Northern Japan. – Composition and interpretation. – J. Fac. Sci. Hokkaido Univ., Ser. Geol. Min., **17**: 513–604.

MIKI, SH., 1956: Endocarp remains of *Alangiaceae, Cornaceae* and *Nyssaceae* in Japan. – J. Inst. Polytechnics, ser. D, **7**: 275–295.

MILDENHALL, D. C., 1980: New Zealand late Cretaceous and Cenozoic plant biogeography: a contribution. – Palaeogeogr., Palaeoclimatol., Palaeoecol. **31**: 197–233.

MOHR, B., 1986: Palynologischer Nachweis eines Mangrovenbiotyps in der Südatlas-Randzone (Marokko) und seine paläoökologische Bedeutung. – Dokum. Nat. **33**: 20–28.

– FECHNER, G., 1986: Eine eozäne Mikroflora (Sporomorphae und Dinoflagellaten-Zysten) aus der Südatlas-Randzone westlich Boumalne du Dades (Marokko). – Berl. geowiss. Abh. A **66**: 381–414.

MOORE, H. E. JR., 1973: Palms in the tropical forest ecosystems of Africa and South America. – In: Tropical forest ecosystems in Africa and South America: a comparative review, pp. 63–88. – Washington: Smithsonian Inst. Press.

MORLEY, R. J., 1977: Palynology of Tertiary and Quaternary sediments in Southeast Asia. – Proc. Indonesien Petrol. Assoc., 6th Ann. Conv. 1977: 255–276.

– 1982: Fossil pollen attributable to *Alangium* LAMARCK (*Alangiaceae*) from the Tertiary of Malesia. – Rev. Palaeobot. Palyn. **36**: 65–94.

MULLER, J., 1964: A palynological contribution to the history of the Mangrove vegetation in Borneo. – In: Ancient Pacific flora, pp. 33–42. – Univ. Hawaii Press.

– 1968: Palynology of the Pedawan and Plateau Sandstone Formations (Cretaceous–Eocene) in Sarawak, Malaysia. – Micropaleontology **14**: 1–37.

– 1970: Palynological evidence on early differentiation of Angiosperms. – Biol. Rev. **45**: 417–450.

– 1974: A composition of Southeast Asian with European fossil angiosperm pollen floras. – Sympos. Origin and Phytogeogr. of Angiosperms. Birbal Sahni Inst., Palaeobot., Spec. Publ. **1**: 49–56.

– 1979: Reflection on fossil Palmenpollen. – 4. Int. palyn. Conf. Lucknow 1976/77, **1**: 568–578.

– 1981: Fossil pollen records of extant angiosperms. – Bot. Rev. **47**: 1–42.

– 1984: Significance of fossil pollen for angiosperm history. – Ann. Missouri Bot. Gard. **71**: 419–443.

MÜRRIGER, F., PFLUG, H. P., 1952: Über eine palynologische Untersuchung des Braunkohlenlagers der Grube Emma bei Marxheim (Untermaingebiet). – Notizbl. Mess. L. Amt Bodenf. **6**/3: 56–66.

NAGY, E., 1962: New pollen species from the Lower Miocene of the Bakony Mountains (Varpalota) of Hungary. – Acta Bot. **8**: 153–163.

– 1969: Palynological elaborations of the Miocene layers of the Mecsek Mountains. – mafi évk. **52**: 1–417.

– 1985: Sporomorphs of the Neogene in Hungary. – Geol. Hung., ser. Palaeontol. **47**: 1–471.

NAIR, P. K. K., 1962: Pollen grains of Indian plants. 3. *Malvaceae* and *Bombacaceae*. – Bull. Nat. Bot. Gard. **63**: 9–30.

NILSSON, S., ROBYNS, A., 1986: World pollen and spore flora 14. *Angiospermae. Bombacaceae* KUNTH., pp. 1–59. – Stockholm: Almqvist & Wiksell.

NOREM, W. L., 1955: Pollen, spores and other organic microfossils from the Eocene of Venezuela. – Micropaleont. **1**: 261–265.

NORRIS, G., JARZEN, D. M., AWAI-THORNE, B. V., 1975: Evolution of the Cretaceous terrestrial palynoflora in the Western Canada. – Geol. Ass. Canada, spec. pap. **13**: 333–364.

OLLIVIER-PIERRE, M. F., 1982: La microflore du Paléocéne et de l'Eocéne des Ganntour (Maroc). – Sci. Geol. Bull. **35**: 117–127.

OLTZ, D. F., 1969: Numerical analyses of palynological data from Cretaceous and early Tertiary sediments in East Central Montana. – Palaeontographica B **128**: 90–166.

PACLTOVÁ, B., 1963: Die palynologische Forschung in den Schichten Chatt-Aquitan im Braunkohlenbecken Modrikamen. – Geol. Prace Zpravy **27**: 121–123. (In Czech.)

– 1966: The results of micropalaeobotanical studies of the Chattien-Aquitanian Complex in Slowakia. – Rozpravy Cesk. Akad. Ved., Rada Mat. Prirod. Ved. **76**: 1–68.

PAIS, J., 1981: Contribuicao pora o contrecimento da vegetacão miocénica da parte ocidental da Bacia to Teje. – Thesis, Lisbonne.

PENNEY, J. S., 1969: Later Cretaceous and Early Tertiary palynology. – In TSCHUDY, R. H., SCOTT, R. A., (Eds.): Aspects of palynology, pp. 331–376. – New York: Wiley Interscience.

POTONIÉ, R., 1934: Zur Mikrobotanik des eozänen Humodils des Geiseltales. – Arb. Inst. Paläobot. Petrogr. Brenngesteine **4**: 25–126.

RAO, A. N., ONG, E. T., 1972: Germination of compound pollen grains. – Grana Palyn. **12**: 113–120.

RAO, A. R., VIMAL, K. P., 1950: Plant microfossils from Palanga Lignite (? Eocene), Bikaner. Curr. Sci. **19**. 82–84.

REITSMA, T., 1970: Pollen morphology of the *Alangiaceae*. – Rev. Palaeobot. Palynol. **10**: 249–332.

ROBYNS, A., 1963 a: Contribution a l'étude monographique du genre *Bombax* s. l. – Grana Palyn. **5**: 73–79.

– 1963 b: Essai de monographie du genre *Bombax* s. l. (*Bombacaceae*). – Bull. Jard. Bot. L'Etat, Bruxelles **33**: 1–315.

RÖGL, F., STEININGER, F. F., 1983: Vom Zerfall der Tethys zu Mediterran und Paratethys. Die neogene Paläogeographie und Palinspastik des zirkum-mediterranen Raumes. – Ann. Naturhist. Mus. A**83**: 135–163.

ROUSE, G. E., HOPKINS, JR., W. S., PIEL, K. M., 1971: Palynology of some Late Cretaceous and Early Tertiary deposits in British Columbia and adjacent Alberta. – Geol. Soc. Ann., spec. paper **127**: 213–246.

RÜFFLE, L., 1986/87: Characters and homologies of ancient angiosperms as related to their origin. – Palaeobotanist (Lucknow, India) **35**: 263–266.

SAAD, S. I., 1962: Pollen morphology of *Ctenolophon*. – Bot. Not. **115**: 49–57.

SAH, S. C. D., 1967: Palynology of an upper Neogene profile from Rusizi valley, Burundi. Tervuren Musée Royal de l'Afrique Centrale, Ann. Sec. Geol. **57**: 1–173.

– KAR, R. K., 1970: Palynology of the Laki Sediments in Kutch. 3. Pollen from the boreholes around Jhulrai Baranda and Panandhro. – Palaeobotanist **18**: 127–142.

– SINGH, R. Y., 1974: Palynological biostratigraphy of the Tura formation in the type area. – In: Symp. Strat. Palynol., Birbal Sahni Inst. Palaeobot., spec. publ. **3**: 76–98.

SAMOILOVIČ, S. R., 1961: Vgl. in ANONYMUS 1961.

SATO, S., 1963: Palynological study on Miocene sediments of Hokkaido, Japan. – J. Fac. Science Hokkaido Univ., Ser. 4, **12**: 1–93.

SCHRANK, E., 1984: Organic-geochemical and palynological studies of a Dakhla Shale Profile (Late Cretaceous) in Southeast Egypt. A. – Berl. geol. Abh. A, **50**: 189–207.

SCHUSTER, R. M., 1974: Plate tectonics and its bearing on the geographical origin and dispersal of angiosperms. – In BECK, C. B., (Ed.): Origin and early evolution of angiosperms, pp. 48–138. – Amsterdam: Elsevier.

SMITH, A. C., 1973: Angiosperm evolution and the relationship of the floras of Africa and America. – In MEGGERS, B. J., AYENSU, E. S., DUCKWORTH, W. D., (Eds.): Tropical forest ecosystems in Africa and South America: a comparative review; pp. 49–61. – Washington: Smithsonian Inst. Press..

SMITH, A. G., HURLEY, A. M., BRIDEN, J. C., 1981: Phanerozoic paleocontinental maps. – Cambridge: Univ. Press.

SNEAD, R. G., 1969: Microfloral diagnosis of the Cretaceous-Tertiary boundary, Central Alberta. – Res. Counc. of Alberta Bull. **25**: 1–148.

SRIVASTAVA, S. K., 1972: Pollen genus *Erdtmanipollis* KRUTZSCH 1962. – Pollen & Spores **14**: 309–322.

STANLEY, E. A., 1965: Upper Cretaceous and Paleocene plant microfossils and Paleocene dinoflagellates and hystrichosphaerids from northwestern South Dakota. – Bull. Amer. Paleont. **49**: 179–383.

STEIN, N., 1978: Coniferen im westl. Teil des malayischen Archipels. – Studien zu ihrer Verbreitung und Ökologie innerhalb der vollhumiden SO-asiatischen Tropen. – Biogeographica **11**: 1–168.

STUCHLIK, L., SHATILOVA, I. I., 1987: Palynological study of Neogene deposits of Southern Poland and Western Georgia. – Acta Palaeobotanica (Krakow) **27** (2): 21–52.

SUC, J. P., 1980: Contribution à la connaissance, du Pliocène et du Pleistocène inférieur des régions mediterrannées d'Europe occidentale par l'analyse palynologique des dépots du Languedoc-Roussillon (Sud de la France) et de la Catalogne (Nord-Est de l'Espagne). – Thesis, Université Sc. Techn. du Languedoc, Montpellier.

TANAI, T., 1972: Tertiary history of vegetation in Japan. – In GRAHAM, A., (Ed.): Floristics and paleofloristics of Asia and eastern North America; pp. 235–255. – Amsterdam: Elsevier.

THANIKAIMONI, G., VASANTHI, G., 1972: *Sarraceniaceae*: palynology and systematics. – Pollen & Spores **14**: 143–155.

– CAVATINI, C., VENKATACHALA, B. S., RAMANUJAM, C. G., KAR, R. K., (Eds.) 1984: Selected tertiary angiosperm pollens from India and their relationship with African Tertiary pollens. – Inst. Franc. Pondicherry, Trav. Sect. Scient. Techniques **20**: 1–3.

THIERGART, F., 1940: Die Mikropaläontologie als Pollenanalyse im Dienst der Braunkohlenforschung. – Schr. Gebiet Brennstoffgeologie **13**: 1–82.

– FRANTZ, U., RAUKOPF, K. K., 1963: Palynologische Untersuchungen von Tertiärkohlen und einer Oberflächenprobe nahe Knysna, Südafrika. – Advanc. Frontiers Plant Sci. **4**: 151–178.

TRALAU, H., 1964: The genus *Nypa* VAN WURMB. – Kungl. Svenska Vetenskapsak. Handlingar **10**: 1–29.

TRAVERSE, A., 1955: Pollen analysis of the Brandon Lignite of Vermont. – U.S. Dep. Inter. Bur. Mines, Rep. Invest. **5151**: 1–107.

TSCHUDY, R. H., 1973: Stratigraphic distribution of significant Eocene palynomorphs of the Mississippi embayment. – U.S. Geol. Survey, prof. pap. **743**-B: 1–24.

– LOENEN, S. D. VAN, 1970: Illustrations of plant microfossils from the Yazoo Clay (Jackson Group, Upper Eocene), Mississippi. – U.S. Geol. Surv., prof. pap., **643**-E: 1–5.

TSUKADA, M., 1964: Pollen morphology and identification 3. Modern and fossil tropical pollen with emphasis on *Bombacaceae*. – Pollen & Spores **6**: 393–462.

VAKHRAMEEV, V. A., DOBRUSKINA, I. A., MEYEN, S. V., ZAKILINSKAJA, E. D, 1978: Paläozoische und mesozoische Floren Eurasiens und die Phytogeographie dieser Zeit. – Jena: VEB G. Fischer.

VENKATACHALA, B. S., KAR, R. K., 1969: Palynology of the tertiary sediments of Kutch 1. Spores and pollen from borehole Nr. 14. – Palaeobotanist **17**: 157–178.

VENKATACHALA, B. S., SHARMA, K. D., 1974a: Palynology of the Cretaceous sediments from the subsurface of Vridhachalam area, Cauvery Basin. – Geophytology **4**: 153–183.

– – 1974b: Palynology of the Cretaceous sediments from the subsurface of Pondicherry area, Cauvery Basin. – New Botanist **1**: 170–200.

WALKER, J. W., WALKER, A. G., 1985: Ultrastructure of Lower Cretaceous angiosperm pollen and the origin and early evolution of flowering plants. – Ann. Missouri Bot. Gard. **71**: 464–521.

WALTER, H., 1962: Die Vegetation der Erde in ökologischer Betrachtung. I. Die tropischen und subtropischen Zonen. – Jena: VEB G. Fischer.

WIGGINS, V. D., 1982: *Expressipollis striatus* n. sp. to *Anacolosidites striatus* n sp. an Upper Cretaceous example of suggested pollen aperture evolution. – Grana **21**: 39–49.

WIJMSTRA, A., 1971: The palynology of the Gujana coastal basin. – Thesis, Amsterdam.

WILSON, M. A., 1978: Palynology of three sections across the uppermost Cretaceous/Paleocene boundary in the Yukon territory and district of Mackenzie, Canada. – Palaeontographica B **166**: 99–183.

WOLFE, J. A., 1972: An interpretation of Alaskan Tertiary floras. – In GRAHAM, A., (Ed.): Floristics and paleofloristics of Asia and Eastern North America, pp. 200–233. – Amsterdam: Elsevier.

– 1975: Some aspects of plant geography of the Northern hemisphere during the Late Cretaceous and Tertiary. – Ann. Missouri Bot. Gard. **62**: 264–279.

– 1976: Stratigraphic distribution of some pollen types from the Campanian and Lower Maestrichtian rocks (Upper Cretaceous) of the Middle Atlantic States. – U.S. Geol. Survey, prof. paper **977**. 1–18.

ZAKLINSKAJA, E. D., 1963: Angiospermous pollen grains and their value for Upper Cretaceous and Tertiary stratigraphy. – Trudy Geol. Inst. Akad. Nauk S.S.R. **74**: 1–258. (In Russian.)

ZIEGLER, P. A., 1982: Geological atlas of Western and Central Europe. – Amsterdam: Shell Dutch Petr. Maatsch.

Address of the author: Dr. habil. WILFRIED KRUTZSCH, Museum für Naturkunde der Humboldt-Universität, Invalidenstrasse 43, DDR-1040 Berlin, German Democratic Republic.

Pl. Syst. Evol. 162, 63–78

—Plant—
Systematics
and
Evolution

Evolution of the Australian flora through the Tertiary

DAVID C. CHRISTOPHEL

Received February 2, 1986

Key words: Angiosperms, *Myrtaceae, Proteaceae, Casuarinaceae, Lauraceae, Nothofagus*, etc. – Tertiary floras of Australia, evolution of the Australian flora.

Abstract: The evolution of the Australian flora through the Tertiary has not been reviewed by a megafossil palaeobotanist for almost a century. Based on material available and published studies, the Australian Tertiary can be considered in three units: Eocene, Oligo-Miocene, and Pliocene. Key taxa in the Eocene include *Casuarinaceae, Proteaceae, Fagaceae, Podocarpaceae,* and *Lauraceae.* Many known Eocene deposits are interpreted as warm, humid rainforests.

Oligo-Miocene floras reflect a climatic deterioration, with greater occurrence of sclerophylly and xerophylly. A reduction of tropical taxa is seen with an increase of *Myrtaceae* and earliest records of *Mimosaceae, Chenopodiaceae,* and *Poaceae.* These trends continue into the Pliocene, which is not well represented in Australia.

Problems needing to be addressed include the discrepancy between the pollen and megafossil records of *Nothofagus,* the origins of Australia's wet tropical taxa, and the role of fire in the early evolution of the Australian flora.

The evolution of the Australian flora, in one context or another, has been the subject of numerous papers. Some of these papers have been written by taxonomists (e.g., BURBIDGE 1960, BARLOW 1981), others by ecologists (e.g., SPECHT 1981, LANGE 1982), but none, since ETTINGSHAUSEN (1888) by megafossil palaeobotanists. As evolution can be defined as change through time, it seems appropriate that the topic of the evolution of the Australian flora should be approached also from the perspective of megafossils. This paper will specifically deal with the evolution of the Australian flora during the Tertiary, and will consider the fossil evidence primarily from the point of megafossils. Excellent work has been done on the Tertiary pollen record of Australia (e.g., MARTIN 1981, HARRIS 1971, TRUSWELL & HARRIS 1982), and this evidence will also be included as needed.

At the outset it seems appropriate to draw attention to a semantic distinction. This is the difference between the evolution of the Tertiary flora and of the Tertiary vegetation of Australia. The former deals with the component plants of the continent, while the latter is related to the actual associations in which those plants are found – that is to say, the community structure. The Tertiary flora can be meaningfully discussed in terms of the fossil taxa and parataxa identified from different Tertiary age deposits. To consider the Tertiary vegetation and its changes, it is necessary to interpret the relationships between components of the fossil floras,

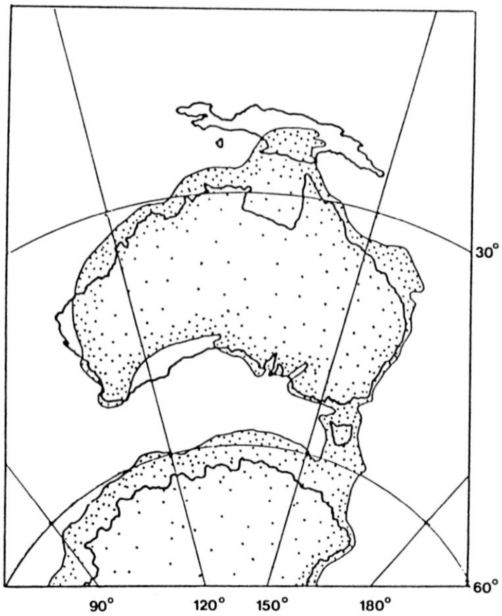

Fig. 1. Map of Australia showing the present land area and also the estimated land surface between 40 – 45 million years BP. The heavy black lines show the present day outlines of the Antarctica, Australia and Papua-New Guinea, and the stippling enclosed by lighter lines shows the hypothesized Eocene land surface

and make judgments on the relative autochthony of those components. In many cases it is true that the thorough documentation of the floras must precede vegetation studies, which must be accompanied by analyses of the depositional history of the localities. That some reviewers of Australia's fossil history (e.g., LANGE 1982) appear to draw few positive conclusions from that history is perhaps due to their considering mainly vegetation and not floristic components as evidenced from both micro- and mega-floras.

A brief examination of Australia's Tertiary geological history must precede any discussion of flora evolution. In general, the topic has been well reviewed by many, including KEMP (1978) and BARLOW & HYLAND (1988). The now classic work by BURBIDGE (1960) and all previous dissertations on the evolution of the flora were notable for the absence of continental drift as an explicative agent, while those following have included it. Important features of Australia's drifting to be considered include firstly its position at the beginning of the Tertiary (see Fig. 1). There is fairly good agreement that during this period a connection still existed with Antarctica which, in turn, was still linked with South America. The southern half of what was to be Papua-New Guinea was located along the eastern coast of northern Australia. By early Oligocene, Australia was separated from Antarctica and the circulation of the resulting cold water currents no doubt had a major effect on the climate and hence the vegetation (e.g., KEMP 1978). A further event to note is the collision of the Australian Plate with the Sunda Plate in the Miocene, at which time Papua-New Guinea was in its present position to the north of Australia, and connected to it by a land continuum (e.g., BARLOW & HYLAND 1988, and SPECHT 1988). The floristic results of that collision are still much in debate, and will be discussed later in this paper. Of internal importance is recent geological evidence which suggests that the eastern highlands of Australia, previously considered of recent origin, have been present through most of the Tertiary (OLLIER 1986).

Throughout this paper, the Tertiary will be considered primarily from the point of view of Eocene, Oligocene/Miocene, and Pliocene floras. The biostratigraphic zonation of the Australian Tertiary is well documented by HARRIS (1985). As may be seen from that treatment, the Oligocene/Miocene boundary is not clear-cut across the various biozonation schemes, and hence the precise placement of some deposits within a specific epoch is not agreed upon. The Paleocene is not extensively considered due to the basic paucity of floras of that age within Australia.

The remainder of the paper will be presented in three parts. The first section will deal with the evolution of key taxa within the Australian flora through the Tertiary. This will include both taxa which are key elements in the extant flora and also those which were major components of the Tertiary flora but are less significant in the extant flora. The second section will examine the floristic patterns apparent from the synthesis of the temporal and spacial distribution patterns of the individual key taxa, and will also comment on possible climatic implications. The final section will attempt to highlight some of the issues currently remaining to be resolved in our understanding of the evolution of the Australian Tertiary flora.

As a final point of introduction, it should be noted that most of the literature cited in this paper is reasonably recent. Still, the classic works near the beginning of the twentieth century (e.g., ETTINGSHAUSEN 1888, DEANE 1902, and JOHNSTON 1885) can not go unacknowledged. However, many of the taxa described by those early researches have either proved to be incorrectly identified or else remain to be verified. For this reason, their data can not be used confidently and, basically, have been omitted. It is unfortunate that some recent authors (e.g., WHITE 1986) still persist in using these identifications and hence risk perpetration of numerous early errors.

Key Australian plant taxa

In considering key Australian taxa, two groups of plants must be listed. The first of these includes those plants which make a dominant, and in many cases unique, contribution to the extant Australian flora. At the family level these include the *Myrtaceae, Mimosaceae, Chenopodiaceae, Poaceae,* and to a slightly lesser extent, the *Proteaceae* and the *Casuarinaceae*, with *Eucalyptus* (*Myrtaceae*), *Acacia* (*Mimosaceae*), and *Banksia* (*Proteaceae*) of particular interest at the generic level.

The second group of plants to be considered are those which have been demonstrated to be important components of the Australian Tertiary flora, and which have remained significant in the modern vegetation or not, e.g., members of *Podocarpaceae, Proteaceae, Myrtaceae, Casuarinaceae, Fagaceae* and *Lauraceae*, and with *Nothofagus* (*Fagaceae*), *Gymnostoma* (*Casuarinaceae*), *Ilex* (*Aquifoliaceae*), *Anacolosa* (*Olaceae*) and *Banksiaephyllum* (*Proteaceae*) as genera of particular interest. All of the above list of plants are seed plants. In addition three groups of ferns – *Lygodium, Gleichenia*, and the tree ferns (*Cyathea* and *Dicksonia*) are also worth mentioning.

Myrtaceae. This angiosperm family is prominent on both lists of key plants. Its members provide the greatest biomass, as well as the highest diversity of any group making up the Australian terrestrial flora (e.g., PRYOR & JOHNSON 1981). The distinctive individuality of *Eucalyptus* trees gives an Australian character to the country's open forests and mallee lands, while *Leptospermum, Melaleuca, Baeckea* and other shrubs of the family are equally distinctive components of the heath communities. In the closed forest, *Acmena* and *Syzygium* play important floristic roles.

Despite the importance of this family to the extant flora, its fossil record is not well documented. The pollen is known from throughout the Tertiary, with pollen definitely considered to have *Eucalyptus* affinities known from the Oligocene/Miocene. The family has traditionally been divided into two subfamilies, the *Leptospermoidae* and the *Myrtoidae*, based on the presence of woody capsular fruits and fleshy berry fruits respectively (JOHNSON & BRIGGS 1981), but more recently this distinction has been suggested to be artificial (JOHNSON & BRIGGS 1984, CHRISTOPHEL & LYS 1986).

The earliest documented foliar remains have been described as two species of *Myrtaciphyllum* from the Eocene of Anglesea, Victoria, but their extant generic affinities have not as yet been determined (CHRISTOPHEL & LYS 1986). We have also recently collected additional species from the Eocene of Golden Grove, South Australia, which are as yet undescribed (Fig. 3 A).

By the Oligocene/Miocene, foliar remains of *Myrtaceae* are quite abundant, with records from Morwell, Yallourn, and Bacchus Marsh in Victoria, from Pioneer in Tasmania and from Kiandra in New South Wales (see Fig. 2 for locations). All of these leaves correspond to morphotypes found in both extant mesic and sclerophyllous *Myrtaceae* (berry fruited and woody-capsular), and hence tell us little about generic evolution in the family yet. None of these reported taxa show any affinities with the leaves of *Eucalyptus*, however. One other Miocene report is worth mentioning and that concerns leaf and fruit impressions from Chalk Mountain in New South Wales (HOLMES & al. 1983). Although cuticular confirmation is not

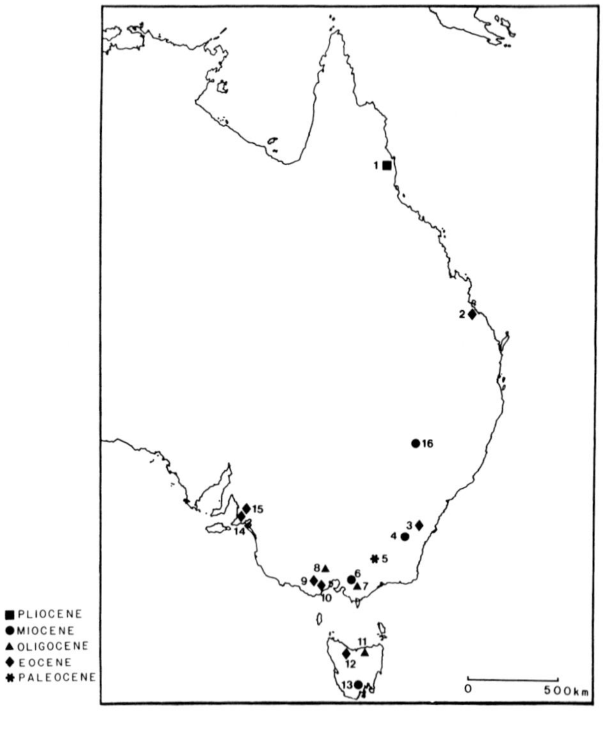

Fig. 2. Map of eastern Australia showing plant megafossil localities considered in the paper. *1* Butcher's Creek, Qld.; *2* Rundle, Qld.; *3* Nerriga, N.S.W.; *4* Kiandra, N.S.W.; *5* Mt Hotham, Vic.; *6* Morwell, Vic.; *7* Yallourn, Vic.; *8* Bacchus Marsh, Vic.; *9* Dean's Marsh, Vic.; *10* Anglesea, Vic.; *11* Pioneer, Tas.; *12* Cethana, Tas.; *13* New Norfolk, Tas.; *14* Maslin Bay, S.A.; *15* Golden Grove, S.A.; *16* Warrumbungle Mts, N.S.W.

possible, these fossils appear definitely eucalyptoid and as such would likely represent our earliest record of that group. By the Pliocene this portion of the family is well represented (e.g., YALLOURN on Fig. 2).

It is worth noting that some very interesting records of leptospermoid fruits have been reported from silcrete floras of northern S. Australia (e.g., LANGE 1978). The diversity of leptospermoid fruit types present in these floras suggests that they might provide good insight into evolution within the family. However, this material is generally found weathering on the land surface, and has thus far not been convincingly dated, with suggestions ranging from Eocene to Pliocene. Unless an age can be established for these fossils, their import is, however, negligible.

Mimosaceae. This family, which for the purposes of this discussion really is represented by only one genus, *Acacia,* is second only to the *Myrtaceae* in its Australian diversity and the fraction of the continent for which its members provide a canopy (SPECHT 1981). Its fossil record, however, is even more enigmatic. *Acacia* pollen is first reported from the Oligocene/Miocene (MARTIN 1981) and the only Tertiary megafossil record for it comes from that age as well (J. G. DOUGLAS, Geological Survey of Victoria, pers. comm.) in Victoria. The Gondwanic distribution (South America, Africa, and Australia) of *Acacia* in the extant flora suggests that the genus is much older than our records currently indicate, and the reason for its absence in the Early Tertiary is not obvious.

Proteaceae. This is another large Gondwanic family with genera such as *Banksia, Grevillea* and *Hakea* being significant components of modern shrublands and open forests, while numerous, small, frequently endemic genera (e.g., *Orites, Austromuellera*) form major elements in the closed forests. The evolution and biogeography of the family is well documented (JOHNSON & BRIGGS 1975, 1981) although little use has been made of the fairly extensive fossil record of the family. The pollen record extends through the entire Tertiary (MARTIN 1981), and shows affinities to numerous genera. The major organ genus for the pollen is *Proteacidites,* with other pollen types including *Beupreaidites* and *Banksiaedites.* The last named genus is particularly interesting in that, as its name implies, it shows affinities to *Banksia.* However, *Banksiaedites arcuatus* has been suggested to be more similar to the closely related *Musgraveinae* by A. MARTIN (1982), and has been recovered from *Musgraveinanthus* flowers by CHRISTOPHEL (1984). *B. arcuatus* is a widespread Eocene pollen type, and it is therefore suggested that the *Musgraveinae*, which are now restricted to a small area of north-eastern Queensland, were once a widespread element in the eastern Australian Tertiary flora.

In the Early Tertiary, a large number of *Proteaceae* pollen types (as many as 35 – 40 species) are found at individual localities. Megafossil deposits whose pollen spectra have such diversity of *Proteaceae* however, do not show an equivalent megafossil diversity for the family. For example, the Eocene Anglesea locality has over 20 species of pollen ascribed to *Proteaceae*, but only three megafossil taxa (CHRISTOPHEL & al. 1987). This problem is particularly perplexing in that the *Proteaceae* are considered to be poor pollen dispersers and hence indicators of local pollen rain. As suggested by A. MARTIN (1982), either the pollination mechanism for the family must have changed through time, or some currently unknown phenomenon must be invoked to explain the discrepancy between pollen and megafossil diversity. As might be expected, the Early Tertiary megafossil record for the family is one of rainforest taxa, with large pinnately lobed leaves (e.g., *Maslinia:* BLACKBURN 1981, and the Anglesia lobed *Proteaceae* aff. *Orites:* CHRISTOPHEL & al. 1987), pinnately compound grevillioid leaves, and *Musgraveinae* inflorescences, which are now known from both Anglesea and Golden Grove. Tertiary leaves of the *Banksiae* have recently been reviewed by HILL & CHRISTOPHEL (1988). A leaf type ascribed to *Banksiaephyllum* by them is also known from Eocene deposits, but in most cases it is found in associations uncharacteristic for *Banksia* and it has been recently suggested by CHRISTOPHEL & GREENWOOD (1988) that these leaves might belong to the *Musgraveinae* inflorescences found in those deposits. True *Banksia* remains, in the form of cones, have been reported from the Eocene of Western Australia (MCNAMARA & SCOTT 1983).

By the Oligocene/Miocene some of the pollen diversity alters, and megafossil remains show a more sclerophyllous assemblage. A large variety of *Banksiaephyllum* leaves are reported from YALLOURN (COOKSON & DUIGAN 1950) and this time their affinities with true *Banksia* are more likely (HILL & CHRISTOPHEL 1988). HILL (pers. comm.) has also reported the presence of megafossils of the family in the Tasmanian mid-Tertiary, possibly including such genera as *Lomatia.* All evidence suggests that the relative importance of the family in the extant Australia flora has been reflected throughout the Tertiary, and the fossil record of it is one of the best we have.

Casuarinaceae. While known as a monogeneric family for over 100 years, *Casuarinaceae* has now been divided into four genera (JOHNSON 1980, 1982). Of these, *Casuarina* and *Allocasuarina* are primarily xerophytic plants and comprise most of the species found in present day Australia. *Gymnostoma,* a third genus, has its centre of diversity in new Caledonia and Papua-New Guinea, with only one un-

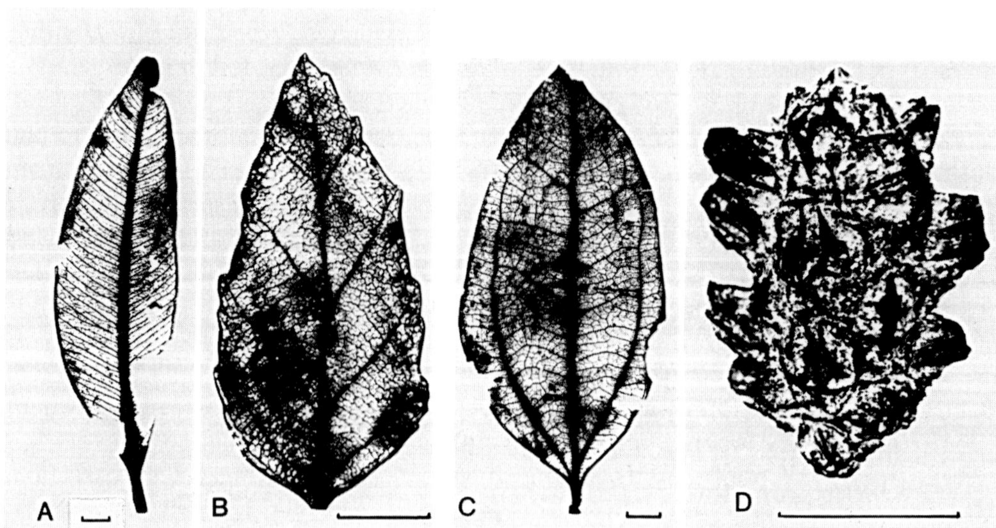

Fig. 3. Mummified leaves and reproductive structures from Australian megafossil localities. *A Myrtaciphyllum* spec. CHRISTOPHEL & LYS from the Eocene Golden Grove locality; *B Nothofagus* leaf from the Bacchus Marsh locality; *C Lauraceae* leaf from the Eocene Golden Grove locality; *D Gymnostoma* seed cone from the Eocene Anglesea locality. – Bars: 1 cm

described species found in restricted populations in the northern rainforests of Queensland. A fourth, as yet undescribed genus completes the family, but is non-Australian and hence not discussed further here.

Pollen of the genus is known throughout the Tertiary, and KERSHAW (1970) has concluded that while some differences for the most divergent species occur, it is not possible to regularly distinguish the pollen from *Gymnostoma* and the other genera. This is unfortunate, in that CHRISTOPHEL (1980) has reported that *Gymnostoma* megafossils are exclusively present in the Paleocene and Eocene deposits examined, while both *Gymnostoma* and *Casuarina/Allocasuarina* are found in the Oligocene, with *Casuarina/Allocasuarina* predominating in the Miocene and younger sediments. A *Gymnostoma* seed cone is shown in Fig. 3 D. If confirmed, this shift of generic composition could prove an interesting biostratigraphic tool and also has implications for climatic interpretation, as extant *Gymnostoma* is found in far more mesic environments than the other two described genera. Tertiary records of *Gymnostoma* are also confirmed from South America (CHRISTOPHEL 1980) and New Zealand (HOLDEN 1983), but the family appeared to become extinct in both of those regions by the end of the Miocene.

***Chenopodiaceae* and *Poaceae*.** Of those families left which are considered key components of the modern vegetation, these two may be considered together. Both form the understorey of various parts of the dry central or Eremaean portion of the Australian flora. Both are known only as microfossils in the Tertiary record, and both are not reported from localities earlier than the Miocene (MARTIN 1981). While it is not surprising that our megafossil record of arid zone plants is meagre, nonetheless, the stock from which they evolved must have been present prior to aridification, and hence should ultimately show up – at least at the family level – in the megafossil record.

***Lauraceae*.** Extant Australian representatives of this family have recently been revised by B. P. HYLAND (C.S.I.R.O., Atherton, Queensland; pers. comm.) who has recognized 107 species of trees – most found in the closed forests of the north and east of Australia. The sole additional representative of the family in Australia is the genus *Cassytha* with 10 – 12 species, but as this is a specialized leafless parasitic twiner, unknown from the fossil record, it will not be included in this discussion.

While representatives of the family play a major role in the make-up of most of the closed forest types in Australia, this is a very minor portion of the whole vegetation (less than 0.5% of the land surface) and hence was not included in the key families of extant plants. However, the family is a major component of the Australian Tertiary flora. HILL (1987) has now recognized that 13 of the 26 parataxa described for the Eocene Nerriga flora (Fig. 2, no. 3) are *Lauraceae*, and that they represent 45% of the flora. CHRISTOPHEL & al. (1987) have identified four major *Lauraceae* taxa in the Eocene Anglesea deposit which, in the best studied lenses represent over 30% of the flora. It is suspected that several more leaf types from Anglesea will also prove to be *Lauraceae*. The Eocene Rundle flora (Fig. 2, no. 2) has also proven rich in *Lauraceae* leaf remains (A. ROWETT, Adelaide University; pers. comm.) and the new Eocene locality from Golden Grove in South Australia has likewise been shown to have the *Lauraceae* as a major component (Fig. 3 C).

Oligocene/Miocene floras in Australia show the family still well represented, as is evidenced from the Kiandra flora (SELKIRK 1972), and from current studies on the Bacchus Marsh flora and the Morwell and Yallourn floras. In all of the floras mentioned above, the simple, pinnate leaf form as well as the acrodromous "*Cinnamomum*" type of leaf are present. The importance of the family in the evolution of Tertiary floras in Australia has been minimized in previous treatments. The most likely reason for this is that pollen of the family does not preserve well, and hence the excellent record of Tertiary pollen studies in Australia which has been a major data source for many of the flora evolution essays, has not shown the family in its true representation. Studies presently under way (CHRISTOPHEL & ROWETT, in press) have failed to find a suite of leaf architectural characters which will allow the identification of extant genera in Australia, but in combination with cuticular features the possibility appears somewhat more hopeful.

Podocarpaceae. The highest diversity of this family in the extant Australian flora is in the cool temperate regions of Tasmania with five of the six Australian genera and half of the Australian species. According to GREENWOOD (1987), this high diversity in Tasmania has biased studies of the Australian Tertiary pollen record to interpret the presence of *Podocarpaceae* as cool temperate indicators. On a world scale, however, the greatest diversity of the family occurs in tropical regions (e.g., New Guinea and New Caledonia) in a wide variety of mesic environments. The megafossil record of the *Podacarpaceae* from southern Australia, in particular the Eocene Anglesea deposit, contains taxa which suggest similar environments to those found in the tropical range of the *Podocarpaceae* including some genera now extinct (*Dacrycarpus, Decussocarpus*) or very restricted in range within Australia (*Prumnopitys*), which have traditionally been included within *Podocarpus* sensu lato (GREENWOOD 1987). The early Tertiary record on continental Australia is primarily one of *Podocarpus* sensu lato while the later Tertiary record shows a greater influence of *Dacrydium* sensu lato. This may well correlate with some of the changing climate/vegetation patterns discussed below. The fossil pollen record supports a strong diversity within the family throughout the Tertiary (MARTIN 1981).

Nothofagus. No discussion of the evolution of the Australian flora could be complete without the inclusion of a section on *Nothofagus*. The evolution of the genus as a whole has been the subject of numerous studies – including recent efforts by HUMPHRIES (1983), ROMERO (1980) and this volume and the elegant treatment by TANAI (1986) based on South American fossils to mention but a few. In Australia, our knowledge of the genus based on megafossils is due almost entirely to the work of R. S. HILL on the Tasmanian fossil flora (HILL 1983 a, b, 1984, 1986). The only report of confirmed megafossils from the mainland of Australia is from the Miocene of Bacchus Marsh, in Victoria (CHRISTOPHEL 1985) (Fig. 3 B). Thorough searches of all the Eocene continental sites shown in Fig. 2 have failed to produce any evidence of the genus, and the *N. brassii* group, which the pollen record suggests dominated in the Early Tertiary, is completely unknown as fossil leaves.

The fossil pollen record is a completely different story, however. Pollen of the genus dominates most deposits from southern and Central Australia throughout the Tertiary (MARTIN 1981). It is less prevalent in northern deposits (e.g., FOSTER 1982) but still present. This major input of the taxon to the fossil pollen rain has led numerous authors to write of the pan-Australian *Nothofagus* forests of the Early Teritary. Examples of this include GALLOWAY & KEMP (1981) and BARLOW (1981) in his introduction to the multi-volumed "Flora of Australia".

The ever growing list of Early Tertiary megafossil localities (Fig. 2) demonstrates the error of these interpretations. HILL's work has shown that *Nothofagus* leaves and fruits are well capable of being preserved in Australian depositional situations. Therefore the argument of poor preservation as an explanation for the absence of leaves and fruits can not be credibly invoked. Several other explanations are worth consideration. The first of these is that areas of *Nothofagus* forest did exist, and that their notorious prolific pollen production was responsible for the dominance in the pollen rain. This assumes that these pockets were out of areas of megafossil deposition, much akin to the "upland" theories of angiosperm origin used to explain their possible early presence before the fossil record shows them. Another possibility is that areas of southern Gondwana (e.g., Tasmania and Antarctica) where megafossils have been documented, were indeed covered with *Nothofagus* forests and that northerly air flow accounted for the pollen presence in continental Australia. Further support for this is the fact that the *Nothofagus* element is considerably diluted in the northern portions of the continent. A final possibility is that *Nothofagus* trees were reasonably widespread, but that they existed as single trees or small populations, and hence were far less likely to appear in any given fossil leaf deposit. Whatever the reality might have been, although the presence and diversity of *Nothofagus* in the Australian Tertiary, as shown by the pollen record can not be denied, the myth that the continent was covered by forests of that genus must be firmly laid to rest once and for all.

Tertiary floras in Australia

As mentioned in the introduction, the Tertiary history of Australia may be considered in three broad chronological sections. The Early Tertiary is represented primarily by Eocene floras. The palynological record of the Paleocene (e.g., MARTIN 1981) tells us that conifers played a significant role in the pollen rain composition, as did some of the older Gondwanic families such as the *Proteaceae*, and to a lesser extent, *Casuarinaceae, Myrtaceae*, and *Fagaceae*.

Eocene. From the palynological perspective, by mid-Eocene, the significance of the coniferous element had dropped markedly, while the contribution of *Nothofagus* had increased strongly and the contribution of the *Proteaceae* had dropped somewhat. Fortunately, Eocene megafossil floras are reasonably plentiful, and the information they provide is somewhat different. The overall picture to emerge is one of a highly variable, perhaps mosaic vegetation in the Eocene. At Anglesea, several discrete clay lenses exhibit a distinctive flora (CHRISTOPHEL & al. 1987) with a diversity (about 100 leaf types) and taxonomic composition similar to that found in the rainforests of northern Queensland today (CHRISTOPHEL & GREENWOOD 1988). The *Lauraceae* form a significant portion of the vegetation.

At Nerriga, the somewhat older Eocene flora shows a dominance of *Lauraceae* (HILL 1987), but with a lower overall diversity. The mid-Eocene Golden Grove flora has a diversity of 40 – 50 leaf types with some overlap apparent, both with the S. Australian Maslin Bay flora (CHRISTOPHEL & BLACKBURN 1978) and the Anglesea flora. The *Lauraceae* are again prevalent, and the *Elaeocarpaceae, Myrtaceae, Proteaceae* and *Podocarpaceae* present. A large number of specimens of both fertile and sterile *Lygodium* add a distinctive element to this flora, as does the common occurrence of *Brachychiton* (*Sterculiaceae*; CHRISTOPHEL & GREENWOOD 1987).

The Maslin Bay flora just mentioned has been the object of several preliminary studies (e.g., CHRISTOPHEL & BLACKBURN 1978, BLACKBURN 1981), but the majority of the taxa are yet to be studied. Nonetheless one can say that most of the higher taxa found at the other Eocene localities are present, e.g., *Gymnostoma* and the *Lauraceae*. The flora exhibits perhaps the greatest diversity of any of the known Eocene localities, but the fact that the leaves are preserved as compressions rather than in the mummified form found at most other Eocene localities (e.g., Fig. 3 D), has resulted in more preparation time and hence slower progress.

Cuticular studies of the Rundle Eocene flora of Queensland provide our only data on the northern Australian Eocene leaf floras (A. ROWETT, Adelaide University; pers. comm.). The general composition at the family level is similar to all other Eocene floras thus far described as is the relative diversity. Remains of the *Lauraceae* again predominate. Pollen studies of this same locality by him have shown a marked reduction in *Nothofagus* pollen as compared with the southern floras. There is also an increased contribution of palm pollen at the Rundle locality.

Recent studies have shown that there is a definite correlation between the physiognomy of leaf litter samples and of canopy samples in Australian rainforests (CHRISTOPHEL & GREENWOOD 1988). Physiognomic comparison of the Eocene Tertiary floras and leaf litter samples from various Australian rainforest types (particularly Simple Notophyll Vine Forest and Simple Mesophyll Vine Forest – sensu WEBB 1959) have shown strong similarities between the fossil floras and these exant forest types.

From all of the above it may be generalized that a complex closed forest system existed across southern Australia during the Middle and Upper Eocene with much local variation. The climate is likely to have been warm temperate to tropical (or more appropriately Microtherm – Megatherm: NIX 1982) and generally moist, supporting a range of taxa similar to those found in northern and eastern coastal Queensland today. While certain genera, e.g., *Gymnostoma* (*Casuarinaceae*), *Bowenia* (*Zamiaceae*), *Musgraveinanthus* (*Proteaceae*), and *Lygodium* (*Schizaeaceae*) occur in common between two or more of the localities, the floras were generally distinctive through time and across space. The concept of a pan-Australian *Nothofagus* forest is contrary to all megafossil evidence at hand from the mainland.

The Cethana Flora reported from Tasmania by HILL (1984) is possibly Eocene. He reported a high angiosperm diversity, but more importantly, described two species of *Nothofagus* leaves from the deposit. His physiognomic analysis of the flora showed a high correspondence with the modern day Tasmanian rainforest – that is more temperate than the mainland flora. If the flora does indeed prove to be Eocene, then the Tasmanian flora had clearly already diversified from the mainland forests – at least those studied thus far.

Oligocene/Miocene. As compared with the Eocene, these floras of Australia show a marked change in composition. The palynofloras have undergone the "*Nothofagus* shift", with the *fusca* and *menzesii* pollen types now more common than the tropical *brassii* type. The high diversity Eocene leaf floras have been replaced by a lower diversity, for more sclerophyllous group of taxa. In the LaTrobe Valley deposits (represented by the Morwell and Yallourn deposits in Fig. 1) a far more microphyllous flora is encountered, with *Myrtaceae* more prevalent and including some sclerophyllous genera such as *Leptospermum*. *Epacridaceae* are also present for the first time, and the *Proteaceae* is represented by true examples of *Banksia*. The *Casuarinaceae* are represented by *Casuarina*/*Allocasuarina*, and among the non-angiosperms, *Callitris* (*Cupressaceae*) is found and the ferns include several species of *Gleichenia*. The mesic elements are not completely gone, however, and in one interseam clay at Morwell, a much larger leafed flora is encountered with abundant *Lauraceae* (D. Greenwood, Adelaide University, pers. comm.).

At Bacchus Marsh a low diversity Oligocene/Miocene megaflora is encountered, which includes the earliest continental record of *Nothofagus* fruits (Christophel 1985). Additional elements include *Elaeocarpaceae* fruits, a small-leafed *Myrtaceae* genus, *Araucaria* and a few small-leaved *Lauraceae*. A similar aged flora from Kiandra in New South Wales also has *Myrtaceae* leaves and some *Lauraceae*.

A final, significant flora of this age is reported from the Chalk Mountain Formation in New South Wales (Holmes & al. 1983). This impression flora is significant as it contains fruits and leaves which appear to be the earliest documentation of eucalyptoid *Myrtaceae* in Australia. The remainder of the flora has not been formally described, but flowers of the *Cunoniaceae* have also been recovered (Keith Holmes, pers. comm.).

Once again the Tasmanian record has been provided by R. S. Hill. In an excellent combined palynology/megafossil study, the Oligocene/Miocene Pioneer flora was described (Hill & MacPhail 1982). Again the flora bears similarities to the extant temperate Tasmanian flora, and *Nothofagus* occurs again.

Looking in broad terms at the Oligocene/Miocene floras of Australia, the generalization may be made that the floras reflect the climatic deterioration presumed to have been brought about by the separation of Antarctica from Australia and the resultant cold, circumpolar currents and the formation of ice caps in Antarctica (Kemp 1978). This is generally expressed by a reduction in leaf size, and an increase in scleromorphy. While Beadle (1966) has suggested that sclerophylly may have originally arisen in response to depauperate soils, its dramatic increase in the Oligocene/Miocene seems likely tied to climatic deterioration. The high diversity forests of the Eocene are replaced by the lower diversity floras whose components include sclerophyllous *Proteaceae* and *Myrtaceae* as well as the switch from the mesic *Gymnostoma* to more xeromorphic genera of *Casuarinaceae* and from the sub-tropical *brassii* types of *Nothofagus* to the small leafed, serrate temperate species. It is also well to remember that although megafossil evidence is lacking, the pollen record of this time clearly shows the first records of the *Chenopodiaceae*, *Mimosaceae* (*Acacia*) and the *Poaceae*. Thus Burbidge's (1960) "Austral" element which ultimately gave rise to her Eremaean flora appears to have had its origins in this time period.

A final event of this time worth remembering is that the collision between the leading edge of the Australian Plate (in the form of New Guinea) and the Sunda Plate occurred in the Miocene. Unfortunately, no northern Australian megafossil floras are known from this time period to serve as bench-marks for the possible floristic changes which followed.

Pliocene. This is another epoch for which we have little megafossil documentation in Australia. The pollen record and its interpretation have been well set out by

MARTIN (1981) and particularly by SLUITER & KERSHAW (1982). In general the pollen record of this time reflects a gradual trend toward the extant flora with some new elements entering (possibly resulting from the Malaysian collision). SLUITER & KERSHAW (1982), using other studies by KERSHAW as well, highlighted one particularly interesting point – namely that the present day N. Queensland rainforests, with their host of primitive angiosperms, are not really old at all. The pollen history of N. Queensland over the last few million years has seen many cycles of rainforest and of drier, more open vegetations. Refugia have obviously been important in this system. Indeed, the present rainforest's age is measured in thousands, and not hundreds of thousands or millions of years! If such diverse and complicated ecosystems as these have changed that rapidly, it is to be expected that older floras could well exhibit massive and rapid changes, which would be complicated to document.

Again, the analysis of the Tasmanian Pliocene flora is the work of HILL. The Regatta Point flora (HILL 1982) shows the presence of primarily modern Tasmanian taxa, but with a few continental genera, such as *Quintinia* and *Elaeocarpus* which have since become extinct on the island. The Pliocene/Pleistocene of Australia is therefore a record of an adjusting toward the modern flora, with numerous Pleistocene fluctuations, perhaps indicating climatic variations mirroring the Northern Hemisphere glaciations.

Discussion and areas of future work

The strong current world interest in the ecology, evolution and conservation of rainforests also has stimulated research on the origins and evolution of Australia's wet tropics. At a recent symposium in Brisbane on that topic it became clear that one major aspect of this evolution is still controversial. This has to do with the effect the collision between the Australian Plate and the Sunda Plate in the Miocene had on the ultimate composition of Australian rainforests. The Indomalaysian intrusive element of the flora has long been considered an important factor by some workers (e.g., THORN 1986, HARTLEY 1986) and at the recent symposium, (BARLOW & HYLAND 1988). Other workers (e.g., WEBB & al. 1986, CHRISTOPHEL & GREENWOOD 1988) have chosen instead to emphasize the autochthonous element in the Australian rainforests. WEBB & al. (1986) have emphasized the interdependent nature of the rainforest communities, and hence the unlikeliness of establishment of large numbes of new taxa by migration. They instead see the diversity of communities and of micro-associations with the existing forest as evidence of differential extinctions and refugia and not of major regional migration. The lack of the major group of Malaysian forest trees, the *Dipterocarpaceae,* in Australia is cited as an example.

CHRISTOPHEL & GREENWOOD (1988) have instead chosen to emphasize the presence of modern taxa in associations similar to those found in the Eocene floras of southern Australia (pre-collision) and to suggest that such fossil evidence is the only positive evidence of autochthony. Not many years ago, the *Ebenaceae* was listed as one of the groups of Indo-Malaysian intrusives, but its documentation in the Australian Eocene (CHRISTOPHEL & BASINGER 1982, BASINGER & CHRISTOPHEL 1985) has caused that interpretation to be reassessed.

Whatever the ultimate influence the Indo-malaysian flora proves to have had on the composition of the extant Austalian rainforest, it will only be positively demonstrated by thorough collection and study of the Tertiary floras of Australia.

A second major area which still needs resolution is the origin and early radiation of some of Australia's key modern day taxa. Of particular import here are *Eucalyptus, Acacia,* and the *Chenopodiaceae,* and to a lesser extent *Cassia* and the *Epacridaceae*. All of these groups show a high diversity in the extant flora, and all have an Austral history or distribution which suggests that they are Gondwanic in origin. None, however, have an Australian pre-Miocene fossil record, and in the case of *Cassia, Acacia* and the *Chenopodiaceae*, there is no megafossil record at all. Clearly it is only by increasing the number of localities and taxa studied from our extensive Tertiary record in Australia that we can hope to shed any light on the origins and radiation of these important taxa.

Another unresolved issue in our studies of the evolution of the Australian flora is the role of fire in the evolution of modern taxa. The evolved fire tolerance of many sclerophyllous elements in the modern flora, including the lignotubers of *Eucalyptus* and the suckering ability of many other plants is well documented. Some plants, such as *Banksia* and *Hakea* (*Proteaceae*) are even credited with requiring fire to release seed.

The charcoal content of a fossil deposit can be measured, and hence some notion of the frequency of fire at a locality gained. However, with the exception of the period of time following human habitation of Australia (a palaeobotanically insignificant time period), there is no reason to suspect any alteration in the frequency of random firestarting events (chiefly lightning strikes). What has altered is the climate, and hence the potential "flammability" of the flora. Rainforest plants are particularly susceptible to fire damage. However, the simple fact remains that few fires get started, or stay burning within a wet rainforest. Therefore, the preponderance of rainforest taxa in the early Tertiary of Australia does not suggest a particular infrequency of fire starts, but rather a climate and vegetation in which those starts would not be likely to develop.

Similarly, the increase in fire resilient taxa in the Mid to Late Tertiary does not indicate an increase in fire starts, but merely a drier climate, with drier vegetation and hence more likelihood of fires staying alight. Given such a situation, natural selection would ensure that fire resilient and tolerant (rather than obligate) taxa would survive. Despite the apparent logic of this interpretation, the role of fire in our flora's evolution continues to be questioned (e.g., Recher & Christensen 1981).

It is only fitting to conclude this essay on the Tertiary evolution of the Australian flora with the enigma of *Nothofagus*. As mentioned earlier, the fervour surrounding the search for the origin and migration route of this genus somewhat parallels that of the angiosperms in general. Despite the best efforts of the researchers into this topic mentioned earlier, the ultimate source of the extant species in the southern floras remains conjecture. As for Australia, the pollen record shows us abundant quantity and diversity, particularly of the tropical *brassii* type in the Early Tertiary, and a switch to the more temperate (but equally prolific) *fusca* and *menzesii* types in the Mid Tertiary. With the exception of Tasmania, the megafossil record stands strangely mute on the diversity, or even the presence of the genus in the fossil record. While it can easily be said that more work is required to document our

Tertiary floras and perhaps solve that problem, it is of paramount importance that the myth of the pan-Australian *Nothofagus* flora be lain to rest, or at least temporarily shelved, pending megafossil evidence of its existence.

As any serious student of the extant Australian flora knows, with hybridization rampant in the floras' dominent components such as *Eucalyptus*, the floristic composition of the continent is truly dynamic. The Tertiary flora of Australia shows evidence of that same dynamism, and it is hoped that our approach to and study of that flora will match it.

I wish to thank S. M. SHAW for preparation of Figs. 2 and 3 and D. R. GREENWOOD for Fig. 1. Some of the original research upon which this paper is based was supported by A.R.G.S. grant E83/15626 to the author.

References

BARLOW, B. A., 1981: The Australian flora: its origin and evolution. – In GEORGE, A. S., (Ed.): Flora of Australia **1**, pp. 25–75. – Canberra: Austral. Govt. Publ. Service.

– HYLAND, B., 1988: The origins of the flora of Australia's wet tropics. – Proc. Ecol. Soc. Austral. **15**: (in press).

BASINGER, J. F., CHRISTOPHEL, D. C., 1985: Fossil flowers and leaves of the *Ebenaceae* from the Eocene of Southern Australia. – Canad. J. Bot. **63**: 1825–1843.

BEADLE, N. C. W., 1966: Soil phosphate and its role in molding segments of the Australian flora and vegetation, with special reference to xeromorphy and sclerophylly. – Ecology **47**: 992–1007.

BLACKBURN, D. T., 1981: Tertiary megafossil flora of Maslin Bay, South Australia: numerical taxonomic study of selected leaves. – Alcheringa **5**: 9–28.

BURBIDGE, N. T., 1960: The pyhtogeography of the Australian region. – Austral. J. Bot. **8**: 75–209.

CHRISTOPHEL, D. C., 1980: Occurrence of *Casuarina* megafossils in the Tertiary of Southeastern Australia. – Austral. J. Bot. **28**: 249–259.

– 1984: Early Tertiary *Proteaceae*: The first floral evidence for the *Musgraveinae*. – Austral. J. Bot. **32**: 177–186.

– 1985: First record of well-preserved megafossils of *Nothofagus* from mainland Australia. – Proc. Roy. Soc. Victoria **97**: 175–178.

– BASINGER, J. F., 1982: Earliest floral evidence for the *Ebenaceae* in Australia. – Nature **296**: 439–441.

– BLACKBURN, D. T., 1978: Tertiary megafossil flora of Maslin Bay, South Australia: a preliminary report. – Alcheringa **2**: 311–319.

– GREENWOOD, D. R., 1987: A megafossil flora from the Eocene of Golden Grove, South Australia. – Trans. Roy. Soc. S. Austral. **111**: 155–162.

– – 1988: A comparison of Australian tropical rainforest and Tertiary fossil leaf-beds. – Proc. Ecol. Soc. Austral. **15**: (in press).

– LYS, S., 1986: Mummified leaves of two new species of *Myrtaceae* from the Eocene of Victoria, Australia. – Austral. J. Bot. **34**: 649–662.

– HARRIS, W. K., SYBER, A. K., 1987: The Eocene flora of the Anglesea Locality, Victoria. – Alcheringa **11**: 303–323.

COOKSON, I. C., DUIGAN, S.L., 1950: Fossil *Banksieae* from Yallourn, Victoria, with notes on the morphology and anatomy of living species. – Austral. J. Scient. Res., Ser. B., **3**: 133–165.

DEANE, H., 1902: Preliminary report on the fossil flora of Pitfield, Mornington, Sentinel Rock (Otway Coast), Berwick and Wonwron. – Rec. Geol. Surv. Victoria **1**: 13–32.

ETTINGSHAUSEN, C. VON, 1888: Contributions to the Tertiary flora of Australia. – Mem. Geol. Surv. New South Wales Palaeontology **2**.

Foster, C. B., 1982: Illustrations of Early Tertiary (Eocene) plant microfossils from the Yamba Basin, Queensland. – Geol. Surv. Qld. Publ. **381**: 1–33.

Galloway, R. W., Kemp, E. M., 1981: Late Cainozoic environments in Australia. – In Keast, A., (Ed.): Ecological biogeography of Australia, pp. 51–80. – The Hague: W. Junk.

Greenwood, D. R., 1987: Early Tertiary *Podocarpaceae*: Megafossils from the Anglesea Locality, Victoria. – Austral. J. Bot. **35**: 111–133.

Harris, W. K., 1971: Tertiary stratigraphic palynology, Otway Basin. – Geol. Surv. S. Austral., Special Bulletin, 67–87.

– 1985: Middle to Late Eocene depositional cycles and dinoflagellate zones in southern Australia. – Spec. Publ., S. Austral. Dept. Mines Energy **5**: 133–144.

Hartley, T. G., 1986: Floristic relationships of the rainforest flora of New Guinea. – Telopea **2**: 619–630.

Hill, R. S., 1983a: Reconstruction of the Oligocene vegetation at Pioneer, northeast Tasmania. – Alcheringa **7**: 281–299.

– 1983b: Evolution of *Nothofagus cunninghamii* and its relationship to *N. moorei* as inferred from Tasmanian macrofossils. – Austral. J. Bot. **31**: 453–465.

– 1984: Tertiary *Nothofagus* macrofossils from Cethana, Tansmania. – Alcheringa **8**: 81–86.

– 1987: Lauraceous leaves from the Eocene of Nerriga, New South Wales. – Alcheringa **10**: 327–352.

– Christophel, D. C., 1988: Tertiary leaves of the tribe *Banksieae* (*Proteaceae*) from south-eastern Australia. – Bot. J. Linn. Soc. **97**: 205–227.

– Macphail, 1982: Reconstruction of the Oligocene vegetation at Pioneer, North East Tasmania. – Alcheringa **7**: 281–299.

Holden, A. M., 1983: Studies in New Zealand Oligocene and Miocene plant macrofossils. – Ph.D. Thesis, Victoria University, Wellington, New Zealand.

Holmes, W. B. K., Holmes, F. M., Martin, H. A., 1983: Fossil *Eucalyptus* remains from the Middle Miocene Chalk Mountain Formation, Warrumbungle Mountains, New South Wales. – Proc. Linn. Soc. New South Wales **106**: 299–310.

Humphries, C. J., 1983: Biogeographical explanations and the southern beeches (*Fagaceae: Nothofagus*). – In Sims, R. W. & al., (Eds.): Evolution, time and space, pp. 335–365. – London, New York.

Johnson, L. A. S., 1980: Notes on *Casuarinaceae*. – Telopea **2**: 83–84.

– 1982: Notes on *Casuarinaceae* 2. – J. Adelaide Bot. Gard. **6**: 73–87.

– Briggs, B. G., 1975: On the *Proteaceae* – the evolution and classification of a southern family. – Bot. J. Linn. Soc. **70**: 83–112.

– – 1981: Three old southern families – *Myrtaceae, Proteaceae* and *Restionaceae*. – In Keast, A., (Ed.): Ecological biogeography of Australia, pp. 427–469. – The Hague: W. Junk.

– – 1984: *Myrtales* and *Myrtaceae* – a phylogenetic analysis. – Ann. Miss. Bot. Gard. **71**: 700–756.

Johnston, R. M., 1885: Descriptions of two new species of Tertiary fossil plants belonging to the genera *Eucalyptus* and *Taxites*. – Pap. Proc. Roy. Soc. Tasmania 1885, 335–337.

Kemp, E. M., 1978: Tertiary climatic evolution and vegetation history in the southeast Indian Ocean region. – Palaeogeogr. Palaeoclimatol. Palaeoecol. **24**: 169–208.

Kershaw, A. P., 1970: Pollen morphological variation within the *Casuarinaceae*. – Pollen & Spores **12**: 145–161.

Lange, R. T., 1978: Carpological evidence for fossil *Eucalyptus* and other *Leptospermeae* (subfamily *Leptospermoideae* of *Myrtaceae*) from a Tertiary deposit in the South Australian arid zone. – Austral. J. Bot. **26**: 221–233.

Lange, R. T., 1982: Australian Tertiary vegetation. – In Smith, J. M. B., (Ed.): A history of Australasian vegetation, pp. 44–89. – Sydney: McGraw Hill.

McNamara, K. J., Scott, J. K., 1983: A new species of *Banksia* (*Proteaceae*) from the Eocene Merlinleigh Sandstone of the Kennedy Range, Western Australia. – Alcheringa **7**: 185–193.

Martin, A. R. H., 1982: *Proteaceae* and the early differentiation of the Central Australian flora. – In Barker, W. R., Greenslade, P. J. M., (Eds.): Evolution of the flora and fauna of arid Australia, pp. 77–84. – Adelaide: Peacock.

Martin, H. A., 1981: The Tertiary flora. – In Keast, A., (Ed.): Ecological biogeography of Australia, pp. 391–406. – The Hague: W. Junk.

Nix, H. A., 1982: Environmental determinants of biogeography and evolution in Terra Australis. – In Barker, W. R., Greenslade, P. J. M., (Eds.): Evolution of the flora and fauna of arid Australia, pp. 47–66. – Adelaide: Peacock.

Ollier, C. D., 1986: The origin of alpine land forms in Australasia. – In Barlow, B. A., (Ed.): Flora and fauna of alpine Australasia: ages and origins, pp. 3–25. – Melbourne: C.S.I.R.O.

Pryor, L. D., Johnson, L. A. S., 1981: *Eucalyptus*, the universal Australian. – In Keast, A., (Ed.): Ecological biogeography of Australia, pp. 499–536. – The Hague: W. Junk.

Recher, H. F., Christensen, P. E., 1981: Fire and the evolution of the Australian biota. – In Keast, A., (Ed.): Ecological biogeography of Australia, pp. 135–162. – The Hague: W. Junk.

Romero, E. J., 1980: Arquitectura foliar de las especies sudamericanas de *Nothofagus* Bl. – Bol. Socied. Argent. Bot. **19**: 289–308.

Selkirk, D. R., 1972: Fossil *Manginula*-like fungi and their classifications. – Proc. Linn. Soc. New South Wales **97**: 141–149.

Sluiter, I. P., Kershaw, A. P., 1982: The nature of Late Tertiary vegetation in Australia. – Alcheringa **6**: 211–222.

Specht, R. L., 1981: Evolution of the Australian flora: some generalizations. – In Keast, A., (Ed.): Ecological biogeography of Australia, pp. 783–805. – The Hague: W. Junk.

– 1988: Origin and evolution of terrestrial plant communities in the wet-dry tropics of Australia. – Proc. Ecol. Soc. Austral. **15**: (in press).

Tanai, T., 1986: Phytogeographic and phylogenetic history of the genus *Nothofagus* Bl. (*Fagaceae*) in the Southern hemisphere. – J. Fac. Sci. Hokkaido Univ. Ser. 4, **21**: 505–582.

Thorn, R. F., 1986: Antarctic elements in Australian rainforests. – Telopea **2**: 611–617.

Truswell, E. M., Harris, W. K., 1982: The Cainozoic palaeobotanical record in arid Australia: fossil evidence for the origin of an arid-adapted flora. – In Barker, W. R., Greenslade, P. J. M., (Eds.): Evolution of the flora and fauna of arid Australia, pp. 67–76. – Adelaide: Peacock.

Webb, L. J., 1959: A physiognomic classification of Australian rain forests. – J. Ecol. **47**: 551–570.

– Tracey, J. G., Jessup, L. W., 1986: Recent evidence for autochthony of Australian tropical and subtropical rainforest floristic elements. – Telopea **2**: 575–589.

White, M. E., 1986: The greening of Gondwana. – Reed Books Pty. Frenches Forests, New South Wales, Australia.

Address of the author: David C. Christophel, Department of Botany, University of Adelaide, Adelaide, South Australia 5000, Australia.

Pl. Syst. Evol. 162, 79–91

Plant Systematics and Evolution

Development and regional differentiation of the European vegetation during the Tertiary

Dieter H. Mai

Received February 2, 1987

Key words: History of vegetation, Tertiary; paleotropical and Arctotertiary geofloras, forest types, herbs, aquatic angiosperm communities, floristic regions of Europe.

Abstract: The Tertiary vegetation of Europe evolved from paratropical to warm-temperate and temperate forms in response to a progressive, non-linear, climatic cooling. Its vegetational forms are composed mainly of two separate ecological units: the evergreen, laurophyll "paleotropical geoflora" and the deciduous, broad-leaved "Arctotertiary geoflora". The development of the Tertiary climate and its interaction with the vegetation are convincingly indicated by the geoflora's migration; the changes in its composition; and the development of the Tertiary forest, swamp, and aquatic plant communities. The "paleotropical geoflora" is characterized in the upper Cretaceous to the upper Miocene by paratropical rain forest, subtropical rain and laurel forests, temperate laurel forests and edaphically-mediated formation of laurel-conifer forests. The "Arctotertiary geoflora" advanced into Europe in waves since the Paleocene and formed the basis for the Tertiary mixed mesophytic forests. These can be divided into warm-temperate rain forests, oak-hornbeam-chestnut or mixed beech-oak-hornbeam forests, and edaphic formations such as bottomland and swamp forests. Beginning in the lower Cretaceous, the hydrophytic vegetation developed independently of the forest vegetation and formed very diverse herbaceous fresh water, swamp, salt water, and coastal formations. Considerable differences in composition allow to separate floral regions and provinces in Eurosiberia. Instead of three ill-defined floral regions in the Paleocene, there are four well-defined floral regions in the Pliocene. A Mediterranean region cannot be recognized, although Mediterranean (eumesogeic) floral elements appear in the Eocene/Oligocene and thereafter. The Mediterranean sclerophyll forests probably arose after the destruction of the laurophyll forests during the Pleistocene.

During 150 years of scientific European Tertiary paleobotany, nearly a million paleontological specimens were collected and investigated. This has resulted in significant insights, as specialists have analyzed fossil floras consisting of leaves, fruits, seeds, pollen, and wood fragments. The irrefutable evidence which Tertiary paleobotany has presented, now is abundant and clear enough to allow broad conclusions to be drawn.

The purpose of this summary article is the presentation of a coherent picture of the development and chorogenesis of the main European vegetation types. The work and conceptions of many important predecessors thereby have been incorporated: Unger, Ettingshausen, Heer, Gardner, Saporta, Engler, Reid & Chandler, Kryshtofovich, Kirchheimer, Szafer, and others. Unanswered questions and unsolved problems will be clearly delineated. The basis of the following comments are two detailed publications by Mai (1981 a, 1985 a).

Laurophyll vegetation

Evergreen rain and laurel forests are the two most important vegetation types in the European Tertiary. This evergreen vegetation can be derived from the "paleotropical geoflora", a term defined by Engler (1882). Synonyms for this term are "Tethys flora" (Reid & Chandler 1933), "Poltawa flora" (Kryshtofovich 1928), "*Mastixioideae* flora" (Kirchheimer 1938), and "boreotropical flora" (Wolfe 1975). According to their characteristics as defined by the authors, all these paleofloras are identical. Unfortunately, no unified terminology has been accepted yet. We will continue to use the "geoflora concept" (Chaney 1959) here, although there are other controversial concepts (Wolfe 1972, 1980).

These floras are based on reliable determinations of a large number of genera, subfamilies, and families, which are completely absent in present-day Europe. Their extant distribution patterns are as follows:

1) Pan(sub)tropical, areas with gaps in Africa; e.g., *Aphananthe* Planch., *Clethra* Gron. (Knobloch & Mai 1986), *Myrica* L. s. l. (Gladkova 1965), *Persea* Boehm., *Theaceae* (Mai 1981 a), *Zanthoxylum* L. (Tiffney 1980).

2) Paleotropical; e.g., *Alangium* Lam. (Mai 1970 a), *Canarium* Stickman (Gregor 1979 b), *Euodia* J. R. & G. Forst. (Tiffney 1980), *Phoenix* L. (Bąkowski 1967), *Toddalia* Juss. (Gregor 1979 a).

3) Southeast Asian-neotropical disjunct (so-called "amphi-Pacific" disjunction); e.g., *Castanopsis* Spach (Mai 1981 a), *Eurya* Thunb., *Leucothoe* D. Don (Knobloch & Mai 1986), *Itea* L. (Mai 1985 b), *Meliosma* Bl. (van Beusekom 1971), *Saurauia* Willd. (Mai 1970 c), *Symplocos* Jacq. (Kirchheimer 1949), *Trigonobalanus* Forman (Mai 1970 b, p. p.).

4) Southeast Asian; e.g., *Dichroa* Lour. (Mai 1985 b), *Gironniera* Gaudich. (Knobloch & Mai 1986), *Illicum* L. (Mai 1970 a), *Manglietia* Bl. (Mai 1971), *Mastixia* Bl. (Kirchheimer 1943), *Nypa* van Wurmb (Lakhanpal 1952, Tralau 1964 a), *Pentaphylax* Gardn. & Chapm. (Knobloch & Mai 1986), *Sarcococca* Lindl. (Gray & Sohma 1964), *Wikstroemia* Endl. (Gregor 1978 a).

5) Canarian-Macaronesian; e.g., *Visnea* L. (Mai 1981 a).

The fact that there are many extinct genera of the European Tertiary flora which are related to the above-mentioned taxa, suggest a primary developmental centre for the European laurophyll vegetation (Mai 1965, 1981 a). Evidence for a tropical-subtropical floral belt, a "broad belt of tropical rainforests" (Chandler 1964) or a circumhemispheric "Gelinden region" (Kryshtofovich 1955) does not exist in the oldest Tertiary. During the old Tertiary, the rich autochthonous paleosubtropical laurophyll flora of the European developmental centre was isolated in the east by a connection between the Tethys and the Arctic Sea ("Turgai strait"; Vinogradov 1967/68). In the south there was the Tethys Sea. In the northwest a land bridge to America was populated and blocked by another flora (Tiffney 1985).

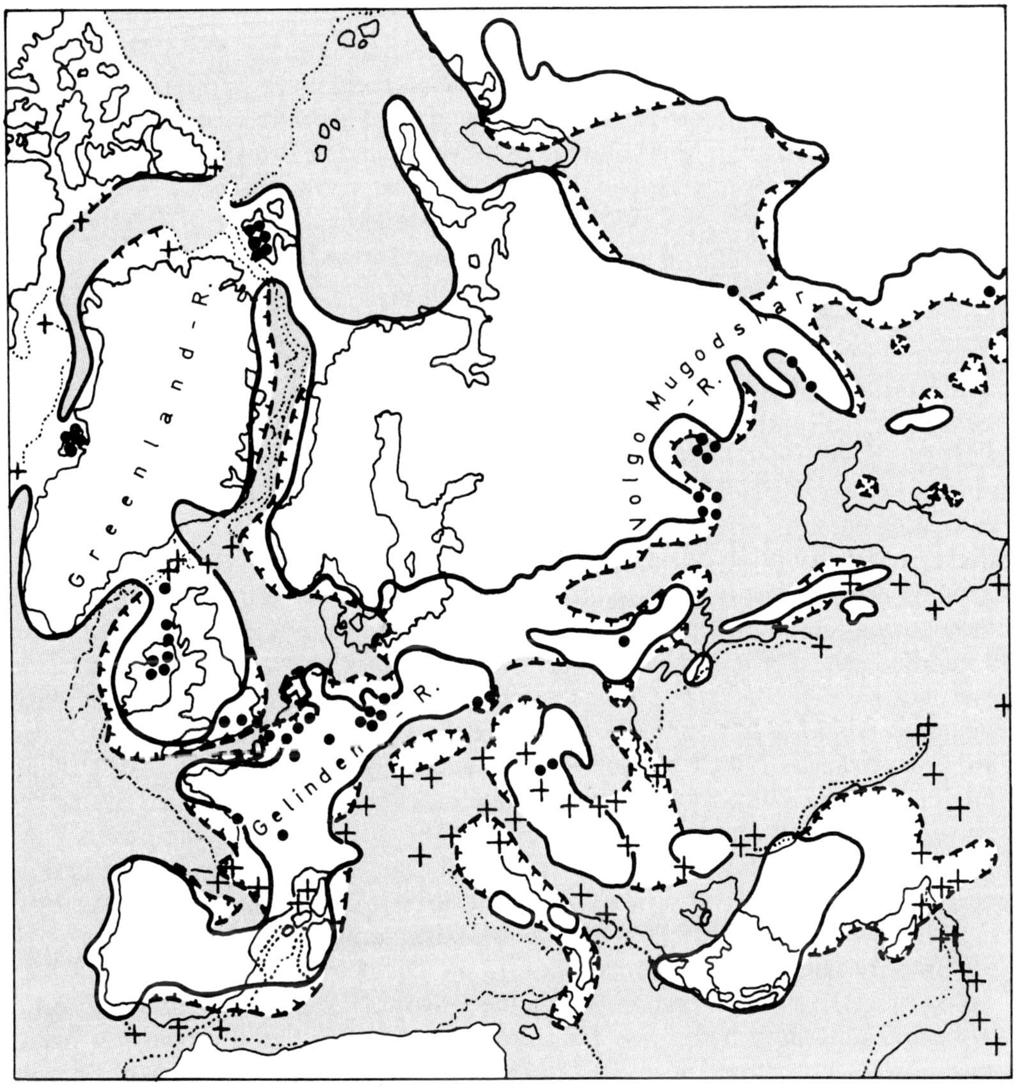

Fig. 1. Paleogeographical-paleocontinental map of W. Eurasia during the Paleocene (65 – 53.6 MY b. p.). —— Present limits of land and sea, ···· Paleocene limits of continental shelfs, ▬▬ Paleocene limits of land and sea, ⊤ Paleocene limits of epicontinental seas, + centres of orogenic activity, ● Paleocene locations of fossil macrofloras; sea grey, and white (after MAI 1987)

During the Paleocene three floral regions can be accepted in Europe (Fig. 1):

1) The Gelinden region (SAPORTA 1881) with an ancient laurophyll flora;

2) The Greenland region (SEWARD 1931) with an Arctotertiary flora;

3) The Volgo-Mugojar region (MAKUL'BEKOV 1977) with the *Ushia-Oxycarpia* flora.

Based upon the few definitive specimens from the lower Paleocene of Greenland and the rich upper Paleocene flora from Gelinden in Belgium, far-reaching generalizations were made (KRYSHTOFOVICH 1955). As a third western Asian floral region, a Volgo-Mugojar province (or region; MAKUL'BEKOV 1977) was defined according to the Upper Paleocene flora of the "Kamyschin type". It remains to be seen whether a useful synthesis of floral development during the oldest Tertiary can be made from this mixture of stratigraphic and regional paleofloristic theories.

The development of the primary European flora definitely began in the later Upper Cretaceous. In this context, attention should be drawn to the new "monograph of fossil fruits and seeds in the Central European Cretaceous" (Knobloch & Mai 1986), in which such significant families as *Mastixiaceae, Theaceae, Sabiaceae, Saurauiaceae,* and *Pentaphylaceae* are unequivocally documented.

The period of most intensive differentiation of this laurophyll vegetation was the Eocene. Even in the newer literature (Chandler 1964, Mai 1965) the description of these floras as being of a "tropical character" has continued. However, when taking into account the diverse recent investigations concerning the boundary between tropical and subtropical vegetation on the Northern hemisphere (Richards 1952, Panfilov 1959, Fosberg & al. 1961, Wang 1961, Wolfe 1979), one has to follow more strict guidelines. Of all the investigated fossil genera, only the genus *Nypa* Van Wurmb is of genuine tropical nature (Tralau 1964 a). However, many subtropical distribution during the Palèogene (often only during the Eocene). At *Sterculia* L., etc.) justify our speaking of "paratropical" conditions (Wolfe 1969) during the developmental optimum of this vegetation, i.e., during the lower to middle Eocene. The formation of a real N. hemisphere belt of this vegetation could not have begun earlier than the Eocene (examples are Japan: Tanai 1972; Pacific N. America: Bones 1979, Scott 1954, Wolfe 1972, 1977, Manchester 1981), nevertheless, a belt necessarily with several interruptions (Tiffney 1985). Only c. 65 of the European genera had an Eurasian, Euramerican, or even pan-subtropical distribution during the Paleogene (often only during the Eocene). At this time we had a uniform evergreen laurel and rain forest vegetation in the middle and southern latitudes of Europe, forming a "Central European region" (Fig. 2) (Mai 1981 a: Table 1). The retreat and gradual decline of this laurophyll vegetation began in Europe already during the Oligocene.

Today we know with certainty that the laurophyll elements of the European Neogene disappeared before the Pleistocene cooling. This decline did not occur continuously, but rather in cyclic intervals under the influence of a repeatedly changing climate (Sittler 1965, Mai 1967, 1970 d, Mai & Walther 1983). This has led to a periodical vicariance of closely related species. The laurel forest tended increasingly toward a temperate appearance, similar to the "evergreen broad-leaved forests" of southern Japan, southern China, the Himalaya, or the Canary Islands. Only a few laurophyll elements from the C. European Pliocene can be definitely classified as relics (e.g., *Daphniphyllum* Bl., *Sabia* Colebr., *Styrax* L.). It was mostly groups presently containing also deciduous species, which advanced far into the Sino-Japanese or N. American deciduous forest areas (Mai 1970 c). S. Europe and especially the Kolchis ("Kolchidian endemism"; Kolakovskij 1956, 1969, Chochieva 1965) remained richer in laurophyll relics.

The decline of wide-spread and abundant humid-subtropical elements in Europe can be explained only by drastic climatic and paleogeographic changes, which took place during the later Upper Miocene (Rögl & Steininger 1983). These changes lead to the disintegration of the late Tertiary Eurasian laurel forest belt and a nearly total extinction of the European laurophyll elements.

In present-day Europe, laurophyll relics exist only in the species-poor laurel

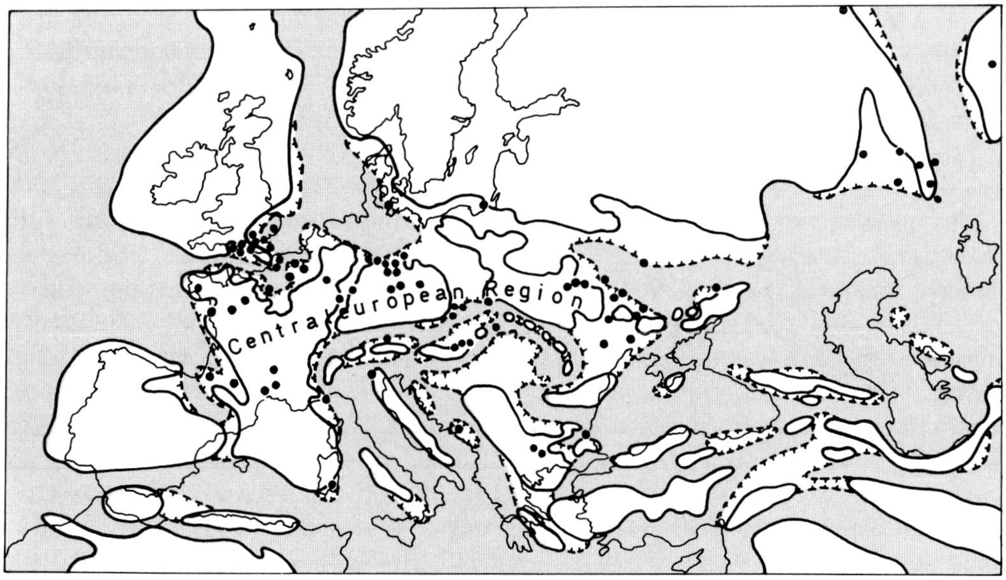

Fig. 2. Upper Eocene (42 – 34 MY b. p.) locations of fossil macrofloras in W. Eurasia (after MAI & WALTHER 1985). For explanations see Fig. 1

forests of the Canary Islands (e.g., *Persea indica, Ocotea foetens, Visnea mocanera;* MEUSEL 1965, OBERDORFER 1965), in the Lusitanic and Kolchidic floristic provinces (e.g., *Laurocerasus officinalis, Rhododendron ponticum;* WALTER 1968), and , mostly as ± xeromorphic derivates, in the so-called Mediterranean "*Laurocerasus*-belt" (*Laurus nobilis, Chamaerops humilis, Styrax officinalis*; SCHMID 1970).

Deciduous vegetation

Deciduous forests, typical for Europe today, are undoubtedly genetically young, and the result of a long development of the deciduous "Arctotertiary geoflora" (GARDNER & ETTINGSHAUSEN 1879, ENGLER 1879). An archaic, probably circumarctic flora had developed during the Upper Cretaceous in the "Greenland region" (KRYSHTOFOVICH 1928, CHANEY 1947). A mesothermic-humid climate and seasonally changing light periodicity formed the impetus for this development (SEWARD & CONWAY 1935). This circumarctic old Tertiary flora contained many typical deciduous elements, which exist today, sometimes in narrowly limited areas:

1) Panholartic, partly with extensions into tropical mountain regions; e.g., *Acer* L. (MAI 1981 a), *Juglans* L. (BERRY 1916), *Prunus* L. (MAI 1984).

2) Eastern Asian/north(east)ern American/European; e.g., *Carpinus* L., *Platanus* L. (MAI 1981 a), *Fagus* L. (TRALAU 1962).

3) Eastern Asian/northeastern American (so-called ASAGRAY disjunction); e.g., *Carya* NUTT. (MAI 1981 b), *Chionanthus* L. (GREGOR 1978 b), *Liriodendron* L. (MAI 1965, SCHILIN 1986), *Nyssa* L. (EYDE & BARGHORN 1963), *Schizandra* MICHX. (GREGOR 1981), *Stewartia* L. (MAI 1981 a).

4) Eurasian; e.g., *Parrotia* C. A. MEYER (TRALAU 1963), *Pterocarya* KUNTH (ILJINSKAJA 1953, TRALAU 1963), *Zelkova* SPACH (TRALAU 1963).

5) Eastern Asian: e.g., *Actinidia* LINDL., *Corylopsis* SIEB. & ZUCC. (TRALAU 1963), *Cercidiphyllum* SIEB. & ZUCC. (JÄHNICHEN & al. 1980), *Eucommia* OLIV., *Phellodendron* RUPR. (TRALAU 1963), *Glyptostrobus* ENDL., *Metasequoia* MIKI, *Pseudolarix* GORD. (FLORIN 1963), *Tapiscia* OLIV. (MAI 1980).

6) Northern American; e.g., *Leitneria* CHAPM. (MAI 1980), *Taxodium* A. RICH. (FLORIN 1963).

Intrageneric endemics arose among this Arctotertiary flora already during the Paleogene in western America, Europe/Greenland/eastern America, and eastern Siberia/Japan (DOROFEEV 1964). More recent data indicate that a strong invasion of Arctotertiary elements into Europe began in the Paleocene (only a few years ago, this invasion was thought to have taken place during the Lower Oligocene). Because many species (e.g., *Corylus macquarrii, Platanus schimperi*, and several red oaks) in C. and W. Europe are nearly or completely identical to the classic Arctotertiary flora of NW. Greenland, we may conclude that this early floral invasion came directly from the north (JOHNSON 1935, KOCH 1963, LAURENT 1912). Only analogous species ("Turgai-flora", KRYSHTOFOVICH 1928, 1935) were found in the Siberian and N. American Paleogene. This flow of Arctotertiary species from the north was checked by the European paleotropical geoflora during the Eocene. Nevertheless, deciduous species also arose during the development of a winter-dry climate from the Eocene subtropical flora (e.g., *Ailanthus* DESF., *Catalpa* SCOP., *Cephalanthus* L., *Koelreuteria* LAXM., *Morus* L., etc.). Thus, we have also to consider a floral invasion from south to north.

In the course of the Lower Oligocene oceanic regression during which a definite climatic cooling and continentalization occurred, temperate and deciduous elements invaded all European forest communities, resulting in very rich "mixed mesophytic forests". After the Middle Oligocene, "migration waves of Turgai floral elements" (ANDREÁNSZKY 1966) moved westwards; nevertheless they did not at all lead to floristic uniformity in the late Tertiary Eurosiberian deciduous forest belt (DOROFEEV 1965).

The exchange between Arctotertiary and paleosubtropical floras as well as their interaction and mutual penetration in C. Europe are the main aspects of the floral and climatic development in the Oligocene/Miocene (MAI 1964, 1967).

Drastic changes in the deciduous forest floras are first detectable in the upper Middle Miocene (Lower Sarmatian; ANDREÁNSZKY 1959, GIVULESCU 1967, GRANGEON 1958, SZAFER 1961). The Mediterranean aridization (RÖGL & STEININGER 1983) caused a certain climatic continentalization in adjacent C. Europe. A strong flow of suboceanic-continental species began. Very clearly and for the first time in floral history, one can observe the appearance of extant European and Near-Eastern species (MAI 1981 a). The exact date (between 13.5 MY and 6.5 MY) is uncertain, because of the prevailing difficulties to parallel the bioprovinces Paratethys and North Sea/Atlantic.

The formation of a Eurasian floral region in the Pliocene (FRENZEL 1968) was accompanied by the development of new European species which up to the present belong to the endemic stock of its humid-temperate flora (e.g., *Fagus sylvatica* L., TRALAU 1962; *Aesculus hippocastanum* L., SZAFER 1954, CHOCHIEVA 1965). Many warm-temperate taxa, which belong to the so-called "eastside elements" (JÄGER 1968), began to take on the character of relics (MAI 1980). The forest communities clearly began to decline more and more dramatically with the following cold periods of the Pleistocene. Therefore, we have to speak of relic fragments of deciduous Arctotertiary forests in present-day S. and C. Europe, Euxina, and Hyrkania (Meusel & Schubert 1971).

Hydrophytic and herbaceous vegetation

Herbaceous aquatic angiosperms, which developed parallel to and independently of woody life forms, are already detectable in the earliest angiosperm floras. The following families have been more or less definitely documented for the Middle and Upper Cretaceous (95 – 65 MY) in Europe: *Sparganiaceae, Zingiberaceae, Araceae, Typhaceae, Nymphaeaceae, Nelumbonaceae,* and *Droseraceae* (Depape & Gauthier 1953, Knobloch & Mai 1986, Němejc 1975).

Most of the extant European families with aquatic members can be traced back to the Eocene/Oligocene with the aid of definite (carpological) evidence. This extraordinary wealth of herbaceous life forms even increased during the Miocene/Pliocene. Only with the floral decline caused by the Pleistocene glaciations, did some aquatic families die out completely in Europe (e.g., *Azollaceae,* Dorofeev in Kryshtofovich 1957; *Cabombaceae,* Tralau 1959; *Pontederiaceae, Saururaceae, Trapellaceae,* Tralau 1964 b; Mai 1985 a).

A remarkable phenomenon is the occurrence in some important families of now extinct genera, which gave the Tertiary hydrophytic vegetation an individual character. Extant genera which are monotypic, species-poor or, by nature of their distribution, relics, are conspicuous in the Tertiary floras because of their extraordinary wealth of species (Dorofeev 1968, 1974, Palamarev 1979, Mai 1985 a).

Clear species differences between the otherwise widely distributed aquatic and swamp plant communities suggest effective paleogeographic limits and separate floral regions and provinces for hydro- and helophytes. Dorofeev (1958, 1965) has analyzed such differences in the Oligocene floras, defining a "European" in contrast to a "Siberian endemism". We prefer the term "European-western Siberian vicariances".

The development of the aquatic and swamp vegetation in western Eurasia during the Miocene and Pliocene is characterized by a strong dependence on the paleogeography, i.e., the complete disintegration of the Tethys and a temporary narrowing of the Mediterranean and small Paratethyan seas. This was accompanied by a stepwise elevation of the European mountain systems, whose erosion products collected in huge deltas. Climatic and terrestrial floral changes also strongly influenced the aquatic vegetation.

On the basis of aquatic and swamp plants, the following Pliocene European floral regions can be characterized (Dorofeev 1958, 1965, Mai 1985 a) (see Fig. 3):

1) Western European floral region.
2) Eastern European (Uralian) floral region.
3) Southeastern European-Pontic floral region.
4) Western Siberian-Kasachic floral region.

This example demonstrates, that special attention should be given to endemic aquatic species when characterizing vegetational regions.

In contrast to aquatic herbaceous flora, which are, as a whole, geohistorically quite old, the evolution of terrestrial herbaceous flowering plants has been shown to be more recent (cf., Seward 1931, Kirchheimer 1957). Only the component

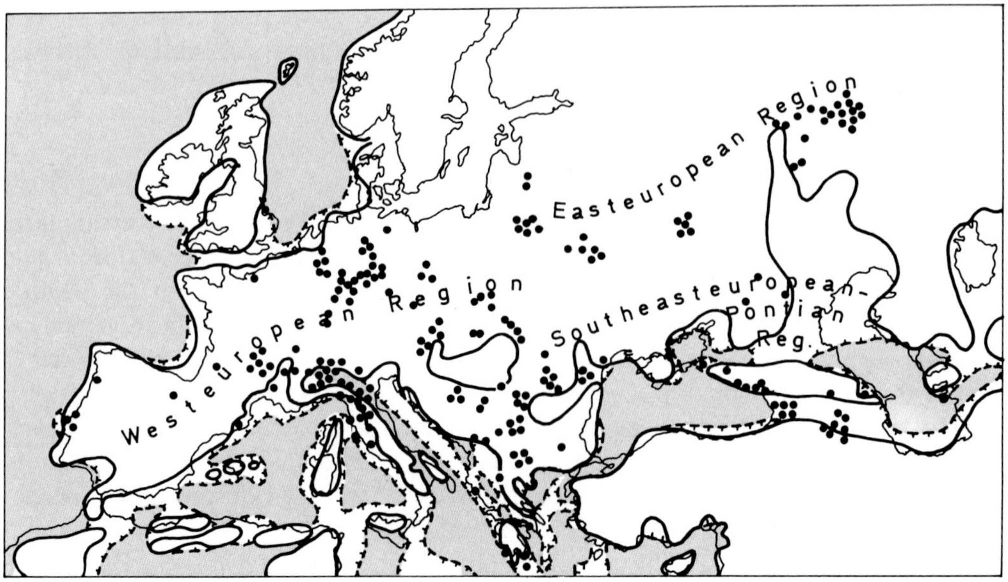

Fig. 3. Pliocene (Upper Pontian to Tegelen, about 6.0 – 1.7 MY b. p.) floristic regions in W. Eurasia. For explanations see Fig. 1

of evergreen ferns, selaginellas, and mosses has existed over long periods of time. Early fossils demonstrate its origin in Mesozoic forests.

The first evidence of herbaceous angiosperm sciophytes and hygrophytes on the forest floor is from the Eocene (e.g., *Alpinia* ROXB., *Decodon* GMEL., *Pachysandra* MICHX., *Solanum* L.). REID & CHANDLER (1933) have estimated that the herbs made up less than 5% of this period's total flora.

A considerable increase of herbaceous forms occurred during the Upper Miocene and Pliocene. In one estimate (KIRCHHEIMER 1957) they comprised 45%, in another (REID & CHANDLER 1933) 43 – 72% of the vegetation of this time. A species-rich herbaceous stratum obviously evolved from herbaceous vines (*Solanum* L., *Humulus* L., *Polygonum* L.), and from shrubs, dwarf- and half-shrubs (*Decodon* GMEL., *Hypericum* L., *Potentilla* L., *Teucrium* L., etc.). Genera occurring in swamps and along shores also have participated in the development of the herbaceous stratum in damp forests (*Carex* L., *Ranunculus* L., *Lysimachia* L., etc.). Furthermore, we find a number of spring-flowering and many more late-flowering summer-green forms at that time (MAI 1981 a).

The demonstration of ecological "specialists" among these herbs, often with considerable sociological fidelity, offers paleobotanical research valuable information for the paleosociological analyses of European Tertiary forests.

Sclerophyll vegetation

Very little fossil material is available to elucidate the origin of the Mediterranean sclerophyll forests. Nonetheless, this vegetation type is so important as to warrant some short comments.

The Mediterranean (eumesogeic) floristic elements were once believed to be connected with fossil floras, described as "xeromorphic" (ANDREÁNSZKY 1959,

1963, BERGER 1953). Recently, these floras have been subject to thorough analyses (KNOBLOCH 1969, GREGOR 1980) with the result that they cannot be regarded as forerunners of Mediterranean vegetation.

Unequivocal but rather rare fossil evidence is available from Tertiary floras in C. Europe in respect to genera and species groups, which today grow under Mediterranean climatic conditions and are characteristic components of the Oleo-Ceratonion (EBERLE 1965). *Cupressus* L., *Nerium* L., *Olea* L., the *Pinus pinaster* group, *Punica* L., *Pyracantha* M. J. ROEM., and often *Tetraclinis* MAST., e.g., were members of the laurophyll floras of the Eocene. *Ceratonia* L., *Cercis* L., *Phillyrea* L., *Pistacia* L., and the *Quercus ilex* group belonged to the "mixed mesophytic forest" flora of the Oligocene. This has led to the conclusion (MAI 1964) that the Mediterranean xeromorphic vegetation has its origin in the Tertiary laurophyll vegetation.

However, as xeromorphoses of laurophyll groups they appeared later and only became more frequent during the Lower Sarmatian (13.5 MY). Their evolution into typical extant Mediterranean taxa may have been due to the drastic narrowing of the Paratethys and the complete disintegration of the Tethys (PALAMAREV 1967) which reached its maximum during the Messinian (6 – 5.5 MY; RÖGL & STEININGER 1983). But even after this "salinity crisis", we have found no fossil evidence for an Oleo-Ceratonion but only for some of its taxa, sporadic among primarily mesophytic forest floras. One must conclude therefore that the genuine Mediterranean mesoxerophytic sclerophyll forest in its present form is a very young phenomenon which was established only after the disappearance of the laurophyll vegetation. As such it is documented from many locations with Mediterranean Pleistocene interglacial floras (PRINCIPI 1938, FRENZEL 1968). This opens a wide field for future research.

References

ANDREÁNSZKY, G., 1959: Die Flora der sarmatischen Stufe in Ungarn. – Budapest: Akad. Kiadó.

– 1963: Das Trockenelement in der jungtertiären Flora Mitteleuropas. – Vegetatio **11**: 113 – 129, 155 – 172.

– 1966: On the upper oligocene flora of Hungary. Analysis of the site at the Wind brickyard, Eger. – Stud. Biol. Hung. **5** (5): 5 – 151.

BĄKOWSKI, Z., 1967: Rośliny eoceńskie z Tatr i Podhala. – Prace Muz. Ziemi Warszawa **10**: 167 – 213.

BERGER, W., 1953: Pflanzenreste aus den obermiozänen Ablagerungen von Wien-Hernals. – Ann. Naturhist. Mus. Wien **59**: 141 – 154.

BERRY, E. W., 1916: The lower Eocene floras of southeastern North America. – U.S. Geol. Surv. Prof. Pap. **91**: 1 – 481.

BONES, T. J., 1979: Atlas of fossil fruits and seeds from North Central Oregon. – OMSI Occasional Pap. Nat. Sci. **1**: 1 – 22.

CHANDLER, M. E. J., 1964: The Lower Tertiary floras of Southern England 4. A summary and survey of findings in the light of botanical observations. – London: British Museum Natural History.

CHANEY, R. W., 1947: Origin and development of natural floristic areas with special reference to North America. Tertiary centers and migration routes. – Ecol. Monographs **17**: 139 – 148.

– 1959: Miocene floras of the Columbia Plateau, 1. Composition and interpretation. – Carnegie Inst. Wash. Publ. **617**: 1 – 134.

COCHIEVA, K. I., 1965: Flora i rastitel'nost' Chaudinskogo gorizonta Gurij. – Trudi Inst. Palaeobiol. Akad. Nauk Gruz. SSR **52**: 219–222.

DEPAPE, G., GAUTHIER, H., 1953: Sur le genre *Pistia* et sa présence dans la flore éocène de la vallée de l'Oued Mgoun (Maroc méridionale). – Ann. Univ. Lyon C **7**: 25–30.

DOROFEEV, P. I., 1958: Novye dannye ob oligocenovoj flore d. Belojarka na r. Tavde v Zapadnoj Sibiri. – Dokl. Akad. Nauk USSR **123**: 543–545.

– 1964: Razvitie tretičnoi flory SSSR po dannym paleokarpologičeskich issledovanij. – Leningrad: Autoref. Diss.

– 1965: O nekotorych problemach istorii flory. – Bot. Ž. **50**: 1509–1522.

– 1968: Ob oligocenovoj flore Zaural'ja. – Paleont. Ž. **2**: 111–119.

– 1974: *Cabombaceae, Nymphaeaceae, Ceratophyllaceae.* – In TAKHTAJAN, A. L., (Ed.): Iskopaemye cvetkovye rastenija SSSR **1**. – Leningrad: Nauka.

EBERLE, G., 1965: Pflanzen am Mittelmeer. Mediterrane Pflanzengemeinschaften Italiens mit Ausblick auf das ganze Mittelmeergebiet. – Frankfurt a. M.: W. Kramer.

ENGLER, A., 1879, 1882: Versuch einer Entwicklungsgeschichte der Pflanzenwelt seit der Tertiärperiode 1–2. – Leipzig: Engelmann.

EYDE, R., BARGHORN, E. S., 1963: Morphological and paleobotanical studies of the *Nyssaceae* 2. The fossil record. – J. Arnold Arbor. **44**: 328–376.

FLORIN, R., 1963: The distribution of conifer and taxad genera in time and space. – Acta Hort. Berg. **20** (4): 121–312.

FOSBERG, F. R., GARNIER, B. J., KUECHLER, A. W., 1961: Delimination of the humid tropics. – Geogr. Rev. **51**: 333–347.

FRENZEL, B., 1968: Grundzüge der pleistozänen Vegetationsentwicklung Nord-Eurasiens. – Wiesbaden: F. Steiner.

GARDNER, J. S., ETTINGSHAUSEN, C. v., 1879: A monograph of the British Eocene flora 1. *Filices.* – London: Palaeont. Soc.

GIVULESCU, R., 1967: Stand unserer Kenntnisse über die sarmato-pleistozäne Flora des Pannonischen Beckens. – Feddes Repert. **74**: 95–108.

GLADKOVA, A. N., 1965: Iskopaemye *Myricaceae* Zapadnoj Sibiri. – Trudi vsegeli. **239**: 142–190.

GRANGEON, P., 1958: Contribution a l'étude de la paléontologie végétale du Massif de Coiron (Ardèche). – Mém. Soc. Hist. Nat. Auvergne **6**: 1–299.

GRAY, J., SOHMA, K., 1964: Fossil *Pachysandra* from Western America with a comparative study of pollen in *Pachysandra and Sarcococca.* – Amer. J. Science **262**: 1159–1197.

GREGOR, H.-J., 1978 a: Die miozänen Frucht- und Samen-Floren der Oberpfälzer Braunkohle I. Funde aus den sandigen Zwischenmitteln. – Palaeontographica B **167**: 8–103.

– 1978 b: *Chionanthus* LINNÉ – ein neues subtropisches Element im europäischen Jung-Tertiär. – Feddes Repert. **88**: 645–653.

– 1979 a: Systematics, biostratigraphy and paleoecology of the genus *Toddalia* JUSSIEU (*Rutaceae*) in the European Tertiary. – Rev. Paleobot. Palynol. **28**: 311–363.

– 1979 b: Erster Nachweis der Gattung *Canarium* STICKMAN 1759 (*Burseraceae*) im europäischen Alttertiär. – Stuttgarter Beitr. Naturk. B **47**: 1–15.

– 1980: Ein neues Klima- und Vegetations-Modell für das untere Sarmat (Mittelmiozän) Mitteleuropas unter spezieller Berücksichtigung floristischer Gegebenheiten. – Verh. Geol. Bundesanst. Wien **3**: 337–353.

– 1981: *Schizandra Geisertii* nova spec. – ein exotisches Element im Elsäßer Pliozän (Sessenheim, Brunssumien). – Mitt. bad. Landesver. Naturk. Naturschutz N. F. **12** (3/4): 241–247.

ILJINSKAJA, I. A., 1953: Monografija roda *Pterocarya* KUNTH. – Trudi BIN. AN. SSSR, Ser. 1, Leningrad **10**: 7–123.

JÄHNICHEN, H., MAI, D. H., WALTHER, H., 1980: Blätter und Früchte von *Cercidiphyllum* SIEBOLD & ZUCCARINI im mitteleuropäischen Tertiär. – Schriftenr. geol. Wiss. Berlin **16**: 357–399.

JÄGER, E. J., 1968, Die pflanzengeographische Ozeanitätsgliederung der Holarktis und die Ozeanitätsbindung der Pflanzenareale. – Feddes Repert. **79**: 157–335.

JOHNSON, T., 1935: *Quercus* L. in the Tertiary Beds of Ireland and Scotland. – Mem. Proc. Manchester Lit. Philosoph. Soc. **79**: 83–98.

KIRCHHEIMER, F., 1938: Beiträge zur näheren Kenntnis der Mastixioideenflora des deutschen Mittel- bis Oberoligozän. – Beih. Bot. Zentralb. B **58**: 303–375.

– 1943: Die Mastixioideen in der Flora der Gegenwart. – Braunkohle **42**: 17–19, 26–30.

– 1949: Die Symplocaceen der erdgeschichtlichen Vergangenheit. – Palaeontographica B **90**: 1–52.

– 1957: Die Laubgewächse der Braunkohlenzeit. – Halle/Saale: Knapp.

KNOBLOCH, E., 1969: Tertiäre Floren von Mähren. – Brno: Moravské Mus.

– MAI, D. H., 1986: Monographie der Früchte und Samen in der Kreide von Mitteleuropa. – Rozpr. ústř. úst. geol. Praha **47**: 1–219.

KOCH, E. B., 1963: Fossil plants from the Lower Paleocene of the Agatdalen (Angmârtussat) Area, Central Nûgssuaq Peninsula, Northwest Greenland. – Grønlands geol. Unders. **38**: 1–120.

KOLAKOVSKIJ, A. A., 1956: K istorii flory Kolchidskogo refugiuma. – Sb. Akademik W. N. Sukačev k 75-let. dnja roždenija, Moskva, Leningrad: 275–285.

– 1969: Predstavitel' oligocenovoj Mastiksevoj flory Evropy v pliocene Kolchidskogo refugiuma. – Bull. Acad. Sci. Georgian SSR **55**: 737–740.

KRYSHTOFOVICH, A. N., 1928: Grenlandskaja tretičnaja flora na Severnom Urale i botaniko-geografičeskie provincii tretičnogo perioda. – Priroda **5**: 499–502.

– 1935: A final link between the Tertiary floras of Asia and Europe (contribution to the age of the Arcto-Tertiary floras of the Northern Holarctic). – New Phytol. **34**: 339–344.

– 1955: Razvitie botaniko-geografičeskich oblastjei severno poluascharija s načala tretičnovo perioda. – Vopr. Geol. Azii, Moskva-Leningrad **2**: 824–844.

– 1957: Paleobotanika. – Leningrad: Gostoptechizdat.

LAKHANPAL, R. N., 1952: *Nipa sahnii* a palm fruit in the Tertiary of Assam. – Palaeobotanist **1**: 289–294.

LAURENT, L., 1912: Flore fossile des schistes de Menat (Puy-de-Dôme). – Ann. Mus. Hist. Nat. Marseille **14**: 1–246.

MAKUL'BEKOV, N. M., 1977: Paleogenovye flory Zapadnogo Kazachstana i Nižnego Povolž'ja. – Alma-Ata: Nauka Kazach. SSR.

MAI, D. H., 1964: Die Mastixioideen-Floren im Tertiär der Oberlausitz. – Paläont. Abh. B **2**: 1–192.

– 1965: Der Florenwechsel im jüngeren Tertiär Mitteleuropas. – Feddes Repert. **70**: 157–169.

– 1967: Die Florenzonen, der Florenwechsel und die Vorstellungen über den Klimaablauf im Jungtertiär der Deutschen Demokratischen Republik. – Abh. zentr. geol. Inst. Berlin **10**: 55–81.

– 1970a: Subtropische Elemente im europäischen Tertiär I. – Paläont. Abh. B **3**: 441–503.

– 1970b: Die tertiären Arten von *Trigonobalanus* FORMAN (*Fagaceae*) in Europa. – Jahrb. Geol. (Berlin) **3**: 381–409.

– 1970c: Funde von *Saurauia* WILLD. im europäischen Alttertiär. – Wiss. Zeitschr. Friedrich-Schiller-Univ. Jena, Math.-Nat. R. **19** (3): 385–392.

– 1970d: Change of climate and biostratigraphy in the continental younger Tertiary of Boreal Province. – Giorn. Geol. Bologna **35** (1): 85–90.

– 1971: Fossile Funde von *Manglietia* BLUME (*Magnoliaceae*). – Feddes Repert. **82**: 441–448.

– 1980: Zur Bedeutung von Relikten in der Florengeschichte. – In: 100 Jahre Arboretum (1879–1979), pp. 281–307. – Berlin: Akademie Verl.

MAI, D. H., 1981 a: Entwicklung und klimatische Differenzierung der Laubwaldflora Mitteleuropas im Tertiär. – Flora **171**: 525–582.

– 1981 b: Der Formenkreis der Vietnam-Nuß [*Carya poilanei* (CHEV.) LEROY] in Europa. – Feddes Repert. **92**: 339–385.

– 1984: Karpologische Untersuchungen der Steinkerne fossiler und rezenter *Amygdalaceae* (*Rosales*). – Feddes Repert. **95**: 299–329.

– 1985 a: Entwicklung der Wasser- und Sumpfpflanzen-Gesellschaften Europas von der Kreide bis ins Quartär. – Flora **176**: 449–511.

– 1985 b: Beiträge zur Geschichte einiger holziger *Saxifragales*-Gattungen. – Gleditschia **13**: 75–88.

– 1987: Neue Früchte und Samen aus paläozänen Ablagerungen Mitteleuropas. – Feddes Repert. **98**: 197–229.

– WALTHER, H., 1983: Die fossilen Floren des Weißelster-Beckens und seiner Randgebiete. – Hall. Jahrb. Geowiss. **8**: 59–74.

– – 1985: Die obereozänen Floren im Weißelster-Becken (Bezirk Leipzig, DDR) und seiner Randgebiete. – Abh. Staatl. Mus. Mineral. Geol. Dresden **33**: 1–220.

MANCHESTER, S., 1981: Fossil plants of the Eocene Clarno Nut Beds. – Oregon Geol. **43**: 75–81.

MEUSEL, H., 1965: Die Reliktvegetation der Kanarischen Inseln in ihren Beziehungen zur süd- und mitteleuropäischen Flora. – In GERSCH, M., (Ed.): Gesammelte Vorträge über moderne Probleme der Abstammungslehre. **1**, pp. 117–136. – Jena: G. Fischer.

– SCHUBERT, R., 1971: Beiträge zur Pflanzengeographie des Westhimalajas. 1–3. – Flora **160**: 137–194, 373–432, 573–606.

NĚMEJC, F., 1975: Paleobotanika 4. – Praha: Academia.

OBERDORFER, E., 1965: Pflanzensoziologische Studien auf Teneriffa und Gomera (Kanarische Inseln). – Beitr. naturk. Forsch. Südwestdeutschland **24**: 47–104.

PALAMAREV, E., 1967: Xerotherme Elemente in der Tertiärflora Bulgariens und Aspekte zum Problem der Formierung der mediterranen Flora auf der Balkanhalbinsel. – Abh. zentr. geol. Inst. Berlin **10**: 167–175.

– 1979: Die Gattung *Stratiotes* L. in der Tertiärflora Bulgariens und ihre Entwicklungsgeschichte in Eurasien. – Fitologija (Sofia) **12**: 3–36.

PANFILOV, D. V., 1959: Položenije granicy meždu tropičeskimi i subtropičeskimi landschaftami v vostočnoj Azii (po materialam issledovanija prirody Junani). – Izvest. AN SSSR, ser. geograf. **3**: 31–41.

PRINCIPI, P., 1938: Le flore del Quaternario. – Ann. Fac. Agr. Forest. R. Univ. Firenze **1**: 3–136.

REID, E. M., CHANDLER, M. E. J., 1933: The London Clay flora. – London: British Museum Natural History.

RICHARDS, P. W., 1952: The tropical rain forests – an ecological study. – Cambridge: Univ. Press.

RÖGL, F., STEININGER, F. F., 1983: Vom Zerfall der Tethys zu Mediterran und Paratethys. Die neogene Paläogeographie und Palinspastik des zirkum-mediterranen Raumes. – Ann. Naturhist. Mus. Wien A **85**: 135–163.

SAPORTA, G., 1881: Die Pflanzenwelt vor dem Erscheinen des Menschen (translation into German by C. VOGT). – Braunschweig: Vieweg u. Sohn.

SCHILIN, P. V., 1986: Pozdnemelovye flory Kazachstana. Sistematičeskij sostav, istorija razvitija, stratigrafičeskoe značenie. – Alma-Ata: Nauka Kazach. SSR.

SCHMID, E., 1970: Die Abgrenzung der Vegetationsgürtel im Mittelmeergebiet. – Feddes Repert. **81**: 203–213.

SCOTT, R. A., 1954: Fossil fruits and seeds from the Eocene Clarno Formation of Oregon. – Palaeontographica B **96**: 66–97.

Seward, A. C., 1931: Plant life through the ages. – Cambridge: Univ. Press.
– Conway, V., 1935: Additional Cretaceous plants from Western Greenland. – Kungl. Svenska Vet. Handl., 3. ser., **15** (3): 1–41.
Sittler, C., 1965: Le paleogene des fosses Rhenan et Rhodanien. Études sedimentologiques et paleoclimatiques. – Mém. Serv. Carte geol. Als. Lorr. Strasbourg **24**: 1–392.
Szafer, W., 1954: Plioceńska flora okolic Czorsztyna i jej stosunek do Plejstocenu. – Prace Inst. Geol. Warszawa **11**: 1–238.
– 1961: Mioceńska flora ze Starych Gliwic na Śląsku. – Prace Inst. Geol. Warszawa **33**: 1–205.
Tanai, T., 1972: Tertiary history of vegetation in Japan. – In Graham, A., (Ed.): Floristics and paleofloristics of Asia and Eastern North America, pp. 235–255. – Amsterdam: Elsevier.
Tiffney, B. H., 1980: Fruits and seeds of the Brandon Lignite, 5. *Rutaceae*. – J. Arnold Arbor. **61**: 1–40.
– 1985: The Eocene North Atlantic land bridge: its importance in Tertiary and modern phytogeography of the Northern hemisphere. – J. Arnold Arbor. **66**: 243–273.
Tralau, H., 1959: Extinct aquatic plants of Europe. On the fossil and recent distribution of *Azolla filiculoides, Dulichium arundinaceum, Brasenia schreberi*, and *Euryale ferox*. – Bot. Not. **112**: 385–406.
– 1962: Die spättertiären *Fagus*-Arten Europas. – Bot. Not. **115**: 147–176.
– 1963: Asiatic dicotyledonous affinities in the Cainozoic flora of Europe. – Kungl. Svenska Vet. Handl. **9**: 1–87.
– 1964a: The genus *Nipa* van Wurmb. – Kungl. Svenska Vet. Handl. **10**: 1–29.
– 1964b: The genus *Trapella* Oliver in the Tertiary of Europe. – Bot. Not. **117**: 119–123.
Van Beusekom, C. F., 1971: Revision of *Meliosma* (*Sabiaceae*), Section *Lorenzanea* excepted, living and fossil, geography and phylogeny. – Blumea **19**: 355–529.
Vinogradov, A. P., 1967/68: Lithological-paleogeographic atlas of U.S.S.R. 4. Paleogene, Neogene, & Quaternary. – Leningrad: Nauka.
Walter, 1968: Die Vegetation der Erde in öko-physiologischer Betrachtung 2. Die gemäßigten und arktischen Zonen. – Jena: VEB G. Fischer.
Wang, C. W., 1961. The forests of China with a survey of grassland and desert vegetation. – Cambridge/Mass.: Harvard Univ. Press.
Wolfe, J. A., 1969: Paleogene floras from the Gulf of Alaska region. – U.S. Geol. Surv., Open-file Rept. 1969: 1–114.
– 1972: An interpretation of Alaskan Tertiary floras. – In Graham, A., (Ed.): Floristics and paleofloristics of Asia and Eastern North America, pp. 201–233. – Amsterdam: Elsevier.
– 1975: Some aspects of plant geography of the Northern hemisphere during the Late Cretaceous and Tertiary. – Ann. Missouri Bot. Gard. **62**: 264–279.
– 1977: Paleogene floras from the Gulf of Alaska Region. – U.S. Geol. Surv. Prof. Paper **997**: 1–108.
– 1979: Temperature parameters of humid to mesic forests of Eastern Asia and relation to forests of other regions of the Northern hemisphere and Australasia. – U.S. Geol. Surv. Prof. Paper **1106**: 1–37.
– 1980: Tertiary climates and floristic relationships at high latitudes in the Northern hemisphere. – Palaeogeography, Palaeoclimatology, Palaeoecology **30**: 313–323.

Address of the author: Dr Dieter H. Mai, Bereich Botanik und Arboretum des Museums für Naturkunde der Humboldt-Universität, Späthstrasse 80/81, DDR-1195 Berlin, German Democratic Republic.

Pl. Syst. Evol. 162, 93–107

Plant Systematics and Evolution

Paleobotanical evidences of the Tertiary history and origin of the Mediterranean sclerophyll dendroflora

EMANUEL PALAMAREV

Received January 25, 1987

Key words: Gymnosperms, angiosperms, *Gnetopsida, Pinopsida, Magnoliopsida, Liliopsida.* – Tethyan and Paratethyan sclerophyll dendroflora, Paleomediterranean and Mediterranean floristic elements.

Abstract: The evolution of important woody plant groups, ancestral to the modern Mediterranean dendroflora, is surveyed. Altogether, the history and phytogeography of 86 fossil species or species-groups is considered. The major part of the Paleomediterranean woody plants appears in Miocene mixed mesophyllous and mesoxerophyllous, evergreen and deciduous forests. The initial formation of basic Mediterranean sclerophyllous woody vegetation types is referred to periods from the Late Sarmatian to the Late Pontian, in geographic areas between 37° and 45° N latitude.

The history of the flora and vegetation of the Mediterranean region as remnants of the ancient Mediterranean of Tethyan floristic subkingdom (TAKHTAJAN 1978) has been intensively studied. Some of the authors have built their hypotheses on the basis of floristic or phytogeographic considerations (STOJANOFF 1925 & 1930, RECHINGER 1950, CONTANDRIOPOULOS 1962, POPOV 1963, MEUSEL 1969 a & b, FAVARGER 1971, GREUTER 1971, etc.), whereas other have also used regional paleofloristic or paleogeographic facts (WULF 1944, HORVAT 1959, CREUTZBURG 1963, ANDREANSZKY 1963 b, PALAMAREV 1967, GREUTER 1970, 1975, PACHOMOV 1976, LOVRIČ 1979, etc.). AXELROD (1975) has pointed to some similarities between the development of the Madro-Tethyan sclerophyll vegetation of California and the Tertiary Tethyan-Mediterranean plant formations.

The present paper gives a detailed representation of the origin and history of the basic ligneous plant species which built the Ancient Tethyan and Paratethyan sclerophyllous communities. The latter develop into the present-day Mediterranean forest vegetation by the end of their Tertiary history. More than 80 megataphofloras have been analysed and compared on the basis of own and literature data. These are distributed in different areas in Europe, Southwestern Asia, Kazakhstan and Western Siberia. They belong to 33 regions (Fig. 1) and a stratigraphic range from the Middle Eocene to the Late Pliocene, i.e. from 45 to 2 million years ago.

The floristic and phytogeographic synthesis includes 86 species or species-groups, belonging to 52 genera and 32 families. These species are considered as ancestral to about 100 extant Mediterranean and also some Submediterranean species

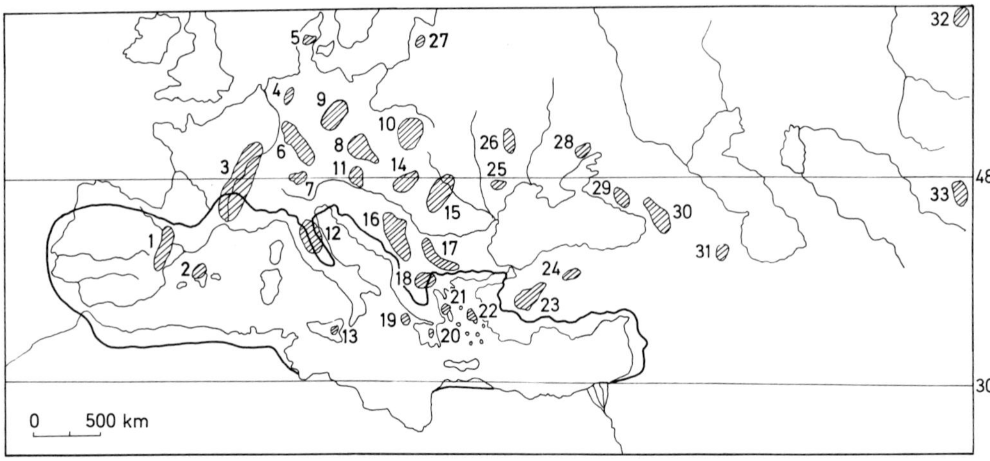

Fig. 1. Geographical location of the Tertiary macrofloras containing Paleomediterranean sclerophyll woody elements in Eurasia, (E Eocene, O Oligocene, M Miocene, P Pliocene), i.e. in Spain (1, O, M), the Balearic Islands (2, M), France (3, O, M, P), the Netherlands (4, P), Denmark (5, M), the Federal Republic of Germany (6, O, M, P), Switzerland (7, M), Czechoslovakia (8, O, M), the Democratic Republic of Germany (9, E, O, M, P), Poland (10, M, P), Austria (11, E, O, M), Italy (12–13, O, M), Hungary (14, O, M); Romania (15, O, M, P), Yugoslavia (16, O, M, P), Bulgaria (17, E, O, M, P); Greece (18–22, M, P), Turkey (23–24, O, M, P), and the Soviet Union (25–33, O, M, P). The bold line marks the recent Mediterranean floristic region

(Table 1). The eco-coenotic features of the paleofloras analysed elucidate the milieu in which the Paleomediterranean xerothermic elements studied appear and develop. Their "chronoareas" (after ZHILIN 1984) during the Tertiary (Table 1) are determined on the basis of their geographic distribution in Eurasia.

The paleofloristic data are presented in four groups according to the first appearance of the paleotaxa. As they are morphogenetically linked with their recent derivatives, but described as separate fossil species, they are here combined into species-groups.

Paleobotanical record

Group 1: species of an Eocene appearance. The data in Table 1 suggest an Eocene origin for nine species (10,4%), including members of the genera *Pinus, Tetraclinis, Periploca, Platanus, Rhamnus, Chamaerops* and *Smilax*. Very early, i.e. already in the Middle Eocene, the area of *Smilax websteri* (*S. excelsa:* the species presented here and afterwards in brackets correspond to recent derivatives) became apparent. Its expansion reached a maximum during the Miocene with an almost continuous European distribution reaching 56° N latitude. The remaining species of this group appear in the Late Eocene and change to some extent during the following periods of the Tertiary. As examples, one can point to *Pinus hepios* (*P. halepensis* s. str.), *P. thomasiana* (*P. pallasiana*) and *Tetraclinis brachyodon* (*T. articulata*) with vast areas of long duration. Throughout the Oligocene and Early Miocene they occupy large regions of W., C., S. and SE. Europe and SW. Asia. *Pinus thomasiana* reaches furthest north with its borderline in Europe and W. Siberia situated at 56–60° N

Table 1. Geochronological ranges and geographical distribution of Paleomediterranean woody plants in the Tethys and Paratethys region, together with their present equivalent taxa. *A* Austria, *Ba* Balearic Islands, *Bu* Bulgaria, *Cz* Czechoslovakia, *Dk* Denmark, *F* France, *Gd* German Democratic Republic, *Gf* German Federal Republic, *Gr* Greece, *H* Hungary, *I* Italy, *N* The Netherlands, *Pl* Poland, *Ro* Romania, *S* Spain, *Sua* Soviet Union (Armenia), *Suaz* Soviet Union (Azerbaijan), *Suf* Soviet Union (Russian Soc. Federal Sov. Republic), *Sug* Soviet Union (Georgia), *Suk* Soviet Union (Kazakhstan), *Sum* Soviet Union (Moldavia), *Suu* Soviet Union (The Ukraine), *Suw* Soviet Union (Western Siberia), *Sw* Switzerland, *T* Turkey, *Yu* Yugoslavia; E Eocene, O Oligocene, M Miocene, P Pliocene. The taxa are listed in alphabetical order.

Ancestral taxon	Tertiary					Present equivalent taxon
	E	O	M	P	Area	
Gnetopsida						
Ephedraceae						
Ephedra campylopoda CAM. foss.				+	Bu	*E. fragilis* DESF. subsp. *campylopoda* (CAM.) A. & G.
Pinopsida						
Cupressaceae						
Cupressus palaeosempervirens KOLAK. & SHAKR.		+	+	+	F-I-Pl-H-Gr-Bu	*C. sempervirens* L.
Juniperus bessarabica NEGRU			+		Sum	*J. macrocarpa* (SIEBTH. & SM.) BALL
J. drupacea-pliocenica RÉR.			+	+	S-F-Pl-Bu-Sug	*J. drupacea* LABILL.
J. excelsa L. foss.				+	Bu	*J. excelsa* L.
J. foetidissima WILLD. foss.			+	+	Bu	*J. foetidissima* WILLD., *J. thurifera* L.
J. oxycedrus L. foss.			+	+	Pl-Bu	*J. oxycedrus* L.
Tetraclinis brachyodon (BRONGN.) MAI & WALTHER	+	+	+		F-Gf-Gd-I-Yu-A-H-Pl-Cz-Ro-Bu	*T. articulata* (VAHL) MAST.
Pinaceae						
Abies ramesii SAP.		+	+		Ba-F	*A. numidica* CARR., *A. cilicica* (ANT. & KOTSCHY) CARR.
A. saportana RÉR.			+	+	S-Pl-Bu	*A. pinsapo* BOISS., *A. cephalonica* LOUD.
Cedrus vivariensis N. BOUL. (= *C. miocenica* LAUBY)			+	+	F-Pl-Bu	*C. libanii* RICH., *C. atlantica* (ENDL.) MAN.
Picea omoricoides WEB. (= *P. palaeoomorica* MÜLL.-STOLL)		+		+	Gf-Pl	*P. omorica* (PANČ.) PURK.
Pinus geanthracis (GOEPP.) REICH.			+	+	Gf-Gd-Pl-Bu-Sug-Suu	*P. peuce* GRISEB.
P. hepios UNG. sensu MAI & WALTHER (incl. *P. halepensis* MILL. foss.)	+	+	+	+	F-I-Gd-A-H-Yu-Ro-Bu-Gr-Suf	*P. halepensis* MILL. sensu stricto
P. laricioides MENZEL			+	+	S-F-Gf-Gd-A-Cz-Ro-Bu	*P. nigra* ARN. sensu lato

Table 1 (continued)

Ancestral taxon	Tertiary					Present equivalent taxon
	E	O	M	P	Area	
P. robustifolia SAP. sensu MAI & WALTHER 1985	+	+	+		S-I-F-Gf-Gd-Yu-GraBu-T	*P. pinaster* AIT.
P. salinarum PARTSCH.			+		Pl-Sum	*P. brutia* TEN.
P. thomasiana (GOEPP.) REICH.	+	+	+	+	Gf-Gd-Dk-Cz-A-Pl-Ro-Bu-Suu	*P. pallasiana* LAMB.
Magnoliopsida						
Aceraceae						
Acer campylopterix-group (*A. campylopterix* UNG., *A. opulifolium-pliocenicum* SAP., *A. obtusilobum* UNG.)			+	+	F-Gf-Yu	*A. opalus* MILL. sensu lato
A. crenatifolium ETT.			+	+	Gd-Cz-A-Pl	*A. hyrcanum* F. & M., *A. reginae-amaliae* BOISS.
A. angustilobum-group (*A. angustilobum* HEER, *A. haselbachensis* WALTHER, *A. berganum* MAI)		+	+		Gf-Gd-Sw-Cz-Bu-T	*A. heldreichii* ORPH.
A. pseudomonspessulanum-group (*A. pseudomonspessulanum* UNG., *A. loclense* HANTKE, *A. pseudocreticum* ETT., *A. hercynicum* MAI)		+	+	+	F-Gf-Gd-Sw-Cz-A-H-Pl-Ro-Bu-Gr-T	*A. monspessulanum* L., *A. orientale* L., *A. creticum* L.
Anacardiaceae						
Cotinus orbiculatus (HEER) BUDANTS. (= *Rhus palaeocotinus* SAP., *R. palaeoradicans* SAP.)		+	+	+	F-Gf-A-H-Pl-Bu-Sug-Suk-Sum-Suu	*C. coggygria* SCOP.
Pistacia acuminata-group (*P. acuminata* C. & E. M. REID, *P. lentiscoides* ANDR. & CZIFF.)			+	+	S-Gf-N-H-Ro-Pl-Yu-Gr-Sua	*P. lentiscus* L.
P. miocenica SAP.			+	+	Ba-F-Bu-Suu-Sum	*P. terebinthus* L., *P. palestina* L.
P. reddita (SAP.) SAP. & MAR.		+	+		F-H-Bu	*P. mutica* F. & M.
Rhus coriaria L. foss.			+		H	*R. coriaria* L.
Apocynaceae						
Nerium oleander-pliocenicum SAP. (= *N. oleander* L. foss.)			+	+	Ba-F-Ro-Bu	*N. oleander* L.
Aquifoliaceae						
Ilex balearica DESF. foss.				+	Ba-F	*I. balearica* DESF.
Asclepiadaceae						
Marsdenia takhtajanii STEPHYRTZA			+		Sum	*M. erecta* (L.) R. BR.
Betulaceae						
Alnus cycladum UNG.			+	+	Gr-Sug	*A. cordata* (LOISEL) DESF.

Table 1 (continued)

Ancestral taxon	Tertiary					Present equivalent taxon
	E	O	M	P	Area	
Carpinus neilreichii-group (*C. neilreichii* Kov., *C. bergeri* Giv., *C. polonica* Zabl., *C. europaea* Negru, *C. sub-orientalis* Sap.)		+	+	+	F-Gf-Gd-Cz-A-H-Pl-Ro-Bu-Gr-Sug-Sum-Suu	*C. orientalis* Mill.
Ostrya kryshtofovichii-group (*O. kryshtofovichii* Baik., *O. tenerrimum* Sap., *O. carpinifolia* Scop. foss.)		+	+	+	F-H-Pl-Bu-Gr-Suk-Sum-Suu	*O. carpinifolia* Scop.
Buxaceae						
Buxus pliocenica Sap. & Mar. sensu Palamarev & Kitanov 1988			+	+	S-F-Gf-Gd-A-Cz-Pl-Ro-Yu-Bu-Gr-T-Sug-Sum-Suu	*B. sempervirens* L. sensu lato
Caprifoliaceae						
Lonicera lipthyana Andr.			+		H	*L. arborea* Boiss.
Viburnum čukurovense Palam.			+		Bu	*V. orientalis* Pall.
V. pseudotinus Sap. & Mar. (= *V. tinus* L. foss.)		+	+		F-H-Ro	*V. tinus* L.
Coriariaceae						
Coriaria longaeva-group (*C. longaeva* Sap., *C. collinsonae* Gregor, *C. myrtifolia* L. foss.)		+	+	+	F-Gf-Ro-Bu	*C. myrtifolia* L.
Ericaceae						
Arbutus elegans Kolak.			+	+	Ro-Bu-Sug	*A. andrachne* L.
A. praeunedo Andr. (= *A. unedo* L. foss.)		+	+		H-Pl-Bu	*A. unedo* L.
Fabaceae sensu lato						
Anagyris foetida L. foss.			+		Pl-Bu	*A. foetida* L.
Ceratonia emarginata Heer		+	+	+	Gf-H-Bu-Sug-Suk	*C. siliqua* L.
Cercis cyclophylla Heer (= *C. virgilliana* Massal.)		+	+	+	F-I-Gf-H-Yu-Suk-Sum	*C. siliquastrum* L.
Colutea macrophylla Heer (= *C. parcefolia* Sap.)			+	+	S-F-Gf-Ro-Bu-Sug	*C. arborescens* L., *C. cilicica* Boiss. & Bal.
Fagaceae						
Castanea pliosativa Kolak. (= *C. sativa* Mill. foss., *C. atavia* Ung. p. p.)			+	+	F-I-Gf-Gd-Dk-Pl Cz-H-Yu-Ro-Bu-Sug-Suu	*C. sativa* Mill.
Quercus cardanii Massal. sensu Palamarev & Kitanov 1988			+	+	F-I-Gd-N-H-Ro-Bu-Gr-Sug	*Q. canariensis* Willd., *Q. hartwissiana* Stev.
Q. equitrojanii Andr. (incl. *Q. trojana* Webb foss.)			+	+	H-Bu-T	*Q. trojana* Webb

Table 1 (continued)

Ancestral taxon	Tertiary					Present equivalent taxon
	E	O	M	P	Area	
Q. grandidentata UNG.			+	+	Gf-Cz-Pl-H-Ro-Yu-T	*Q. aegilops* L., *Q. brantii* LINDL.
Q. kubinyi-group (*Q. kubinyi* (ETT.) CZECZOTT, *Q. saperi* (MENZEL) HUMMEL)			+	+	F-A-H-Pl-Ro-Bu-Sug-Suu-Suw	*Q. libanii* OLIV.
Q. mediterranea UNG. sensu PALAMAREV & PETKOVA 1987		+	+	+	S-F-I-Gf-A-H-Ro-Yu-Bu-Gr-T-Sug	*Q. coccifera* L. sensu lato
Q. palaeocerris SAP. (= *Q. cerris-fossilis* KOLAK.)			+	+	F-Gd-H-Ro-Bu-Suf-Sug	*Q. cerris* L.
Q. praeilex SAP. (= *Q. praecursor* SAP., *Q. ilex* L. foss.)		+	+	+	S-F-H-Ro-Pl-Bu-Gr-T-Sug	*Q. ilex* L., *Q. baloot* GRIFF.
Q. pseudoalnus ETT.			+		A-H-Yu	*Q. alnifolia* POECH.
Q. pseudocastanea GOEPP. (= *Q. mioaxelrodii* KASAPLIGIL)		+	+	+	F-I-Gd-Pl-Cz-H-A-Ro-Bu-Gr-T-Sug-Suk-Suu	*Q. castaneifolia* CAM., *Q. afares* POMEL.
Q. pubescens L. foss.			+	+	Pl	*Q. pubescens* L.
Q. sosnowskyi-group (*Q. sosnowskyi* KOLAK., *Q. suber-pliocenica* N. BOUL. *Q. pseudosuber* SANTI foss., *Q. cerrisicarpa* KOLAK.)			+	+	F-Pl-Ro-Bu-T-Sug	*Q. suber* L.
Q. venturii MASSAL. (= *Q. boissieri* REUT. subsp. *pesmenii* KASAPLIGIL)			+		I-H-Ro-T	*Q. infectoria* OLIV., *Q. faginea* LAM.
Lauraceae						
Laurus nobilis L. foss.		+	+	+	F-H-Yu-Ro-Bu-Suk	*L. nobilis* L.
Moraceae						
Ficus europaea NEGRU			+		Pl-Sum	*F. carica* L.
Myrtaceae						
Myrtus rectinervis SAP. (= *M. bossniaca* ENGEL., *M. aquensis* SAP.)		+	+		F-Pl-Yu-Sug	*M. communis* L.
Oleaceae						
Fraxinus ornus L. foss.			+	+	F	*F. ornus* L.
F. praedicta HEER (= *F. grossidentata* LAURENT)			+	+	F-Gf-Pl-Ro-Suk	*F. angustifolia* VAHL sensu lato
Jasminum contiguum C. & E. M. REID				+	N	*J. fruticans* L.
Olea notii-group (*O. notii* UNG., *O. oleastroides* ZABL., *O. moldavica* NEGRU)		+	+		I-Gr-Cz-Pl-Gf-Bu-Sum	*O. oleaster* HOFFM. & LINK
Phillyrea lanceolata N. BOUL.			+		F-Sug	*P. angustifolia* L.

Table 1 (continued)

Ancestral taxon	Tertiary					Present equivalent taxon
	E	O	M	P	Area	
P. pschechensis KUTUZK. (= *P. latifolia* L. foss.)			+		Suf-H	*P. latifolia* L.
Periplocaceae						
Periploca kryshtofovichii KORN. (incl. *P. graeca* L. foss.)	+	+	+	+	I-Ro-Bu-Sua-Suf-Sug-Suk	*P. graeca* L.
Platanaceae						
Platanus academiae-group (*P. academiae* GAUDIN, *P. schimperi* SAP., *P.leariloba* KOLAK.)	+	+	+	+	I-Gr-Bu-Sug-Suu	*P. orientalis* L.
Punicaceae						
Punica palaeogranatum-group (*P. palaeogranatum* KUTUZK., *P. tertiaria* GREGOR, *P. granatum* L. var. *planchonii* SAP.)		+	+		S-F-Gf-Pl-Bu-Suf-Sug	*P. granatum* L.
Rhamnaceae						
Paliurus ovoideus-group (*P. ovoideus* (GOEPP.) HERR, *P. thurmanii* HEER, *P. orbiculatus* SAP.)			+		F-Gf-Pl-Ro-Bu-Gr-Sug-Suk-Sum	*P. spina-christii* MILL.
Rhamnus colubrinoides ETT. (= *R. alaternoides* HEER, *R. alaternus* L. foss.)	+		+		Gf-A-H-Pl-Bu-Sua	*R. alaternus* L.
Rosaceae						
Cerasus mahaleb L. foss.			+		Pl	*C. mahaleb* L.
Prunus insitita L. var. *pliocenica* MÄDLER				+	F-Gf-Gd-Pl	*P. insitita* L.
P. palaeocerasus ETT.			+	+	Cz-Suu	*P. cerasus* L.
Pyracantha acuticarpa C. & E. M. REID (incl. *P. coccinea* L. foss.)			+	+	Gf-N-Dk-Cz-Pl-Ro-Sug-Sum-Suu	*P. coccinea* L.
Styracaceae						
Styrax maximus-group (*S. maximus* (WEB.) KIRCHH., *S. pristinum* ETT., *S. pseudoofficinalis* BAIK.)			+	+	Gf-Gd-F-N-A-H-Pl-Sug-Suu	*S. officinalis* L.
Tamaricaceae						
Tamarix kryshtofovichii TAKHT. & KUTUZK.			+		Sua	*T. ramosissima* LED.
Thymelaeaceae						
Daphne kimmerica KOLAK.			+	+	Pl-Suaz-Sug	*D. laureola* L.
D. oeningensis A. BR.			+		Gf-A-Pl-Suu	*D. oleoides* SCHREB.
Ulmaceae						
Celtis lacunosa-group (*C. lacunosa* KIRCHH., *C. japetii* UNG., *C. begonioides* GOEPP., *C. vulcanica* KOV., *C. cernua* SAP.)		+	+	+	F-Gf-Cz-Pl-H-A-Bu-Suk-Sum	*C. australis* L.

Table 1 (continued)

Ancestral taxon	Tertiary					Present equivalent taxon
	E	O	M	P	Area	
Verbenaceae						
Vitex paucidenticulata KUTUZK.			+		Suf	*V. agnus-castus* L.
Liliopsida						
Arecaceae						
Chamaerops helvetica FRIEDR. (incl. *C. humilis* L. foss.)	+	+	+	+	Gd-Pl-Gr-Sug-Sum	*C. humilis* L.
Ruscaceae						
Ruscus subaculeatus BERGER			+		A	*R. aculeatus* L.
Smilacaceae						
Smilax hastata (BRONGN.) SAP. sensu PALAMAREV & PETKOVA 1987		+	+	+	S-I-F-Gf-Cz-H-Ro-Yu-Bu-T-Sug-Sum	*S. aspera* L.
S. websteri WESSEL sensu PALAMAREV & PETKOVA 1987	+	+	+	+	S-F-I-Gf-Gd-Dk-Cz-H-Ro-Yu-Bu-Suu	*S. excelsa* L.

latitude. The most eastern partial area of this species is situated in W. Siberia during the Late Oligocene (GORBUNOV 1958). The area of *Pinus robustifolia* (*P. pinaster*) is the most diffuse. Its partial areas exist during the following stages: a) Upper Eocene – Middle Oligocene in C. Europe (GDR, MAI & WALTHER 1985), b) Upper Oligocene in W. Europe (France, SAPORTA 1865) and c) Upper Oligocene in SE. Europe (Bulgaria, PALAMAREV 1962).

The above mentioned Eocene species significantly reduce their areas and gradually disappear during the Late Miocene and Early Pliocene, to be replaced by their present-day derivatives with the S. European flora.

Group 2: species of an Oligocene appearance. 23 species (26,8%) are placed in this group (Table 1). They are from the genera *Abies, Cupressus, Picea, Pinus, Acer, Arbutus, Carpinus, Celtis, Ceratonia, Cercis, Coriaria, Cotinus, Laurus, Myrtus, Olea, Ostrya, Pistacia, Punica, Quercus, Viburnum* and *Smilax*. *Pistacia reddita* (*P. mutica*) and *Punica palaeogranatum* (*P. granatum*) from the Lower Oligocene are the earliest. It is worth noting that the most ancient find of *Punica palaeogranatum* is from the early Oligocene flora of the Rhodope region in Bulgaria (PALAMAREV & PETKOVA unpubl.).

The remaining species apparently start their evolution from the Middle and Late Oligocene. Some aspects concerning the development of their areas are of interest. There are taxa with large and continuous areas, including regions of almost the whole of Europe (except the north), as well as parts of the Caucasus, Asia Minor and Kazakhstan, e.g., *Acer pseudomonspessulanum*-group (*A. monspessulanum*-group), the *Carpinus neilreichii*-group (*C. orientalis*), *Cercis cyclophylla* (*C. siliquastrum*), *Cotinus orbiculatus* (*C. coggygria*), *Quercus mediterranea* (*Q. coccifera*

s. l.), *Q. pseudocastanea* (*Q. castaneifolia, Q. afares*), and *Smilax hastata* (*S. aspera*). Alternatively, there are species having restricted and discontinuous areas, e.g., *Abies saportana* (*A. pinsapo, A. cephalonica*), *A. ramesii* (*A. numidica, A. cilicica*), *Picea omoricoides* (*P. omorica*), *Arbutus praeunedo* (*A. unedo*), *Coriaria longaeva*-group (*C. myrtifolia*), *Myrtus rectinervis* (*M. communis*), *Viburnum pseudotinus* (*V. tinus*). The *Celtis lacunosa*-group (*C. australis*), *Ceratonia emarginata* (*C. siliqua*), the *Olea notii*-group (*O. oleaster*), the *Ostrya kryshtofovichii*-group (*O. carpinifolia*) and *Quercus praeilex* (*Q. ilex, Q. baloot*) occupy an intermediate position in this respect, being disjunct, i.e. SW. European-Caucasian, E. European-Kazakhstanian and Caucasian-Kazakhstanian. Besides, the Caucasian part of these areas is always younger than its other parts. This could be due to the existence of an European and a Kazakhstanian Paleogene centres of dispersal for semixerophytic or xerophytic elements, expanding to the East and West, respectively. It seems that the Caucasian part of these areas survived for a longer period in comparison to other parts which disappeared already before the Pliocene.

It is typical of the major part of the Oligocene Paleomediterranean species that their northermost borderline reached 50 – 53° N latitude during the different stages of their development.

Group 3: species of a Miocene appearance. This group is the most numerous, including 49 species or 57% (Table 1). It is represented by the genera *Abies, Cedrus, Juniperus, Pinus, Acer, Alnus, Anagyris, Arbutus, Buxus, Cerasus, Colutea, Daphne, Ficus, Fraxinus, Lonicera, Marsdenia, Nerium, Paliurus, Phillyrea, Pistacia, Prunus, Pyracantha, Quercus, Rhus, Styrax, Tamarix, Viburnum, Vitex* and *Ruscus*. Part of the species are of an Early Miocene appearance, i.e. *Pinus laricioides* (*P. nigra*), *Acer crenatifolium* (*A. hyrcanum, A. reginae-amaliae*), *Alnus cycladum* (*A. cordata*), *Colutea macrophylla* (*C. arborescens*), *Fraxinus praedicta* (*F. angustifolia* s. l.), *Paliurus ovoideus*-group (*P. spina-christii*), *Pistacia miocenica* (*P. terebinthus, P. palaestina*), *Quercus grandidentata* (*Q. aegilops, Q. brantii*), *Styrax maximus*-group (*S. officinalis*). Among these species, *Pinus laricioides, Fraxinus praedicta, Paliurus ovoideus* and *Styrax maximus* stand out with their quite extensive Tertiary areas; *Styrax maximus*, e.g., can be qualified as European – Southwest-asian – Kazakhstanian. The northern borderline of the species mentioned above coincides more or less with the northern boundary of C. Europe.

In contrast, some Early Miocene species have restricted or diffuse areas, e.g., *Acer crenatifolium* (C. Europe), *A. campylopterix*-group (*A. opalus* s. l., W. – SE. Europe), *Alnus cycladum* (S. Balkan – Georgia), and *Pistacia miocenica* (SW. – E. Europe – Georgia).

The major part of the species discussed appear during the Middle and particularly the Late Miocene. A more expanded European (E) or European – Southwestasian (ESWA) area is characteristic only of *Buxus pliocenica* (*B. sempervirens*-group): ESWA, the *Pistacia acuminata*-group (*P. lentiscus*): ESWA, *Pyracantha acuticarpa* (*P. coccinea*): ESWA, *Quercus cardanii* (*Q. canariensis*): ESWA, *Q. palaeocerris* (*Q. cerris*): ESWA, and *Castanea pliosativa* (*C. sativa*): ESWA. Concerning the subgroup of the Middle and Late Miocene Paleomediterranean elements, the more diffuse areas characterize *Cedrus vivariensis* (*C. libanii, C. atlantica*), *Juniperus drupacea-pliocenica* (*J. drupacea*), *Pinus salinarum* (*P. brutia*), *Anagyris foetida-*fossilis (*A. foetida*), *Arbutus elegans* (*A. andrachne*), *Daphne kimmerica* (*D. laureola*), *D. oeningensis* (*D. oleoides*), *Nerium oleander-pliocenica* (*N. oleander*), *Phillyrea* *lanceolata* (*P. angustifolia*), and *Prunus palaeocerasus* (*P. cerasus*). These species exhibit three types of disjunction, namely the SW. European – E. European type, the E. European – SW. Asian type and the C. European – Caucasian type. The smallest area has been found in *Viburnum čukurovense* (*V. orientalis*), regarded as a Bulgarian Miocene endemic.

Group 4: species of a Pliocene appearance. This is the smallest group with only five species or 5,8% (Table 1) and consists of the following taxa: *Ephedra campylopoda*-fossilis (*E. fragilis* subsp. *campylopoda*), *Juniperus excelsa*-fossilis (*J. excelsa*), *Ilex balearica*-fossilis (*I. balearica*), *Jasminum contiguum* (*J. fruticans*), and *Prunus insitita* var. *pliocenica* (*P. insitita*). A specific feature is their restricted Tertiary areas: the first four species have partial areas in Bulgaria, SW. France and the Netherlands, while the fifth has a SW. and C. European area.

The limited number of the Paleomediterranean elements having a Pliocene appearance could be attributed to the fact that the transformation of the major part of the pre-Pliocene species into their recent Mediterranean derivatives mostly took place during the Pliocene.

Discussion

The appearance of ligneous sclerophyll Paleomediterranean species began during the second part of the Eocene. This process took place within the boundaries of the zonal vegetation composed of the subtropical woody vegetation (of the Dryophylleto-Daphnogenetum-sempervirentifruticosum type) which occupied large territories of the entire Tertiary Tethys region in Eurasia. These sclerophyll species played a secondary role, forming restricted azonal communities influenced by relatively dry edaphic and microclimatic conditions. However, the role of these Paleomediterranean elements increased during the Oligocene, when they formed specific tree and shrub coenoses in different areas of the Tethys. They arose, both from various subtropical and warm-temperate broad-leaved evergreen and deciduous communities, as well as from coniferous mixed coenoses. These communities are mainly composed of *Lauraceae, Fagaceae, Magnoliaceae, Theaceae* and some Conifers. Thus, Oligocene xerothermic centres are formed on the Iberian, Apenninian and Balkan Peninsulas, as well as in some C. European and French regions (SAPORTA 1862, 1865; PRINCIPI 1916, PANTIČ 1956, GRANGEON 1958, ANDREANSZKY 1966, PALAMAREV 1967, FERNANDEZ MARRON 1971, 1979, etc.). The species of the genera *Quercus* (sectt. *Ilex* and *Heterobalanus*), *Arbutus, Pistacia, Ceratonia, Acer* (sect. *Goniocarpa*), *Periploca, Smilax* and *Pinus* (sect. *Banksia*) are best represented in these centres. They probably formed restricted and isolated formation types of Querceta mediterraneae, Querceta praeilicis and Pineta hepios.

The evolution of sclerophyllous coenoses of a Paleomediterranean type appears irregular during the subsequent phases of the Neogene. The phenomenon is more distinct during the Burdigalian of Balearic Islands, Greece and Kazakhstan (UNGER 1867, ARENES & DEPAPE 1956, KORNILOVA 1960); during the Sarmatian and Tortonian on territories of the Iberian, Apenninian, and Balkan Peninsulas (PRINCIPI 1926, MENENDEZ AMOR 1955, PANTIČ 1956, BERGER 1957, PANTIČ & MIHAJLOVIČ 1976, 1977; VELITZELOS & SCHNEIDER 1979, VELITZELOS & PETRESCU 1981, VELITZELOS & al. 1981, 1985, 1986; PALAMAREV & PETKOVA 1987), as well as in Hungary, Moldavia, and pre-Caucasus (ANDREANSZKY 1959, KUTUZKINA 1964,

1974; NEGRU 1972, STEPHYRTZA 1974, 1982); during the Late Aragonian in Asia Minor (MÄDLER & STEFFENS 1979); During the Messinian – Late Pontian period this developed further in different regions of the Tethys and Paratethys, as is supported by the concentration of Paleomediterranean species in some local paleofloras. Communities of a Mediterranean sclerophyllous type during this period contained evergreen oaks, species of *Arbutus, Ceratonia, Olea, Phillyrea, Pistacia, Pyracantha* and *Pinus hepios*. PANTIČ & MIHAJLOVIČ (1977, 1979) published data about the development of maquis-like coenoses in Yugoslavia during the Pannonian. HEIMANN & al. (1975) reported data about the existence of such vegetation in Greece during the Messinian, while TAKHTAJAN (1956), KOLAKOVSKYI (1964), and RATIANI (1972) have demonstrated the existence of sclerophyllous coenoses of the Mediterranean type during the Pontian and Akchagylian in Armenia and Georgia.

The process of evolution of sclerophyllous vegetation of the type discussed is supported also by the data from the Pannonian flora of Romania (GIVULESCU 1962), the Late Pontian flora of Bulgaria (STEFANOFF & JORDANOFF 1935, KITANOV 1984), and the Early Pliocene floras of France (DEPAPE 1922) and Turkey (KASAPLIGIL 1977, 1981).

The areas of the Paleomediterranean species appearing during the Miocene reach 47 – 49° N latitude during their evolution. This undoubtedly is an indication that the present-day border of the Mediterranean floristic region if a post-Miocene event. The drastic reduction of the areas probably begins during the Messinian, continues in the Pliocene and reaches the present extent during the Pleistocene.

When comparing the dynamics of the distribution areas of Tertiary ancestors and their modern derivatives, two opposite trends become obvious. On the one hand, there are cases where the areas of the Tertiary species are more extensive than those of the derivatives, e.g., in *Tetraclinis brachyodon – T. articulata, Pyracantha acuticarpa – P. coccinea, Quercus cardanii – Q. canariensis, Q. kubinyi – Q. libanii, Q. pseudocastanea – Q. castaneifolia*, and *Styrax maximus – S. officinalis*. On the other hand, the areas of the Tertiary ancestors are more restricted than those of their recent derivatives, comparing *Juniperus bessarabica – J. macrocarpa, J. foetidissima* fossilis – *J. foetidissima, J. oxycedrus* fossilis – *J. oxycedrus, Anagyris foetida* fossilis – *A. foetida, Cerasus mahaleb* fossilis – *C. mahaleb, Ficus europaea – F. carica, Fraxinus ornus* fossilis – *F. ornus, Jasminum contiguum – J. fruticans, Marsdenia takhtajanii – M. erecta, Quercus pubescens* fossilis – *Q. pubescens, Tamarix kryshtofovichii – T. ramosissima*, and *Ruscus subaculeatus – R. aculeatus*.

KOLAKOVSKYI (1974) offers considerations concerning the altitudinal zonation of the vegetation of the Ancient Mediterranean area during the Neogene. According to the principle of actualismus, one can assume some of the Paleomediterranean conifers to have been an integral part of the sclerophyllous coastal, lowland or hilly phytocoenoses, which we now prefer the call "paleomediterranean sclerophyllous tree and shrub communities" (instead of the term "pre-maquis" introduced earlier: PALAMAREV 1967). Such Conifer elements are *Pinus hepios, P. salinarum, P. robustifolia, Juniperus bessarabica, J. drupacea-pliocenica* and *Tetraclinis brachyodon*. Other species probably have formed mixed mesoxerophyllous to mesophyllous forests at higher elevations, e.g., *Abies saportana, A. ramesii, Cedrus vivariensis, Cupressus palaeosempervirens, Picea omoricoides, Pinus geanthracis, P.*

laricioides, P. thomasiana. This type of paleocoenoses probably formed a montane Mediterranean vegetation belt.

The evergreen mesoxerophyllous oaks of the sectt. *Ilex* and *Heterobalanus* obviously have played an important role in the formation of the sclerophyllous coastal, lowland and hilly forests. The paleobotanical data suggest that they have given rise to the oak-chestnut woods and the so-called "oak-maquis" (Menitzkyi 1984) later on.

The Late Tertiary sclerophyllous coenoses also contain deciduous oaks (from sectt. *Quercus* and *Cerris*). These oaks are considered by Menitzkyi (1984) to appear first in the Tethyan subtropical flora at the beginning of the Miocene as montane elements and then to descend to the lower belts. According to this author, the dispersal of the numerous deciduous oaks into S. Europe and the Caucasus has taken place after the disappearance of the Himalayan sector of the Tethys during the latest Miocene (Rögl & Steininger 1983).

Some Paleomediterranean species groups have paticipated in the composition of the azonal riparian vegetation, e.g., members of *Nerium, Periploca, Platanus, Tamarix, Vitex,* and *Smilax.*

The evolution of the sclerophyllous woody communities composed of Paleomediterranean elements is closely connected with the aridity of the climate during the different phases of the Neogene. This aridity had a cyclic character and mainly affected regions of the Alpine-Himalayan Orogenic Belt. Papp & Steininger (1979) report data about a considerable aridity in the Tethys between 30° and 40° N latitude during the Messinian – Late Pontian period, probably as a result of the extensive Messinian regressive phases and corresponding increases in land surface. The dispersal of the xerophytic element probably was of a cyclic nature, just as the climate. This is supported by the mesophytic nature of the Pliocene. S. European floras, reconstituted after the "xeric period" from the Sarmatian to the Messinian. The last "xeric wave" effecting Paleomediterranean elements probably occurred during the latest Romanian (resp. Piacenzian) and the Pleistocene. These arid phases stimulated the formation of Neogene xerothermic centres on the Iberian, Apenninian and Balkan Peninsula, as well in Asia Minor. The composition of the paleofloras suggests a warm-temperate climate with a humidity of the present W. Mediterranean type. Its summer aridity is moderated by the relatively high air humidity, favoured by persisting air circulations. Such a type of climate made possible the long co-existence of laurophyllous, mesoxerophyllous and xerophyllous woody plants in the Neogene Tethyan and Paratethyan vegetation.

In the Late Pliocene the differences between the W. and E. Mediterranean floristic areas are already apparent, differences which further increase during the Pleistocene.

Andreanszky (1963 a, b) in his hypothesis about the origin of the so-called "Atlantic xerophilic element" including the Paleomediterranean floristic element, postulates the presence of a xerothermic centre on the territory of the Maroccanian and Spanish Meseta (between 32 and 40° N latitude) from which the dispersal of the xerothermic ligneous element had started. However, there is no paleobotanical evidences for such a centre. Moreover, the paleofloras analysed by us support the presence of several Euroasiatic centres of different ages (from the Burdigalian to the Akchagylian) on the territories from the Iberian Peninsula to Kazakhstan, belonging to different paleofloristic phytochoria (Iberian, Ligurian, Pannonian,

Karpathian, Balkanian, Krimea-Caucasian, Asia Minor, etc.). The considerations above mentioned makes it clear that the ANDREANSZKY's "Atlantic xerophilous element" should be regarded as a Tertiary Tethyan and Paratethyan sclerophyllous element.

In conclusion we would like to point out that the paleobotanic data indicate the presence of Paleomediterranean elements during the Middle and Late Paleogene in Eurasia which have been of subordinate significance for the composition and structure of the vegetation. Only during the Late Miocene and Pliocene, a semiarid climatic zone was formed between 37° and 45 °N latitude with more prominent sclerophyllous woody coenoses. Their direct descendants nowadays constitute much of the woodland vegetation of the Mediterranean floristic area with their two basic coenotic units, Querceta ilicis and Oleo-Ceratonion.

References

ANDREANSZKY, G., 1959: Die Flora der sarmatischen Stufe in Ungarn. – Budapest: Akademia kiado.

– 1963a: Das Trockenelement in der alttertiären Flora Mitteleuropas auf Grund paläobotanischer Forschungen in Ungarn. – Vegetatio **11**: 95–111.

– 1963b: Das Trockenelement in der jungtertiären Flora Mitteleuropas. – Vegetatio **11**: 155–172.

– 1966: The Upper Oligocene flora of Hungary. – Stud. biol. Hung. **5**: 3–151.

ARENES, J., DEPAPE, G., 1956: La flore burdigalienne des îles Baléares. Rev. Gen. Bot. (Paris) **63**: 1–43.

AXELROD, D., 1975: Evolution and biogeography of Madrean-Tethyan sclerophyll vegetation. – Ann. Missouri Bot. Gard. **62**: 280–334.

BERGER, W., 1957: Untersuchungen an der obermiozänen Flora von Gabbro in der Toskana. – Paläontogra. Ital. **51**: 3–96.

CONTANDRIOPOULOS, J., 1962: Recherches sur la flore endemique de la Corse et sur ses origin. – Ann. Fac. Sci. Marseille **32**: 1–35.

CREUTZBURG, N., 1963: Die paläogeographische Entwicklung der Insel Kreta vom Miozän bis zur Gegenwart. – Kritika Hronika **15–16**: 336–344.

DEPAPE, G., 1922: Recherches sur la flore de la vallée du Rhône. – Ann. Sci. Nat. Bot., ser. 10, **4**: 3–265.

FAVARGER, C., 1971: Relations entre la flore mediterranéene et celle des enclaves à végétation submediterranéene d'Europe centrale. – Boissiera **19**: 149–177.

FERNANDEZ MARRON, M. T., 1971: Estudio paleoecologico y revision sistematica de la flore fosil del Oligoceno español. – Fac. Cienc. Univ. Compl. Sec. Biol. (Madrid) **152**: 1–177.

– 1979: Essai de resolution de problemes stratigraphiques de la limite paleogene-neogene par les études de macroflore. – Ann. Geol. Pays Hell. **1**: 403–412.

GIVULESCU, R., 1962: Die fossile Flora von Valea Neagra. – Paläontogr. **110**: 128–187.

GRANGEON, P., 1958: Contribution à l'étude de la flore stampienne de Limagne. – Travaux Labor. Géol. Mineral. Univ. Clermont **24**: 43–99.

GREUTER, W., 1970: Zur Paläogeographie und Florengeschichte der südlichen Ägäis. – Feddes Repert. **81**: 233–242.

– 1971: Betrachtungen zur Pflanzengeographie der Südägäis. – Opera Bot. **30**: 49–64.

– 1975: Historical phytogeography of the southern half of the Aegean area. – In JORDANOV, D., (Ed.): Problems of Balkan flora and vegetation, pp. 17–21. – Sofia: Academic Press.

GORBUNOV, M. G., 1958: Tertiary *Pinaceae* from Western Siberia. – Bot. Ž. (Moscow) **43**: 337–352.

Heimann, J., Jung, W., Braune, K., 1975: Schichtenfolge und Flora des Messinian in Nord-Korfu (Griechenland). – Mitt. Bayer. Staatssammlungen **15**: 169–177.

Horvat, J., 1959: Die Pflanzenwelt Südosteuropas als Ausdruck der erd- und vegetationsgeschichtlichen Vorgänge. – Acta Soc. Bot. Polon. **28**: 381–405.

Kasapligil, B., 1977: A Late-Tertiary conifer hardwood forest from the vicinity of Güven village. – Bull. Min. Research Explor. Inst. Turkey **88**: 25–33.

– 1981: Past and present oaks of Turkey. – Phytologia **49**: 95–146.

Kitanov, G., 1984: The Pliocene flora from Gotse Delcev region. – Fitologija **25**: 41–70.

Kolakovskyi, A. A., 1964: The Pliocene flora from Kodor. – Suchumi: Suchumi Bot. Gard.

– 1974: Altitudinal zonation of the forest vegetation of the Kolchis during the Tertiary. – Ann. Forest Inst. (Tbilisi) **21**: 98–115.

Kornilova, V. S., 1960: The Lower Miocene flora from Kushuk. – Alma-Atha: Academic Press Kazakh. SSR.

Kutuzkina, E. F., 1964: The Sarmatian flora from Armavir. – Paleobotanika (Leningrad) **5**: 148–229.

– 1974: Some members of the Mediterranean flora in Upper Sarmatian from Krasnodar region. – Bot. Ž. (Moscow & Leningrad) **59**: 251–259.

Lovrič, A. Z., 1979: Derive des continents et phytogeographie adriatiques. – Rapp. Proced. Comm. Int. Explor. Sci. Mer. Mediterr. (Monaco) **25**: 61–63.

Mädler, K., Steffens, P., 1979: Neue Blattfloren aus dem Oligozän, Neogen und Pleistozän der Türkei. – Geol. Jahrb. **33**: 3–33.

Mai, D. H., Walther, H., 1985: Die obereozänen Floren des Weißelster Beckens und seiner Randgebiete. – Abh. Staatl. Mus. Miner. Geol. **33**: 5–176.

Menendez Amor, J., 1955: La depression ceretana y sus vegetales fosiles. – Mem. Real. Acad. Sci. (Madrid), Ser. Cienc. Nat., **18**: 3–232.

Menitzkyi, J. L., 1984: The Asiatic oaks. – Leningrad: Nauka.

Meusel, H., 1969 a: Die Beziehungen zwischen der mediterranen und holarktischen Flora. – In V. Symp. Flora Europaea (Sevilla): 149–157.

– 1969 b: Die Beziehungen in den Florendifferenzierungen von Eurasien und Nordamerika. – Flora **158**: 537–564.

Negru, A. G., 1972: The Early Sarmatian flora of the Northern Moldavia. – Kishinew: "Shtiinza".

Pachomov, M. M., 1976: Paleogeographic and floristic community of the ancient Mediterranean. – In Palynology in USSR (Moscow): 112–116.

Palamarev, E., 1962: Contribution to the knowledge of the fossil flora from Samokov. – Mitt. Bot. Inst. (Sofia) **10**: 15–21.

– 1967: Xerotherme Elemente in der Tertiärflora Bulgariens und Aspekte zum Problem der Formierung der mediterranen Flora auf der Balkanhalbinsel. – Abh. Zentr. Geol. Inst. (Berlin) **10**: 165–175.

– Kitanov, G., 1988: The fossil macroflora of the Beli Brjag Coal Basin. – In Miscellany in memoriam of Prof. N. Stojanov. – Sofia: Academic Press, 183–205.

– Petkova, A., 1987: The Sarmatian macroflora of Bulgaria. – Les fossiles de Bulgarie **8**: 3–275.

Pantič, N., 1956: Biostratigraphie des flores tertiaires des Serbie. – Ann. Geol. Penins. Balk. **24**: 199–321.

– Mihajlovič, D., 1976: Flores neogenes provenant du continent balkanique et leur importance au point de vue de la paleoclimatologie, paleobiogeographie et biostratigraphie. 1. – Ann. Geol. Penins. Balk. **40**: 104–125.

– – 1977: Flores neogenes provenant du continent balkanique et leur importance au point de vue de la paleoclimatologie, paleobiogeographie et biostratigraphie. 2. – Ann. Geol. Penins. Balk. **41**: 159–171.

PANTIĆ, N., MIHAJLOVIĆ, D., 1979: Flores neogenes provenant du continent balkanique et leur importance au point de vue de la paleoclimatologie, paleobiogeographie et biostratigraphie. 3. – Ann. Geol. Penins. Balk. **43**: 239–257.

PAPP, A., STEININGER, F. F., 1979: Paleogeographic implication of Late Miocene deposits in the Aegean region. – Ann. Geol. Pays Hell. **2**: 966–969.

POPOV, M. G., 1963: Bases of florogenetic. – Moscow: Academic Press.

PRINCIPI, P., 1916: La flora oligocenica di Chiavon e Salcedo. – Mem. Descr. Carta geol. Ital. **10**: 1–130.

– 1926: Flora sarmaziana di Polenta in provincia di Forli. – Atti Soc. Lig. Sci. Lett. **5**: 5–26.

RATIANI, N. N., 1972: Some data about the Pliocene flora from "Malki Shirak". – Comm. Akad. Sci. Grus. SSR (Tbilisi) **68**: 241–244.

RECHINGER, K. H., 1950: Grundzüge der Pflanzenverbreitung in der Ägäis. – Vegetatio **2**: 55–119.

RÖGL, F., STEININGER, F. F., 1983: Vom Zerfall der Tethys zu Mediterran und Paratethys. – Ann. Naturhist. Mus. Wien **85**: 135–163.

SAPORTA, G., 1862: Etudes sur la végétation du Sud-Est de la France à l'époque tertiaire. – Ann. Sci. Nat. Bot. **17**: 191–311.

– 1865: Etudes sur la végétation du Sud-Est de la France à l'époque tertiaire. – Ann. Sci. Nat. Bot. **4**: 5–264.

STEFANOFF, B., JORDANOFF, D., 1935: Studies upon the Pliocene flora of the plain of Sofia. – Ann. Acad. Sci. Bulg. **29**: 3–150.

STEPHYRTZA, A. G., 1974: The Early Sarmatian flora from Bursuk. – Kishinew: "Shtiinza".

– 1982: Some species of the genera *Pyracantha, Pistacia, Paliurus* and *Berchemia* in the Early Sarmatian flora from Bursuk. – Bot. Ž. (Moscow & Leningrad) **67**: 1221–1230.

STOJANOFF, N., 1925: Über die Herkunft des xerothermen Elements in der Flora der Balkanhalbinsel. – Jahrb. Univ. Sofia **3**: 217–248.

– 1930: Versuch einer Analyse des relikten Elements in der Flora der Balkanhalbinsel. – Engl. Bot. Jahrb. **43**: 369–418.

TAKHTAJAN, A. L., 1956: The fossil flora from Agarka in Armenian SSR. – Bot. Ž. (Moscow & Leningrad) **41**: 652–657.

– 1978: Floristic regions of the earth. – Leningrad: Nauka.

UNGER, F., 1867: Die fossile Flora von Kumi auf der Insel Euboea. – Denkschr. K. Akad. Wiss., Math.-Nat. Kl. **27**: 27–90.

VELITZELOS, E., SCHNEIDER, H., 1979: Jungtertiäre Pflanzenfunde aus dem Becken von Vegora in West Mazedonien. Eine Fächerpalme (*Chamaerops humilis* foss.). – Ann. Geol. Pays Hell. **2**: 796–799.

– PETRESCU, J., 1981: Seltene pflanzliche Fossilien aus dem Braunkohlenbecken von Vegora. – Ann. Geol. Pays Hell. **2**: 767–777.

– – SIMEONIDIS, N., 1981: Tertiäre Pflanzenreste aus der Ägäis. Die Macroflora der Insel Lesvos. – Ann. Geol. Pays Hell. **2**: 500–514.

– GREGOR, H. J., 1985: Neue paläobotanische Befunde im Neogen Griechenlands. – Doc. Naturae **25**: 1–4.

– – 1986: Geologische Daten zu den fossilführenden Fundstellen Lava, Prosilion und Likudi nebst Bemerkungen zu deren Frucht- und Samenflora. – Doc. Naturae **29**: 34 40.

– KNOBLOCH, E., 1986: Die pliozäne Flora von Skoura bei Sparta auf dem Peloponnes. – Doc. Naturae **29**: 21–29.

WULF, E. W., 1944: Historical plant geography. – Moscow, Leningrad: Academic Press.

ZHILIN, S. G., 1984: Basic stages of the formation of temperate woody flora during the Oligocene–Early Miocene period in Kazakhstan. – Leningrad: Nauka.

Address of the author: Dr EMANUEL PALAMAREV, Institute of Botany, Bulgarian Academy of Sciences, 1113 Sofia, Bulgaria.

Pl. Syst. Evol. 162, 109–114

—Plant—
Systematics
and
Evolution

History of the montane forests of the northern Andes

T. Van der Hammen

Received February 17, 1987

Key words: Vegetation, flora and phytogeographical elements in the northern Andes; historical changes since the Cretaceous and Tertiary.

Abstract: From the vascular plant genera of the tropical Andean montane forests about 65% have a tropical-neotropical, 15% tropical-Andean, 5% amphi-pacific (Malayo-American), 5% Austral-antarctic and 10% Holarctic distribution. The explosive evolution of Andean centred taxa probably started in the lower Tertiary from tropical Gondwana stock, when in the area of the present day Andes, locally there may have been hills up to 1 000 m. With the final upheaval between the Late Miocene and Late Pliocene, extensive areas with a montane climate were created and populated from the low mountain flora by evolutionary radiation. Taxa of Austral-antarctic origin migrated along the Andean chain to the tropics. At the same time the formation of the Panama isthmus connected N. and S. America, and enabled many plants to enter the tropical Andes from the north. First to cross were tropical to subtropical and warm-temperate species from the Tertiary Laurasian flora that had migrated southward during the period of Miocene cooling, taking refuge in the montane tropics. Because the same happened in SE. Asia there is a good number of amphi-pacific taxa known as fossils from Tertiary Laurasia but largely extinct in the present-day Holarctic (e.g., *Trigonobalanus*). Later to cross were cool-temperate to cold elements that still exist in the Holarctic (e.g., *Alnus* and *Quercus*).

To understand the history of the montane forests of the northern (tropical) Andes, it is necessary to know the general historical background of the South American continent, and of the flora and vegetation of tropical S. America, in relation to temperate S. America and temperate N. America. For that reason we will first give some background information on the earlier history, then treat shortly the vegetation and phytogeography of the Andes, followed by an overview of the Late Tertiary and Quaternary history of the area, and finally general conclusions.

Historical survey

The northern Andes are part of the neotropics. The neotropics correspond to the greater part (northern tropical to sub-tropical part) of S. America, while the southern part of the continent belongs to the Austral-antarctic floral area. The history of the present flora starts in the time of Gondwana land, when Africa, S. America, Australia, and Antarctica were still united. The floras of tropical Africa and S. America are still very similar in the Lower to Middle Cretaceous, forming clearly a well defined floral province (HERNGREEN 1975). In the Upper Cretaceous the continents are moving apart, and from that period on the floral differences between the two continents are increasing. In the Maestrichtian and Paleocene the two areas still have elements in common, but from then on floral differences increase fastly, resulting in the present-day situation. Southern, temperate, S. America, was still connected with Antarctica/Australia until well in the Tertiary. There are, therefore, marked floristic relations with Australia/New Zealand. The Atacama desert area maintained the two floral areas relatively separated. As to the N. and S. American continents, there may have been some connection in the Cretaceous via island chains, but during a major part of the Tertiary, the two continents were separated, apparently not permitting a major interchange of floral elements. During the Pliocene the Panamanian land bridge is formed, preceded by a period of island chains (possibly since the Upper Miocene). This caused an increasing interchange of species between the two continents.

The formation of the Panamanian land bridge takes place in the same period as the final upheaval of the Andes took place.

The history of the Andean geosynclinal area may be traced back into the Mesozoic, but for our purpose it is important to know that tectonic movements in the area during the Maestrichtian and Lower Tertiary (Paleocene-Eocene) caused the rise of large areas above sealevel, and the formation of terrestrial sedimentary basins, hill ridges or low mountains. This type of movements continued during the Oligocene and further increased during the Miocene. The exact elevation of the hills and low mountains that existed during the Tertiary is not exactly known, but there are no indications of the existence of a high mountain vegetation. Hence, we think that the elevations were in general not higher than 1 000 m. Towards the end of the Miocene tectonic movements increase, and hills gradually higher may have come into existence. The final upheaval of the northern Andes to its present elevations, took place during the Pliocene. Then zones with distinctly different climates were created: montane tropical and tropical high mountain (páramo) climates. From then on, the proper montane forests of the northern Andes as we know them today and the open páramo flora and vegetation, are being formed. To know about what properly happened, how these most curious and rich flora and vegetation came into being, it is necessary to study the present phytogeography of the floral elements, and the proper history of the area, its climate, flora and vegetation.

Present vegetation, flora, and phytogeography

In the northern Andes we may distinguish the following zonation of the vegetation (approximate altitudes): 0 – 1 000 m Lower tropical forest, 1 000 – 2 300 m Subandean forest (lower montane), 2 300 – 3 500 (3 200/3 600) m Andean forest (upper montane), 3 500 (3 000/3 500) – 3 900 (3 500/3 900) m High Andean forest. Above 3 200 – 3 900 m we find the open páramo vegetation (sub-páramo, páramo and super-páramo). Until the nival zone (>4 800/5 000 m) there may be considerable differences in altitude of zone limits, according to local circumstances (temperature, humidity, wind protection), as indicated.

The transition of the Andean forest to the subandean forest is characterized by the disappearance of a good number of genera or species of tropical origin, like *Acalypha, Alchornea*, and *Cecropia*. The limit corresponds approximately with the lowermost boundary of the (occasional) occurrence of night frost. A number of genera present in the montane forest and páramo zone came apparently from northern or southern temperate areas (Holarctic/Austral-antarctic), where they have their principal distribution. The percentage of generic elements of temperate origin increases going upward from subandean via Andean to páramo, from <5% to 50%. An analysis of the Andean forest between 2 500 and 3 500 m in the Colombian Cordillera Oriental (130 genera found) showed that approximately 85% is of tropical origin, 10% is of northern (Holarctic) origin and 5% of southern (Andean-Austral-antarctic) origin. Northern genera include *Quercus, Myrica, Juglans, Alnus, Ribes, Berberis, Vaccinium*, and (probably) *Rhamnus* and *Viburnum*. Southern genera are *Weinmannia, Podocarpus, Drimys, Gaultheria*, and *Pernettya*.

When we make a more detailed analysis of the 85% genera of tropical origin, we find that 35% are from the low (neo-)tropical zone, 15% are tropical Andean, 20% pan-tropical, 10% are found in America and SE. Asia (amphi-pacific) and 5% in America and Africa. Some examples of these groups are (Van der Hammen & Cleef 1983/84): neotropical: *Panopsis, Remigia, Ladenbergia, Hieronima, Geonoma* and *Freziera*. Andean: *Mauria, Gaiadendron, Bucquetia, Vallea, Polylepis, Aragoa*, and *Cinchona*. Pantropical: *Cyathea, Phyllanthus, Croton, Ficus* (principally Malaysia), *ugenia, Piper*, and *Ternstroemia*. Tropical amphi-pacific: *Meliosma, Saurauia, Hedyosmum, Turpinia, Bocconia, Styrax, Symplocos, Murraya, Dicksonia*, and *Trigonobalanus*. America/Africa: *Guarea, Lantana, Lippia, Miconia, Hypericum* (woody section).

A very interesting example of the amphi-pacific group is the genus *Trigonobalanus* (*Fagaceae*), a montane element with two species in the Indo-Malayan area and one species recently discovered in the Cordillera Oriental of Colombia (Lozano & al. 1979/80). In this respect it is interesting to note that *Trigonobalanus* is found in the subtropical to warm temperate fossil Tertiary floras of both Europe and N. America. As several other elements of the amphi-pacific group are also found in the same fossil floras, a common history of migration from N. to S. America seems probable (Van der Hammen & Cleef 1983). We will come back to this interesting problem at the end of this article. Gentry (1982) making a general analysis of the neotropical flora, arrived at some very interesting results. Of the neotropical flora (of Gondwanan origin), approximately half the species are Amazonian centred, and half are Andean centred. To this last group belong especially epiphytes (like *Bromeliaceae*) and elements from the understory. As the neotropical flora is aproximately twice as rich in species as the tropical African or tropical SE. Asian flora, these 50% Andean centred elements apparently are the principal cause of this richness, and apparently closely related to the long early history of the Andean orogeny.

Development of vegetation and flora since the Late Tertiary

Let us turn now to the Late Tertiary and Quaternary history of the area. As we said before, during the Tertiary, and especially during the Miocene, hill ridges to low mountains were already present in the area. During that time much of the Andean centred diversification of the flora probably took place in these hills and mountains up to 1 000 m, or possibly locally somewhat higher. Sediments from the probably Lower Pliocene Tequendama member of the Tilatá formation (Cordillera

Oriental; present elevation of sediments 2 600 – 2 800 m), lying approximately horizontally and unconformably on folded and eroded older Tertiary sediment, contain a tropical flora from c. 500 m elevation. Certain elements, however, seem to indicate the presence in the area of hills already higher than 1 000 m: *Weinmannia* and *Podocarpus*, both being elements of southern origin (VAN DER HAMMEN & al. 1973).

Sediments from a next higher portion of the Tilatá formation, show that the area in the mean time had been elevated at least 1 000 m, as the fossil pollen flora indicate an elevation of deposition of c. 1 500 m. The hills present in the surrounding area may have had already an elevation of 2 500 m or more. Noteworthy is the apparently simultaneous appearance of two tropical amphi-pacific elements: *Hedyosmum* and *Symplocos*. Apparently these amphi-pacific elements were amongst the first to cross the Panamanian land bridge. During the sedimentation of a next higher section of the Tilatá formation a pollen flora is found that indicates deposition at an elevation of c. 2 300 m (some 400 m below the present elevation), thus showing that the gradual upheaval of the Cordillera continued. A new element immigrated from the north is *Myrica,* and there are clear indications that páramo-like vegetation was already present on the top of the mountains of the Cordillea Oriental.

In the next and highest part of the Tilatá formation, a pollen flora is found that indicates that the area is already at its present elevation. It even seems to be higher or colder than at present, as there is, at 2 800 m, a dominating open to shrubby vegetation of High-Andean and páramo elements (*Polylepis, Aragoa,* etc.). The generic composition of this pollen flora is 50% of tropical or local origin, and 50% of northern and southern temperate origin, like the present páramo flora. There are indications, however, that the upper forest limit of that time was not at the same elevation as today, but lower: Subandean elements in this pollen flora (blown up from the montane forest) are represented with a higher percentage than today, and some herbs of the Andean forest zone seem to occur in the páramo vegetation. This could be caused by the fact that a number of present-day Andean to High Andean woody species were not yet evolved or were not yet adapted to present-day forest limit temperatures.

From the Late Pliocene and the Quaternary we have a continuous record of the vegetation history: in that time a series of lake sediments of many hundreds of metres was formed, in the high plane of Bogotá (elevation c. 2 580 m). (VAN DER HAMMEN & GONZALEZ 1960, VAN DER HAMMEN 1974, HOOGHIEMSTRA 1984).

In the Late Pliocene we see a gradual enrichment of the flora with new elements, like *Styloceras* and *Juglans*. Towards the beginning of the Pleistocene, *Alnus* appears, and soon forms a very important element of the palynoflora (VAN DER HAMMEN & al. 1973, VAN DER HAMMEN & CLEEF 1983). A long sequence of strong climatic changes, glacials and interglacials follows, with changes of the altitudinal tree line (mainly temperature induced) of 1 200 m or more, and fluctuations of humidity. This has probably a considerable influence on the evolution and distribution of species, the effect being strongest for the páramo flora (interglacial "refugia" in the highest areas). Approx. 1 million years ago (or slightly less), there is the first appearance and successive extension of *Quercus,* one of the last woody immigrants from the North. They become especially important during the Last Glacial, and today the oak, that tends to be a dominant species, covers large areas especially between 2 000 and 3 000 m (but locally lower and up to 3 500 m) on the inner slopes of the North Andean Cordilleras, at both sides of the Magdalena and Cauca valleys.

The Last Interglacial – Last Glacial – Holocene climatic vegetational sequence is known in considerable detail. From these data it has become clear that there is a rather complicated sequence of climatic changes, of temperatures and of rainfall, creating phases with a situation different from the present-day one. There is a Middle Pleniglacial very wet and cool period and an Upper Pleniglacial very dry and cold phase. During the first period, the páramo-zone was probably narrow and very wet, lake levels were high and glaciers had their largest extension; this led to local contact of glaciers and forest. High-Andean forest of *Polylepis* (and probably *Escallonia*) formed a very broad zone, not known in present-day vegetation; below it *Quercus*-forests abounded.

During the dry-cold Upper Pleniglacial, the páramo-zone was very broad and dry, lake levels were very low or lakes disappeared and the upper forest line was locally at 2 000 m or lower. On the slopes to the large interandean valleys (Magdalena valley), montane forest may have disappeared locally almost completely, being replaced by xerophytic vegetations (Van der Hammen 1981).

It will be clear that all this represents very profound changes in the position and composition of the montane forests. While the temperatures high in the tropical Andean mountains descended 8 °C or more during glacial times, in the tropical lowlands this seems to have been much less (possibly some 2 °C). Recent data on the glacial position of oak-forests in the Western Cordillera (Monsalve 1985) show that they were then (at 1 500 m altitude) at a position at least more than 500 m (and may be even 1 000 m) lower than today. This seems to confirm a downward compression of the altitudinal vegetation zones. This would mean that the thermal gradient, being approx. 0.6 today, may have been in the order of 0.9 during the coldest and driest parts of the Last Glacial. This could mean a causal relation between the facts observed and the theoretical thermal gradient for dry and wet air. As to a few known sites of present-day *Quercus* forest at very low levels, near to 1 000 m, these might now, in view of the historical data, be considered as glacial relic stands.

Conclusions

We have seen that the montane forests of the northern tropical Andes are built up from many different floristic elements. Hills up to a 1 000 m may have existed already during the Lower Tertiary. Here the explosive evolution of Andean centred taxa started probably, from the tropical flora of Gondwanan stock and already developed to a neotropical flora. 50% of the neotropical species are Andean centered, showing the extraordinary influence on evolution of the Andean orogeny. When the final upheaval took place, between the Late Miocene and Late Pliocene, extensive areas with a montane climate were formed. These were then populated by adaptation evolution from the flora of the earlier hills and low mountains. Elements of Austral-antarctic origin migrated along the Andes-chain to the tropics. As at the same time the isthmus of Panama was formed, connecting N. and S. America, floral elements from the north entered the tropical Andes. Woody Holarctic elements adapted to the new circumstances, but speciation was limited.

First to cross the Panamanian land bridge were tropical to subtropical and warm-temperate species, that were elements from the Tertiary flora of the Laurasian area, that had migrated southward during the period of Miocene cooling, taking refuge in the montane tropics. The same happened in SE. Asia, so there is a good number of amphipacific taxa, part of which we know as fossils from Tertiary Laurasia, but became largely extinct in the present-day Holarctic. One of the most impressive examples is *Trigonobalanus*, with 2 species in SE. Asia and one recently discovered in Colombia. Next to cross were cool-temperate to cold elements, like *Alnus* and *Quercus* (and of course many taxa of open vegetation types) that still exist in the Holarctic.

The final result are tropical Andean montane forests with a generic composition of 65% of tropical-neotropical distribution, 15% tropical Andean, 5% amphipacific (Malayo-American), 5% Austral-antarctic and 10% Holarctic.

There is certainly still a lot to be investigated before we know more precisely the history of separate taxa. Further palynological study will give us certainly more information, and it may be expected that especially long sections will do so. But it is also to be expected that further development, on a modern basis, of macropaleobotany, will give important new contributions, that cannot be obtained in any other way. For the time being, however, we hope with this overview, to have given at least an impression of the history of a larger plant formation in the neotropics, as related to the very complex historical background of the area.

References

GENTRY, A. H., 1982: Neotropical floristic diversity: phytogeographical connections between Central and South America, Pleistocene climatic fluctuations, or an accident of the Andean orogeny. – Ann. Missouri Bot. Gard. **69**: 557–593.

HERNGREEN, W. F., 1975: Palynology of Middle and Upper Cretaceous strata in Brazil. – Meded. Rijks Geol. Dienst New. Ser. **26**: 39–91.

HOOGHIEMSTRA, H., 1984: Vegetational and climatic history of the High Plain of Bogotá, Colombia: a continuous record of the last 3.5 million years. – Diss. Botan. **79**: 1–368. (Also in: The Quaternary of Colombia **10**, Amsterdam).

LOZANO, C., HERNANDEZ, J. S., HENAO, J. D., 1979/80: El género *Trigonobalanus* Formen en el neo-trópico 1 & 2. – Caldasia **12**: 517–538; **13**: 9–44.

MONSALVE, J. G., 1985: A pollen core from the Hacienda Lusitania. – Pro Calima (Basel) **4**: 40–44.

VAN DER HAMMEN, T., 1974: The Pleistocene changes of vegetation and climate in tropical South America. – J. Biogeogr. **1**: 3–26.

– 1981: Environmental changes in the Northern Andes and the extinction of *Mastodon*. – Geol. Mijnb. **60**: 369–372.

– GONZALES, E., 1960: Upper Pleistocene and Holocene climate and vegetation of the "Sabana de Bogotá" (Colombia, South America). – Leidse Geol. Meded. **25**: 261–315.

– WERNER, H., VAN DOMMELEN, H., 1973: Palynological record of the upheaval of the Northern Andes: a study of the Pliocene and Lower Quaternary of the Colombian Eastern Cordillera and the early evolution of its High-Andean biota. – Rev. Palaeobot. Palynol. **16**: 1–122. (Also in: The Quaternary of Colombia **2**, Amsterdam).

– CLEEF, A. M., 1983: *Trigonobalanus* and the tropical amphi-pacific element in the North Andean forest. – J. Biogeogr. **10**: 437–440.

– – 1983/84: Datos para la historia de la flora andina. – Rev. Chilena Hist. Nat. **56**: 97–107.

Address of the author: TH. VAN DER HAMMEN, Hugo de Vries Laboratorium, Universiteit van Amsterdam, Kruislaan 318, NL-1098 SM Amsterdam, The Netherlands.

Pl. Syst. Evol. 162, 115–132

—Plant—
Systematics
and
Evolution

Late Quaternary vegetation history of the Mexican highland

HERBERT STRAKA and DIETER OHNGEMACH

Received November 12, 1986

Key words: Vegetation history: Quaternary, Pleistocene, Postglacial. – Vegetation of Mexico.

Abstract: Results of pollenanalytical investigations in the Highlands of Mexico (area of Puebla – Tlaxcala) are presented together with a survey of the last 35 000 years of vegetation history. After 3 pine periods from 35 000 to about 7 000 – 8 000 B. P., according to high and late Pleistocene condition, the Holocene vegetation history becomes more diversified: periods of alder, pine, fir and mixed oak forests alternate and end in a period of cultivated landscape.

Within the Mexico Project supported by the "Deutsche Forschungsgemeinschaft" (DFG), pollen samples from the Mexican highland (i.e., the Puebla-Tlaxcala area) were analysed, at the Botanical Institute of the University of Kiel, in an attempt to clarify the vegetational history of the last 35 000 years. These analyses were carried out by Dr D. OHNGEMACH; the text and interpretations are by Prof. H. STRAKA (see OHNGEMACH & STRAKA 1983; see also KLAUS 1973; LAUER 1981; OHNGEMACH 1973, 1977; STRAKA & OHNGEMACH 1975, 1978).

Puebla is situated c. 110 km east of the capital of Mexico; Tlaxcala lies c. 30 km north of Puebla. The area of investigation lies at about 19° lat. N and 99° long. W. The DFG project area is limited in the west by the volcanoes Popocatépetl (5 452 m) and Ixtaccihuatl (5 286 m), the volcano Malinche (4 461 m) rises in the centre of the basin. The area of investigation also includes numerous volcano craters and maars. The highest summit in Mexico, the volcano Pico de Orizaba (or Citlaltépetl, 5 700 m) lies outside, in the east. Table 1 presents a summarized classification of vegetation types. For relevant literature see OHNGEMACH & STRAKA (1983: 21).

Exploring the drilling sites for the planned pollen analyses, we were successful in the maar crater of **Jalapasquillo** (Fig. 1, Point 1, 2 400 m). Here, Dr R. WEBER made three machine drillings, one of them to a depth of 19 m. Samples from the cores have provided data for the pollen diagram in Fig. 3. The sediments were mainly clayish gyttja, sandwiched between layers of volcanic sands. The pollen diagram spans the period from c. 35 000 to c. 8 000/7 000 B. P. (Fig. 2).

Table 1. Extant vegetation belts in Central and eastern Mexico

Climatic belts	Vegetation belts
Arid regions of the highland basins 2200 – 2400 m	Arid formations with *Juniperus deppeana, Pinus cembroides, Yucca* spp., *Agave* spp., *Opuntia* spp.
4000 – 4800 m	Alpine grasslands (zacatonales) with bunch grasses and herbs
Tierra helada 3200 – 4000 m	High mountain conifer forest of *Pinus hartwegii*
Tierra fria 1800 – 3200 m	Mixed conifer forests a) above 2700 m with pine, fir and alder b) up to 2700 m with oaks predominant
Tierra templada 800 – 1800 m	Semi-deciduous mixed mountain forests with a) *Quercus, Liquidambar, Alnus, Fagus, Ulmus, Tilia, Carpinus*, etc. b) *Podocarpus, Oreopanax*, etc. c) Tree ferns, many epiphytes
Tierra caliente 0 – 800 m	a) Tropical rain forests b) Coastal forests with palms

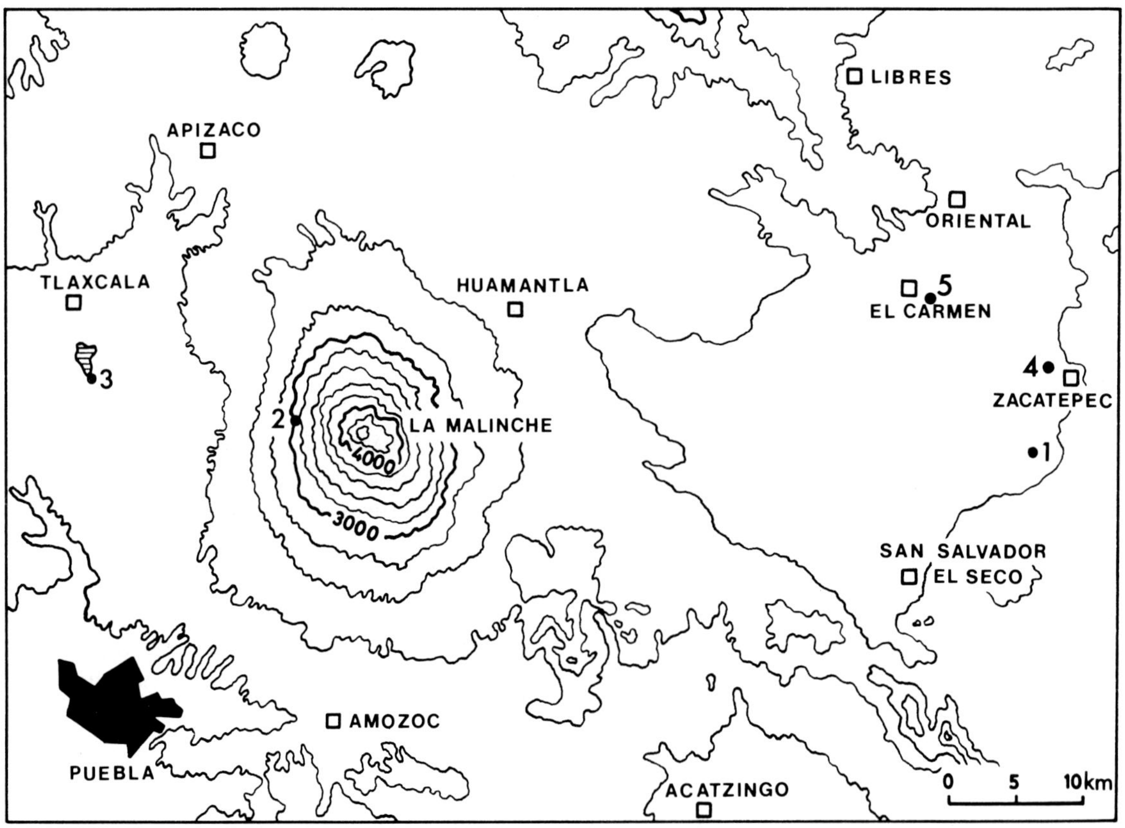

Fig. 1. Location of the points from where the probes for the pollen diagrams were obtained: 1 Maar from Jalapasquillo, 2 Tlaloc crater, 3 Lake Acuitlapilco, 4 Oriental I, 5 Oriental II – IV

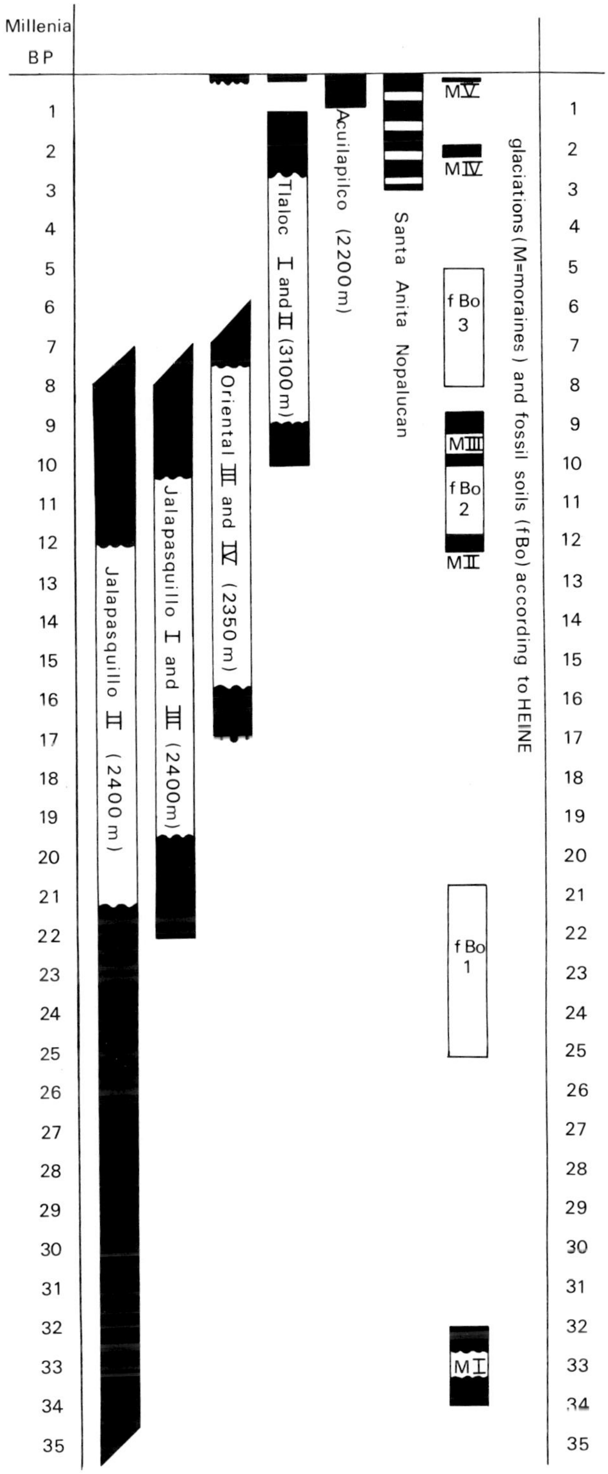

Fig. 2. Comparison of the historical extension of the individual profiles. Together with the moraines (M), fossil soil formations (f Bo), and glaciation periods according to HEINE (1973 a, b, 1974, 1975, 1976, 1978, 1983 a – d; 1984 a, b, 1985)

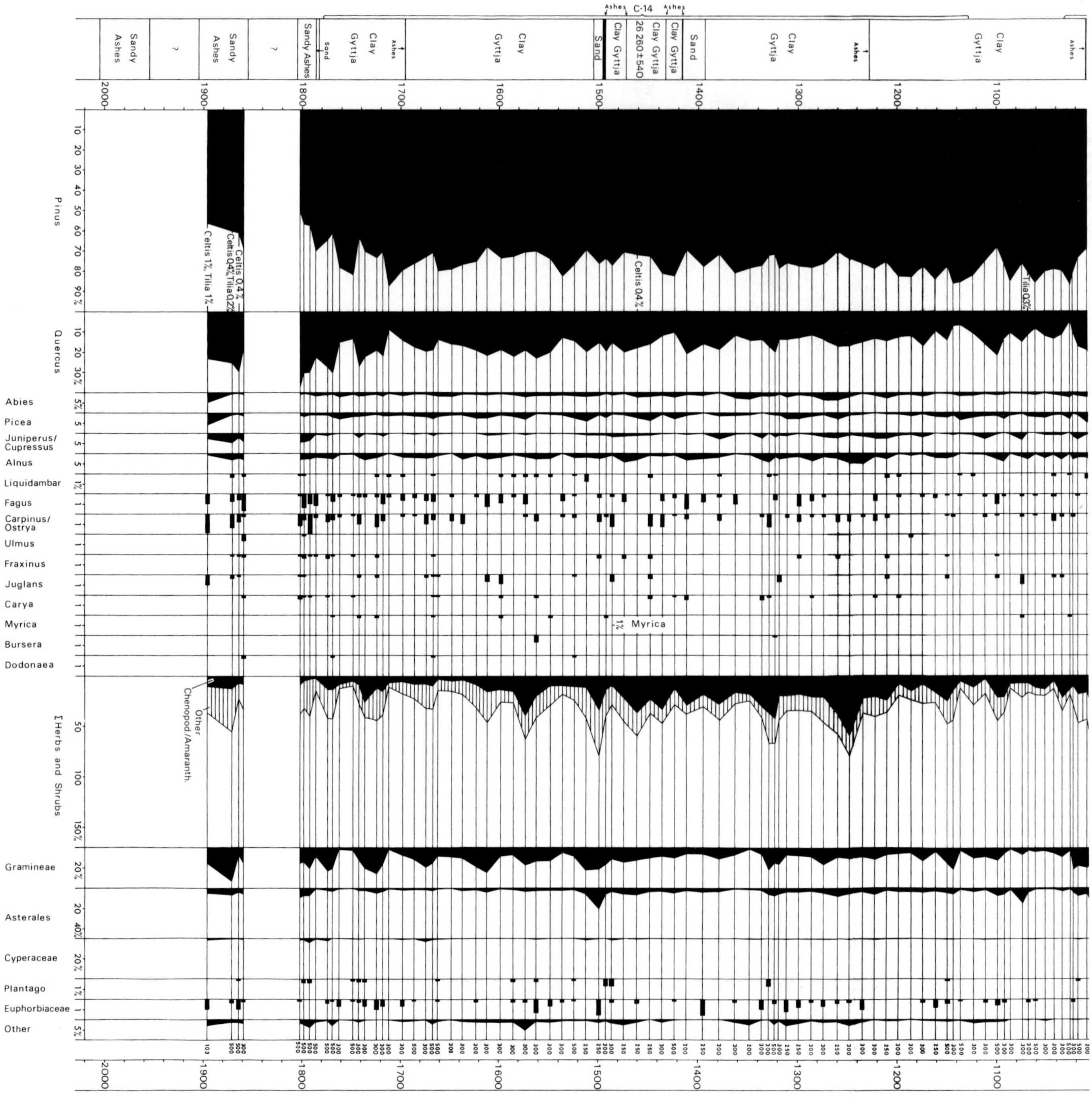

Fig. 3. Pollen diagram Jalapasquillo II; calculation based on arboreal pollen sum = 100% (analyses by OHNGEMACH). – Zone I: 1 900 – 1 718 cm, pine period, relatively rich in *Quercus*, c. 35 000 – 31 000 B. P. Zone II: 1 712 – 751 cm, pine period, average *Quercus* values, c. 31 000 – 18 000 B. P. Zone III: 742 – 289 cm, pine period, poorer in *Quercus*, c. 18 000 – 8 500 B. P. Zone IV: 289 – 170 cm, pine period, no *Quercus*, c. 8 500 – 8 000/7 000 B. P.

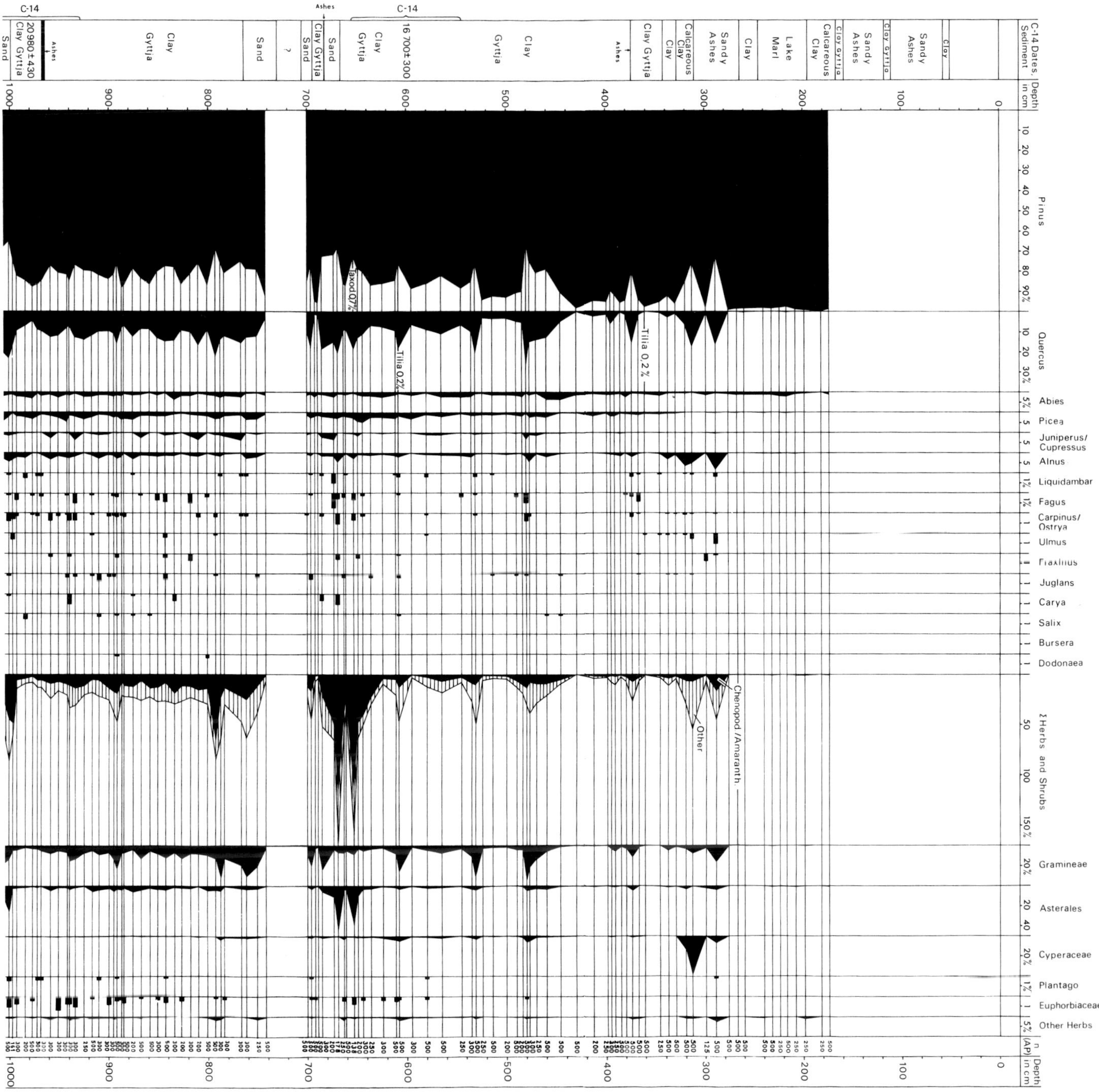
C-14
Ashes
C-14
C-14 Dates, Sediment
Depth in cm
20 980±430
16 700±300
Sand
Clay Gyttja
Clay
Gyttja
Calcareous Clay
Sandy Ashes
Lake Marl
Pinus
Quercus
Abies
Picea
Juniperus/ Cupressus
Alnus
Liquidambar
Fagus
Carpinus/ Ostrya
Ulmus
Fraxinus
Juglans
Carya
Salix
Bursera
Dodonaea
Taxod 0.7%
Tilia 0.2%
Chenopod /Amaranth.
Other
ΣHerbs and Shrubs
Gramineae
Asterales
Cyperaceae
Plantago
Euphorbiaceae
Other Herbs
n (AP)

Table 2. Glacier advances and climate in the Mexican highland during the last 35 000 years according to HEINE (1973 a, b, 1974, 1975, 1976, 1978, 1983 a – d, 1984 a, b, 1985)

Years B. P.	Glacier advances	Climates (compared to present)
100 – 150	M V	cooler and more humid
		warmer and drier
2 000	M IV	somewhat cooler and more humid
		warmer and drier
9 000 – 10 000	M III	humid and cold
		warmer and drier
12 000	M II	humid and cold
21 000		much colder and drier
		cooler (and drier?)
33 000	M I	humid and cold

Pines (*Pinus* spp.) dominate along the whole length of the diagram. Between 1 896 cm and 289 cm, they are accompanied by oaks (*Quercus* spp.). With some difficulty one can recognize three ill-defined zones:

Zone I: 1 896 – 1 718 cm, c. 35 000 – 31 000 B.- P. (C 14 data). In this zone, the portion of *Quercus* pollen is generally higher than 20%, in some cases above 30%. *Pinus* pollen, therefore, makes up 50 – 82%. There are also slightly higher pollen percentages for fir (*Abies*), spruce (*Picea*), juniper/cypress (*Juniperus/Cupressus*[1]), and alder (*Alnus*). Deciduous trees like beech (*Fagus*) and hornbeam/hop-hornbeam (*Carpinus/Ostrya*[1]) apparently grew close to the maar.

Occasionally we have found pollen from *Liquidambar, Ulmus, Fraxinus*, walnut (*Juglans*), *Carya*, and lime (*Tilia*). Perhaps, part of these pollen grains had been carried by long-distance transport. It is noteworthy that pollen of *Podocarpus, Weinmannia, Oreopanax*, and *Clethra* – trees and shrubs from mixed mountain forests – are completely lacking. Non-arboreal pollen (NAP) occurs at variable percentages, from as high as 57% to as low as 10%, and is less frequent than in the following zones.

Zone II: between 1 712 cm and 751 cm, c. 31 000 – 18 000 years B. P. The portion of *Quercus* pollen decreases, being generally less than 20%, but never below 5%. *Abies, Picea, Alnus*, and *Juniperus/Cupressus* remain at low values and their pollen curves show interruptions. The remaining arboreal pollen (AP) and NAP occurs somewhat more frequent than in Zone I.

Zone III: between 742 cm and 289 cm, c. 18 000 – 8 500 B. P. *Quercus* pollen continues to decrease slowly; it rarely reaches 15%, usually is below 10%, and sometimes lacks altogether. Therefore, the relative values for *Pinus* pollen are higher. The remaining AP types occur as in Zone II. The NAP curve shows a double maximum at the beginning of Zone III, mainly due to *Chenopodiaceae/Amaranthaceae*, but also to *Compositae-Tubuliflorae*. Near the end of Zone III there is a

[1] Both genera have the same pollen type.

last maximum of *Quercus* pollen and an increase in the pollen from *Alnus* and *Ulmus*. The end of Zone III is characterized by a sudden decrease in the NAP curve and near absence of pollen from *Quercus, Picea, Juniperus/Cupressus, Alnus, Fagus, Carpinus/Ostrya, Fraxinus, Ulmus, Liquidambar*; even earlier *Juglans* drops out.

Zone IV: 276 – 174 cm, c. 8 500 – 8 000/7 000 B. P. *Pinus* is dominant, with *Abies* comprising only 0.2 – 2.2%. According to other pollen diagrams (one of which will be discussed later), the boundary between Zone III and Zone IV apparently coincides with the end of the cold period, i.e., the transition from the late pleistocene conditions to the protoholocene conditions.

HEINE (1973 a, b; 1975; 1976; 1978; 1983 a – d; 1984 a, b; 1985; cf., also HEINE & HEIDE-WEISE 1973 a, b) has studied the moraines in our area; his compilation (somewhat simplified) is shown in Table 2. He differentiates between five main moraine stages, M I – M V, between which there are drier and usually warmer periods of soil formation. Our Zone I can be paralleled with M I, but there are no indications of M II and M III in the pollen diagrams.

The pollen diagrams **Oriental I – V** (Points 4 and 5 on the map, Fig. 1, elevation of c. 2 350 m) corresponds to the pollen diagrams Jalapasquillo II, I and III. However, those from the **Tlaloc** secondary crater on the Malinche Volcano (locality 2 on the map Fig. 1, elevation c. 3 100 m) are younger (Fig. 2), from c. 10 000 – 1 000 B. P. At the edge of the Malinche Commission well two profiles were taken and supplemented by a 130 cm-deep drilling with the Dachnowsky probe. We present both pollen diagrams, which complement each other Whereas the pollen diagrams Jalapasquillo (Fig. 3), Oriental and Acuitlapilco (Fig. 9) were calculated on the base sum "AP = 100%", the pollen diagrams Tlaloc I and II (Figs. 4 and 8) were calculated as so-called "Iversen diagrams" with their base sums = 100% established on the total terrestrial plant pollen (AP and the terrestrial part of the NAP, without aquatic plants). In the original publication (OHNGEMACH & STRAKA 1983), the pollen diagrams Tlaloc I and II are also presented for comparison with the base arboreal pollen (AP) sum = 100%.

Tlaloc II (Fig. 4) extends further down than Tlaloc I and demonstrates a first period between 300 – 204 cm which corresponds to alpine grasslands (Spanish: "zacatonales"). The NAP portion is quite high, about 35 – 75% of the terrestrial plant pollen. During this period one has to postulate a woodless vegetation around the Tlaloc crater, i.e., at an altitude of more than 3 000 m on the Malinche. The NAP is mainly due to grasses (*Gramineae*) and *Compositae*, and in smaller but regular amounts to *Chenopodiaceae/Amaranthaceae, Plantago, Artemisia, Cyperaceae, Thalictrum* and other *Ranunculaceae, Arceuthobium* (*Loranthaceae*), *Geraniaceae, Umbelliferae* (rarely including *Hydrocotyle*), *Caryophyllaceae, Rosaceae,* and *Cruciferae*. There must have been a species-rich herb/grassland.

Among the *Gramineae* there was also pollen from *Maydeae*. Recognition of crop maize pollen (*Zea mays* subsp. *mays*) is difficult, because the *Maydeae* also contain a number wild grasses, i.e., the genera *Tripsacum* and *Zea* s. l., with *Z. perennis* (= *Euchlaena perennis*), *Z. diploperennis* (newly discovered), and *Z. mays* subsp. *mexicana* (= *Euchlaena mexicana* = "teozintle" or "teosinte"). With no SEM available, first attempt to differentiate these pollen types according to their size was made (Fig. 5). From the diagram it can be seen that *Tripsacum* with only 38 – 58 μm is well-separated, whereas *Z. mays* subsp. *mexicana* with 48 – 110 μm overlaps with *Z. mays* subsp. *mays*, 60 – 116 μm. (The second curve of *Zea mays*

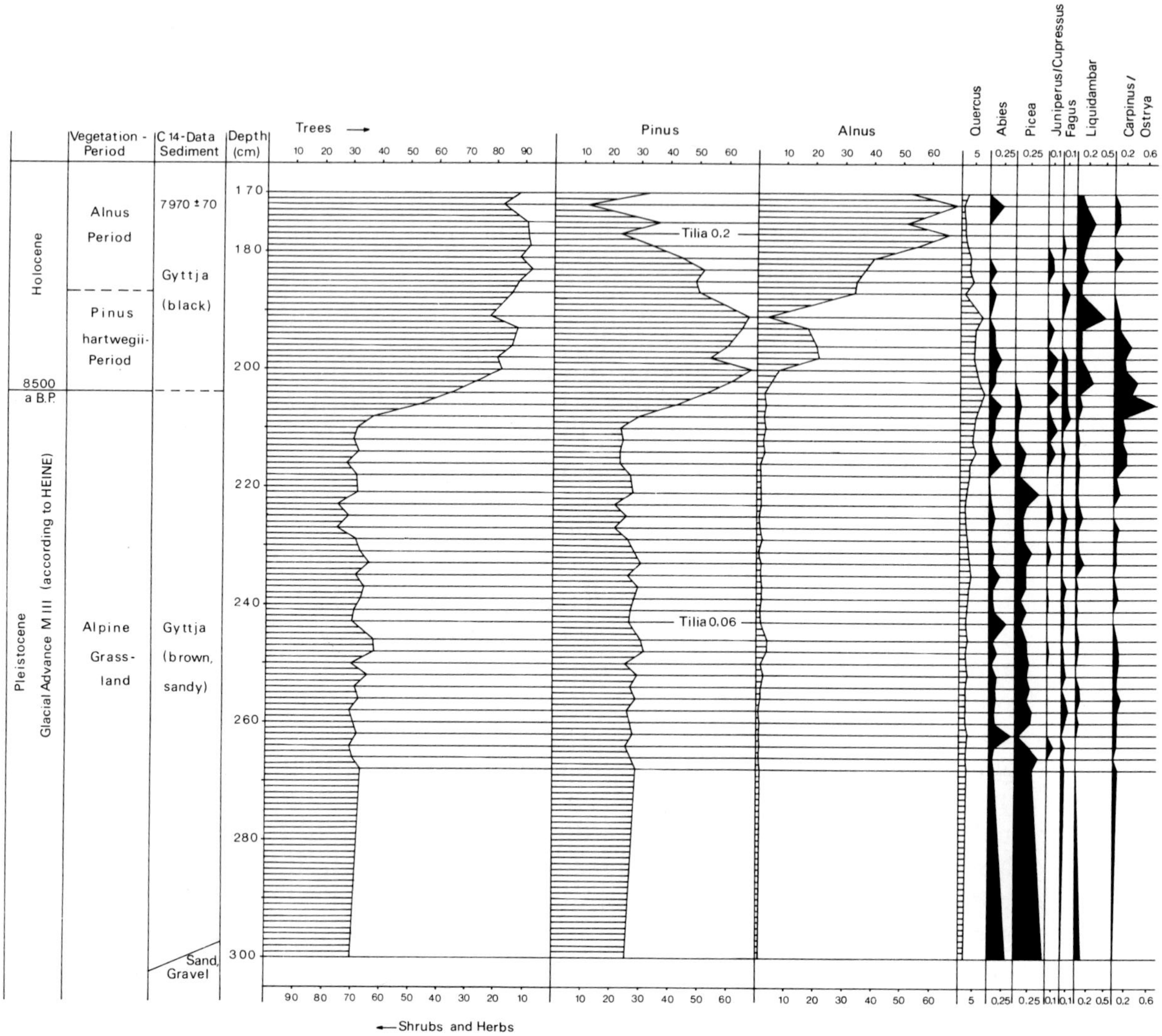

Fig. 4. Pollen diagram Tlaloc II; calculations based on terrestrial plant pollen sum = 100% until c. 8 500 B. P. Period II: 203 – 182 cm, *Pinus hartwegii* period. Period III: 182 – 170 cm,

subsp. *mays* with 77 – 142 µm comes from a 12-year-old preparation where the pollen has obviously swollen in the glycerine.) Later SEM results were disappointing, but not unexpected. *Tripsacum* pollen grains do indeed show microspinules in groups on slightly raised areas (Fig. 6), whereas all *Zea* pollen types (*Zea mays* subsp. *mexicana* and subsp. *mays*) show microspinules in regular arrangements (Fig. 7). However, the crop maize pollen cannot be differentiated. During period I the following maize types occurred: 300 cm, one *Tripsacum*, three *Zea* s. l.; 252 m, one *Zea* s. l.; 243 cm, one *Tripsacum*, 11 *Zea* s. l.

The AP from period I is definitely due to long-distance transport from lower altitudes and mainly comes from *Pinus* (corresponding to the pollen diagram Jalapasquillo), *Quercus*, and *Alnus*. Pollen from *Abies, Picea, Juniperus/Cupressus, Liquidambar, Fagus, Carpinus/Ostrya, Tilia, Fraxinus*, and *Salix* occurs in low

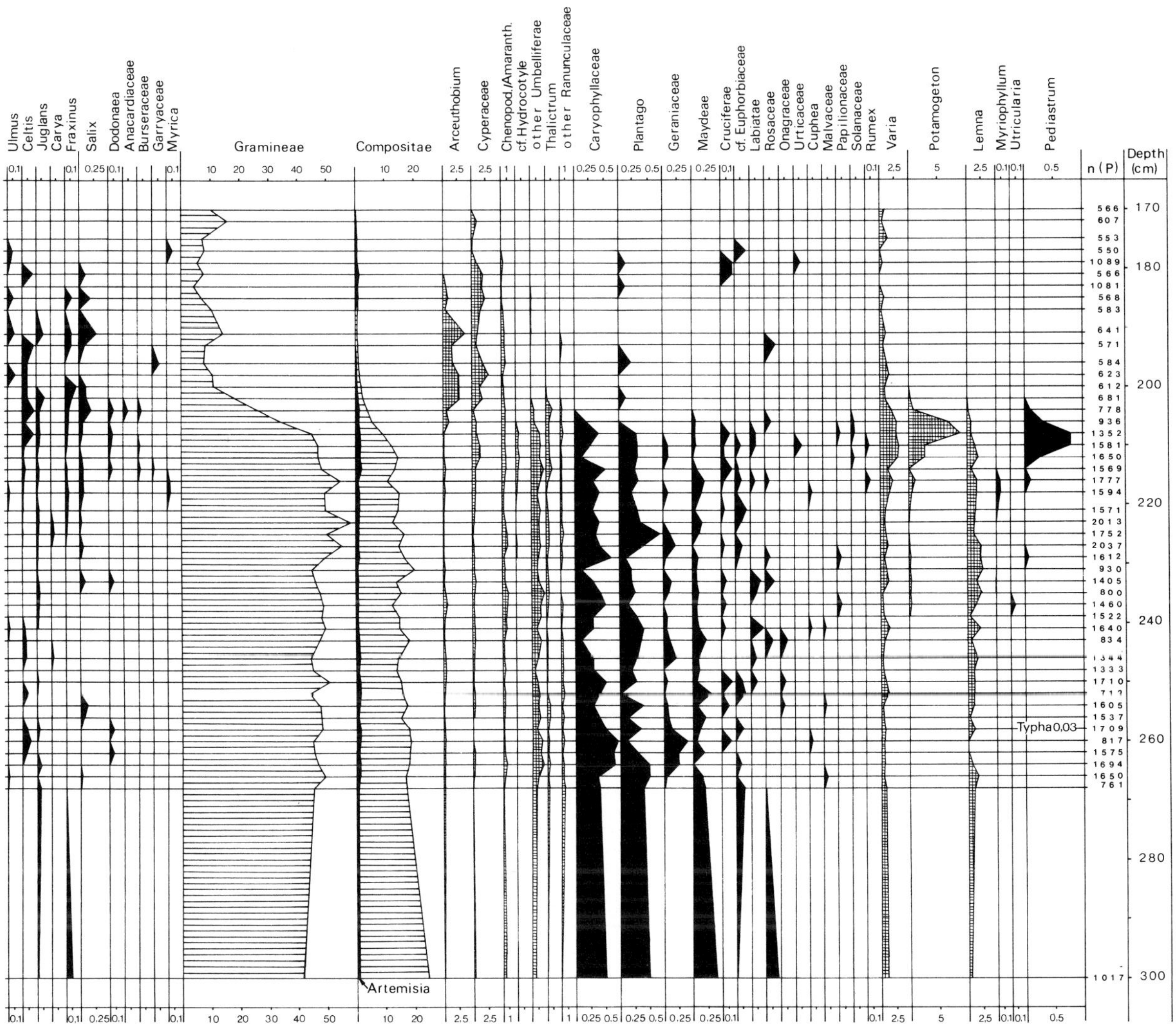

(analyses by OHNGEMACH). – Period I: 300 – 204 cm, alpine grasslands (zacatonales) period; alder period

amounts and sporadically. *Picea* and *Abies* probably occurred on the cliffs of the Malinche. Period I corresponds to the glaciation during the moraine M III (HEINE; Table 2). The climate was more humid and cooler than today.

The end of period I comes at 204 cm with a sudden decline in the NAP curve (especially due to a decline in the *Gramineae*) and the disappearance of *Picea*. Simultaneously, the curve of *Pinus* pollen rises. Several C^{14} data place this period at c. 8 500 B. P. At this elevation on the Malinche cliffs this corresponds to the transition from late glacial to post glacial conditions (HEINE & OHNGEMACH 1976). (At lower elevations in central Europe and N. America this transition took place at c. 10 250 B. P.). Period I corresponds to Zone III of Jalapasquillo.

Period II in the Tlaloc II pollen diagram corresponds to that of *Pinus hartwegii*, reaching the highest altitudes of all pines in Mexico. There are about 60 pine species

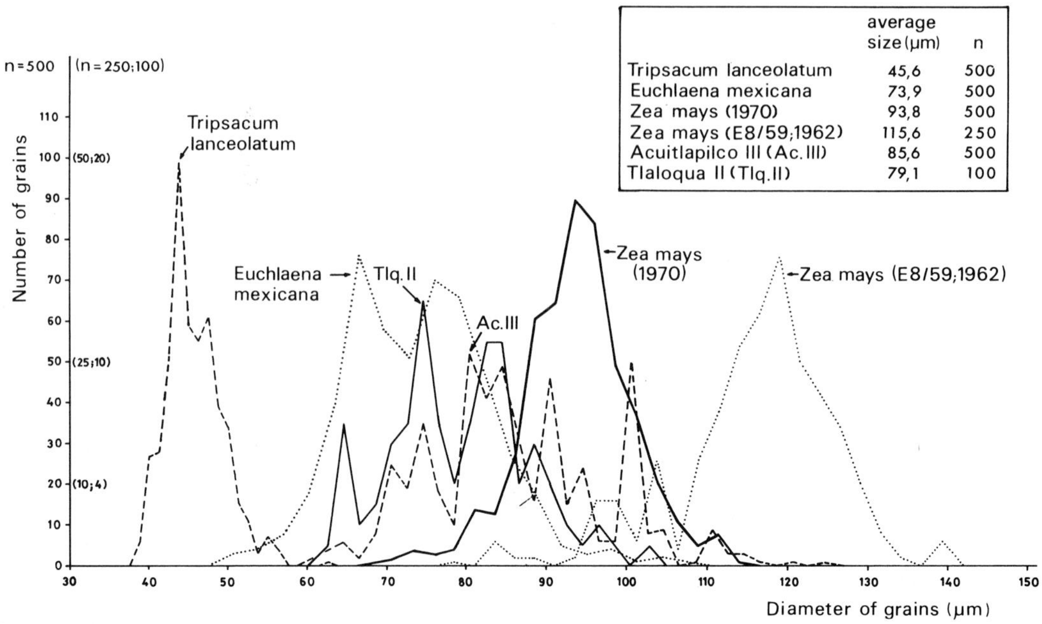

Fig. 5. Size distribution of extant and fossil *Maydeae* pollen grains. *Zea mays* = *Zea mays* subsp. *mays, Euchlaena mexicana* = *Zea mays* subsp. *mexicana*; Tlq Tlaloc, Ac Acuitlapilco

Fig. 6

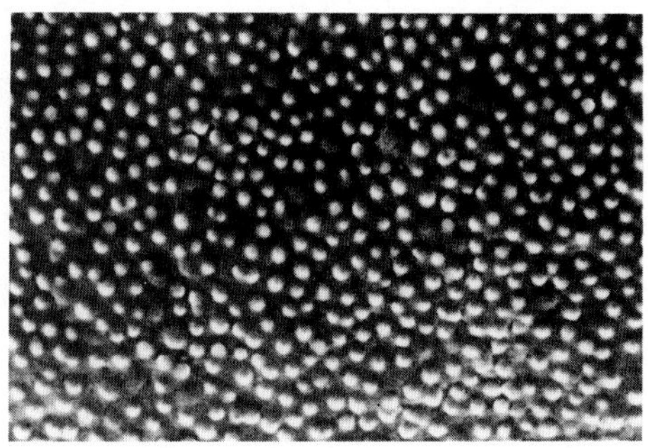

Fig. 7

Fig. 6. Microsculpture of the pollen surface of *Tripsacum dactyloides*; enlarged ×5000 (photograph by B. FRIEDRICH)

Fig. 7. Microsculpture of the pollen surface of *Zea mays* subsp. *mexicana*; enlarged ×5000 (photograph by B. FRIEDRICH)

in Mexico which cannot be separated on the basis of pollen morphology. However, there is a method for definite identification: *Pinus hartwegii* is often and abundantly parasitized by *Arceuthobium cryptopodium* (*Loranthaceae*). Its pollen is easily recognized and reaches values from 2–4% during period II.

Period III is characterized by *Alnus*, most likely *A. firmifolia* which today grows together with *Abies religiosa* and *Pinus montezumae* around the Tlaloc crater.

The pollen diagram **Tlaloc I** (Fig. 8) begins somewhat later than Tlaloc II but extends to more recent layers. Here, period I of the alpine grasslands (zacatonales) reaches from 284 – 256 cm. At 275 cm one pollen grain from the *Zea* s. l. type was identified with SEM, as were two *Tripsacum* and one *Zea* s. l. pollen types at 264 cm.

Period II of *Pinus hartwegii* spans the period from 252 – 227 cm. *Arceuthobium* pollen occurs at 2 – 4%. Period III at 223 – 189 m also corresponds to an *Alnus* period (probably *Alnus firmifolia*). According to Heine: 7 970 + 70 B. P.

Period IV at 185 – 150 cm is that of a pine advance. As *Arceuthobium* pollen was also found here, *Pinus hartwegii* is most likely to have occurred. At 150 – 90 cm, the material is devoid, or almost devoid, of pollen ("period V"). According to Heine: 900 – 800 B. C.

Period VI at 90 – 95 cm corresponds to a period of a sparse pine and mixed woods. The amount of NAP is high and *Alnus* and *Abies* pollen also occur. – Period VII is that of *Abies religiosa* the only fir species in Mexico. This cooler and more humid period probably corresponds to the glacial advance of the moraine M IV (Heine; Table 2), dated at c. 2 000 B. P. More recent datings by Heine seem to indicate 230 – 800 A. D.

Period VIII at 35 – 0 cm is characterized by a period of pine and mixed woods. At 30 cm, 0.5% *Maydeae* pollen was found; at 25 cm, 10 *Zea* s. l. pollen grains were found by SEM. At that time, there were probably maize crops located at some distance from the area of investigation. The pollen diagram Tlaloc I ends at c. 1 000 B. P.

The pollen diagrams from the banks of the **Acuitlapilco Lake** near Tlaxcala (Fig. 1, locality 3; c. 2 200 m) span the last 500 to 1 000 years. Five profiles were drilled with the Dachnowsky probe. Their pollen analysis resulted in comparable data; the most complete diagram is that from Acuitlapilco IV (Fig. 9).

The deepest section a, at 596 cm, corresponds to the period of mixed pine woods. This is located in the lower 55 cm of the Acuitlapilco II pollen diagram and has been dated after C 14 at c. 1 000 B. P. *Pinus* dominates with 82%, *Quercus* reaches 14%, *Alnus* makes up 3%, and *Abies* 1%. The sum of the NAP (calculated on the basis of arboreal pollen sum = 100%) is 57%. There must have been sparse woods with abundant undergrowth, especially *Gramineae, Compositae*, and *Chenopodiaceae/Amaranthaceae*. The 1.3% *Maydeae* pollen (> 60 µm) indicates a rather large area of maize crops.

Section b extends from 591 – 493 cm and represents an oak mixed wood period with 20 – 43% *Quercus* pollen. *Pinus* pollen, which is always over-represented, reaches 42 – 77%; *Alnus* makes up 4.3 – 18%, *Abies* 0 – 1.6%. High NAP values (29 – 93.6%) indicate sparse woods with abundant undergrowth. In the Mexico City basin (situated to the west of our study area) Sears (1952) identified an oak period corresponding to our section b. He interprets it as humid. This is supported by peat deposits in our Acuitlapilco profiles.

The last section c from 493 – 0 cm corresponds to a period of crop landscapes. The NAP sums usually lie at more than 100% (50 – 458%), which leads to the conclusion that enormous areas had been cleared. *Gramineae* and *Compositae* are very frequent and there are many infrequent pollen types. Evidently, there were periods of expanded maize cultivation: *Maydeae* pollen from 50 – 60 µm makes up 0 – 52%, and that over 60 µm makes up 0 – 37%. With SEM, four pollen grains of the *Tripsacum* type and 10 of *Zea* s. l. were recognized. Pollen grains from weeds

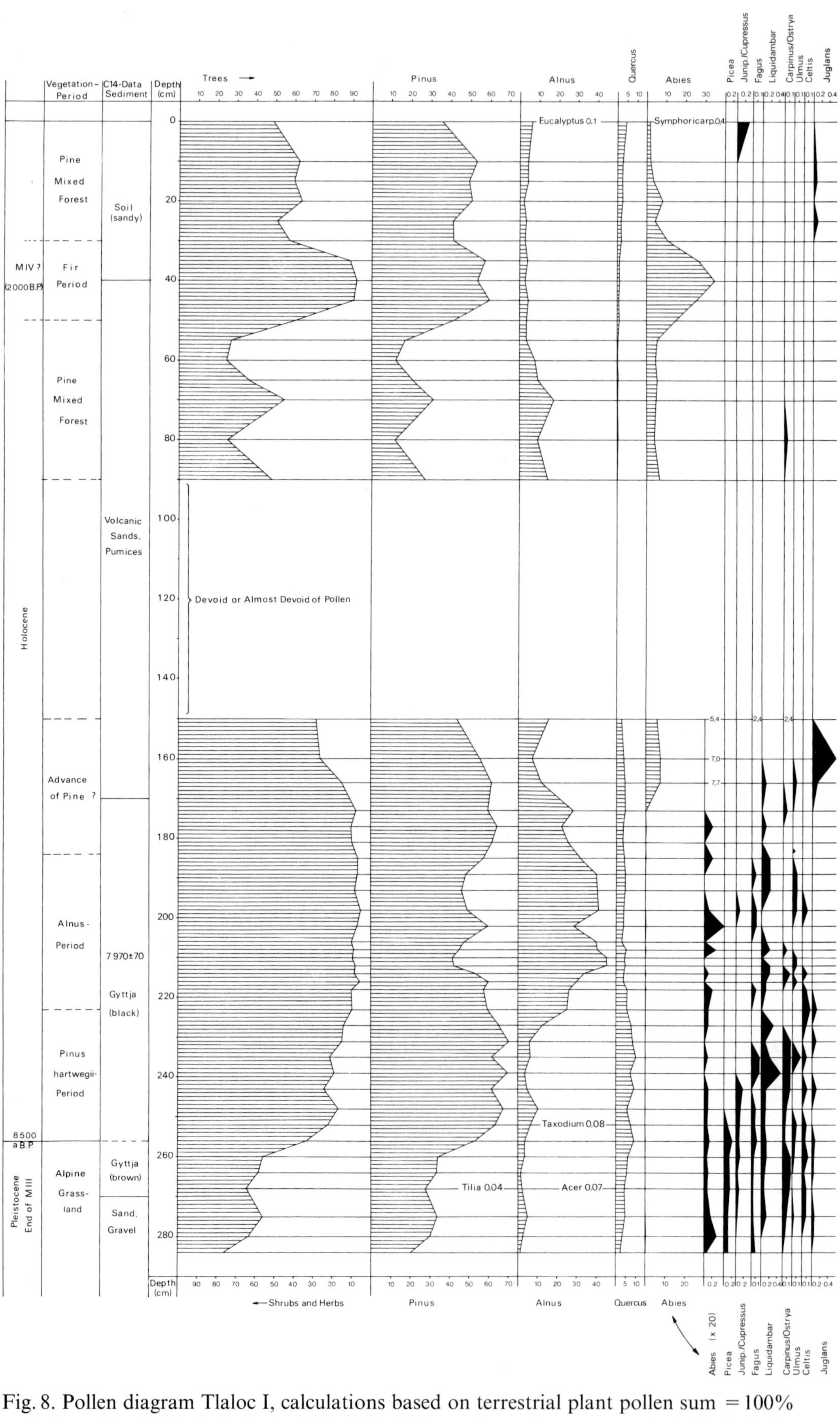

Fig. 8. Pollen diagram Tlaloc I, calculations based on terrestrial plant pollen sum = 100%
period, until c. 8 500 B. P. Period II: 252 – 227 cm, *Pinus hartwegii* period. Period III: devoid of or poor in pollen. Period VI: 90 – 55 cm, period of sparse mixed pine woods.
period

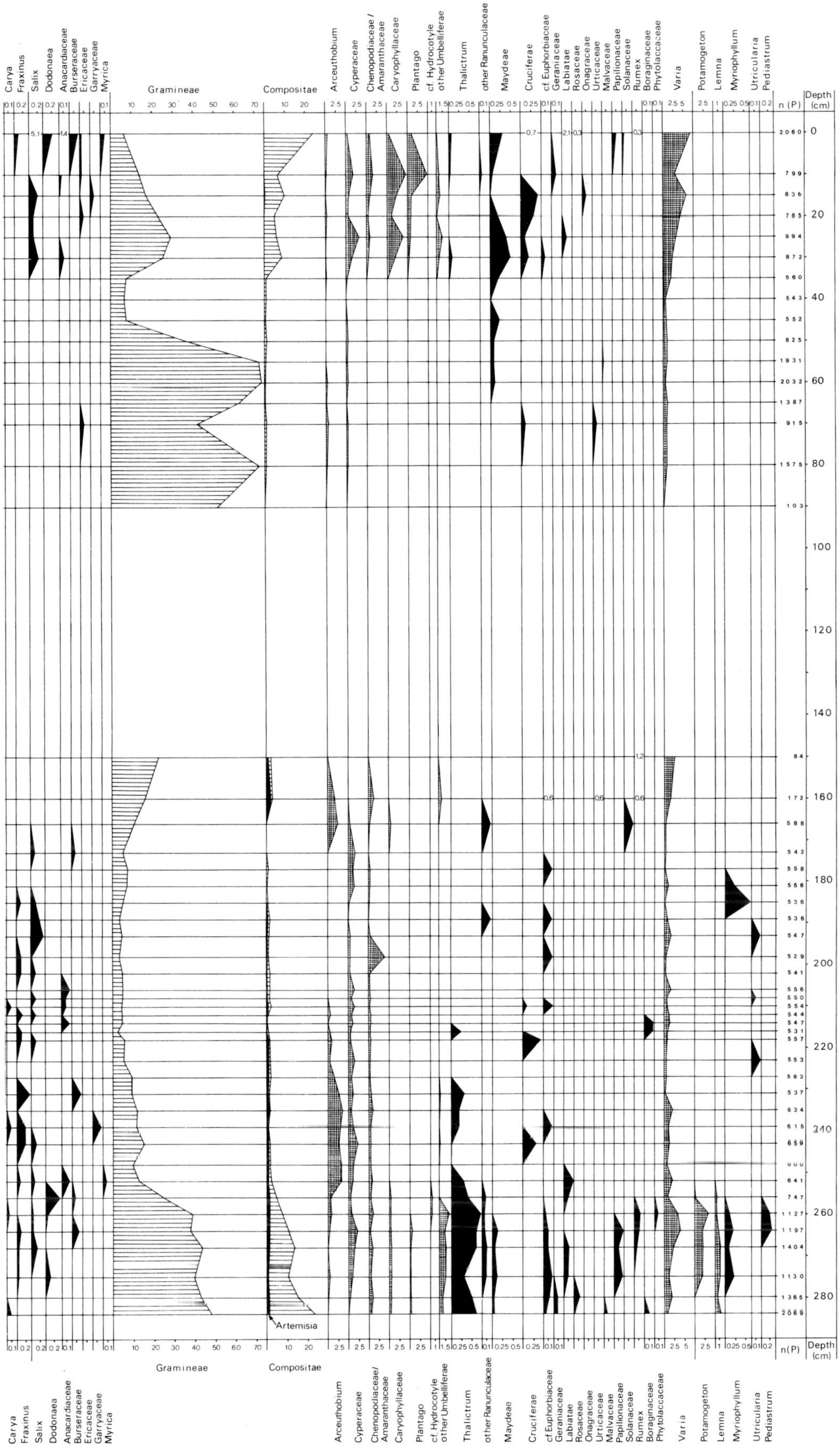

(analyses by Ohngemach). – Period I: 284 – 256 cm, end of alpine grassland (zacatonales) 223 – 189 cm, alder period. Period IV: 185 – 150 cm, pine advance. Period V: 150 – 90 cm, Period VII: 50 – 35 cm, fir period, c. 2000 B. P. Period VIII: 35 – 0 cm, mixed pine wood

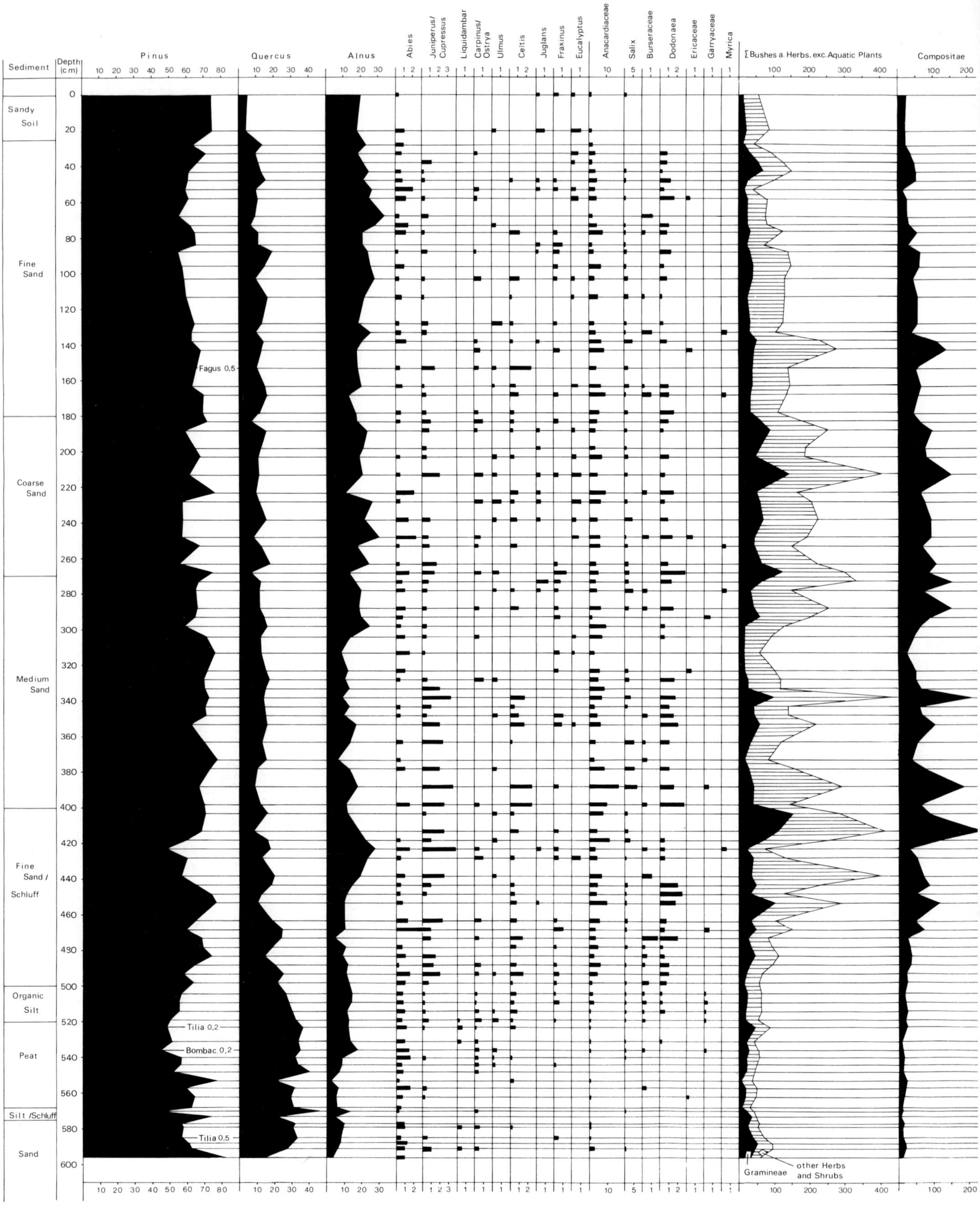

Fig. 9. Pollen diagram Acuitlapilco IV; calculations based on AP sum = 100% (analyses 591 – 493 cm, mixed oak wood period. Section c: 493 – 0 cm, cultivated landscape period. *Maydeae*. Subsection c_3: 128 – 0 cm, poorer in *Maydeae*, richer in *Alnus*

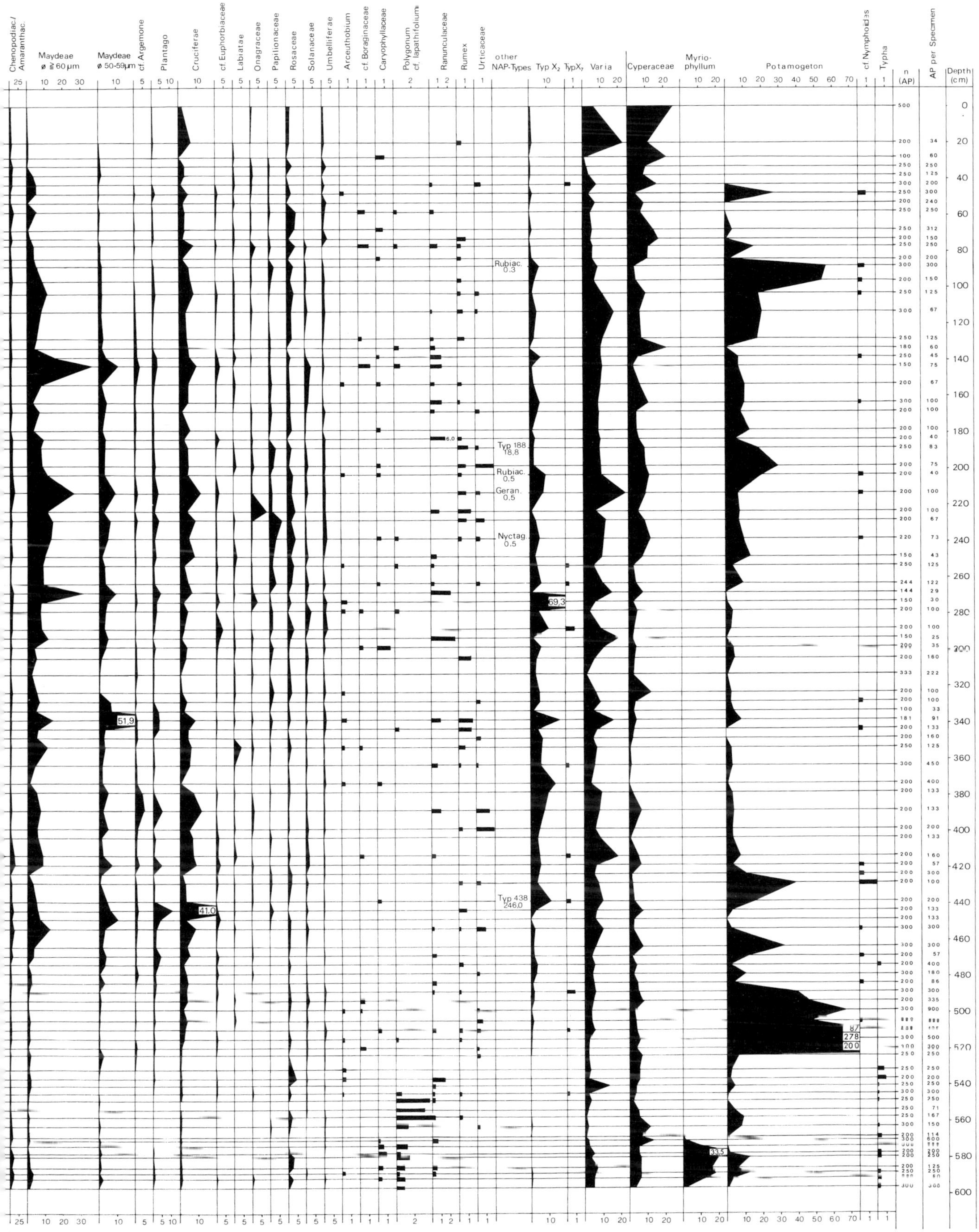

by OHNGEMACH). – Section a: 596 cm, end of mixed pine wood period. Section b: Subsection c_1: 493 – 273 cm, poorer in *Maydeae*. Subsection c_2: 268 – 133 cm, richer in

Table 3. Overview of the vegetation history of the last 35 000 years in the Puebla-Tlaxcala area

Time	Lower elevations (2 200 – 2 400 m)	Upper elevations (3 100 m)
	Acuitlapilco IV, III, II, and I (2 200 m); Santa Ana Nopalucan (2 200 m); (Oriental III, 2 350 m)	Tlaloc II and I (3 100 m)
Today		
	Crop landscape period c_3 poorer in *Maydeae*, richer in alder c_2 richer in *Maydeae* c_1 poorer in *Maydeae*	
Years B. P.		Period VIII Mixed pine forest period
	Mixed oak forest period b_2 richer in alder b_1 poorer in alder	
	Mixed pine forest period	
1 000 – 2 000		
	Jalapasquillo I – III (2 400 m); Oriental III and IV (2 350 m)	Period VII, fir period (M IV); Period VI, mixed pine forest period; (Period V, devoid of or poor in pollen); Period IV, pine advance; Period III, alder period; Period II, *Pinus hartwegii* period
7 000 – 8 000		
	Period IV, pine period (no *Quercus*)	
8 500		
	Period III, pine period, poorer in *Quercus*	Period I, alpine grasslands period (M III)
18 000		
	Period II, pine period, average *Quercus* pollen values	
31 000		
	Period I, pine period, relatively rich in *Quercus* (M I)	
35 000		

found at 263 cm are cf. *Argemone, Plantago, Polygonum* cf. *lapathifolium, Rumex, Urticaceae*, and *Solanaceae* (and perhaps potato and tomato as crop plants). The arboreal pollen is represented by *Pinus* (47 – 77%), *Quercus* (3.5 – 25%), *Alnus* (6.5 – 35%), and other rare types. One can differentiate between three subsections: c_1 493 – 273 cm, poorer in *Maydeae*; c_2 268 – 133 cm, richer in *Maydeae*, with three very high maxima; c_3 128 – 0 cm, poorer in *Maydeae*, richer in *Alnus*.

The relevant results of our studies are presented as a **summary** in Table 3 and can be compared with our Table 2 after HEINE, demonstrating the glaciation of the higher mountain peaks and their moraines. M I corresponds well to Zone I from Jalapasquillo. No corresponding Zone was found for M II. M III probably corresponds to period I from Tlaloc (a period of grasslands or zacatonales). M IV corresponds to period VII from Tlaloc (fir period); no corresponding period was found for M V.

The results of our pollen analyses allow the first detailed reconstruction of the late Quaternary vegetation history of the basin and volcanoes of the Puebla-Tlaxcala area. Other analyses from Mexico City basin (CLISBY & SEARS 1955, SEARS 1952, WATTS & BRADBURY 1980) are not so detailed, but generally correspond to our results.

References

CLISBY, K. H., SEARS, P. B., 1955: Palynology in Southern North America 3. Microfossil profiles under Mexico City correlated with the sedimentary profiles. – Bull. Geol. Soc. America **66**: 511–520.

HEINE, K., 1973 a: Die jungpleistozänen und holozänen Gletschervorstöße am Malinche-Vulkan, Mexiko. – Eisz. Gegenw. **23/24**: 46–62.

– 1973 b: Variaciones más importantes del clima durante los ultimos 40 000 años en Mexico. Deducciones indirectas de observaciones geomorfológicas. – Comunicaciones 7: 51–58.

– 1973 c: Zur Glazialmorphologie und präkeramischen Archäologie des mexikanischen Hochlandes während des Spätglazials (Wisconsin) und Holozäns. – Erdkunde **27**: 161–180.

– 1975: Studien zur jungquartären Glazialmorphologie mexikanischer Vulkane. Mit einem Ausblick auf die Klimaentwicklung. – Das Mexiko-Projekt der Deutschen Forschungsgemeinschaft 7. Wiesbaden: F. Steiner.

– 1976: Schneegrenzdepressionen, Klimaentwicklung, Bodenerosion und Mensch im Zentralmexikanischen Hochland im jüngeren Pleistozän und Holozän. – Z. Geomorph. N. F. Suppl. **24**: 160–176.

– 1978: Neue Beobachtungen der mittelwisconsinzeitlichen Vergletscherungen und Böden mexikanischer Vulkane. – Eisz. Gegenw. **28**: 139–147.

– 1983 a: Ein außergewöhnlicher Gletschervorstoß in Mexiko vor 12 000 Jahren. – Catena **10**: 1–25.

– 1983 b: Outline of man's impact on the natural environment in Central Mexico. – Jahrb. Gesch. Lateinamerikas **20**: 121–131.

– 1983 c: Mesoformen der Periglazialstufe der semihumiden Randtropen, dargestellt an Beispielen der Cordillera Neovolcanica, Mexiko. – Abh. Akad. Wiss. Göttingen, math.-phys. Kl. 3. F., **35**: 403–424.

– 1983 d: Das Verhältnis von Relief- und Bodenentwicklungsphasen im Jungquartär in Zentralmexico und in der Kalahari. – Z. Geomorph. N. F. Suppl. **48**: 145–153.

– 1984 a: Comment on "Pleistocene glaciation of volcano Ajusco, Central Mexico, and comparison with the standard Mexican glacial sequence" by SIDNEY E. WHITE and SALVATORE VALASTRO. – J. Quatern. Res. **22**: 242–246.

– 1984 b: The classical Late Weichselian climatic fluctuations in Mexico. In MÖRNER, N.-A., KARLEN, W., (Eds.): Climatic changes on a yearly to millennial basis, pp. 59–115. – Reidel Publ. Comp..

– 1985: Neue absolute Daten zur holozänen Gletscher-, Vegetations- und Klimageschichte zentralmexikanischer Vulkane. – Regensb. geogr. Schr. **19/20**: 79–92.

Heine, K., Heide-Weise, H., 1973a: Jungquartäre Förderfolgen des Malinche-Vulkans und des Popocatépetl (Sierra Nevada, Mexiko) und ihre Bedeutung für die Glazialgeologie, Paläoklimatologie und Archäologie. – Münster. Forsch. Geol. Paläontol. **31/32**: 303–322.

– – 1973b: Secuencias de erupciones en el volcán de la Malinche y en la Sierra Nevada durane los ultimos 40000 años. – Comunicaciones **7**: 7–8.

– Ohngemach, D., 1976: Die Pleistozän/Holozän-Grenze in Mexiko. – Münster. Forsch. Geol. Paläontol. **38/39**: 229–251.

Klaus, D., 1973: Die eiszeitlichen und nacheiszeitlichen Klimaschwankungen im zentralmexikanischen Hochland und ihre Ursachen. – Erdkunde **27**: 180–192.

Lauer, W., 1981: Klimawandel und Menschheitsgeschichte auf dem mexikanischen Hochland. – Akad. Wiss. Lit. Mainz, Abh. math.-nat. Kl., Jg. 1981, Nr. 2. – Mainz.

Ohngemach, D., 1973: Análisis polinico de los sedimentos del Pleistoceno Reciente y del Holoceno en la región Puebla-Tlaxcala. – Comunicaciones **7**: 47–49.

– 1977: Pollen sequence of the Tlaloqua crater (La Malinche volcano, Tlaxcala, Mexico). – Bol. Soc. Bot. México **36**: 33–40.

– Straka, H., 1983: Beiträge zur Vegetations- und Klimageschichte im Gebiet von Puebla-Tlaxcala. Pollenanalysen im Mexiko-Projekt. (Resumen en español.) Mit einem Anhang von W. Lauer: Synopse der spät- und postglazialen Landschaftsentwicklung im zentralmexikanischen Hochland. – Das Mexiko-Projekt der Deutschen Forschungsgemeinschaft 18. – Wiesbaden: Steiner. (Additional literature cited therein).

Sears, P. B., 1952: Palynology in Southern North America. I. Archaeological horizons in the basins of Mexico. – Geol. Soc. America Bull. **63**: 242–254.

Straka, H., Ohngemach, D., 1975: Pollen analysis in the Puebla-Tlaxcala region (Central Mexican highlands). – Conf. 12th intern. Botan. Congr. Leningrad.

– – 1978: La historia de la vegetación de Puebla-Tlaxcala durante el Cuaternario tardio. – Conf. 2nd Symp. of the "working group Mexico" of the DFG. – Comunicaciones **15**: 189–204.

Watts, W. A., Bradbury, J. P., 1980: Late Pleistocene and Holocene Palaeo-environments and human activity in the West-Central Mexican plateau. – Evidence from Lake Patzcuaro, Michoacan, and from the Cuenca de Mexico. Manuscript.

Address of the authors: Prof. Dr Dr h. c. Herbert Straka and Dr Dieter Ohngemach, Institute of Botany, University of Kiel, Olshausenstrasse 40, D-2300 Kiel, Federal Republic of Germany.

Pl. Syst. Evol. 162, 133–163

—Plant Systematics and Evolution

Mediterranean pines and their history

WILHELM KLAUS †*

Received May 13, 1987

Key words: Gymnosperms, *Pinaceae, Pinus*; *Rzedowskiae* and *Resinosae*: subsectt. nov. – Cone and seed morphology, vegetative characters, Tertiary fossils, phytogeography. – Flora of the Mediterranean region, C. and N. America.

Abstract: The assembly of Mediterranean pines in the sense of MIROV is inhomogeneous in respect to morphological, geographical and evolutionary affinities. Considering new or neglected characters (vegetative and particularly reproductive, cone scales, apophyses, mucros, seeds: Figs. 1–3) in extant populations and fossils, three groups are recognized. The group of coast and island pines extends from the Canary islands to the Himalaya region and is closely related to Caribbean and C. American taxa. This complex evidently has originated from haploxyl ancestors of sect. *Parrya* during the Mesozoic (Upper Jurassic/ Lower Cretaceous) in the NW. Tethys area (Fig. 13). *P. rzedowskii* can be regarded as an extant survivor of this first phase of differentiation. The extremely variable *P. canariensis* (together with *P. roxburghii* in sect. *Sula*) marks a transitional phase towards the more advanced diploxyl species of sect. *Pinea* (with *P. pinea, P. halepensis* & *P. brutia,* and subsect. *Oocarpae*) and sect. *Pinaster* (with *P. pinaster*, etc.) (Figs. 4–12). – The second group consists of diploxyl mountain pines from the areas surrounding the Mediterranean. They are classified as members of the Eurasiatic sect. *Pinus* subsect. *Sylvestres*, have differentiated along the northern Parathetys area, and exhibit close links with E. Asiatic taxa. The third group includes the haploxyl mountain pines *P. cembra* and *P. peuce* which can be regarded as western outposts of the circumpacific centred sect. *Strobus* with a pre-Tertiary origin. – As an appendix, an improved classification scheme is presented for the pine groups discussed.

According to MIROV (1967: 237 ff.) the assembly of Mediterranean pines (slightly extended by me) includes the following species:

Subg. *Haploxylon*

P. cembra	Alps to Carpathians
P. peuce	Balkans

* After the untimely and most regrettable death of Prof. W. KLAUS (13. 11. 1987) his raw manuscript was extensively revised and edited by F. EHRENDORFER.

Subg. *Diploxylon*

P. canariensis	Canary Islands
P. pinea	Portugal to Turkey
P. nigra	widely distributed
P. heldreichii	E. Mediterranean (rare in Italy)
P. montana	Pyrenees to Carpathians
P. sylvestris	from Spain to Turkey and Asia
P. halepensis	widely distributed
P. brutia	E. Mediterranean
P. pithyusa	Caucasus and E. Mediterranean
P. pinaster	widely distributed
P. eldarica	endemic in Transcaucasia

In this contribution I will try to present new or neglected characters of these Mediterranean pines (particularly cones; cf. KLAUS 1980 and Figs. 1–3), fossil material (e.g., Fig. 8) and geographical considerations. My comparative analysis results in the general systematic and evolutionary conclusion that these Mediterranean species represent an extremely heterogeneous assembly. According to their relationships, one can recognize at least three groups, of which only the first will be discussed in detail.

1. Mediterranean shore and island pines. They are the most important for the understanding of the history of pines in that area. As many features and also fossil evidence indicate, they consist mainly of relic pines from the Tertiary (and possible older periods). This is supported, e.g., by fossil cones of *Pinus canariensis* and *P. roxburghii* from the Austrian Miocene basins, located between the extremely disjunct extant areas of these two species. *P. rzedowskii*, a *Haploxylon* pine of sect. *Parrya*, recently discovered by MADRIGAL & CABALLERO (1969), can be regarded as close to the ancestors of *P. canariensis*. Similar cone features can also be seen in *P. tropicalis* from Cuba. On the other hand, the Caribbean pines point with many features towards *P. pinaster*. Species of subsect. *Halepenses* exhibit affinities with *P. canariensis* and *P. oocarpa*. The more isolated *P. pinea* may be positioned between *Cembroides* nutpines and vallate members of *Leiphyllae* (*Lumholtziae*). But no relationships of this group can be traced to *Gerardianae* or to any pine growing today east and north of the Himalaya region (China, Japan, Thailand, Philippines, Korea).

2. Mountain pines from the areas surrounding the Mediterranean. They hardly reach the Mediterranean coasts. Their original distribution evidently was on the C. European and Asiatic mainland northwards of the Mediterranean area. *P. sylvestris, P. nigra, P. heldreichii* and its subsp. *leucodermis* are such pines which have close relatives in E. Asia. Their fossil ancestors (*P. thomasiana, P. hampeana*, etc.) show a wide Tertiary distribution from C. and N. Europe to Siberia, which later was ± disrupted by the elevation of the Eurasiatic mountain systems and the gradual disappearance of the Paratethys.

***3. Haploxylon* pines of subsectt. *Strobi* and *Cembrae*.** There exists neither morphological nor fossil evidence for considering these species as Mediterranean pines. Their evolution appears to be circum-Pacific centred: *P. cembra* is so closely linked with *P. sibirica* (the latter having larger and more primitive cones) that its origin must be sought in the east and towards the main distribution centre of the *Cembrae* (i.a., with *P. sibirica, P. koraiensis, P. armandii, P. pumila, P. albicaulis*). *P. armandii* is a nutpine, but with long peduncles, and may be the connecting link to subsect. *Strobi* which it meets at its eastern extension (with *P. griffithii* in Burma). *P. peuce* is a small cone relative of *P. griffithii*, occurring as an Eurasian relic pine in the mountains of the Balkan Peninsula. Fossils suggest that *P. peuce* and *P. armandii* have been present in the N. European and Asiatic area during the later Tertiary. *P. peuce* has retreated since and today only survives on a few humid mountain systems.

Material, methods, and acknowledgements

My interest in pines was aroused in the early seventies by the *Pinus mugo* problem of the E. Alps. This taxon is important for subfossil pollen determinations in the peat bogs and the reconstruction of the forest history by pollen diagrams. Sampling excursions in the Austrian mountain regions demonstrated a considerable variability of flower colour and cone morphology due to introgressive hybridization (KLAUS 1979). On Monte Baldo in N. Italy, the type locality of *P. mugo*, populations on the humid western slopes towards lake Garda frequently have cones with uncinate apophyses and yellowish green flowers, whereas on the dry eastern slopes cones with flat apophyses dominate, frequently combined with red flowers. The search for ancestral types led to the Pyrenees in the west and the Balkan Mountains in the east. *P. uncinata* collections in the Pyrenees present some difficulties due to hybridization with *P. sylvestris* at lower elevations, but in the higher Pyrenees *P. uncinata* is a pretty stable tree-like monocorm species with unique cone morphology. Excursions in the Balkan mountains (Rila and Pirin) offered rather uniform red flowering and shrubby, polycorm *P. mugo* with flat apophyses.

P. cembra was sampled near the south-westernmost occurrence of that species in the French Alpes Maritimes. In Yugoslavia and Bulgaria several other species like *P. peuce, P. heldreichi,* and its subsp. *leucodermis, P. nigra*, and in Greece *P. halepensis* and hybrids with *P. brutia* could be studied. Pure stands of the latter species were found on an excursion to Crete. To get acquainted with *P. pithyusa*, considered by some authors as a subspecies of *P. brutia*, an excursion to the island of Büyük Ada (Turkey) was organized. *P. pinaster* from the lower Pyrenees, *P. pinea* from Spain (south of Barcelona), and *P. halepensis* (between Valencia and Madrid) were sampled especially for the study of cone morphology.

The cones of cultivated *P. canariensis* from Crete had shown such a variability that it appeared desirable to analyze the full width of variation of that species on the Canary Islands. On Teneriffa, a transsect from low altitudes at the dry SE. slopes (above Los Cristianos) and the cloudy wet N. side (Orotava valley), across the famous forest of Esperanza and to the highest elevations near the Caldera of the Teide (about 2300 m), presented an impressive series of different climatic races. The excessive variability of *P. canariensis* could not be explained without knowledge of the pines of Mexico and C. America which have the richest pine flora of the world. During the following 15 years, eight sampling excursions to Mexico, Guatemala, Santo Domingo, the Bahamas, Florida and the U.S.A. east coast, and to California from San Diego up to the Canadian border, resulted in a wealth of morphological data on cones, conelets, male flowers, seedling growth, and other features. Further material from E. Asia, the Himalaya region, China, Korea, Vietnam, Thailand, Taiwan, the Philippines, and Japan I owe to skilled foresters and botanists from these countries.

Most of the cones of extant populations studied are (if not stated otherwise) from the private collection KLAUS, which will eventually be placed at WU.

Examination of well preserved fossil pine cones was restricted to Austrian collections, mainly from the Tertiary of the Vienna basin (Upper Miocene) and Styria (Middle Miocene and Upper Austrian). In these cases the study of umbo and mucro morphology has been possible. Otherwise, photographs and drawings from the paleobotanical literature do not sufficiently show the details necessary for determination.

An effort to elucidate the history of such a complex assembly of pines as that of the Mediterranean area, could have never been successful without the generous help of skilled colleagues, botanists, geneticists and foresters in Austria as well as in foreign countries. For important suggestions concerning taxonomy, evolution and plant geography as well as the possibility to use SEM, TEM, etc. at the Institute of Botany, University of Vienna. I am grateful to Univ.-Prof. Dr F. EHRENDORFER. In C. Mexico, Prof. J. RZEDOWSKI, Prof. L. VELA, and Prof. MADRIGAL have helped with determinations and introductory excursions. In San Luis Potosi, Prof. F. MEDELLIN with his skilled assistants at the Instituto de Zonas Deserticas, and in N. Mexico Prof. M. A. CAPO-ARTEAGA have introduced me to the pines of Coahuila and Nuevo Leon. Guidance to the Torrey pines and other Californian species I owe to my esteemed friend D. W. NICOL. On the East Coast Dr E. LITTLE jr. from the Smithsonian Institution of Washington as well as Dr MEYER, Director of the Washington Arboretum, have been very helpful.

Selected features and relationships of Mediterranean coast and island pines

The following catalogue contains a selection of features apparently important for elucidating the relationships and evolution of Mediterranean pine species, reaching from S. Europe to the Himalaya region in the east, and to the Caribbean, C. and N. America in the west (Figs. 1 – 3).

Vegetative sphere (Fig. 1) analyses have revealed a number of important characters, overlooked until now.

"Grass stage" of seedling. *P. tropicalis* from Cuba which currently is quoted as a remarkably disjunct member of the Eurasiatic subsect. *Sylvestres* is an example for that unique "grass stage" of seedlings in some pine species. It is due to retarded stem elongation combined with excessive subterranean growth for several years, e.g., in *P. palustris* (Florida), *P. elliottii* subsp. *densa, P. engelmannii*, and some other Mexican pines. This character underlines the true relationship of *P. tropicalis* with other Caribbean pines and not with subsect. *Sylvestres.*

Bracts recurved at tip. This unique feature has been observed without exception in all true Mediterranean pines, in spite of their different taxonomic classification, and also reappears in many Caribbean and American pines. An example is *P. elliottii* which exhibits that character clearly and in exactly the same manner as *P. canariensis* or *P. pinaster*. No C. European pine produces recurved bracts, and also not *P. roxburghii,* which differs in that respect from *P. canariensis* and the other Mediterranean pines.

Colour of spring shoots bluish pruinose. *P. canariensis* is the only pine in the Mediterranean area, Europe and Asia which exhibits this character, whereas it is rather common among C. and N. American *Diploxylon* pines. In the *Haploxylon* sect. *Parrya* I have observed bluish pruinose shoots in *P. nelsonii* only. The feature points clearly to relationships of *P. canariensis* with the Caribbean and C. American pines.

Primary leaves for several years on the seedling. Young plants continue to produce primary leaves for several years, until they reach a height of half a

---

Fig. 1. Selected vegetative features of systematic significance for pines. H Himalaya region, M Mediterranean region, C Caribbean and C. American region, E Europe, NE N. Europe; – – – haploxyl species below this line

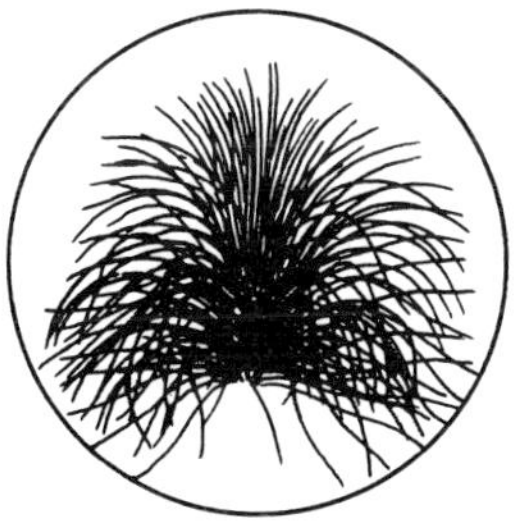

Grass stage of seedling

C *P. tropicalis*
C *P. caribaea*
C *P. cubensis*
USA *P. palustris*
some Mexican pines

No *Sylvestres*!
No species of Europe and Asia

Recurved bracts

M *P. canariensis*
M *P. pinea*
M *P. halepensis*
M *P. brutia* & subsp.
M *P. pinaster*
USA *P. elliottii*
C *P. oocarpa*

No other European and Asiatic species (not *P. roxburghii*)

Bluish pruinose spring shoots

M *P. canariensis*
C *P. tropicalis*
C *P. caribaea*
C *P. cubensis*
USA *P. elliottii*
USA *P. palustris*
– – – – – – – – – –
C *P. nelsoni*

No other species of Europe and Asia

Several years primary leaves on seedlings

M *P. canariensis*
H *P. roxburghii*
M *P. pinea*
M *P. halepensis*
M *P. brutia*
– – – – – – – – – –
C *P. rzedowskii*
C all subsect. *Cembroides*

No other diploxyl species of Europe and Asia

Stem shoots with primary leaves

M *P. canariensis*
M *P. pinea*
some C. American pines

No other species of Europe or Asia

meter or more. The typical bundles of needles (secondary leaves) appear very delayed. This feature is rather common in all species of the *Haploxylon* sect. *Parrya* (e.g., *P. rzedowskii*, Mexico), but more rare in *Diploxylon* pines, where it occurs, e.g., in all true Mediterranean pines and in *P. roxburghii* (Himalaya). None of the subsect. *Sylvestres* from Europe or Asia exhibit that feature, suggesting the lack of closer relationships with the true Mediterranean pines, whose links with the *Haploxylon* sect. *Parrya* are underlined.

Shoots with primary leaves on old stems. In the Mediterranean area this feature is restricted to *P. pinea*, but occasionally also appears in *P. canariensis*. On *P. pinea* trees one can sometimes see the whole trunk and old branches covered by such stem shoots. No other European or Asiatic pine exhibits this character which sporadically also occurs in the C. American species *P. rzedowskii, P. nelsonii, P. caribaea,* and *P. serotina.*

Cone and apophyses (Fig. 2) are very important for the systematics of *Pinus* and have been documented and described in another contribution (KLAUS 1980 a).

Excentromucronate umbo (Figs. 2, 4–8, and 11). The majority of Eurasiatic *Diploxylon* pines as well as some species of the C. American subsect. *Cembroides* develop a small prickle on the umbo, which is located above the horizontal keel. All true Mediterranean, but also a few C. and N. American pine species (*P. occarpa* from Guatemala, *P. tropicalis* from Cuba and *P. resinosa* from northern U.S.A.) belong to this excentromucronate umbo type. Among Mediterranean pines, a surprising variation of this character exists in *P. canariensis* (Figs. 4, 6, and 7), from denticulate-mucronate via tectoid to almost centromucronate. *P. roxburghii* shows a pretty similar umbo and mucro development. A completely flattened or even depressed umbo, often with a reduced and strongly excentric (perexcentric) mucro, is developed in subsect. *Halepenses* (Fig. 11).

Centromucronate umbo (Figs. 2 and 10). There is only one *Diploxylon* pine in the Mediterranean region and in Eurasia which exhibits this feature, i.e., *P. pinaster*. (Only *P. pinaster* subsp. *mesogeensis* from the Pyrenees is slightly excentromucronate and reminiscent of some almost centromucronate variants of *P. canariensis.*) But the bulk of American *Diploxylon* pines from the east and west coast shows centromucronate umbos. Among sect. *Parrya*, centromucronate umbos occur in *P. nelsonii* (Mexico), but also in subsectt. *Balfourianae* (America) and *Gerardianae* (Asia), but not in subsectt. *Cembroides* and *Rzedowskiae*.

Duplomucronate umbo (Figs. 2 and 9). This is characterized by an excentric mucro and an additional central protuberance on the horizontal keel, i.e., a combination of a centro- and an excentromucronate umbo. Diplomucronate umbos are almost restricted to the Caribbean and C. America, where they are most clearly developed in *P. caribaea* subsp. *hondurensis*, slightly reduced in *P. caribaea* subsp. *bahamensis*, occasional in *P. oocarpa* (Guatemala), *P. herrerai* and *P. pringlei* (S. Mexico) , and *P. tecucumani* (Guatemala), but rare in *P. elliottii* var. *densa* (S. Florida). Rather unexpectedly, this feature can be observed also in true Mediterranean pines. Some *P. halepensis* races from Spain (between Valencia and Madrid) show slightly duplomucronate apophyses on some cones (Fig. 9 f), and also a few *P. pinaster* populations from Spain, the Pyrenees and S. France (Fig. 9 b). In *Haploxylon* pines this feature is unknown, only conelets of *P. nelsonii* sometimes have a few duplomucronate scales (Fig. 5 d). As in other characters, this species is an apparent link to *Diploxylon.*

Polygonal umbo delimitation (Figs. 2, 7, and 8). The majority of pines develop a more or less oval to round umbo field. But in the Mediterranean area, many pines exhibit a rhomboidal or polygonal (pentagonal or even hexagonal) umbo field, e.g., *P. canariensis, P. pinaster, P. halepensis,* and *P. brutia.* This is most clearly developed in *P. halepensis* cones from Greece, similar to the fossil *P. salinarum* PARTSCH from the northern Miocene Paratethys shore (Wielicka). In C. America also *P. oocarpa* (Mexico, Uruapan) mostly shows polygonal umbos.

Vallum around the umbo (Figs. 2 and 7c, d). Few pine species develop a ring-shaped wall around the umbo, especially when the ripening of the cones continues for three seasons. In the Mediterranean area *P. pinea* is the only pine with that feature, but some variants of *P. canariensis* exhibit a similar tendency (Fig. 7d). *P. lumholtzii* (Mexico) develops a clear vallum and excentromucronate umbo, well comparable with *P. pinea.* Traces of a vallum can be observed in the haploxyl *P. pinceana* (subsect. *Cembroides*) and in the diploxyl *P. oocarpa, P. tecucumanii* and *P. tropicalis,* but in the other European and Asiatic pines the character is almost lacking. Again, a feature of C. American pines reappears in true Mediterranean pines.

Loss of basal cone scales, combined with excentromucronate apophyses (Fig. 27). According to my own observations, the following species exhibit that character combination.

a) Mediterranean area: *P. canariensis* (W. Canary Islands); *P. canariensis* subsp. *prisca* (Fig. 8, Miocene fossil cone from the Paratethys shore); *P. pinea* (Spain, Portugal, etc.; fossil pollen grains common in Miocene Paratethys deposits).

b) C. to N. America: *P. rzedowskii* (S. Mexican *Haploxylon* pine); *P. tropicalis* (W. Cuba); *P. resinosa* (NW. U.S.A.).

c) Asia: *P. roxburghii* (Himalaya region; Miocene fossil cones found in the Paratethys area of the Vienna basin).

Loss of basal cone scales does not occur in any other Eurasiatic pine species except *P. roxburghii.* Together with excentromucronate apophyses it connects *P. pinea* and *P. canariensis* and suggests their common ancestry. *P. pinaster* subsp. *mesogeensis* exhibits some indications of these features, but cannot be classified as a typical excentromucronate pine. It forms somewhat of a link between *P. canariensis* and the centromucronate *P. pinaster* which does not loose its basal cone scales. *P. canariensis* subsp. *prisca,* the Miocene fossil from the Vienna basin, is another clear example for the loss of basal cone scales.

Outside the Mediterranean area only *P. roxburghii* from the Himalayas and the American *P. resinosa, P. tropicalis,* and *P. rzedowskii* loose their basal scales in ripe cones.

Seed characters (Fig. 3) have already received some attention and are of obvious importance for the systematics and dispersal ecology of pines.

Adnate seed wing. Seeds with undetachable wings are rare among *Diploxylon* pines. Their occurrence in *P. canariensis* and *P. roxburghii* points towards a close relationship of these two species but also towards a similar dispersal mechanism. When such seeds are released, they will not loose their wings, even during long distance flights. Therefore, such pines have the capacity of island hoppers, as suggested by their occurrence and spread during the Cretaceous and Tertiary throughout the Tethys (Fig. 13). The only other pine outside of subg. *Haploxylon*

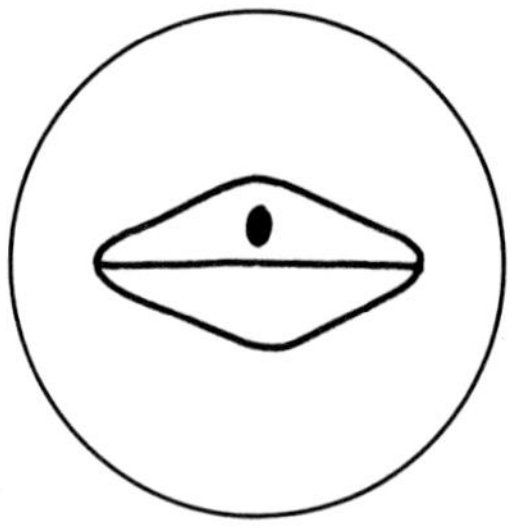

Excentromucronate apophyses of cones

M *P. canariensis*
M *P. pinea*
M *P. halepensis*
M *P. brutia*
C *P. tropicalis*
C *P. oocarpa*
USA *P. resinosa*
– – – – – – – – – –
C *P. rzedowskii*
C *P. pinceana*
C *P. cembroides*

All European and Asiatic diploxyl pines

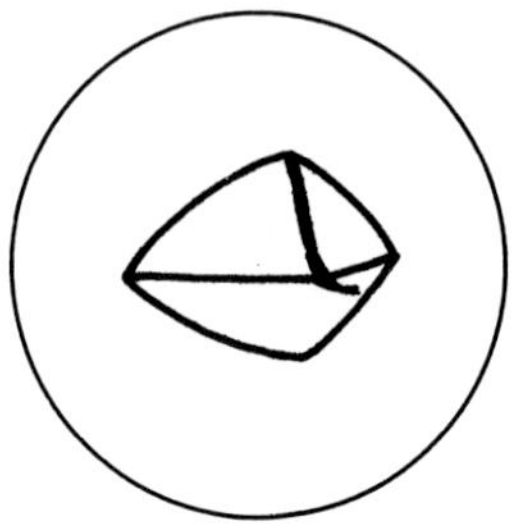

Centromucronate

M *P. pinaster*
C *P. occidentalis*
USA *P. elliottii*
– – – – –
C *P. nelsonii*

No other European and Asiatic diploxyl pines

Most other U.S.A. and Central-American diploxyl pines

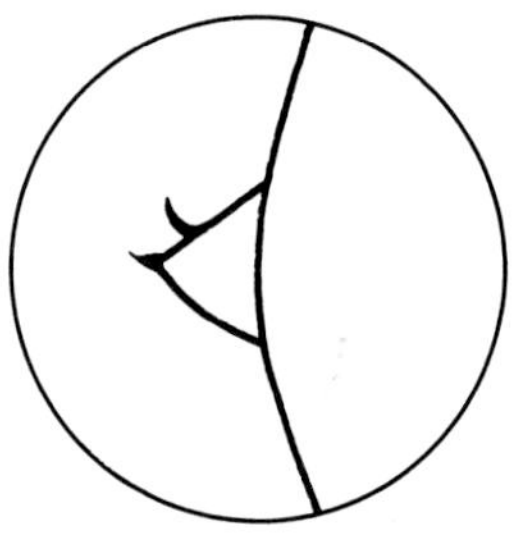

Duplomucronate

M *P. halepensis* (Spain)
M *P. pinaster* (France)
C *P. caribaea* subsp. *hondurensis*
C *P. caribaea* subsp. *bahamensis*
USA *P. elliottii* subsp. *densa* (rarely)
C *P. herrerai*
C *P. oocarpa* (Guatemala)
C *P. tecucumanii*
C *P. pringlei*

No other European, Asiatic or U.S.A. pines

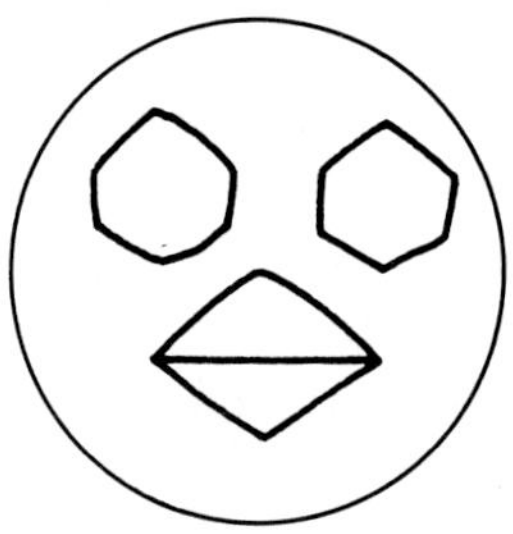

Delimitation of umbo

H *P. canariensis*
H *P. roxburghii*
M *P. halepensis* (Greece)
M *P. pinaster*
NE *P. salinarum* († fossil)
C *P. oocarpa*

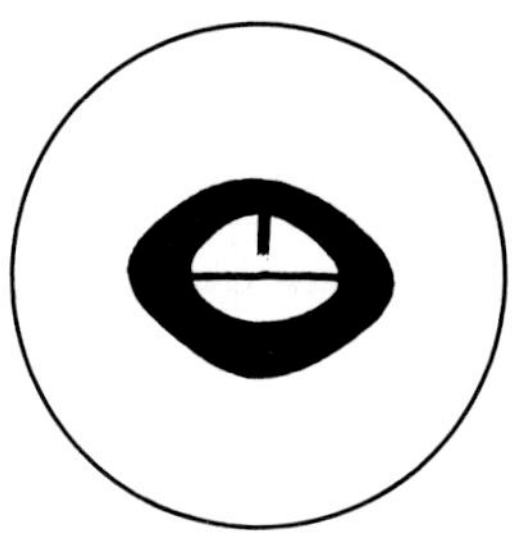

Vallum & excentromucronate

M *P. pinea*
M *P. canariensis* (partly)
C *P. lumholtzii*

Occasionally other American and Asiatic pines

Fig. 2. Selected features of apophysis morphology of pines. For abbreviations see Fig. 1

Loss of basal cone scales

H *P. roxburghii*
M *P. canariensis* & fossils
M *P. pinea*
M *P. pinaster* subsp. *mesogeensis*
C *P. tropicalis*
USA *P. resinosa*
C *P. caribaea* & subsp.
USA *P. eliottii*
C *P. cubensis*

- - - - - - - - - -

C *P. rzedowskii*
C *P. nelsonii*

No species of Asia and Central/North Europe

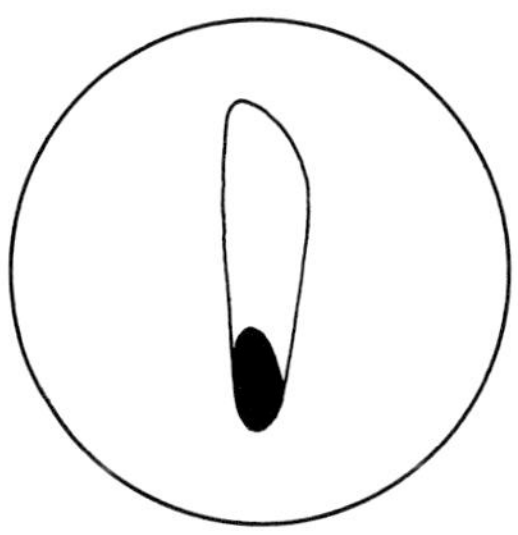

Adnate seed wings

H *P. roxburghii*
M *P. canariensis*
C *P. caribaea*

- - - - - - - - - -

Subsect. *Strobi*
Subsect. *Balfourianae*

No species of Europe and Asia

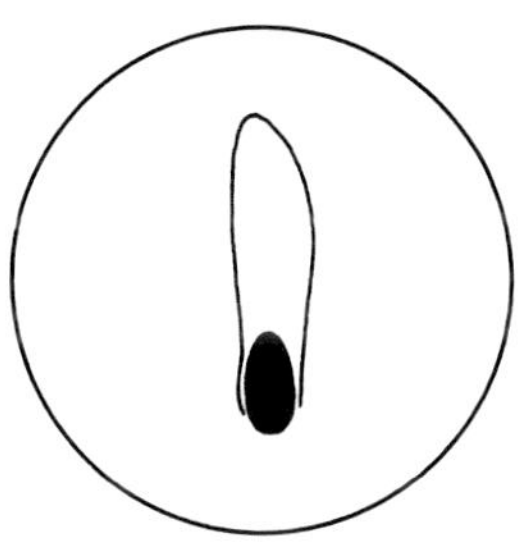

Articulate seed wings

M *P. halepensis*
M *P. brutia*
M *P. pinaster*
E *P. sylvestris*
E *P. nigra*
E *P. uncinata*
E *P. mugo*
E *P. heldreichii* & subsp. *leucodermis*

Most other Asiatic, European and American diploxyl pines

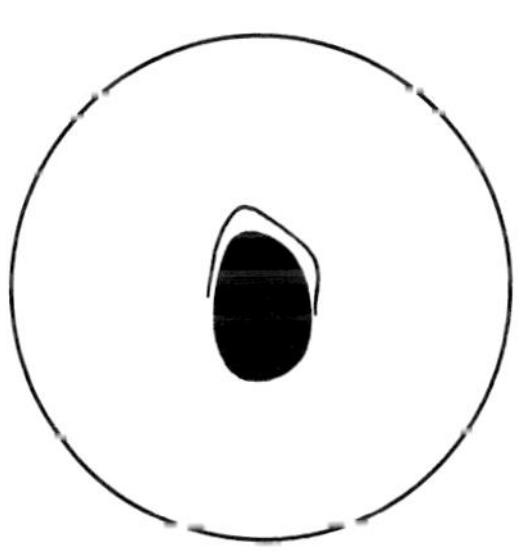

Nut-like seeds

M *P. pinea*
USA *P. torreyana*

- - - - - - - - - -

C *P. pinceana*
C *P. cembroides*
C *P. nelsonii*

P. gerardiana (Asia)
P. bungeana (China)
subsect. *Cembrae* (Asia, Europe, U.S.A.)

Fig. 3. Selected cone and seed features of pines. For abbreviations see Fig. 1

with this unique feature of adnate seed wings is *P. caribaea* with a similar dispersal ecology.

Articulate seed wing. Detachable seed wings have such a wide occurrence among *Diploxylon* pines, that relationships or evolutionary trends cannot be deduced from this feature. It is interesting to note that one *Haploxylon* pine of Mexico, closely related to subsect. *Cembroides* nut pines, also has seeds with articulate wings, i.e., *P. rzedowskii.*

Nut-like seeds. *P. pinea* is the only true Mediterranean pine which produces enlarged nut-like seeds with an easily detachable (articulate) reduced wing. Among *Diploxylon* pines this is a very rare feature, otherwise occurring only in the S. Californian *P. torreyana,* whereas it is common in subg. *Haploxylon* among subsect. *Cembrae* and sect. *Parrya* pines. From the latter section, especially *P. gerardiana* of SW. Asia, is often compared with *P. pinea*, but C. American species of subsect. *Cembroides* also develop nuts.

Systematics, evolution, and fossil occurrence of Mediterranean pines and their relatives

Analyses of the selected features of Mediterranean pine species discussed in the last chapter clearly indicate their close relationships with C. American and Caribbean pines and with one single species of the Himalayan region. In contrast, all the subsect. *Sylvestres* species, including *P. nigra, P. heldreichii* with subsp. *leucodermis, P. sylvestris,* the *P. mugo* complex, and the whole assembly of extant Asiatic pines appear to be of remote or different origin, and will not be discussed further.

Thus, this chapter will present data on the systematics, evolution and fossil occurrence for the more important Mediterranean pines and their American relatives.

Pinus rzedowskii MADRIGAL & CABALLERO (Figs. 4, 12). This is a 3- to 4-needled *Haploxylon* pine (for details see MADRIGAL & CABALLERO 1969), producing cones with a dorsal umbo (therefore belonging to sect. *Parrya*). The recurved basal needle sheeths put the species close to *P. pinceana* in subsect. *Cembroides.* Concerning classification the difficulty arises, that subsect. *Cembroides* is reserved for nut pines, whereas *P. rzedowskii* has an articulate seed wing. With that combination of morphological features the species does not fit in any of the subsections of sect. *Parrya* and should be better classified in its own subsection:

***Pinus* sect. *Parrya* MAYR subsect. *Rzedowskiae* W. KLAUS, subsect. nova:**

Folia 3 – 4, vagina ut in subsectione *Cembroide*, sed semina minora, alis longis secedentibus. Strobilus squamis apophysibus excentromucronatis.

Species typica (et unica): *P. rzedowskii* MADRIGAL & CABALLERO.

P. rzedowskii is particularly remarkable as the only *Haploxylon* pine yet found with three important features which also occur in the diploxyl *P. canariensis:* (a) loss of basal scales in mature cones; (b) very clear and uniformly developed excentromucronate apophyses; (c) winged seeds. Its mature cones are elongated, of medium to large size, and similar to the geologically oldest fossil pine cone, *Pinus andreae* COEMANS (1867) from the Lower Cretaceous of Belgium.

Thus, *P. rzedowskii* suggests an early evolutionary line from haploxyl sect. *Parrya* to diploxyl sect. *Sula*, both already with a dorsal umbo on the apophyses. In that connection it should be mentioned, that *P. cembroides*-like fossil pollen grains have been found in Miocene deposits of Austria (KLAUS 1984). The pollen grains of *P. rzedowskii* are similar to those of subsect. *Cembroides*, whereas those of *P. canariensis* and *P. roxburghii* are different.

A look on the map showing the distribution of species with the combination of features mentioned above (cf., Fig. 12 and p. 146 f.) suggests a first phase of evolution along the northern shore of the Proto-Atlantic, approximately during Upper Jurassic to Lower Cretaceous time. After opening the Mediterranean from Spain to the Himalayan region, the first sectt. *Parrya, Sula,* and *Pinea* pines evidently have entered the Tethys area.

Pinus canariensis SMITH (Figs. 4 – 8, 12). Variation of extant populations. It is of apparent evolutionary significance, that *P. canariensis* exhibits an extremely high morphological variability among pines. On the island of Teneriffa this was studied in the various climatic races, which occur there under different conditions of altitude, temperature and humidity.

With respect to cone morphology, a race with excentromucronate denticulate umbo (Fig. 7 a) occurs mainly on the dry south slopes, at an altitude of approx. 1 600 m. Flat, smooth apophyses with deeply depressed and polygonal umbo can be found only at low altitudes, e.g., at approx. 700 m above Los Cristianos. These forms exactly fit the majority of fossil cones of the Upper Miocene of the Vienna basin (Fig. 8 b, c) with their clearly excentric mucro. Similar conditions I have observed on the basal scales of *P. roxburghii* (Fig. 5 a, b) and on the haploxyl Mexican *P. pinceana* (in populations from the Carneros Pass in Coahuila, but not in those from S. L. Potosi, Queretaro and Hidalgo). At higher altitudes on the Canary Islands (mostly above 2 000 m), the cones of *P. canariensis* become smaller and occasionally exhibit a vallum surrounding a *P. pinea*-like umbo (Fig. 7 d). Similar specimens have been found in the forest of Esperanza (about 1 700 m), a cloudy and wet habitat. These features indicate that *P. canariensis* and *P. pinea* might have genetical connections.

In the forest of Esperanza and also above Orotava, one can frequently encounter cones with extremely protuberant, bent downward and uncinoid umbos on basal scales (Fig. 6 a, b). These umbos are excentromucronate to tectoid and resemble *P. roxburghii* where comparable features are developed much more prominently

---

Fig. 4. Comparison of cones and apophyses from diploxyl *Pinus canariensis* and haploxyl *P. rzedowskii*. *a P. canariensis*, cone (Esperanza forest, Teneriffa), nat. size, *b* apophysis with typical excentromucronate umbo, × 3. *c P. rzedowskii*, cone (Cerro de Chiqueritos, Mpo. de Coalcoman, Michoacan, Mexico); *d* detail of apophysis with umbo and excentric mucro in the upper umbo field, almost identical with *P. canariensis*, × 3.

Fig. 5. Comparison of cones and apophyses of diploxyl *Pinus roxburghii* and *P. pinaster* with haploxyl *P. nelsonii*. *a P. roxburghii* cone, with typical uncinoid apophysis elongation and terminal excentromucronate umbo (in that respect resembling *P. canariensis*: Fig. 10 *a, b*), nat. size; *b* detail of umbo to show the rudimentary excentric mucro (arrow), × 3. *c Pinus pinaster* cone (Italy, Livorno) strongly asymmetric, with elongated, centromucronate apophyses, and occasionally duplomucronate (arrow), nat. size. *d P. nelsonii*, magnified conelet (vicinity of Realejo, Mexico), × 2; among subsect. *Cembroides* the only species with centromucronate apophyses, occasionally duplomucronate and with transitions towards excentromucronate (arrows)

Fig. 4

Fig. 5

(Fig. 5a, b). *P. roxburghii* apparently has evolved under the similar and very humid conditions of the Monsoon belt.

Another variety of *P. canariensis* (Teneriffa, Granadilla) is characterized by apophyses which are excentromucronate, sculptured and shaped like *P. tropicalis* (Fig. 7i) with cones pretty long and narrow, almost indistinguishable from Cuba populations (Fig. 7h).

The most surprising forms of *P. canariensis* have cones resembling *P. pinaster* (Fig. 6a, c). This pine, quite differently classified by CRITCHFIELD & LITTLE (1966: sect. *Sylvestres*) and VAN DER BURGH (1973: sect. *Pinaster* subsect. *Australes*) is characterized by centromucronate, protuberant apophyses (Fig. 10a, b). But that feature is only valid for Italian populations (e.g., Pistoia, Livorno); in France, Spain and especially in the Pyrenees, the mucro can be reduced to a central tectum without prickle, sometimes with a little vallum (Fig. 10c) as in *P. pinea* and *P. canariensis* (Fig. 6a). Thus, it looks as if variants of *P. canariensis* also transgress the limits of sect. *Pinaster*.

Fossils. Three-needled bundles of leaves frequently described as fossil *P. canariensis* as well as other fossils cited in the literature are dubious. A few pictures of SAPORTA (1862 – 1874) from southern France (Oligocene and Miocene) and other fossil pines described by LINDLEY & HUTTON (1931 – 1937) from S. Spain (Murcia Pliocene) are reminiscent of *P. canariensis*, but their umbo morphology has not been studied. These features have been neglected in paleobotanical reports in spite of the fact that ENDLICHER already 1847 recognized the importance of the cone scale morphology for taxonomic purposes. Problematic photographs and pictures from Neogene fossils are an insufficient basis for distribution maps or conclusions about evolutionary trends (cf. KRÄUSEL 1919, DEPAPE 1922, STUDT 1926, SUNDING 1970).

Many specimens of determinable fossil cones of *P. canariensis* have come from the Upper Miocene of the Vienna basin (Pannon E); one from the Miocene of Burgenland and one from the Middle Miocene (Badenien) of Styria (cf. list of fossils in KLAUS 1984b: 44f.). The majority of the Upper Miocene cones morphologically correspond to the more xeromorphic variants of the living species (see Figs. 7 and 8), whereas the Middle Miocene cones have more protuberant apophyses and are apparently closer to more mesomorphic forms. All fossils are from sediments of the Paratethys shore and are thus relatively young relative to the estimated age of *P. canariensis*.

Conclusions. The different variants of *P. canariensis* contain features of almost all Mediterranean shore pines, even if classified in entirely different groups (e.g., subsectt. *Pineae, Halepenses, Sylvestres,* and *Australes*). Furthermore, there are the close relationships to *Haploxylon* pines of C. America (sect. *Parrya* subsect. *Rzedowskiae*) as well as to *Diploxylon* pines of the Caribbean (e.g., subsectt. *Oocarpae* and *Australes*). All this suggests that *P. canariensis* is an old (Lower Cretaceous) relic from an ancient Mediterranean evolutionary centre. Evidently, its roots go back towards the *Haploxylon* sect. *Parrya* (with its subsectt. *Cembroides, Rzedowskiae* and *Nelsoniae*), and its further differentiation has led to the excentromucronate subsectt. *Pineae* and *Halepenses* as well as the centromucronate subsect. *Australes*. The first phase of evolution of these Proto-Mediterranean pines might have occurred during the Upper Jurassic/Lower Cretaceous at the northern shores of a small Protoatlantic (Fig. 13). After the Upper Cretaceous opening of the

Mediterranean, in a second phase, the sect. *Sula* ancestors of *P. canariensis* with adnate seed wings evidently have followed the Tethys coast to the east, branching off under different climatic conditions forerunners of subsect. *Halepenses* and sect. *Pinaster* with articulate seed wings, and of subsect. *Pineae* with nut-like seeds. Subsect. *Canarienses* eventually reached the Himalaya region in the Upper Cretaceous or Lower Tertiary. The sharp southward bend of the eastern Tethys coast crossing the equator apparently formed a barrier for further extension to the south and east, and the loss of a shore finally led to the origin of *P. roxburghii* as a Himalayan relic pine. After the separation of the Atlantic coast lines the Mediterranean and closely related Caribbean and C. American pines developed independently, retaining some common features as a sign of common descent.

Pinus pinea **L.** (Figs. 7 c, 8). Among Mediterranean pines, *P. pinea* is considered by many authors to be an enigmatic and isolated species. In the system of LITTLE & CRITCHFIELD (1969) it is regarded as the only member of subsect. *Pineae*. GAUSSEN (1960) combines *P. pinea* with subsect. *Cembroides* into his sect. *Parryanoides*. But there are several neglected features which throw some light on its relationship and history. *P. pinea* develops recurved bracts (cf. Fig. 7 c), as *P. canariensis, P. pinaster, P. halepensis* and *P. elliotii* from N. America. Seedlings develop for many years bluish pruinose primary leaves (as members of subsectt. *Halepenses, Canarienses, Cembroides, Rzedowskiae*, etc.). A distinct feature is the loss of basal cone scales, also occurring in *P. canariensis*. The vallum around the apophysis umbo appears to be unique; but it also occurs in some races of *P. canariensis* (see Fig. 7 c, d) and among the Mexican pines *P. lumholtzii* and *P. oocarpa* which are also excentromucronate.

The Asiatic *P. gerardiana* is often quoted as an ancestor for *P. pinea*. But *P. gerardiana* develops centromucronate apophyses (see Fig. 7 e, f, g) like *P. bungeana* and exhibits no loss of basal cone scales. Its pollen grains are similar to those of the sect. *Strobus* concerning size and morphology, but differ from those of subsectt. *Cembroides, Rzedowskiae* and especially *P. pinea*. Chemical data also militate against a relationship with Asiatic pines. MIROV (1951) found maderene (beta-caryophyllene), a bicyclic sesquiterpene in *P. bungeana* as well as in *P. pinceana*, a Mexican nut-pine of subsect. *Cembroides*. The presence of cembrene in *P. pinea*

Fig. 6. Cone variation of *Pinus canariensis* in the direction of *P. roxburghii* and *P. pinaster*. *a P. canariensis*, cone (from humid locality above Orotava-Valley, Teneriffa), apophyses with elongated umbo, × 1.5; *b* same cone, basal apophyses: elongated and excentromucronate umbo (arrow) as in *P. roxburghii* (Fig. 5 b), × 2. *c P. pinaster*, cone (France, Montagne Noir) with tectoid, almost centromucronate umbo as in *P. canariensis* (arrows on *a* and *c*), but also with excentromucronate and duplomucronate umbos (arrows), × 2

Fig. 7. Variation of apophysis and mucro in *Pinus canariensis* compared with *P. pinea, P. tropicalis* and *P. gerardiana*. *a P. canariensis* (Teneriffa, lower altitude of dry south side), apophysis with deeply depressed umbo and excentric mucro, × 3; and *b* (Teneriffa, Vilaflor) umbo with polygonal delimitation and excentric mucro (cf. similar fossil cone, Fig. 8 *b, c*), × 3, *c P. pinea*, apophysis with vallum encircling the excentromucronate umbo, × 3; similar to *d P. canariensis* (Teneriffa, 1 800 – 2 000 m), × 3. *e – g P. gerardiana*, apophysis entirely different, umbo clearly centromucronate without vallum × 1.5; basal uncinoid cone scales, × 1.5, and open cone, $^2/_3$ nat. size (Kew Herb.). *h P. tropicalis* (Cuba), apophysis with polygonal, excentromucronate umbo, × 3; similar to *i P. canariensis* (Teneriffa, Granadilla), × 3

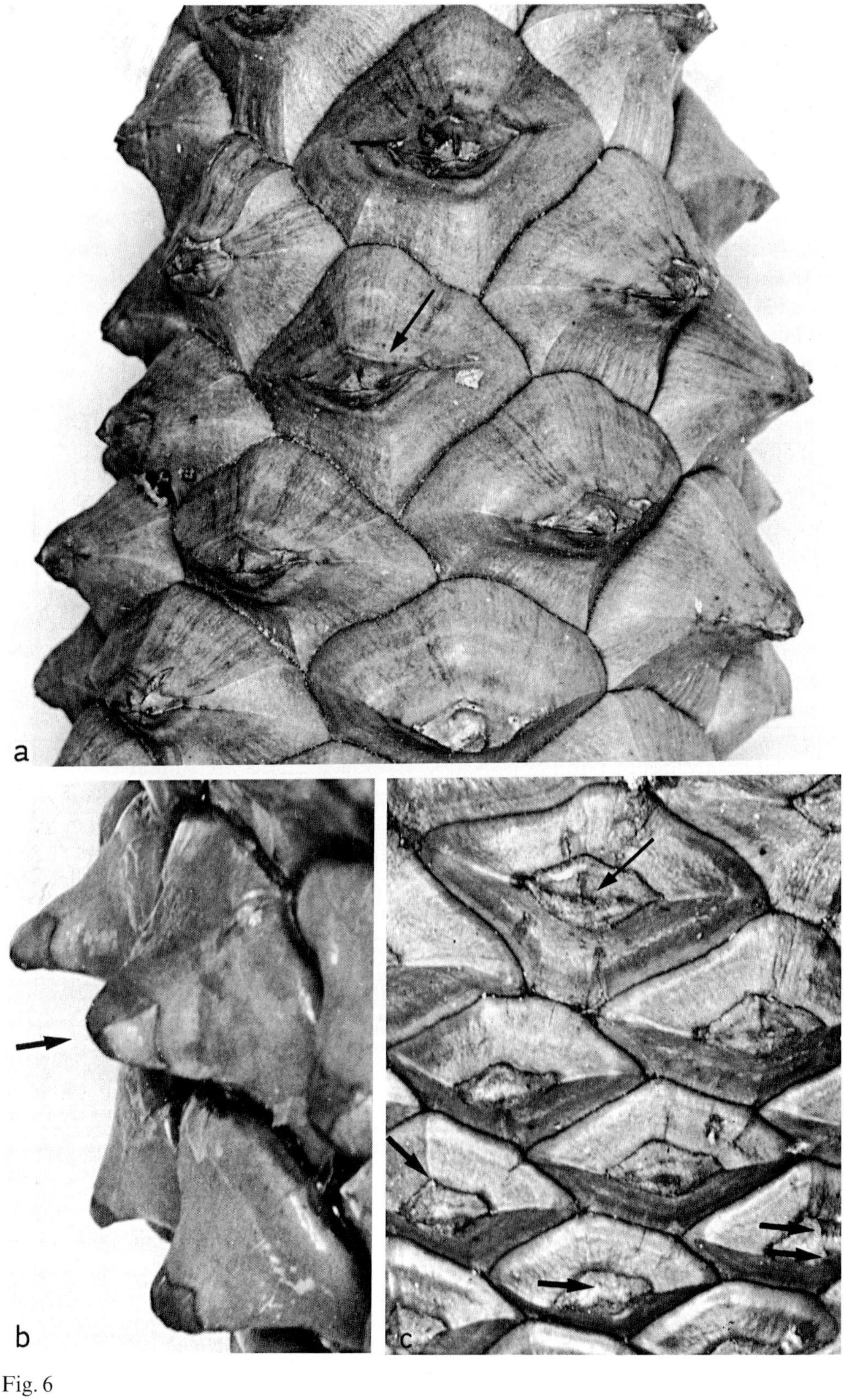

Fig. 6

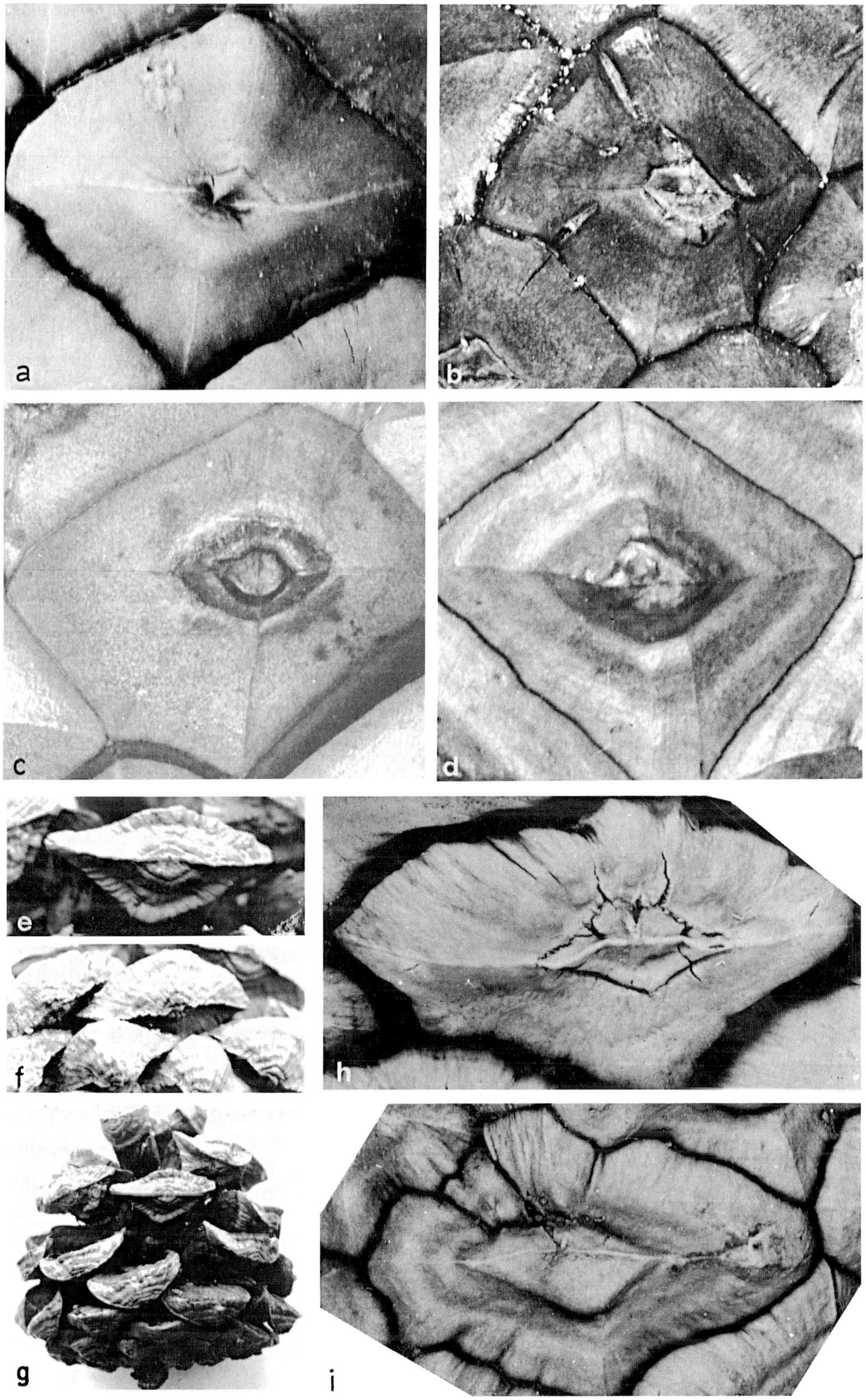

Fig. 7

is no argument for its Asiatic origin, as this compound is widespread among *Haploxylon* pines, also American ones (e.g., in *P. albicaulis*).

Sometimes *P. pinea* is compared with *P. torreyana* from California (MIROV 1967: 560). Indeed, both are similar in cones size and shape, and in nut-like seeds, being the only *Diploxylon* species with these features. But their umbo morphology is entirely different, *P. torreyana* is centromucronate, *P. pinea* excentromucronate and vallate.

According to the present analysis there are so many similarities of *P. pinea* with *P. canariensis* and subsect. *Halepenses*, but also with *P. pinaster* (sesquiterpene and caryophyllene) (see Fig. 7, and the following pages), that it appears definitely related with these Mediterranean pines. Common ancestry is obvious with sect. *Parrya* subsect. *Cembroides* in the western Mediterranean area, where these pines might have been located during the Cretaceous, surviving now in C. America. Already RIKLI (1943) has proposed the origin of *P. pinea* in the western Mediterranean (somewhere on the Iberian Peninsula), where it is found in larger grooves and higher altitudes than anywhere else. The small paleobotanical record does not contradict such an interpretation. MENENDEZ-AMOR (1951) found a pine cone in Pliocene deposits near Malaga and called it "*P. pseudipinea*". Fossil pollen has been found frequently in Austrian Miocene deposits (KLAUS 1984 a).

Pinus tropicalis MORELET (Figs. 7 h, 12). Another important relic pine, related in its cone morphology with *P. canariensis* and *P. oocarpa*, is *P. tropicalis* from the Isle of Pine in W. Cuba. Its classification among the Eurasiatic subsect. *Sylvestres* appears not convincing and already SHAW (1914) mentiones the unique leaf anatomy, which does not occur in any other species of that subsection. The cone analysis reveals similarities with some variants of *P. canariensis* (see Fig. 7 h, i), particularly *P. oocarpa* from Guatemala (Fig. 11 g), but also with *P. cubensis* and *P. caribaea*. Its basal scales are excentromucronate (as in *P. cubensis*), the apical scales centro- or duplomucronate (as in *P. caribaea*). Entirely different from all subsect. *Sylvestres* species is the "grass stage" of the seedlings (Fig. 4), which otherwise appears in *P. palustris* and some Caribbean and Mexican pines. Thus, provisionally, *P. tropicalis* is excluded from subsect. *Sylvestres* and united with *Oocarpae*.

***Pinus halepensis* MILL. and *P. brutia* TEN.** (Figs. 9 f, 10 e, g, and 11 c – h). VAN DER BURGH (1973) has presented good reasons to remove these and related species and subspecies as subsect. *Halepenses* from subsect. *Sylvestres*. Cone morphology supports such a procedure. All *Halepenses* have many features in common with the other Mediterranean coast and island pines: several years primary leaves on seedlings, recurved bracts on spring shoots (Fig. 1), excentromucronate cones, and rectangular delimitations of the umbo fields (Fig. 2). But they also differ in some respect from *P. pinea* and *P. canariensis*. The cones do not loose their basal scales but often remain closed and persist for more than one year on the tree, resembling in some respect the closed cone pines of N. America. They are smaller, mainly have flat or even depressed apophyses and a perexcentromucronate umbo. The

---

Fig. 8. Fossil cone of *Pinus canariensis* subsp. *prisca* from Upper Miocene of Vienna Basin, Austria (Pannon E, Guntramsdorf). *a* Holotype, nat. size; note loss of basal cone scales; *b* – *c* radially cracked, flat apophyses with deeply depressed umbos of polygonal delimitation and excentric mucros in the upper part (arrow), × 3

a
b
c

more eastern species exhibit flat to deeply depressed umbo grooves (as in conelets of *P. roxburghii*). The mucro is much reduced in size and in some cases hardly discernible. *P. halepensis* differs from the other species (*P. brutia, P. pithyusa, P. eldarica*) in its curved peduncles and hanging ripe cones. That species also exhibits some geographical differentiation. In Spain, the cones show more protruding umbos with clearly developed mucros above the horizontal keel. Sometimes even a fork-like division of the mucro (i.e., a duplomucronate condition) is found. Otherwise, in the Mediterranean only *P. brutia* and *P. pinaster* rarely develop indications of that feature (Fig. 9 b). In Greece *P. halepensis* produces more flat umbos, and in *P. brutia* the umbo even may become depressed. When looking for *Pinus* spp. with cones similar to *Halepenses*, the best comparison is with *P. oocarpa* from Guatemala. Its cones remain closed for long time, never loose their basal scales, are of medium size, excentromucronate (and sometimes duplomucronate). *Halepenses* thus are forming a link between subsectt. *Canarienses, Pineae* and *Oocarpae*.

Fossils of *P. halepensis* are met frequently in the Paratetehys area, mainly from the Miocene. Since the Upper Oligocene time the species apparently has changed only slightly. *Pinus salinarum* Partsch (Wielicka in Poland) shows the typical polygonally delimitated umbo areas. The Miocene *P. halepensis* findings from Austria (Badenien: Styria, Pannonien: Vienna basin) are so well preserved, that they can be compared with extant geographic races, e.g., the one from Spain (Badenien) and from Greece (Pannonien). *P. brutia* fossils are also known from the Vienna basin (Klaus 1984 a), Yugoslavia (*P. saturni* Unger), Romania and the coast of the Black Sea. The Miocene distribution thus followed the Tethys and Paratethys shore line, located in C. Europe at that time. A *P. oocarpa*-like fossil cone from the older Miocene of Austria proves that species similar to extant C. American taxa still occurred in the European Mediterranean area at that time.

***P. pinaster* Ait.** (Figs. 5, 6, 9, and 10). In recent classification systems *P. pinaster* is placed in a central position of sect. *Pinaster* subsect. *Australes* among otherwise exclusively American pines. Its distribution in the Mediterranean area somewhat differs from other species, as it is restricted to the west, ranging from Morocco, Portugal and Spain, to S. France and Italy (other occurrences are probably due to human plantation). Besides anatomical features (van der Burgh 1973), particularly cone morphology points to a separation from the other true Mediterranean pines: it is the only species with a centromucronate umbo, a feature it shares with all the American subsect. *Australes* species.

All this might suggest that *P. pinaster* reached the W. Mediterranean as a single species and independent from the other true Mediterranean pines, now surviving there as a relic. The question is, whether *P. pinaster* really is so distinct from the other Mediterranean and European pines. Little & Critchfield (1969) list it among subsect. *Sylvestres* on karyological reasons, as it has two pairs of heterobrachial chromosomes (rather than a single), as all other species of this subsection (Saylor 1961, 1964). Whether this is a significant feature will have to be a matter of further consideration, as *P. halepensis, P. brutia, P. tropicalis* and *P. resinosa* are also included in subsect. *Sylvestres* by these authors.

But there are also many features which link *P. pinaster* with the other true Mediterranean pines. The recurved bracts on spring shoots (Fig. 1) occur in all of them (incl. *P. canariensis*), independent of their different classification. The typical *P. pinaster* of Livorno has cones which do not loose their basal scales when ripe, but subsp. *mesogeensis* corresponds with the other Mediterranean pines in this respect. The centromucronate umbo is clearly developed in all Italian specimens of *P. pinaster* (Livorno, Pistoia: Fig. 10 a, b). But in S. France, the Pyrenees and Spain (*P. pinaster* subsp. *mesogeensis*: Fig. 10 c), the reduced mucro becomes ± excentric and reminiscent of *P. canariensis* (Fig. 6 a), and occasional duplomucronate umbos (Fig. 9) even point in the direction of *P. caribaea, P. oocarpa* and *P. halepensis*. Furthermore, the polygonal umbo delimitation is also recognizable in *P. canariensis, P. halepensis* of Greece and the fossil *P. salinarum* cone from Poland. Finally, the unique and large male flowers of *P. pinaster* are only comparable with *P. canariensis* and *P. roxburghii*.

In conclusion one can say that there are so many links between *P. pinaster* and the other Mediterranean pines (subsectt. *Canarienses, Halepenses, Pineae*) that it appears hardly possible to separate them entirely. Common ancestors are suggested, presumeably among the haploxyl subsectt. *Cembroides* and *Nelsoniae*, or later among the diploxyl subsect. *Canarienses*. *P. pinaster* may have evolved during the divergence of excentromucronate (subsectt. *Canarienses, Halepenses, Oocarpae, Pinae*) and centromucronate (subsect. *Australes*) *Diploxylon* pines. The first indication of such a divergence already occurs within the *Haploxylon* subsect. *Cembroides*, with *P. cembroides* and relatives, excentric and tending towards subsectt. *Rzedowskiae* and *Canarienses*, but with *P. nelsonii* (subsect. *Nelsoniae*) becoming centromucronate. *P. nelsonii* also differs from all other sect. *Parrya* pines of the world by its conelets sometimes with duplomucronate apophyses (Fig. 5 d) and the loss of the basal cone scales (shared with *P. rzedowskii*). Thus, an Upper Mesozoic evolution on the northern shore of the Proto-Atlantic could be envisaged for *P. pinaster*.

The fossil record of *P. pinaster* is poor and does not help much in the problems discussed. Teixeira (1944) reports a "*P. praepinaster*" from the Pliocene of Rio Maior. Among many Upper Tertiary deposits of Austria rich in *Pinaceae* pollen, I have never found any grain reminiscent of *P. pinaster*. This is strange, as pollen of almost all other Mediterranean pines occur frequently in these deposits. On the

Fig. 9. Duplomucronate apophyses in Mediterranean and Caribbean pines, indicating relationships. *a Pinus pinaster* (Italy, Livorno), arrows indicate duplomucronate umbos, most are centromucronate, × 4; *b P. pinaster* subsp. *mesogeensis* (French Pyrenees), apophyses commonly with additional small and extremely excentric mucro besides the centric mucro on the keel. *c P. caribaea* subsp. *hondurensis* (Guatemala, dept. Izabal), side view of most typical duplomucronate umbo with centric and excentric mucro (arrows), × 4. *d–e P. caribaea* subsp. *bahamensis* (Grand Bahama Island, Freeport), side and front view of duplomucronate apophyses with strongly reduced mucros (arrows), × 3. *f P. halepensis* (Spain, Valencia), occasionally with duplomucroante umbos (arrows)

Fig. 10. Variation of mucro development among related Mediterranean pines. *a–b P. pinaster* (Italy), typical mucro a centric, short prickle on the keel bent upwards, × 2. *c P. pinaster* subsp. *mesogeensis* (French Pyrenees), mucro strongly reduced to a tectum or not reaching the horizontal keel, with a tendency towards excentricity (arrow), resembling *d P. canariensis*, especially its conelet, × 2. *e P. halepensis* (Spain), umbo similar to *P. pinaster* in some variants, × 3. *f P. pinaster* (France), variant with the centric mucro entirely absent, the remaining tectum getting excentric and thus resembling some apophyses of *g P. brutia* (Crete) with excentric mucros (arrows)

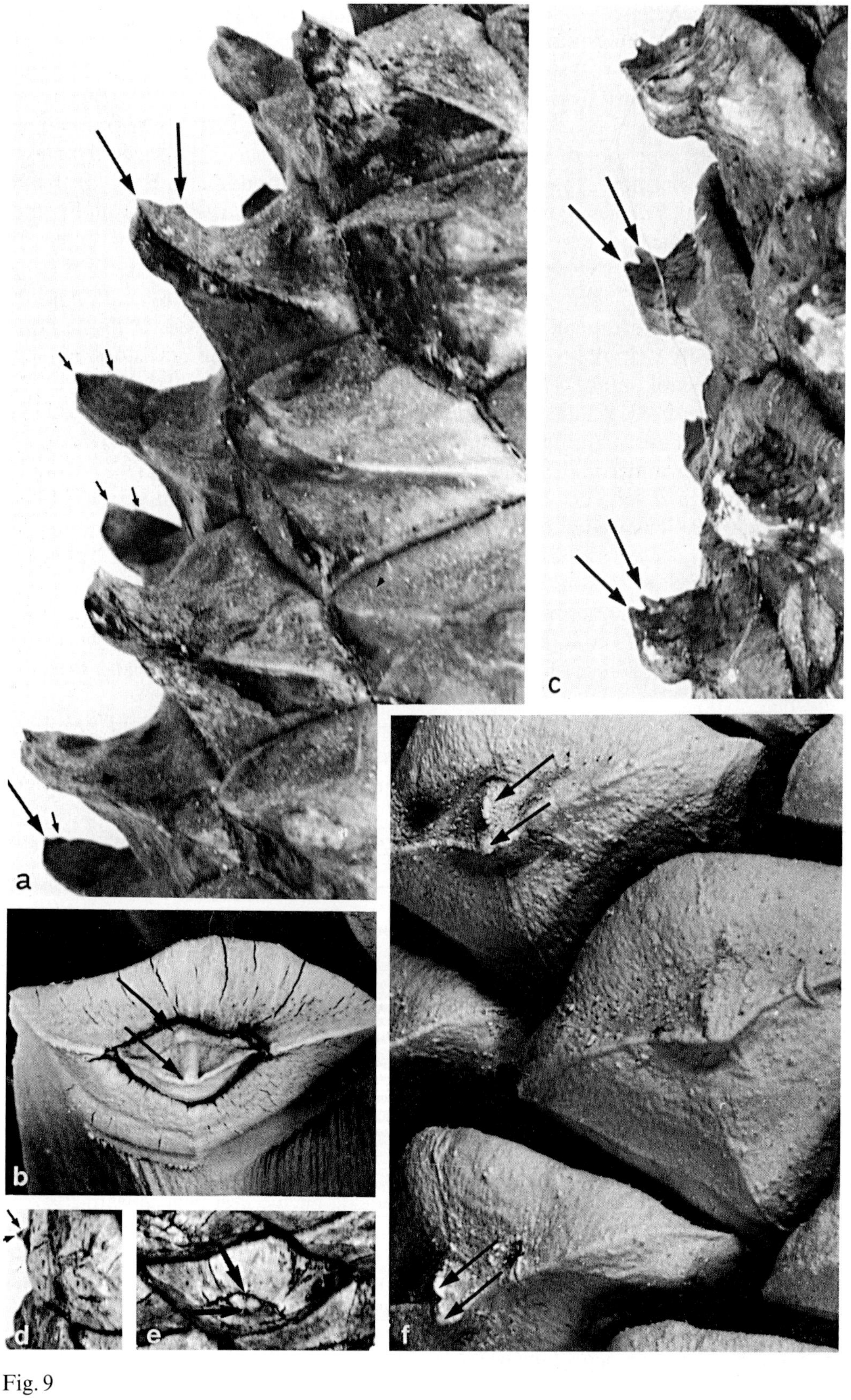

Fig. 9

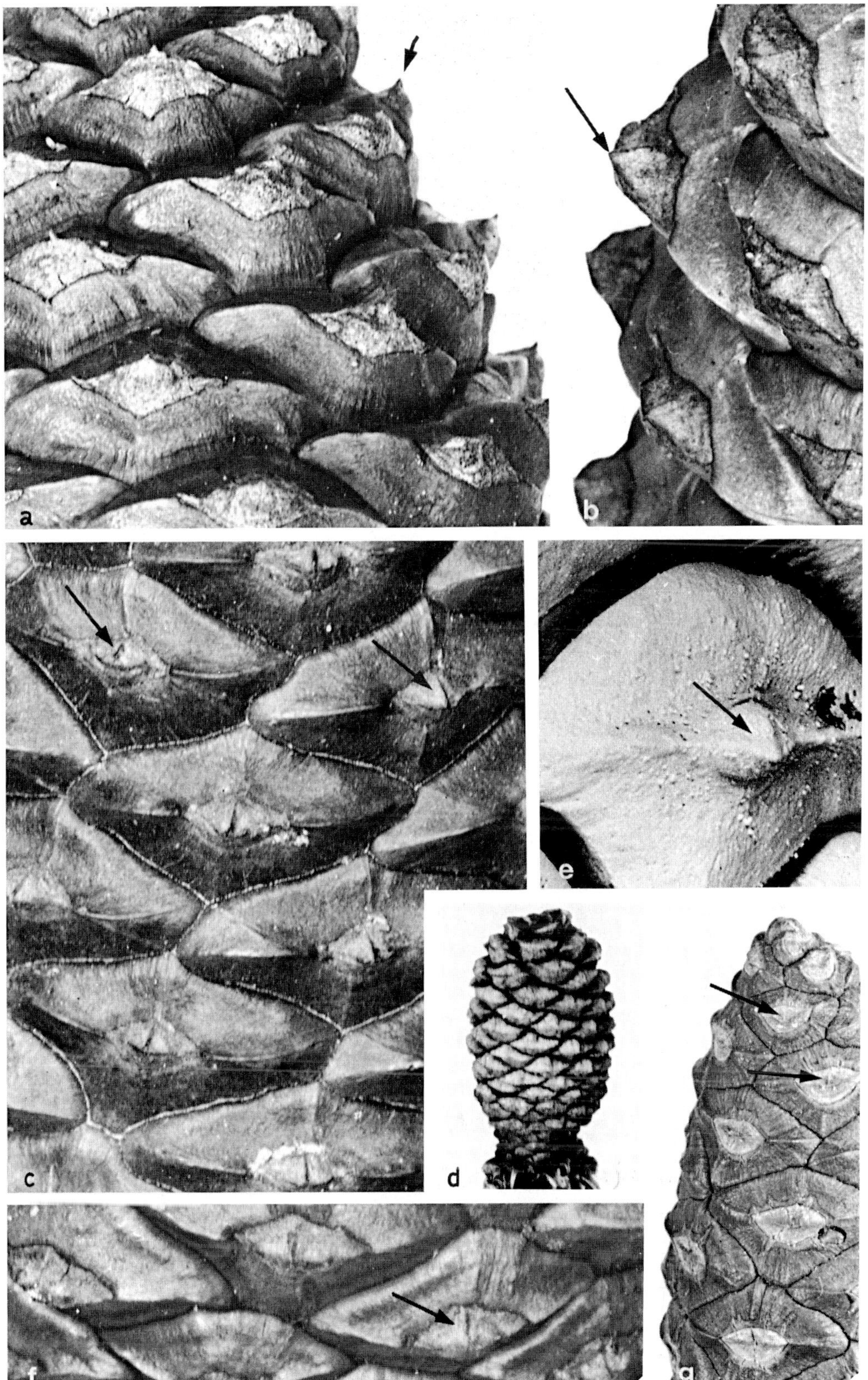

Fig. 10

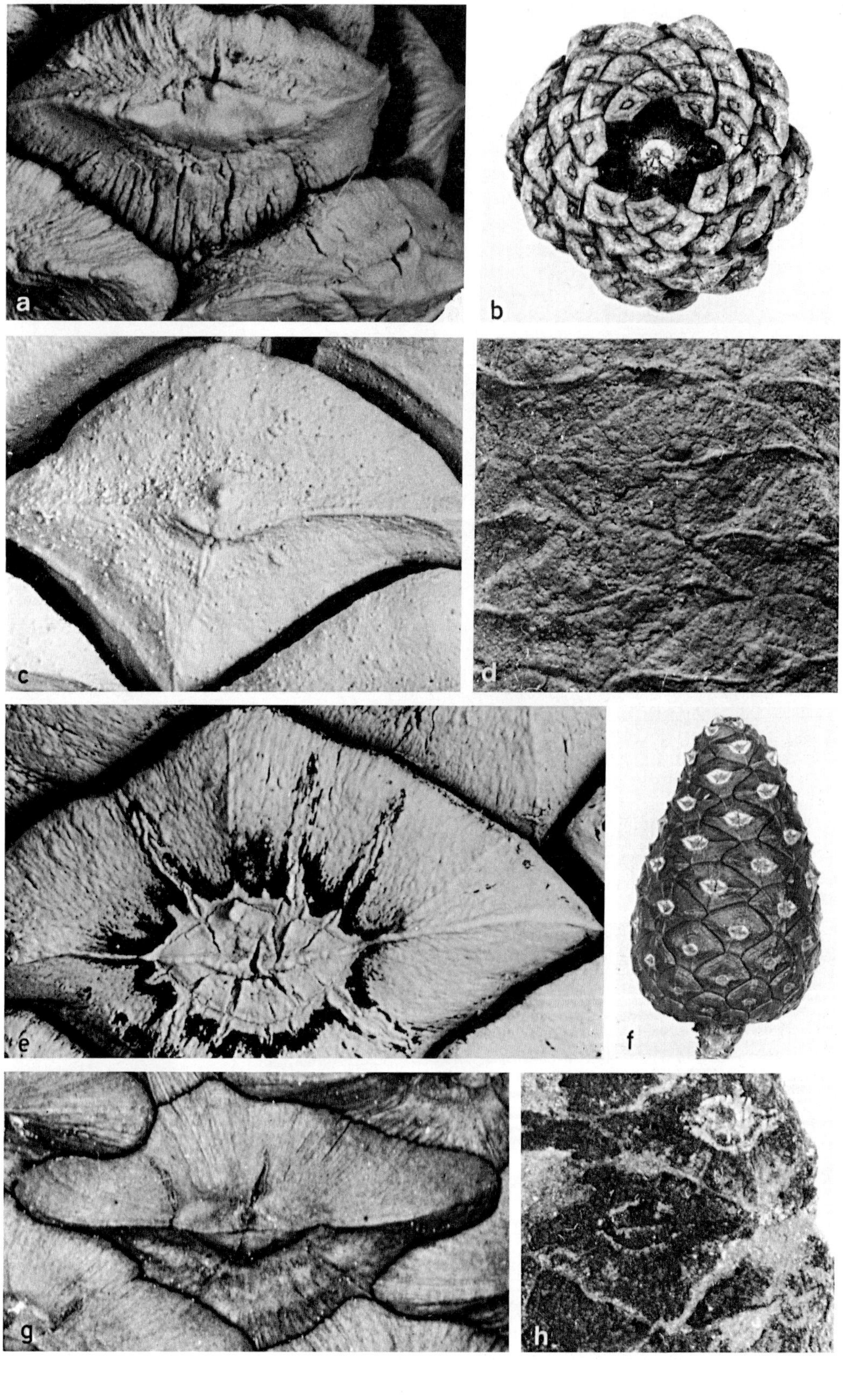
a
b
c
d
e
f
g
h

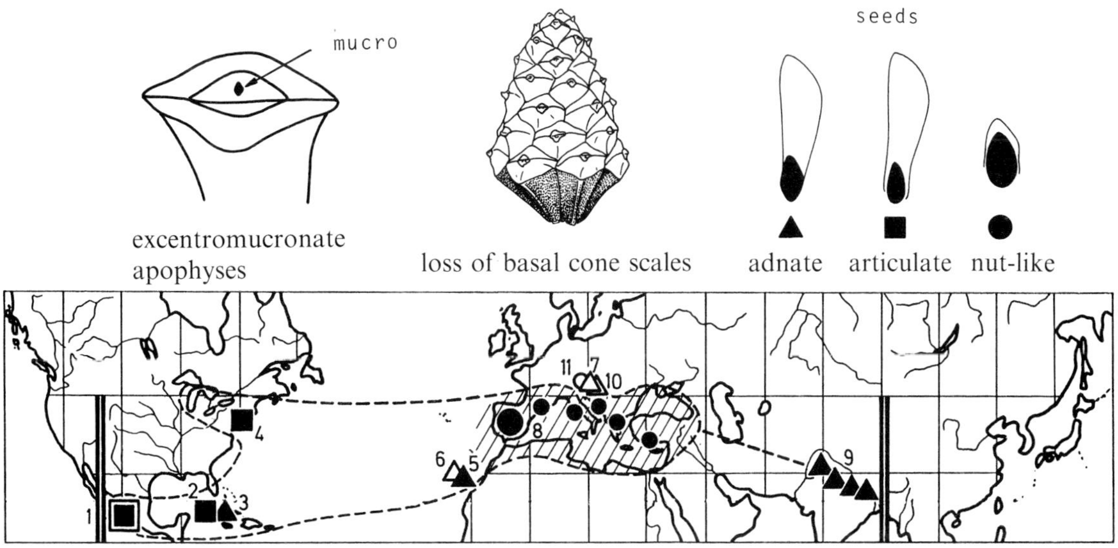

Fig. 12. Distribution of Mediterranean and related *Pinus* species, showing a combination of the following characters: cones loosing basal scales, excentromucronate apophyses, and three different types of seed wing attachments; //// area of Mediterranean shore pines, – – – – area of related species. 1 *P. rzedowskii* (Mexico), 2 *P. tropicalis* (Cuba), 3 *P. caribaea* (Cuba), 4 *P. resinosa* (USA), 5 *P. canariensis*, 6 *P. canariensis* (fossil), 7 *P. canariensis* subsp. *prisca* (fossil), 8 *P. pinea*, 9 *P. roxburghii*, 10 *P. roxburghii* (fossil), 11 *P. pinea* (fossil pollen)

other hand, it is quite certain that other, perhaps more advanced centromucronate pines of the American subsect. *Australes* did occur in Europe and Asia, e.g. *P. spinosa* Herbst (Mai 1965, Klaus 1979, Unger 1847, Kilpper 1968), corresponding in all details with the extant *P. taeda* of the Florida/Georgia swamp district.

Pinus resinosa Ait. (Figs. 11 a, b, and 12). This pine of northeastern N. America usually is placed in subsect. *Sylvestres*, but it exhibits the characteristic syndrome of loss of basal cone scales, excentromucronate apophyses, and articulate seed wings (Fig. 12). The first character is significant for several Mediterranean and American pines, but lacks in Eurasiatic members of subsect. *Sylvestes*. The strongly reduced umbo and mucro resemble *P. brutia*. Also, the bluish purple colour and the shape of male flowers in *P. resinosa* are entirely different and reminiscent of Caribbean and Mexican pines. The reason for all these similarities may perhaps

Fig. 11. Excentromucronate apophyses in Mediterranean and related pines. *a – b P. resinosa*, extremely reduced excentric mucro, reminiscent of *P. brutia* (cf., *e, f, h*) and cone basis showing loss of basal scales, a common feature of *P. canariensis*, *P. pinea* and many Caribbean and American pines. *c P. halepensis* (extant), excentric mucro, × 3; *d P. halepensis* (fossil from Middle Miocene of Styria, Austria), similar shape of apophysis and excentric mucro. *e P. brutia*, typical flat umbo with perexcentric mucro and radially arranged fissures (cf., *P. canariensis*). *f P. brutia*, cone shape reminiscent of some closed cone pines of America, e.g., *g P. oocarpa* (Guatemala), and almost identical with *P. tropicalis* (Fig. 7 *h*) and some *P. canariensis* variants (Fig. 7 *i*); *h P. brutia* (fossil, Upper Miocene of Vienna basin, Austria), comparable apophyses, × 3

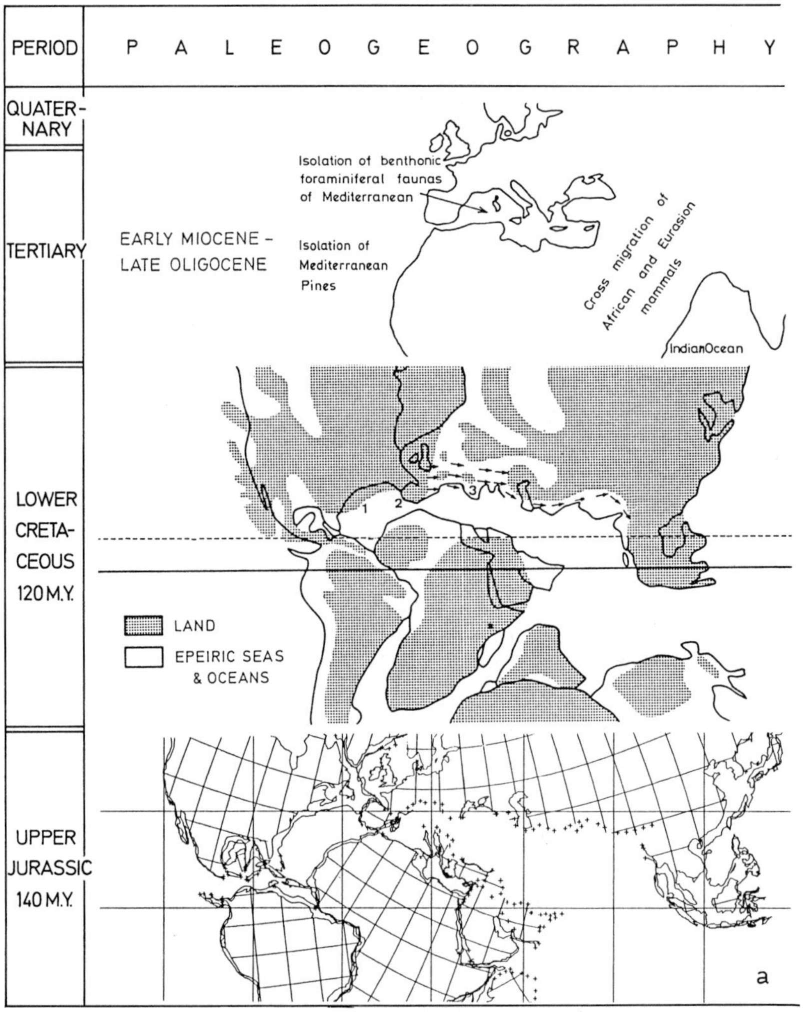

Fig. 13. *a* Paleogeographical conditions during the Upper Jurassic (I), Lower Cretaceous (II), and Middle Tertiary (III) as related to *b* the relationships and evolution of Mediterranean shore pines. Three phases of differentiation can be distinguished. (I) Upper Mesozoic differentiation of haploxyl sect. *Parrya* along N. Proto-Atlantic shore into subsectt. *Cembroides, Nelsoniae*, and *Rzedowskiae*. (II) Differentiation of diploxyl sect. *Sula* (*3*) eastward of Spain along Mediterranean islands until southern limit of pine distribution (– – – –), and distribution of sectt. *Pinea* (*2*) and *Pinaster* (*1*) along Atlantic shore. (III) After closure of Mediterranean basin, evolution of American, Caribbean and Mediterranean pines continued separately

be common ancestry with subsectt. *Canarienses, Pineae,* and *Oocarpae,* whereas the strongly reduced cone and mucro size could be an adaptation to a cold climate. The occurrence of *P. resinosa* far remote from the present Mediterranean coast is not so surprising, considering the common coast line of N. America and the Mediterranean during the Lower Cretaceous.

Even if the relationships of *P. resinosa* need more study, it appears advisable to separate it taxonomically more clearly from sect. *Sylvestres*. Therefore, it is suggested to introduce a new subsection:

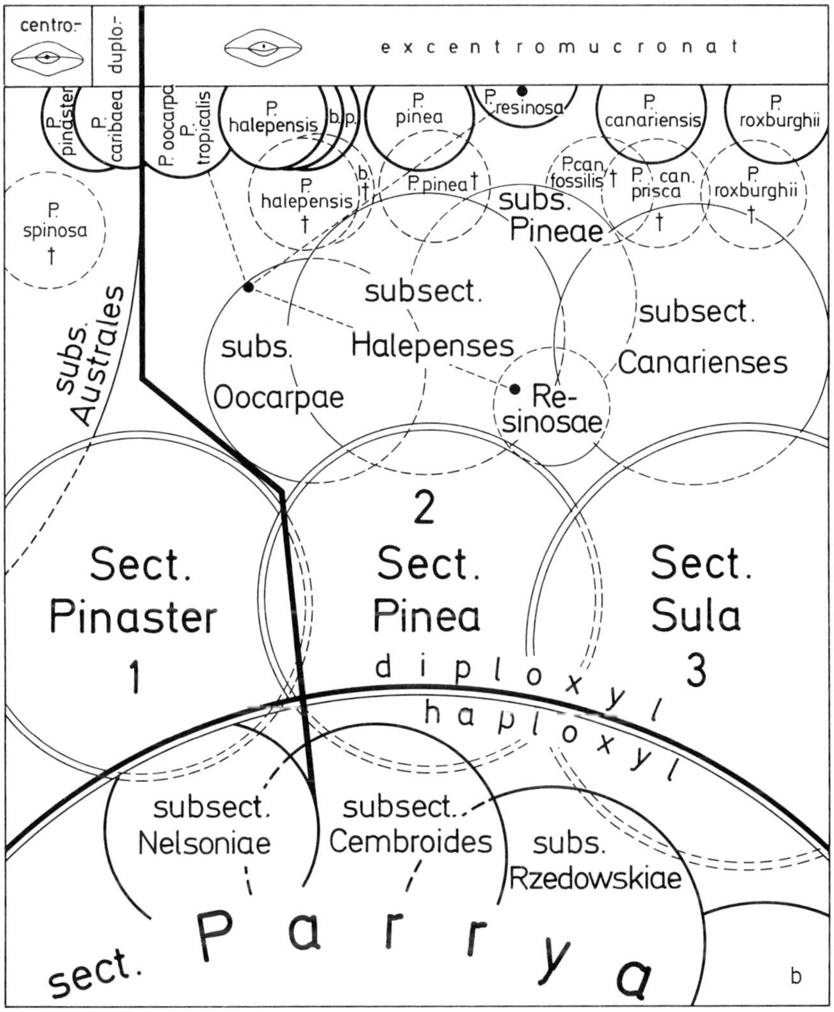

Fig. 13 (continued)

Sect. *Pinus* subsect. *Resinosae* W. Klaus, subsect. nova

Differt a subsect. *Sylvestri* strobilis squamis basalibus non persistentibus, umbonibus perexcentromucronatis (parvimucronatis), strobilorum masculorum colore glaucopurpurato.

Species typica (et unica): *P. resinosa* Ait.

Classification of *Pinus* species mentioned in this report

As an appendix of this comparative study a slightly modified systematic arrangement of the taxa and fossils considered is presented, using the systems of Shaw (1914), Little & Critchfield (1969), Hudson (1969), and van der Burgh (1973) as a basis. For *P. rzedowskii* a new subsection among sect. *Parrya* has been created, and *P. resinosa* also has been separated from subsect. *Sylvestres* in its own subsect. *Resinosae*. *P. tropicalis* has been transferred from subsect. *Sylvestres* to subsect. *Oocarpae*, and the latter has been included among sect. *Pinea*.

Subg. *Strobus* (*Haploxylon*)
- sect. *Strobus*
  - subsect. *Cembrae*
    - *P. cembra*
    - *P. sibirica*
  - subsect. *Strobi*
    - *P. peuce*
- sect. *Parrya*
  - subsect. *Cembroides*
    - *P. cembroides*
    - *P. pinceana*
    - *P. maximartinezi*
  - subsect. *Rzedowskiae*
    - *P. rzedowskii*
  - subsect. *Nelsoniae*
    - *P. nelsonii*

Subg. *Pinus* (*Diploxylon*)
- sect. *Leiophyllae*
  - subsect. *Lumholtziae*
    - *P. lumholtzii*
- sect. *Sula*
  - subsect. *Canarienses*
    - *P. canariensis*
    - *P. roxburghii*
- sect. *Pinea*
  - subsect. *Pineae*
    - *P. pinea*
  - subsect. *Oocarpae*
    - *P. oocarpa* (var. div.)
    - *P. tropicalis*
    - *P. ooconica* (only fossil)
  - subsect. *Halepenses*
    - *P. halepensis*
    - *P. brutia* (subspp. div.)
- sect. *Pinaster*
  - subsect. *Australes*
    - *P. pinaster* (subspp. div.)
    - *P. cubensis*
    - *P. caribaea* (incl. subsp. *hondurensis* et subsp. *bahamensis*)
    - *P. elliottii* (incl. var. *densa*)
    - *P. remorata*
- sect. *Pinus*
  - subsect. *Resinosae*
    - *P. resinosa*
  - subsect. *Sylvestres*
    - *P. sylvestris*
    - *P. uncinata*
    - *P. mugo*
    - *P. nigra*
    - *P. heldreichi* (incl. subsp. *leucodermis*)

References*

AROBBA, D., 1979: Determinazione di "*Pinus halepensis*" MILLER e "*Pinus pinaster*" AITON sulla base di differenze palinologiche. – Archivo Bot. Biogeog. Italiano **55**: 38–58.

COEMANS, E., 1867: Description de la flore fossile du premier étage du terrain Cretacé du Hainaut. – Mém. Brussels Acad. Roy. Sci. Belg. **36**: 1–20.

CRITCHFIELD, W. B., LITTLE, E. L., 1966: Geographic distribution of the pines of the world. – U.S. Dept., Agric. Forest Service, Misc. Publ. **991**. – Washington D.C.

CZECZOTT, H., 1954: The past and present distribution of *Pinus halepensis* MILL. and *Pinus brutia* TEN. – 8th Int. Bot. Congr. Paris **8** (Sec. 2, 4–6): 196–197.

DEPAPE, G., 1922: Flore pliocène de la Vallée du Rhone. – Ann. Sci. Nat. (ser. 10) **4**: 73–265.

ENDLICHER, ST. L., 1847: Synopsis Coniferarum. – St. Gallen.

FARJON, A., 1984: Pines. – Leiden: E. J. Brill.

GAUSSEN, H., 1960: Les gymnospermes actuelles et fossiles. – Trav. Lab. Forest. Toulouse **2**: 1. sect. 6, Genre *Pinus*. 272 p. – Toulouse.

GIVULESCU, R., 1961: Neue *Pinus*-Zapfen aus dem Boroder Becken (Rumänien) mit einer Übersicht der fossilen Kiefern-Zapfen Rumäniens. – Geologie, Berlin **10**: 850–855.

* A number of references used for this contribution are listed here, even if author and editor were unable to cite them in the text.

Gorbunov, M. G., 1958: Tertiary pines (*Pinus* of Western Sibiria). – Bot. Ž. Akad. Wiss. U.S.S.R. (Moskau) **43**: 337–352.

Gregor, H. J., 1980: Funde von *Pinus canariensis* Ch. Smith fossilis aus dem Neogen von La Palma (Kanarische Inseln). – Vieraea **9**: 57–64.

Hudson, R. H., 1960: The anatomy of the genus *Pinus* in the relation to its classification. – J. Inst. Wood Sci. **6**: 26–46.

Kilpper, K., 1968: Koniferenzapfen aus den tertiären Deckschichten des niederrheinischen Hauptflözes. 2. Teil, Genus *Pinus* L. – Palaeontographica B **123**: 213–220.

Klaus, W., 1977: Der Fund einer fossilen Aleppo-Kiefer (*Pinus halepensis* Mill.) im Pannon des Wiener Beckens. – Beitr. Paläont. Österr. **2**: 59–69.

– 1977: Neue fossile Pinaceen-Reste aus dem österreichischen Jungtertiär. – Beitr. Paläont. Österr. **3**: 105–127.

– 1978: On the taxonomic significance of tectum sculpture characters in alpine *Pinus* species. – Grana **17**: 161–166.

– 1978: Neue fossile Föhrenreste aus dem Raum von Linz. Zur Stammesgeschichte der Schwarzföhre, *Pinus nigra* Arn. – Jahrb. OÖ. Mus.-Ver. **123**/1: 20; –222.

– 1979: Blaue, rote und gelbgrüne männliche *Pinus*-Blüten. – Pl. Syst. Evol. **133**: 95–97.

– 1980a: Neue Beobachtungen zur Morphologie des Zapfens von *Pinus* und ihre Bedeutung für die Systematik, Fossilbestimmung, Arealgestaltung und Evolution der Gattung. – Pl. Syst. Evol. **134**: 137–171.

– 1980b: Über eine bemerkenswerte fossile *Pinus*-Art aus dem Unter-Miozän von Niederösterreich. – Ann. Naturhist. Mus. Wien **83**: 169–180.

– 1982: Die Kanarenkiefer (*Pinus canariensis* Smith ssp. *prisca* n. ssp.) und weitere Kiefernreste aus dem Jung-Tertiär von Stoob im Burgenland (Austria). – Biolog. Station Neusiedlersee, BFB-Bericht **44**. – Illmitz, Austria.

– 1982: Ein *Pinus canariensis* Smith-Zapfenfund aus dem Ober-Miozän (Pannon) des Wiener Beckens. – Ann. Naturhist. Mus. Wien **84 A**: 79–84

– 1984a: *Pinus brutia* Ten. ssp. *vindobonensis* n. ssp. und weitere fossile Mediterrankiefern aus Österreichs Miozänablagerungen. – Ann. Naturhist. Mus. Wien **86 A**: 41–54.

– 1984b: Zur Mikroflora des Unter-Sarmat am Alpen-Südostrand. – Beitr. Paläont. Österr. **11**: 291–400.

– 1985: Ein fossiler Zapfen der Aleppo-Kiefer aus dem Mittel-Miozän der Steiermark, Österr. – Ann. Naturhist. Mus. Wien **87**: 115–120.

– 1987: Einführung in die Paläobotanik **1**. – Wien: Deuticke.

Kräusel, R., 1919: Die fossilen Koniferenhölzer (unter Ausschluß von *Araucarioxylon* Kraus). – Palaeontographica B **62**: 185–275.

Lindley, J., Hutton, W., 1931–1937: The fossil flora of Great Britain I–III.

Little, E. L., Critchfield, W. B., 1969: Subdivision of the genus *Pinus* (Pines). – U.S. Dept. Agric., Forest Service, Misc. Publ. **1144**: 1–51.

Madrigal Sanchez, X., Caballero Deloya, M., 1969: Una nueva especie mexicana de *Pinus*. – Inst. Nac. Invest. Forestales. Bol. Tecnico **26**: 1–11.

Mai, D. H., 1965: Eine pliozäne Flora von Kranichfeld in Thüringen. – Abb. zentr. geol. Inst., Berlin **1**: 37–64.

– 1986: Über Typen und Originale tertiärer Arten von *Pinus* L. (*Pinaceae*) in mitteleuropäischen Sammlungen. – Ein Beitrag zur Geschichte der Gattung in Europa. – Feddes Repert. **97**: 9–10, 571–605.

Maleiffe, W., 1929: *Pinus pithyusa* und *Pinus eldarica*. Zwei Relikt-Kiefern der taurisch-kaukasischen Flora. – Mitt. Deutsch. Dendr. Ges. **41**: 138–150.

Menendez-Amor, J., 1951: Una piña fosil nueva del Pliocene de Málaga. – Bol. Roy. Soc. Esp. Hist Nat. **49 B**: 193–195.

Meusel, H., 1965: Die Reliktvegetation der Kanarischen Inseln in ihren Beziehungen zur süd- und mitteleuropäischen Flora. – Gesammelte Vorträge über moderne Abstammungslehre **1**: 117–136.

MIROV, N. T., 1951: Composition of gum turpentines of pines 14. A report on three Mexican pines: *Pinus ayacahuite, P. cembroides*, and *P. pinceana*. – Amer. Pharm. Assoc. J. Sci. Ed. **41**: 673–676.

– 1955: Relationship between *Pinus halepensis* and other *Insignes* pines of the Mediterranean region. – Bul. Res. Council Israel, Sect. D, Botany **5 D**: 65–72.

– 1967: The genus *Pinus*. – New York: Ronald Press.

PALIBIN, I. V., 1935: Stages of development of Caspian flora from the Cretaceous period on. – Soviet. Bot. **3**: 10–50.

PAPAJOANNOU, J., 1936: Über die Artbastarde zwischen *Pinus brutia* TEN. und *Pinus halepensis* MILL. in Nordostchalkidiki (Griechenland). – Forstwiss. Zentralbl. **58**: 194–205.

PRAGER, E. M., FOWLER, D. P., WILSON, A. C., 1976: Rates of evolution in *Coniferes* (*Pinaceae*). – Evolution **30**: 637–649.

PAGE, C. N., 1974: Morphology and affinities of *Pinus canariensis*. – Notes Roy. Bot. Gard. Edinburgh **33**: 317–323.

RIKLI, M. A., 1943: Das Pflanzenkleid der Mittelmeerländer. – Bern: Huber.

RZEDOWSKI, J., 1964: Una especie nueva de pino piñonero del Estado de Zacatecas (México). – Ciencia Méx. **23**: 17–20.

– VELA, G., 1966: *Pinus strobus* var. *chiapensis* en la Sierra Madre del Sur de México. – Ciencia Méx. **24**: 211–216.

SAPORTA, G., 1862–1874: Etudes sur la vegetation du Sud-Est de La France a l'époque tertiare. – Ann. Sci. Nat. Bot., ser. 4, **16**: 309, 348; **17**: 191–311 (1862); **19**: 5–124 (1863); ser. 5, **3**: 5–152 (1865); **8**: 5–136 (1867); **9**: 5–62 (1868); **15**: 277–351 (1872); **17**: 5–44 (1873); **18**: 23–146 (1874).

– 1865: Etudes sur la vegetation de la France à l'époque tertiaire. – Ann. Sci. Nat. Bot. Ser. 5. **3**.

SAYLOR, L. L., 1961: A karyotype analysis of selected species of *Pinus*. – Silvae Genetica **10**: 77–84.

– 1964: Karyotype analysis of *Pinus*-group *Lariciones*. – Silvae Genetica **13**: 165–170.

SHAW, G. R., 1914: The genus *Pinus*. – Publ. Arnold Arbor. **5**: 1–96.

STUDT, W., 1926: Die heutige und frühere Verbreitung der Koniferen und die Geschichte ihrer Arealgestaltung. – Mitt. Inst. Allg. Bot. Hamburg **6**: 167–308.

STUR, D., 1867: Beiträge zur Flora der Süßwasserquarze, der Congerien- und Cerithienschichten im Wiener und Ungarischen Becken. – Jahrb. K. K. Geol. Reichsanst. Wien **17**.

SUNDING, P., 1970: Elementr io Kanarioyuenes flora, og teorier til forklaving av floraens opprinnelse. – Blyttia **28**: 229–259.

– 1979: Origins of the Macaronesian flora. – In BRAMWELL, D., (Ed.): Plants and islands, pp. 13–40. – London: Academic Press.

TEIXEIRA, C., 1944: Pinheiro fosil do Pliocénico de Rio Maior. – Soc. Broteriana Bol., Ser. 2, **19**: 201–221.

UNGER, F., 1847: Cloris protogaea, Beiträge zur Flora der Vorwelt. – Leipzig.

– 1852: Iconographia plantarum fossilium. – Denkschr. Akad. Wiss. Wien, math.-nat. Kl. **4**: 71–118.

– 1860: Sylloge plantarum fossilium **I–III**. – Denkschr. Akad. Wiss. Wien **19**: 1–48. Ibid. **22**: 1–36 (1864); ibid. **25**: 1–76 (1866).

– 1868: Die fossile Flora von Radoboj in ihrer Gesamtheit und nach ihrem Verhältnis zur Entwicklung der Vegetation der Tertiärzeit. – Denkschr. K. K. Akad. Wiss. Wien, math.-nat. Kl. **29**.

VAN DER BURGH, J., 1973: Hölzer der niederrheinischen Braunkohlenformation **2**. Hölzer der Braunkohlengruben „Maria Theresia" zu Herzogenrath, „Zukunft West" zu-

Eschweiler und „Victor“ (Zülpich Mitte) zu Zülpich. Nebst einer systematisch-anatomischen Bearbeitung der Gattung *Pinus* L. – Rev. Paläobotan. Palynol. **15**: 73–275.
ZABLOCKI, J., 1928: Tertiäre Flora des Salzlagers von Wieliczka. Erster Teil. – Acta Soc. Bot. Polon. **5**: 174–208.

Address of the editor: Prof. F. EHRENDORFER, Institut für Botanik der Universität Wien, Rennweg 14, A-1030 Wien, Austria.

Pl. Syst. Evol. 162, 165–191

—Plant—
Systematics
and
Evolution

Paleobotanical evidence on the early radiation of nonmagnoliid dicotyledons

PETER R. CRANE

Received April 17, 1987

Key words: Angiosperms, *Chloranthaceae, Platanaceae, Trochodendrales, Fagaceae, Cercidiphyllum.* – Paleobotany, phylogeny, floral structure, leaf architecture.

Abstract: Paleobotanical studies indicate that several isolated and systematically depauperate groups of extant woody dicotyledons originated in the Mid Cretaceous. The *Chloranthaceae* had probably differentiated into insect-pollinated (*Chloranthus* and *Sarcandra*) and wind-pollinated (*Ascarina* and *Hedyosmum*) forms by the end of the Albian, and leaves referable to the *Trochodendrales* are known from the Albian and Cenomanian. In the latest Cretaceous and Early Tertiary, extinct representatives of the *Trochodendrales* included *Nordenskioldia* and the *Joffrea-Nyssidium* complex. The *Platanaceae* also differentiated before the end of the Albian and initially had insect-pollinated, unisexual flowers with five carpels or stamens. Some of these features persisted in the platanoid lineage until the Early Tertiary, and during the Paleocene and Eocene the *Platanaceae* included forms with elliptical, palmate and pinnate foliage. The history of the *Platanaceae* suggests that several features of the reproductive morphology of extant taxa may have arisen in association with a trend toward wind pollination. In the Mid Cretaceous, platanoid foliage partially intergrades with pinnate *Sapindopsis* and pedate *Debeya-Dewalquea* leaves suggesting a close relationship between *Platanaceae* and *Rosidae* and *Fagaceae* respectively. The *Chloranthaceae*, *Trochodendrales*, and *Platanaceae* all occupy a somewhat intermediate position between the *Magnoliidae* and *Hamamelidae* and are of considerable interest with respect to their role in the initial radiation of nonmagnoliid ("higher") dicotyledons.

In the last three decades paleobotanical investigations have clarified the timing and pattern of early angiosperm evolution, and have identified the Mid Cretaceous as the interval during which the group underwent major diversification. Although it is possible that the angiosperm lineage diverged from related gymnosperm groups as early as the Triassic (CRANE 1985, DOYLE & DONOGHUE 1986) current knowledge of Mesozoic floras makes it unlikely that the angiosperms were diverse or abundant prior to the Cretaceous. The earliest unequivocal angiosperm pollen is reported from the Hauterivian (BRENNER 1984, HUGHES & MCDOUGALL 1987), and through the Hauterivian to Cenomanian interval (approx. 131 – 91 myr B.P.) angiosperm leaves, pollen and other reproductive structures increase dramatically in their variety and abundance in fossil floras (BRENNER 1963, DOYLE 1969, MULLER 1970, DOYLE & HICKEY 1976, HICKEY & DOYLE 1977, CRANE 1987, FRIIS & CREPET 1987). This

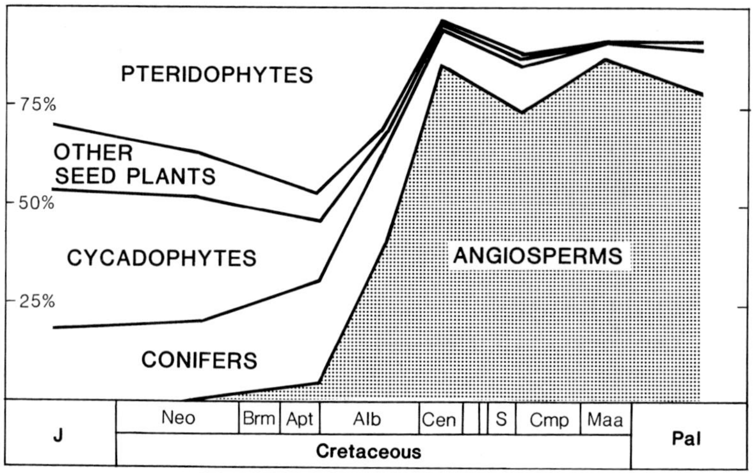

Fig. 1. Floristic change through the Cretaceous. Based on mean species number (expressed as a percentage) in 70 Jurassic, Cretaceous and Paleocene floras. Similar patterns are obtained using mean genus number (expressed as a percentage), or genus or species number without conversion into percentages. Cycadophytes includes *Bennettitales* and true cycads. See CRANE (1987) for lists of data and full explanation. J Jurassic; Neo Neocomian; Brm Barremian; Apt Aptian; Alb Albian; Cen Cenomanian; S Santonian; Cmp Campanian; Maa Maastrichtian; Pal Paleocene

clear chronological pattern is thought to reflect a major ecological and systematic radiation of flowering plants (DOYLE & HICKEY 1976, HICKEY & DOYLE 1977) that correlates with the decline of many gymnosperm and pteridophyte groups (Fig. 1) (CRANE 1987). During this interval many of the features characteristic of extant angiosperms appear for the first time (Fig. 2) and by the early Cenomanian at least three of the six dicotyledonous subclasses (*Magnoliidae, Hamamelidae, Rosidae*) were well established (WOLFE & al. 1975, DOYLE & HICKEY 1976, HICKEY & DOYLE 1977). Paleobotanical work is now clarifying the relationships of the fossil groups involved in the Mid Cretaceous angiosperm radiation and has demonstrated that the lineages clearly leading to several groups of extant woody dicotyledons had already diverged at a very early stage. In this paper I briefly review the fossil history of three of these groups, the *Chloranthaceae*, *Platanaceae* and *Trochodendrales*, in the context of the initial radiation of nonmagnoliid dicotyledons.

The **provenance of specimens** illustrated is as follows: BM, British Museum (Natural History), herbarium; British Museum (Natural History), paleobotanical collections; F, Field Museum of Natural History, herbarium; IU, Indiana University, paleobotanical collection; PP, Field Museum of Natural History, paleobotanical collections; UCMP, University of California, Berkeley, Museum of Paleontology; USNM, U.S. National Museum, Washington, D.C.

Chloranthaceae

The *Chloranthaceae* include the extant genera *Ascarina, Chloranthus, Hedyosmum,* and *Sarcandra*, and have attracted attention because of their simple floral structure and the absence of vessels in the secondary xylem of *Sarcandra* (SWAMY 1953,

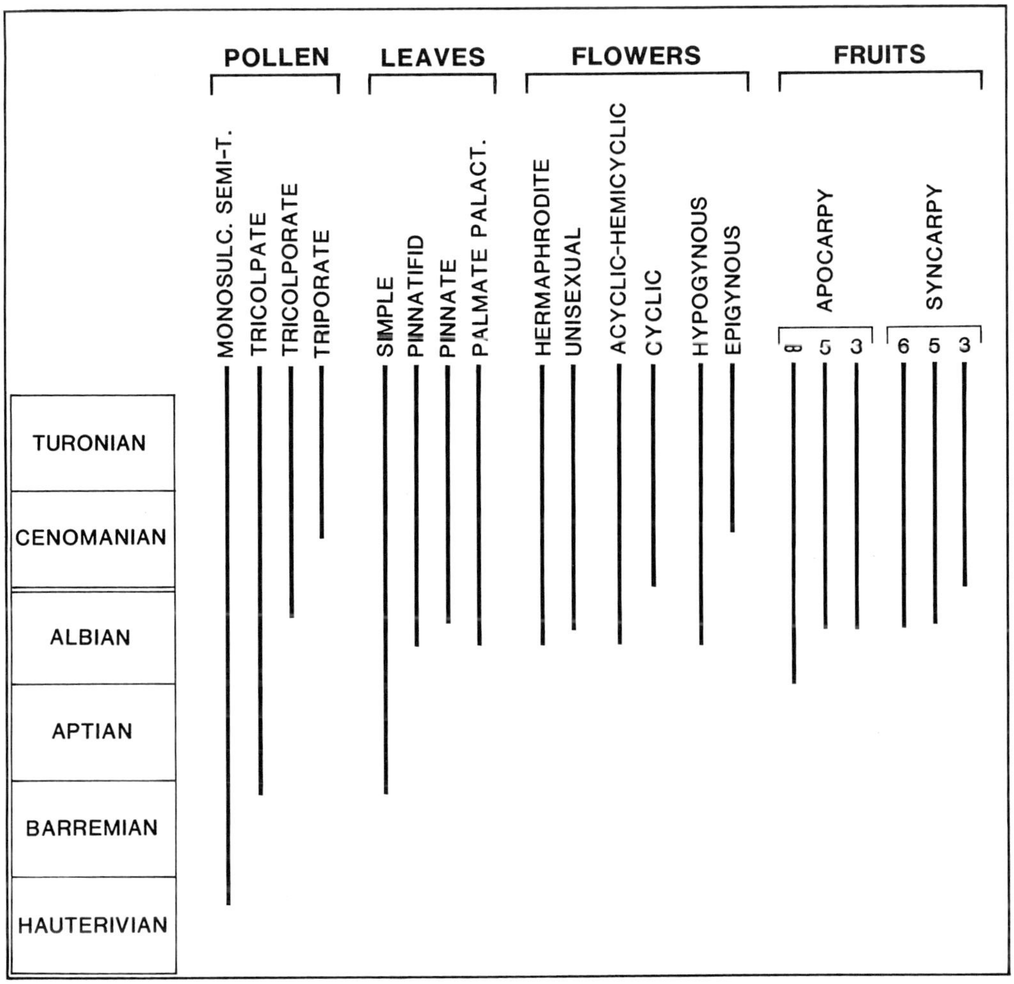

Fig. 2. Appearance of major angiosperm pollen, leaf, flower and fruit characters through the mid-Cretaceous. Monosulc. Semi-T, monosulcate semitectate; Palact, palinactinodromous. Based on DOYLE (1978), DOYLE & HICKEY (1976), BRENNER (1984), FRIIS & CREPET (1987), CRANE (1987), HUGHES & McDOUGALL (1987). Note that monosulcate semi-tectate pollen occurs as early as the Triassic (CORNET 1979) but the relationships of these grains are currently uncertain

SWAMY & BAILEY 1950, ENDRESS 1987 b). *Chloranthus* and *Sarcandra* have bisexual flowers, brightly colored androecia, small stigmatic surfaces and small anther thecae relative to the size of the staminal filament. All of these features are apparently associated with an entomophilous pollination syndrome (ENDRESS 1986 b, 1987 b). *Ascarina* and *Hedyosmum* have inconspicuous unisexual flowers with large dry papillose stigmas apparently associated with an anemophilous pollination syndrome (ENDRESS 1987 b). Together with disintegration of the innermost layers of the ovary wall at anthesis, and the presence of slime cells (ENDRESS 1987 b), these features may be synapomorphies in the context of the *Magnoliidae* as a whole. Present knowledge therefore suggests that *Ascarina* and *Hedyosmum* together form a clade, but whether *Chloranthus* and *Sarcandra* form a monophyletic group is less certain

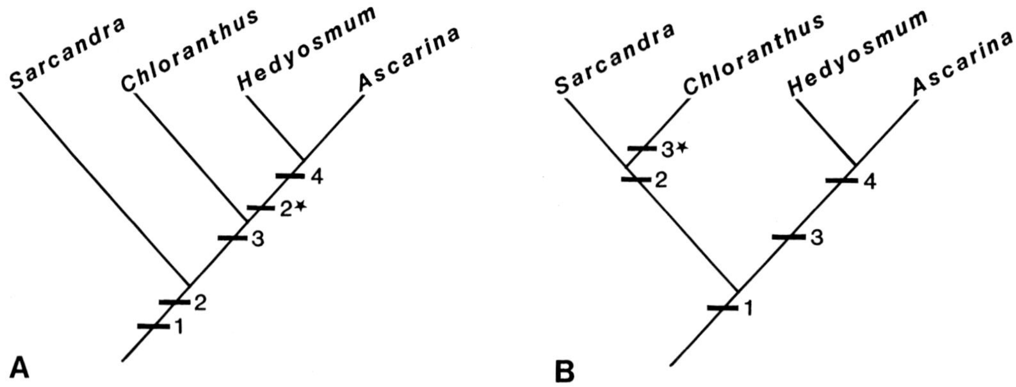

Fig. 3. Alternative hypotheses of relationships in extant *Chloranthaceae*. 1 Chloranthaceous synapomorphies, e.g., flowers lacking a perianth (except possibly *Hedyosmum*, see ENDRESS 1987 b), unicarpellate gynoecium with a single pendent orthotropous ovule (ENDRESS 1987 b), stamens anterior with respect to gynoecium (ENDRESS 1987 b); 2 swollen, brightly colored androecia and other characters apparently associated with entomophilous pollination (ENDRESS 1987 b); 3 vessels; 4 synapomorphies of *Ascarina* and *Hedyosmum*, e.g., unisexual flowers, disintegration of the inner ovary wall, presence of slime cells (ENDRESS 1987 b). *A* Single origin of vessels (3) requires either loss of entomophilous pollination syndrome (as shown, 2 and 2*) or independent acquisition of these features in *Chloranthus* and *Sarcandra*; *B* single origin of the entomophilous pollination syndrome (2) and monophyly of *Chloranthus*, *Sarcandra* clade requires double origin of vessels (as shown, 3 and 3*) or loss of vessels in *Sarcandra* (e.g., YOUNG 1981)

(Fig. 3). If vessels evolved only once within the family then *Chloranthus* would be resolved as more closely related to *Ascarina* and *Hedyosmum* (Fig. 3 A).

Systematically, the *Chloranthaceae* are included within the *Magnoliidae* but similarities to certain *Hamamelidae*, particularly the *Trochodendrales* (simple sometimes unisexual flowers, chloranthoid leaf teeth, absence of vessels in *Sarcandra, Tetracentron, Trochodendron*), have led to the suggestion that the group occupies an intermediate position between these two subclasses (WALKER & WALKER 1984, ENDRESS 1987 b). WALKER & WALKER (1984) included the family in their *Piperiflorae* with the *Lactoridaceae, Piperaceae* and *Saururaceae*. The *Piperaceae* and *Saururaceae* were interpreted as more closely related to each other than either is to the *Lactoridaceae*; and the *Chloranthaceae* were interpreted as the sister group to the other three families. ENDRESS (1987 b) also acknowledged resemblances to the *Piperales* but placed the family in the *Laurales* based on similarities to the *Trimeniaceae* and *Amborellaceae*.

Paleobotanically the *Chloranthaceae* first attracted attention because of similarities between pollen of extant *Ascarina* and the fossil grain *Clavatipollenites hughesii* COUPER, which is widespread in the Mid Cretaceous (COUPER 1958, KUPRIANOVA 1967, DOYLE 1969, MULLER 1981, 1984). Detailed S.E.M. and T.E.M. investigations of early angiosperm pollen grains have corroborated these observations and demonstrated strong similarities between *Clavatipollenites* and *Ascarina*, and between the fossil grain *Asteropollis* and pollen of *Hedyosmum* (WALKER & WALKER 1984). Other fossil grains (*Stephanocolpites*) have been compared with those of extant *Chloranthus* (WALKER & WALKER 1984). Studies of Cretaceous angiosperm reproductive structures have also yielded well preserved chloranthoid androecia from the Upper Albian of eastern North America and the Upper Santonian/Lower Campanian of southern Sweden (FRIIS 1985, FRIIS & al., 1986, CRANE & al. in prep.).

Upper Cretaceous androecia are three-lobed with two pairs of marginal pollen sacs on the central lobe and a single pair of pollen sacs on the outer margin of the two lateral lobes (Fig. 4 C). Androecia of this kind are known only in *Chloranthus* (Fig. 4 A), and thus plants with some, but not necessarily all of the diagnostic characters of this genus were present at least by the Campanian. The single androecium known from Lower Cretaceous consists of three obovoid stamens that are slightly fused at the base (Fig. 4 D). Each filament is swollen and bears two pairs of pollen sacs. The individual stamens resemble those of *Sarcandra* (Fig. 4 B) and some extant *Winteraceae* (ENDRESS 1986 b, 1987 b), but the staminal aggregation is more like that in *Chloranthus*. The fossil androecium thus exhibits features seen in two chloranthaceous genera, and combines the presumed primitive condition of two pairs of pollen sacs per filament with a more specialized floral arrangement in which the stamens are fused (even if only slightly) into a three-lobed androecium. This fossil provides good evidence of the *Chloranthaceae*, and perhaps the lineage leading to extant *Chloranthus*, before the end of the Early Cretaceous. The swollen filaments, small pollen sacs, small pollen and the apparent presence of pollenkitt are also indicative of insect pollination (FRIIS & al. 1986).

Based on the presence of *Clavatipollenites*-type pollen in well dated sediments (HUGHES & al. 1979, HUGHES & McDOUGALL 1987) the origin of the *Chloranthaceae* can be inferred to have been as early as the Barremian-Hauterivian. WALKER & WALKER (1984) suggested that the abundance of *Clavatipollenites* and *Asteropollis* in some Mid Cretaceous palynofloras may reflect the first appearance of wind pollination in the angiosperm clade: a conclusion consistent with pollination in extant *Ascarina* and *Hedyosmum*. If this is accepted, then the *Chloranthaceae* were not only widespread in the Cretaceous vegetation of the Northern Hemisphere

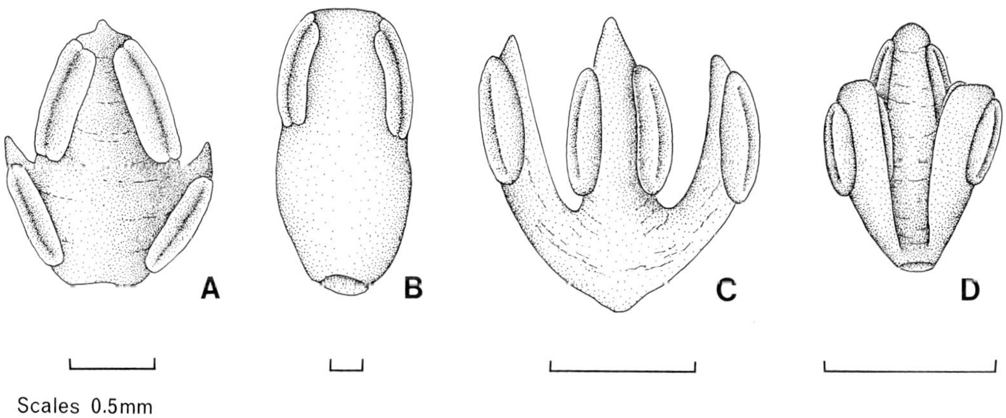

Fig. 4. Androecia of fossil and extant *Chloranthaceae*. *A* Extant *Chloranthus officinalis* BL. (F 411034). *B* Extant *Sarcandra chloranthoides* GARDNER (after ENDRESS 1987 b). *C* Fossil androecium Santonian/Campanian, Scania, Sweden (after FRIIS & al., 1986). *D* Fossil androecium, Albian, Maryland, U.S.A. (after FRIIS & al. 1986)

(MULLER 1981), but had also developed both anemophilous and entomophilous pollination syndromes at an early stage in the angiosperm radiation.

Trochodendrales

As most recently circumscribed (ENDRESS 1986a), the *Trochodendrales* comprise four extant genera, *Trochodendron* and *Tetracentron* included in the *Trochodendraceae*, and *Cercidiphyllum* and *Euptelea* included in the monotypic families *Cercidiphyllaceae* and *Eupteleaceae* respectively. Systematically the *Trochodendrales* occupy an analogous position in the *Hamamelidae* to that of the *Chloranthaceae* in the *Magnoliidae*. They exhibit similarities to certain *Magnoliidae* (particularly *Chloranthaceae*), including vesselless wood (*Tetracentron, Trochodendron*), leaves with chloranthoid teeth (*Cercidiphyllum, Tetracentron, Trochodendron*), and "oil cells" (*Tetracentron, Trochodendron*) (ENDRESS 1986a). The wide range in carpel number in the *Trochodendrales* is also similar to the floral plasticity exhibited by the *Magnoliidae* as a whole (ENDRESS 1986a, b). Usually, however, the *Trochodendrales* are placed within the *Hamamelidae* on the basis of their triaperturate pollen grains and other features (ENDRESS 1986a).

The *Trochodendrales* have an extensive fossil record and leaves with the architectural characteristics of this group (actinodromous venation frequently associated with chloranthoid teeth) first appear in the Mid Cretaceous (UPCHURCH & WOLFE 1987). Leaves of "*Populus*" *potomacensis* WARD, *Populophyllum renifome* FONTAINE and *Menispermites potomacensis* BERRY from the Mid-Late Albian of the Potomac Group may be closely related to the *Trochodendrales,* and similar foliage is known from Mid Cretaceous sediments in Eurasia (e.g., KVAČEK 1983), as well as elsewhere in North America (e.g., LESQUEREUX 1892, BELL 1956, BERRY 1921a, SPICER pers. comm.). The reproductive structures of these probable early *Trochodendrales* are unknown, but two later extinct representatives of the group, *Nordenskioldia* and the *Joffrea-Nyssidium* complex, are becoming increasingly well known from studies of Early Tertiary material.

Nordenskioldia. *Nordenskioldia borealis* HEER was first described from the Paleocene of Greenland (then believed to be Miocene) based on infructescences bearing sessile schizocarp-like fruits (HEER 1870). *Nordenskioldia* is now known to be widespread in the latest Cretaceous and Early Tertiary of the Northern Hemisphere (BROWN 1962, KRYSHTOFOVICH 1956, 1958, KRASSILOV 1976, CHANDRASEKHARAM 1974, CRANE & al. 1986a). In North America *Nordenskioldia* persisted into the Miocene (leaves of "*Populus*" *heteromorpha* KNOWLTON and associated in infructescences, Latah flora KNOWLTON 1926, *Carpolithus* spec. Whitebird flora BROWN 1936). The relationships of *Nordenskioldia* remained uncertain until KRYSHTOFOVICH (1956, 1958) recognized the leaves associated with the infructescences and suggested the possibility of a close relationship to *Trochodendron* and *Tetracentron.* Recent studies in the Paleocene of North America (CRANE & al. 1986a, 1988b) support this conclusion, and the discovery of silicified material with well preserved anatomy has clarified many aspects of the structure of this plant.

Infructescences are long, branched (Fig. 9A) and bear numerous sessile fruits (Figs. 5A, B and 7A, B). Each fruit consists of a whorl of about 15 achene-like fruitlets (Figs. 5A and 7D) partially embedded in a shallow cup-like receptacle (Figs. 5A and 7F). Distally the fruit is bilaterally symmetrical (Fig. 5A), but at

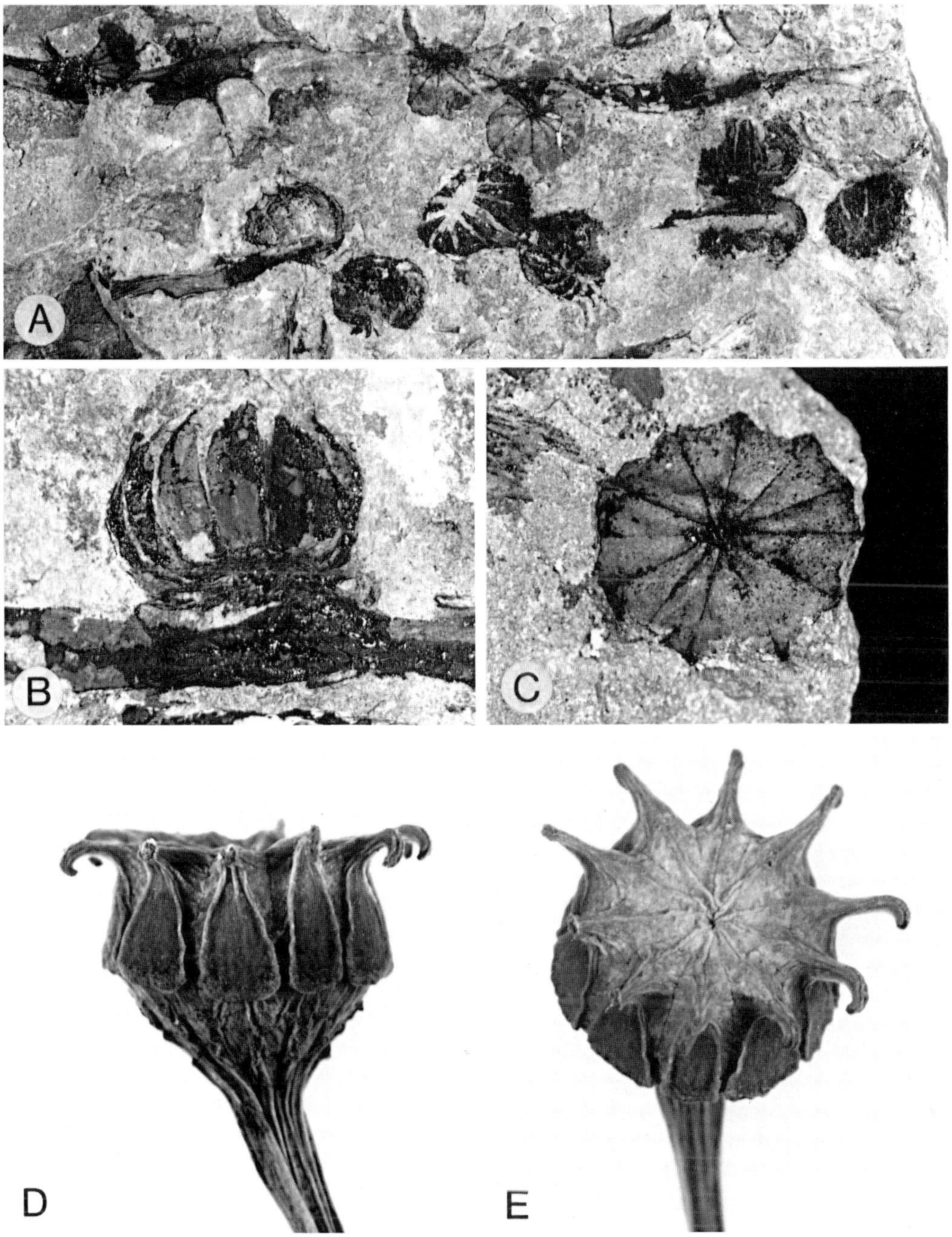

Fig. 5. Fruits of fossil *Nordenskioldia borealis* HEER and extant *Trochodendron aralioides* SIEB. & ZUCC. *A* Portion of two *Nordenskioldia* infructescence axes with several attached fruits, note sediment filling between the fruitlets and the persistent collar and central column from which the fruitlets have been shed (PP 34522), × 1.75; *B* lateral view of a fruit showing infructescence axis and separating fruitlets (PP 34527), × 5; *C* basal collar showing positions of attachment of 12 fruitlets (PP 34529), × 5; *D Trochodendron aralioides* SIEB. & ZUCC., lateral view of young fruit showing carpels and semi-inferior gynoecium, × 7; *E Trochodendron aralioides* SIEB. & ZUCC., oblique apical view of young fruit showing radially arranged carpels and recurved styles, × 7. Fossil material from Tongue River Member, Fort Union Formation, Paleocene (near Melville, Montana)

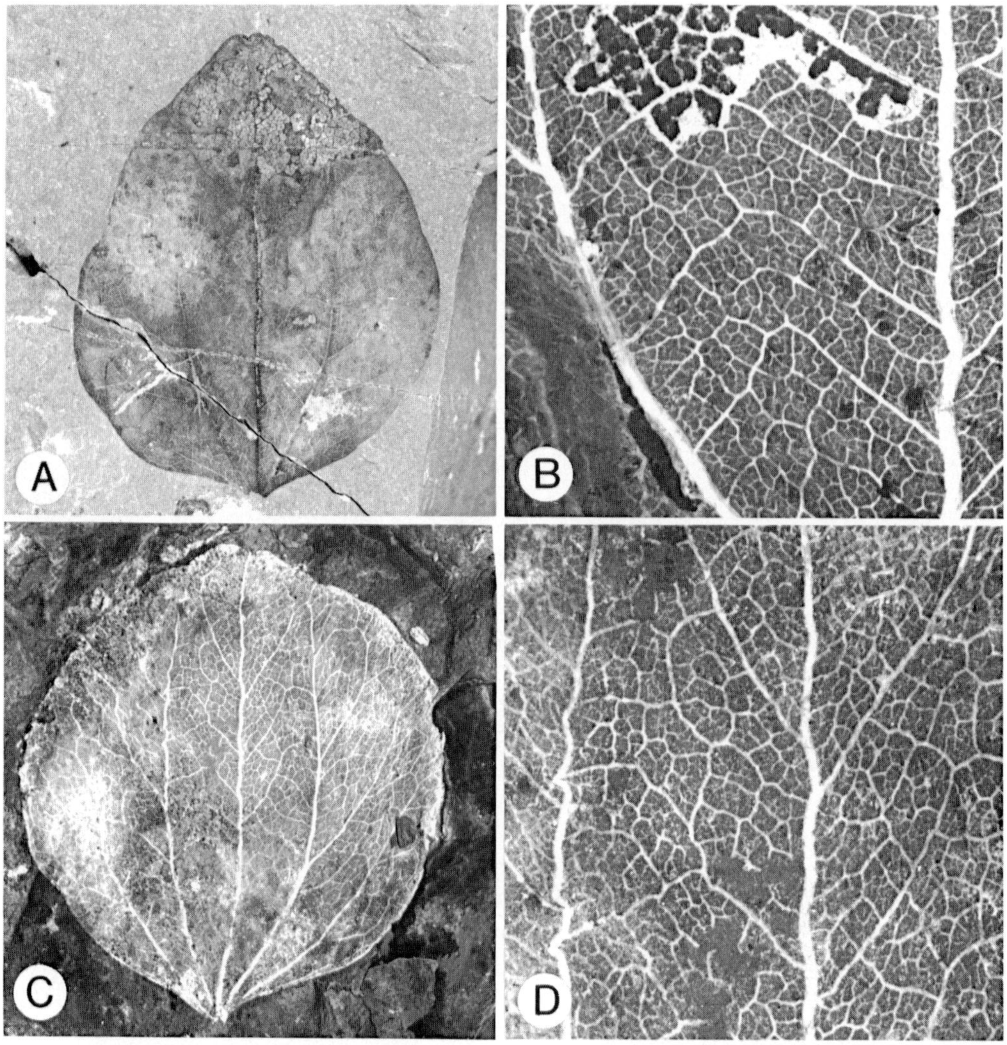

Fig. 6. Fossil leaves ["*Cocculus*" *flabella* (NEWBERRY) WOLFE] associated with *Nordenskioldia borealis* HEER. *A* Small leaf showing actinodromous venation and entire undulate margin (PP 34548), × 3; *B* detail of venation showing third, fourth, and fifth order veins delimiting polygonal areolae with branched freely ending veinlets (cf. WOLFE 1966, HICKEY 1977) (IU 15722-7040), × 6; *C* leaf with actinodromous venation (PP 34538), × 1.5; *D* detail of leaf in *C* with poorly preserved fine venation, × 5; *A* from Tongue River Member, Fort Union Formation, Paleocene (near Melville, Montana); *B* – *D* from Sentinel Butte Member, Fort Union Formation, Paleocene (near Almont, North Dakota)

the attachment to the receptacle the arrangement of the fruitlets is radially symmetrical (Fig. 5 C). Each fruitlet is wedge-shaped in transverse section with two longitudinal bands of fibers along the outer margin (Fig. 7 C, D). Several less prominent bands of fibers form ridges on the radial walls (Fig. 7 C) and laterally compressed specimens show that these ridges occasionally branch and are slightly oblique to the long axis of the fruitlet (Fig. 7 G). The endocarp consists of a mass of obliquely oriented fibers. Each fruitlet contains a single flattened seed with a

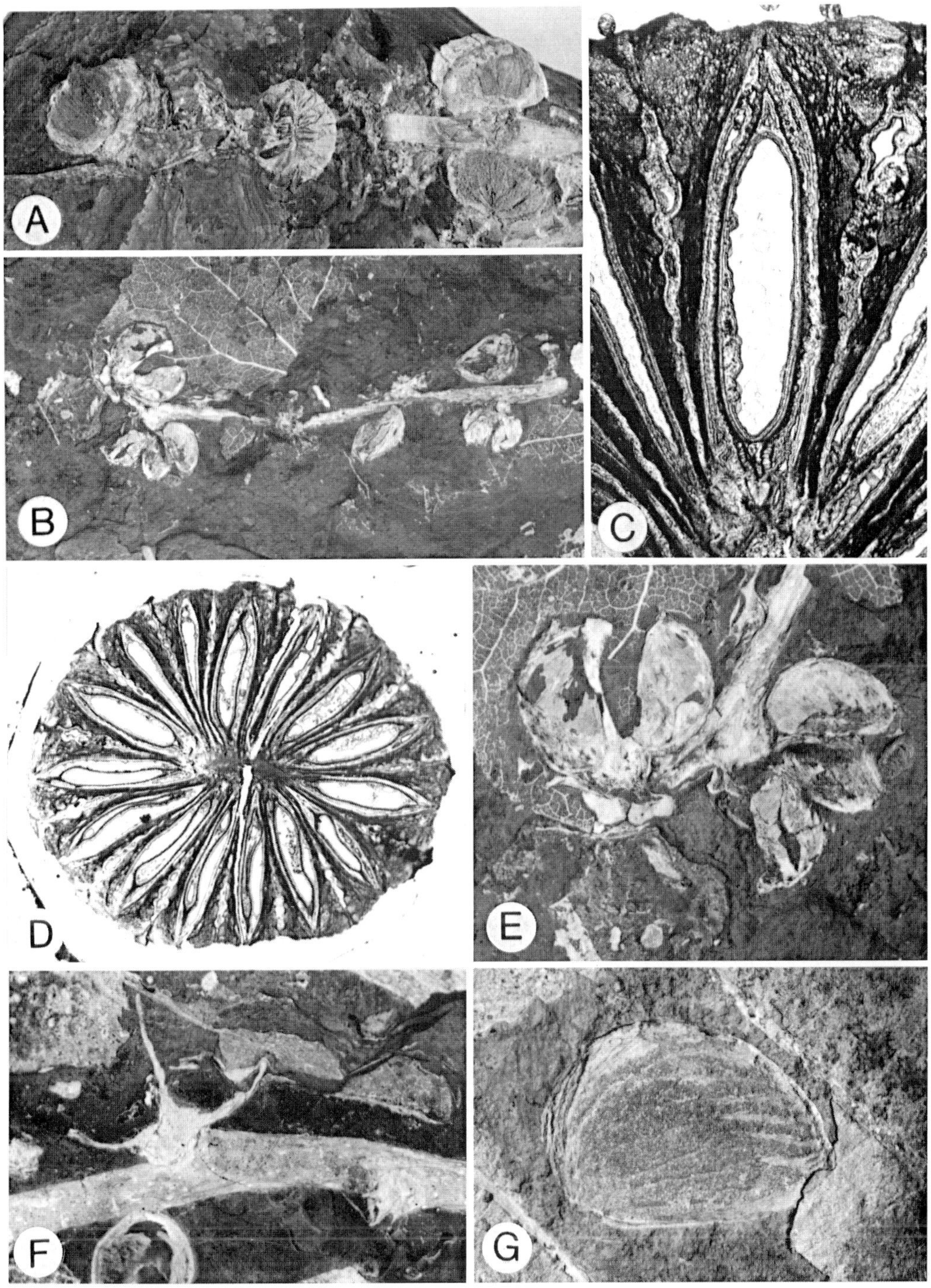

Fig. 7. Infructescences, fruits and fruitlets of *Nordenskioldia borealis* HEER. *A* Portion of infructescence showing four attached fruits (PP 34536), × 2; *B* portion of infructescence axis showing two partially fragmented fruits and associated fruitlets (PP 34537), × 1; *C* transverse section through a silicified fruit in *D* showing a wedge-shaped fruitlet containing a single seed, × 15; *D* transverse section through a silicified fruit showing 15 fruitlets radiating from a central axis (PP 34535), × 5; *E* two partially fragmented fruits from *B* showing separating fruitlets, × 2.5; *E* portion of infructescence axis showing section through persistent collar and central column of a fruit from which the fruitlets have been shed (PP 34425), × 2.5; *G* isolated fruitlet showing vascular strands on the radial walls and attachment scar (PP 34210), × 4. Fossil material from Sentinel Butte Member, Fort Union Formation, Paleocene (near Almont, North Dakota)

narrow flange along each margin (Fig. 7 C). At maturity fruitlets were dispersed individually (Fig. 7 E, G) but the remains of the receptacle persisted on the infructescence axis as a shallow cup often with a central column around which the fruitlets were arranged (Figs. 5 A and 7 E, F). In one specimen this cup shows small elliptical scars that may indicate the position of stamens and hence bisexual flowers. Unfortunately the consistent occurrence of this feature has not yet been established and the stamens, pollen, perianth and sexuality of *Nordenskioldia* flowers are currently unknown.

Shoots of the *Nordenskioldia* plant have been identified by anatomical similarities to infructescence axes (CRANE & al. 1986 a, 1988 b). Both have secondary xylem in which the axial elements consist entirely of tracheids separated by predominantly uniseriate rays. Vegetative axes are clearly differentiated into long and short shoots (CRANE & al. 1988 b). Leaves of the *Nordenskioldia* plant ["*Cocculus*" *flabella* (NEWBERRY) WOLFE 1966] have been identified by consistent association with the infructescences at numerous localities (KRYSHTOFOVICH 1956, 1958) and have previously been assigned to a variety of extant genera. The leaves are simple, with actinodromous venation and a margin that is more or less entire or has irregularly developed marginal crenations (CRANE & al. 1988 b) (Fig. 6 A, C).

Nordenskioldia is clearly related to *Trochodendron* and *Tetracentron*. The actinodromous leaves and the presence of short shoots are similarities to *Tetracentron* and *Cercidiphyllum*, while the wood anatomy is like that of *Tetracentron* and *Trochodendron*. The syncarpous, semi-inferior gynoecium, the position of the styles and the number of carpels is very similar to that of *Trochodendron* (Fig. 5 D, E). Taken together these features indicate that *Nordenskioldia* is closely related to the *Trochodendraceae*, but more detailed resolution of relationships will require additional morphological and anatomical information.

***Joffrea-Nyssidium* complex.** Fossil leaves similar to those of extant *Cercidiphyllum* are possibly the most abundant element in latest Cretaceous and Early Tertiary floras from middle and high latitudes in the Northern hemisphere (Fig. 8) (BROWN 1939, SCHLOEMER-JÄGER 1958, ILJINSKAJA 1974, CHANDRASEKHARAM 1974, TANAI 1981, CRANE 1984 a). BROWN (1939) first noted the association of these leaves with distinctive elongated infructescences and small winged seeds at numerous localities in North America. Based on the combined information from these three different organs BROWN recognized certain differences from the extant genus but considered the similarities sufficiently strong to assign the fossil remains to *Cercidiphyllum*. Subsequent work has confirmed the association noted by BROWN and has demonstrated winged seeds in permineralized fruits belonging to this complex (*Jenkinsella* REID & CHANDLER 1933, CHANDLER 1961, CRANE 1984 a).

The most fully understood of the Early Tertiary *Cercidiphyllum*-like plants is *Joffrea speirsii* CRANE & STOCKEY known from seedlings, shoots, leaves, pistillate inflorescences, infructescences, seeds, and possibly staminate inflorescences from the Paleocene of southern Alberta, Canada (CRANE & STOCKEY 1985). Similarities to extant *Cercidiphyllum* include opposite phyllotaxy, leaf architecture, distinct long and short shoot differentiation of the vegetative axes, the general organization of the pistillate flowers and inflorescences, the winged seeds with a distinctive "hairpin loop" of the raphe and possibly the organization of the staminate flowers (CRANE & STOCKEY 1985, 1986). Differences from extant *Cercidiphyllum* include

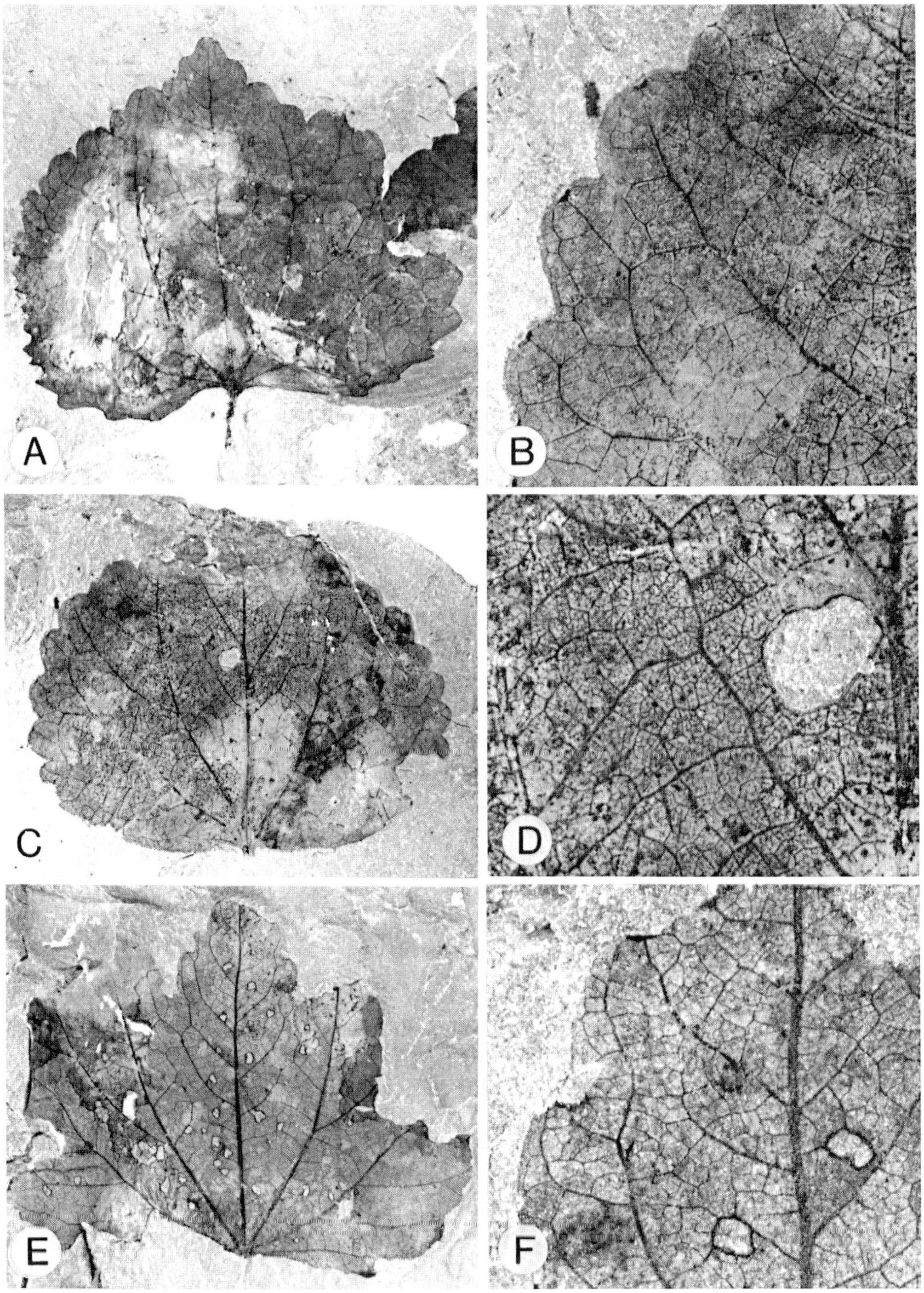

Fig. 8. Leaves of fossil *Cercidiphyllaceae* [*Trochodendroides arctica* (HEER) BROWN] – *Nyssidium-Joffrea* complex. *A* Leaf showing actinodromous venation and crenate margin (PP 34516), × 1; *B* margin of leaf in *C* showing glandular (chloranthoid) teeth and details of venation, × 3; *C* leaf showing actinodromous venation, glandular teeth and well preserved fine venation (PP 34519), × 1; *D* detail of fine venation from leaf in *C* showing fourth, fifth and sixth order veins delimiting polygonal areolae with branched freely ending veinlets (cf. WOLFE 1966, HICKEY 1977), × 5; *E* lobed leaf with broken margin note similarities in leaf architecture to *A* and *C* (PP 34520), × 1; *F* detail of margin from central lobe of leaf in *E* showing poorly preserved fine venation and glandular teeth, × 5. Fossil material from Tongue River Member, Fort Union Formation, Paleocene (near Melville, Montana)

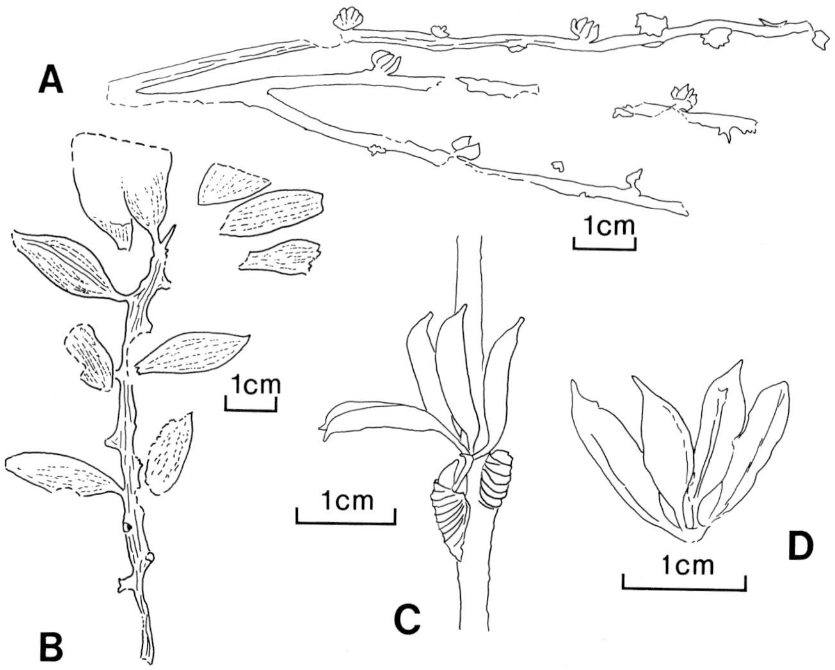

Fig. 9. Infructescences of extant and fossil *Trochodendrales*. *A Nordenskioldia borealis* HEER portion of infructescence with numerous sessile fruits (PP 34546); *B Nyssidium* infructescence (PP 34528); *C* infructescence of extant *Cercidiphyllum japonicum* SIEB. & ZUCC.; *D* fossil infructescence similar to extant *Cercidiphyllum* (UCMP-9309): *A, B* from the Tongue River Member, Fort Union Formation, Paleocene (near Melville, Montana); *D* John Day Formation, Oligocene (Pentecost Ranch, John Day Basin, Oregon)

the form of the short shoots, the frequent occurrence of two carpels per flower, the large number of flowers on an elongated inflorescence (Fig. 9 B), details of the anatomy of the follicle wall, the large number of seeds per follicle, anatomy of the seed coat and possibly the large size and number of the bracts in each staminate inflorescence (CRANE & STOCKEY 1985).

Comparisons of *Joffrea* with similar fossil plants from the Early Tertiary of Europe and Asia suggest that there was considerable diversity in phyllotaxy, shoot growth, inflorescence position and the number and crowding of follicles in each infructescence, among extinct *Cercidiphyllaceae* (CRANE & STOCKEY 1986). This diversity is also manifested in the variation that occurs in the fossil foliage of this group and has led to considerable systematic difficulties. Variation in the leaf margin has led to confusion with the foliage of the *Nordenskioldia* complex, and variants with distinctly trilobed laminae (Fig. 8 E, F) have sometimes been referred to "*Ampelopsis*" *acerifolia* (NEWBERRY) BROWN (1962).

Ecologically, the abundance of the *Joffrea-Nyssidium* complex in many fossil floras from fluvial depositional environments, combined with the production of large quantities of small seeds, suggests that these *Cercidiphyllum*-like plants were early colonizers of disturbed habitats during the early Tertiary (CRANE 1987). The occurrence of large numbers of fossil seedlings of both *Joffrea* and extinct *Platanaceae* at the Joffre Bridge locality, Alberta (STOCKEY & CRANE 1983, CRANE &

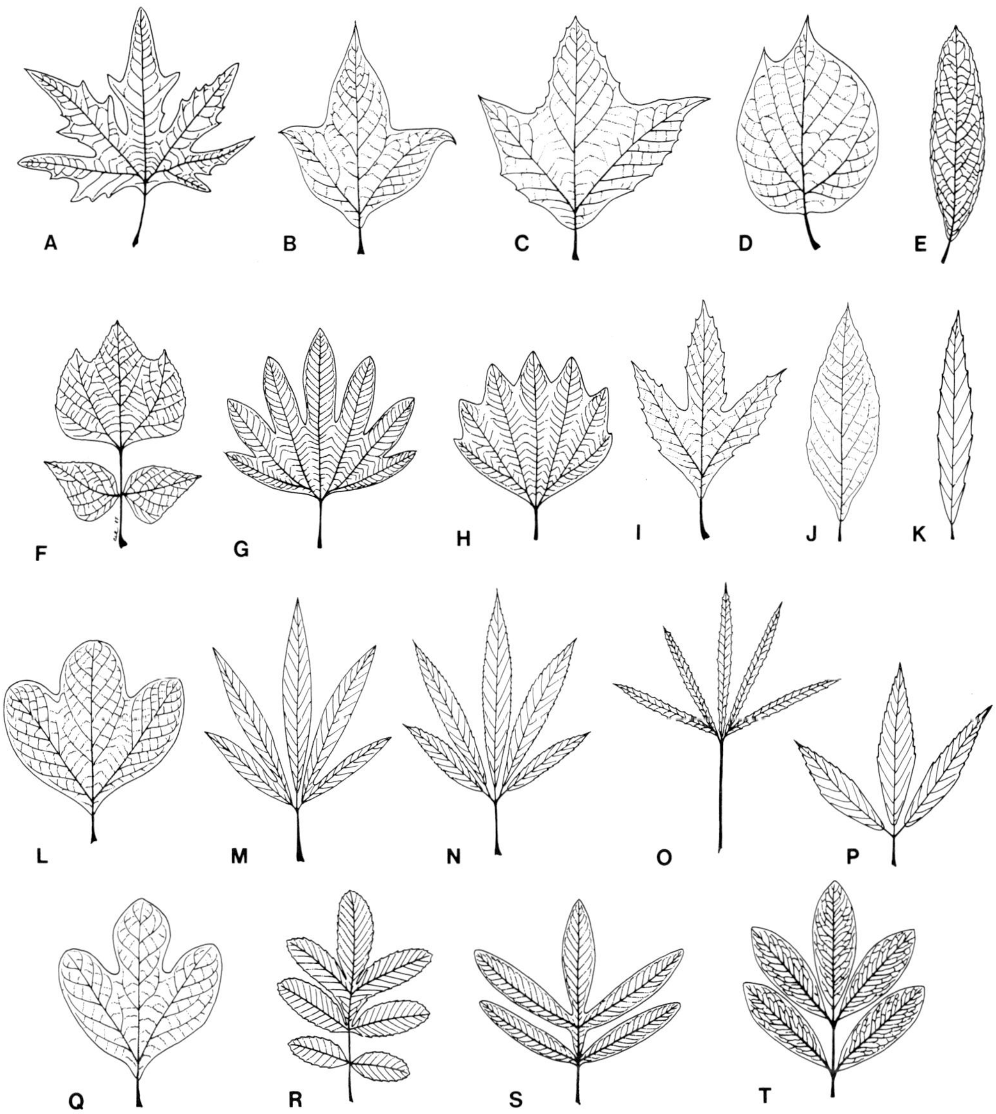

Fig. 10. Foliar variation in extant and fossil *Platanaceae*, and related forms (not to scale). *A Platanus orientalis* L., extant (after MANCHESTER 1986); *B* – *D Platanus mexicana* MORIC., extant (after NEE 1981): *E Platanus kerrii* A GAGNEPAIN, extant (after MANCHESTER 1986); *F Platanites hebridicus* FORBES, Paleocene (after CRANE & al. 1988 a); *G Macginitiea angustiloba* (LESQUEREUX) MANCHESTER, Eocene (after MANCHESTER 1986); *H Macginitiea whitneyi* (LESQUEREUX) MANCHESTER, Eocene-early Oligocene (after MANCHESTER 1986); *I Platanus schimperi* (HEER) SAPORTA & MARION, Paleocene (CRANE, unpubl.); *J Dicotylophyllum flexuosa* (NEWBERRY) WOLFE (after CRANE & al. 1988 b); *K Dryophyllum* spec. Eocene (CRANE unpubl.); *L* Generalized *Araliopsoides*, "*Sassafras*" type, Albian (HICKEY & DOYLE 1977); *M Dewalquea westerhausiana* (RICHTER) RÜFFLE, Upper Cretaceous (after RÜFFLE & KNAPPE 1977); *N Dewalquea insignis* HOS. & V. D. MARCK, Upper Cretaceous (after RÜFFLE & KNAPPE 1977); *O Dewalquea pulchella* KNOWLTON, Cenomanian (after KNOWLTON 1917); *P Dewalquea fraxinifolia* JOHNSON & GILMORE, Upper Cretaceous (after JOHNSON & GILMORE 1921); *Q* Generalized *Araliaephyllum* type, Albian (after HICKEY & DOYLE 1977); *R Sapindopsis belviderensis* BERRY, Albian (after BERRY 1921 b); *S Sapindopsis magnifolia* FONTAINE, Albian (after HICKEY & DOYLE 1977); *T Sapindopsis* spec. Albian (after HICKEY & DOYLE 1977)

STOCKEY 1985) provides striking evidence that these plants were important Early Tertiary weeds prior to the radiation of many herbaceous groups that today dominate that niche. The *Joffrea-Nyssidium* complex persisted into the Late Eocene (CRANE 1984a) and Oligocene (BECKER 1973). Fossil plants with compact inflorescences closely resembling extant *Cercidiphyllum* are first recorded from the Early Oligocene of North America and Europe (Fig. 9 C, D) (JÄHNICHEN & al. 1980).

Platanaceae

The *Platanaceae* comprise a single extant genus divided into two subgenera (LEROY 1982). Subg. *Platanus* includes approximately seven species characterized by the possession of stipules and typical palmate palinactinodromous platanoid leaves (Fig. 10 A – D). The center of distribution for the subgenus is in southwestern North America and Mexico, with one disjunct species (*Platanus occidentalis* L., Fig. 10 A) in eastern North America and another (*P. orientalis* L.) in eastern Europe and western Asia. Subg. *Castaneophyllum* includes only *Platanus kerrii* GAGNEPAIN, a species native to southeastern Asia, that differs from other members of the family in lacking stipules and having elliptical, pinnate veined leaves (Fig. 10 E). The *Platanaceae* have long been known to extend back to the Mid Cretaceous (LESQUEREUX 1892) but only recently has it become clear that this group may have played an important role in the Mid Cretaceous radiation of dicotyledons (DOYLE & HICKEY 1976, HICKEY & DOYLE 1977, DILCHER 1979, CRANE & al. 1986 b).

Mid Cretaceous *Platanaceae*. During the Mid Cretaceous the *Platanaceae* are known from leaves, wood, infructescences and both staminate and pistillate flowers (KNAPPE & RÜFFLE 1975, DILCHER 1979, UPCHURCH 1984 a, b, SCHWARZWALDER & DILCHER 1988, CRANE & al. 1986 b, FRIIS & al. 1988, MANCHESTER pers. comm.). Mid Cretaceous platanoid foliage is characterized by palinactinodromous primary venation but exhibits considerable variation in leaf architecture and cuticular structure (Fig. 10 L, Q) (SCHWARZWALDER & DILCHER 1987, UPCHURCH 1984 a). Mid Cretaceous infructescences consists of elongated axes bearing numerous sessile, more or less spherical heads of numerous flowers (Fig. 11 C). Flowers are unisexual with stamens and carpels in fives surrounded by prominent perianth parts (CRANE & al. 1986 b). The stamens have short filaments, long pollen sacs and produced very small tricolpate reticulate pollen grains [*Tricolpites minutus* (BRENNER) DETTMAN] approx. 8 – 10 µm in polar length (CRANE & al. 1986 b, FRIIS & al. 1988). The carpels are oblong in shape and differ from those of extant *Platanus* in lacking the mass of hairs that are present at the base of each fruitlet (Fig. 12 D – E).

Platanoid-*Sapindopsis* complex. In addition to palmately lobed platanoid leaves, pinnatifid or pinnate leaves assigned to the fossil genus *Sapindopsis* (Fig. 10 R – T) are a further common component of Mid Cretaceous leaf floras (BERRY 1921 b, DOYLE & HICKEY 1976, HICKEY & DOYLE 1977), and it has been suggested that foliage of this kind was probably produced by early representatives of the subclass *Rosidae* (HICKEY & WOLFE 1975, DOYLE & HICKEY 1977). *Sapindopsis* foliage exhibits considerable variation in gross morphology, venation and the form of the margin (BERRY 1921 b, HICKEY & DOYLE 1977) and many specimens exhibit interesting similarities to the leaves of Mid Cretaceous platanoids. Some forms have trilobed terminal leaflets with palinactinodromous primary venation closely resembling that in contemporaneous platanoid leaves (Fig. 10 R) (HICKEY & DOYLE 1977). Other variants exhibit details of tooth venation resembling that seen in some platanoid foliage. This partial intergradation in the morphology of platanoid and *Sapindopsis* leaves is paralleled by similarities in cuticular structure (UPCHURCH 1984 a) and has been used to suggest a close phylogenetic relationship between these two groups of Mid Cretaceous angiosperms. By extension a close, possibly sister group, relationship has been inferred between the *Hamamelidae* and *Rosidae* (HICKEY & WOLFE 1975, DOYLE & HICKEY 1976, HICKEY & DOYLE 1977).

The reproductive structures of *Sapindopsis* plants are unknown but at some Mid Cretaceous localities where platanoid leaves are either absent or rare, *Sapindopsis* foliage is associated with inflorescences of the platanoid type. These observations have led to the suggestion that in reproductive structure, as well as leaves, the platanoid and *Sapindopsis* groups may have exhibited significant similarities (HICKEY & DOYLE 1977). Recent work in the Potomac Group of eastern North America has provided some support for this conclusion (CRANE & al. 1986 b). At the West Brothers locality *Sapindopsis* foliage dominates the leaf flora, and in addition to platanoid inflorescences sieving of unconsolidated sediments has yielded other more or less spherical inflorescences that are sessile along the inflorescence axis (CRANE & al. 1986 b). Each spherical head consists of numerous small flowers in which the central floral organs are arranged in fives. It remains to be determined whether these organs are stamens or carpels but as in the platanoid flowers they are surrounded by prominent perianth parts. These reproductive structures differ from the platanoid inflorescences that occur at the same locality in their size, the smaller number of flowers per inflorescence, and the presence of clumps of pollen grains with distinct endoapertures (CRANE & al. 1986 b). Cuticles isolated from these inflorescences more closely resemble those from *Sapindopsis* foliage rather than the cuticles isolated from the rare platanoid leaves at this locality (UPCHURCH pers. comm.). Although these results are preliminary the data presently available indicate that the reproductive structures of at least some Mid Cretaceous platanoid and *Sapindopsis* plants were similar.

Late Cretaceous and early Tertiary *Platanaceae*. Following their first appearance in Mid Cretaceous sediments the *Platanaceae* persist as an important component of the angiosperm fossil record throughout the Late Cretaceous and Early Tertiary. Few of these plants are well understood although detailed information is now available on several *Platanaceae* from the Upper Cretaceous (Santonian-Campanian) of southern Sweden and North Carolina (FRIIS 1985, FRIIS & al. 1988), and the Early Tertiary of Europe (CRANE & al. 1988 a) and western North America (MANCHESTER 1986).

Staminate inflorescences from the Santonian-Campanian of Europe and North America conform to the structure of Mid Cretaceous flowers in having well developed perianth parts and five stamens with short filaments and long pollen sacs in each flowers (FRIIS 1985). Flowers from southern Sweden have especially prominent perianth parts and stamens in which the connective is expanded distally to form a pronounced conical extension above the pollen sacs (FRIIS 1985, FRIIS & al. 1988). Dispersed platanoid stamens from the same locality are more like those of the extant genus in having a peltate distal expansion of the connective (FRIIS 1985). Flowers from North Carolina have a similar peltate connective and also

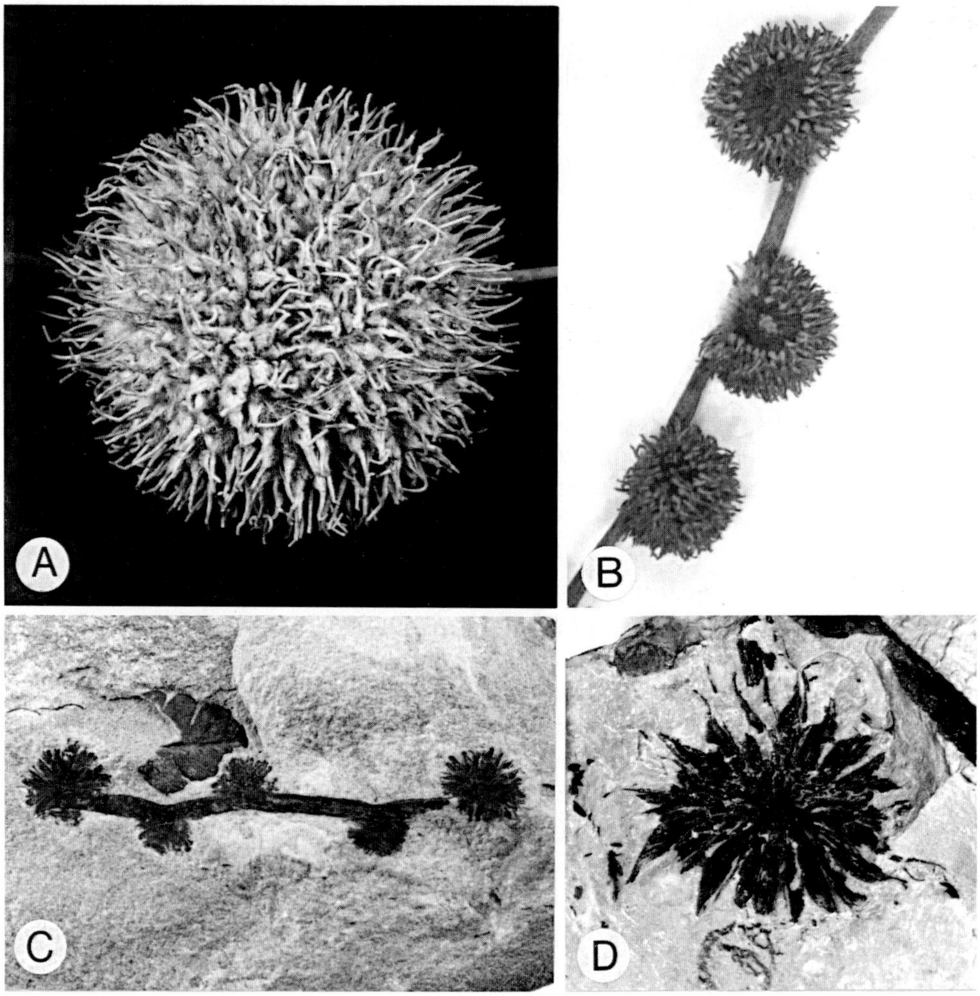

Fig. 11. Extant and fossil platanoid infructescences. *A Platanus orientalis* L., × 2; *B Platanus kerrii* GAGNEPAIN (BM, Kerr, April 28, 1932), × 2; *C "Sparganium" aspensis* BROWN, Aspen Shale (see BROWN 1933, late Albian-Cenomanian) Wyoming (USNM 41147), × 2; *D Macginicarpa*-like infructescence, Wind River Formation (Middle Eocene), Wyoming (PP 10151), × 2

differ from the Swedish material in having less prominent perianth parts. In all of this Late Cretaceous staminate material the pollen grains are smaller than those of the extant genus. Pistillate reproductive structures associated with the staminate material are less well preserved, but the limited information available suggests that the fruitlets were also derived from carpels with five flowers and lacked the mass of basal hairs seen in the extant genus (FRIIS & al. 1988).

The most completely understood of Early Tertiary *Platanaceae* is the "*Macginitiea angustiloba* plant" known from the middle to late Eocene of North America on the basis of leaves (Fig. 10 G), petiole and stem anatomy, pistillate inflorescences, fruits, staminate inflorescences and pollen (MANCHESTER 1986). In several respects, this plant is more similar to the Late Cretaceous *Platanaceae* than the extant genus.

Flowers are unisexual with well developed perianth parts and carpels or stamens in fives. The fruitlets show no indication of the prominent hairs that occur on those of the extant genus (Fig. 12). The pollen is also smaller (12 – 16 µm) than that of extant *Platanus*. As in the staminate flowers from Sweden the stamens have distally elongated connectives but are unusual in that the connectives of the five stamens in an individual flower are held together by a mass of hairs. The androecium from individual flowers adheres together even after it is shed from the inflorescence (MANCHESTER 1986).

Leaves of *Platanites hebridicus* FORBES from the Paleocene of northwest Scotland provide interesting information on the foliar variability among Early Tertiary elements of the platanaceous clade (CRANE & al. 1988 a). Each leaf is pinnate and consists of a shallowly three-lobed terminal leaflet and two smaller asymmetric lateral leaflets (Fig. 10 F). In all architectural features the terminal leaflet closely resembles the leaves of extant *Platanus* subg. *Platanus*, but the presence of distinct lateral leaflets is unknown in this group. Staminate reproductive structures associated with *Platanites hebridicus* resemble those of extant *Platanus* in having stamens with a peltate distal expansion of the connective, and pollen grains within the size range of those of the extant species (16 – 22 µm). The number of stamens per flower is unknown. Pistillate reproductive structures associated with *P. hebridicus* are

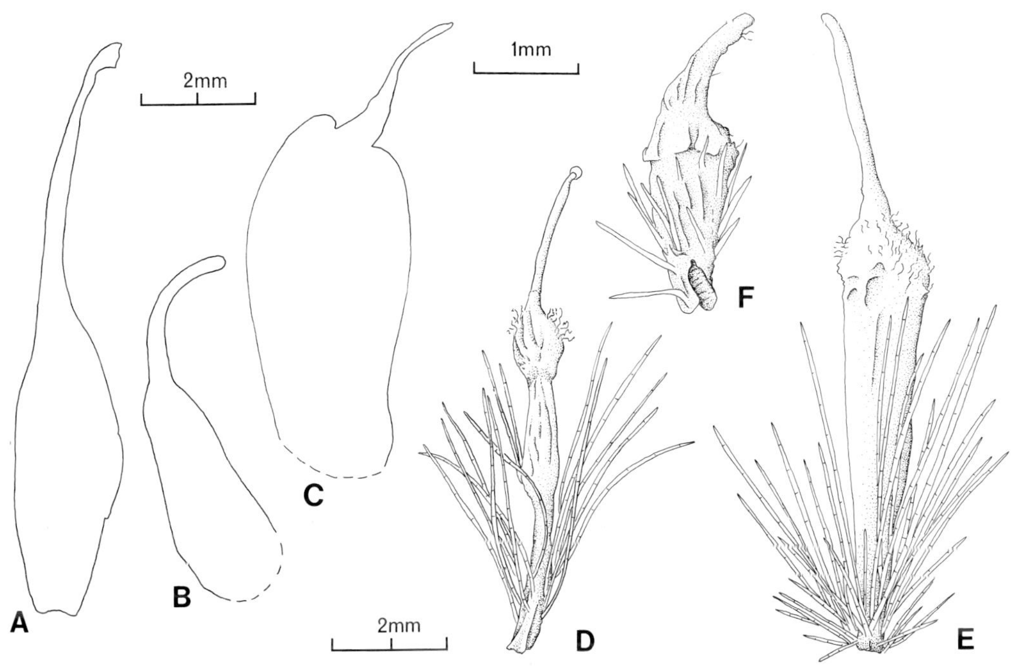

Fig. 12. Fruitlets of extant and fossil *Platanaceae*. *A* Fruitlet associated with *Platanus schimperi* (HEER) SAPORTA & MARION (PP 34558), from Woolwich & Reading Formation, Paleocene (Cold Ash Quarry, Berkshire, England); *B* fruitlet associated with *Platanites hebridicus* FORBES (v. 25058), from Interbasaltic Leaf Beds, Paleocene (Mull, Scotland); *C* Fruitlet of *Macginicarpa glabra* MANCHESTER (IU 5146), from Clarno Formation, Middle Eocene (West Branch Creek, Oregon) (after MANCHESTER 1986); *D Platanus orientalis* L.; *E Platanus mexicana* MORIC.; *F Platanus kerrii* GAGNEPAIN (BM, Kerr, April 28, 1932). All fruitlets except *F* to same scale (2 mm)

poorly understood but the number of achenes per fruiting head and achene shape resembles that in other Early Tertiary *Platanaceae* such as *Macginitiea* (Figs. 11 D and 12 B, C). The number of achenes per flower is unknown.

Macginitiea and *Platanites* both document the considerable foliar variation in the platanoid clade during the early Tertiary. Additional material from the Paleocene of North America indicates that other compound leaves previously assigned to *Cissus marginata* (LESQUEREUX) BROWN (1962), and *Negundo fremontensis* BERRY (1931) may also be platanaceous. In addition certain simple Paleocene leaves [Fig. 10 J, *Dicotylophyllum flexuosum* (NEWBERRY) WOLFE] previously assigned to *Quercus* (BROWN 1962) and *Meliosma* (HICKEY 1977) may also be referable to the family. The foliar variability of the *Platanaceae* during the Early Tertiary thus appears to have encompassed morphologies ranging from simple to palmate to pinnate (Fig. 10 F – J)

Chronological trends in the platanoid lineage. Although our current understanding of the fossil record of the *Platanaceae* is too poor for detailed cladistic analysis several chronological trends in floral structure seem to be clear. Primitive platanoid flowers were unisexual, delimited by a distinct perianth, and each contained five carpels or stamens. This basic plan seems to have persisted at least until the Eocene although throughout the Upper Cretaceous and Early Tertiary there was evidently considerable variability in other characters including the development of the perianth, the structure of stamens and leaf morphology.

In general, throughout the history of this clade there seems to be a trend toward increasing pollen size but while it is likely that the Mid Cretaceous representatives were insect-pollinated this is less certain for some Early Tertiary fossils (e.g., *Platanites*). Within the platanoid clade there has apparently been a trend toward wind-pollination although curiously the staminodes present in some pistillate flowers of the extant genus (CRONQUIST 1981) have not been detected in fossil material. The trend to anemophily was apparently accompanied by reduction in the size of the perianth parts. Extant *Platanus* seems to represent the culmination of this trend in which the perianth is so reduced that it no longer constrains the arrangement of carpels and stamens during floral development (see ENDRESS 1987 a), thus accounting for the highly variable number of achenes in extant platanaceous infructescences. Possibly associated with this process was a decrease in the number of heads per infructescence, an increase in the size of each fruiting head (Fig. 11), an increase in the number of achenes per fruiting head (Fig. 11) and a concomitant change in the shape of platanaceous achenes from the ellipsoidal forms of the Early Tertiary to the obovoid forms of the extant genus (Figs. 11 and 12). Possibly the characteristic dispersal hairs associated with the achenes of extant *Platanus* (Fig. 12 D – F) also became well developed at this time.

In the fossil record it is difficult to assess the time of first appearance of this suite of characters associated with extant *Platanus*, but fruiting heads with densely crowded obovoid achenes that unequivocally have a distinct tuft of hairs at the base are first recognized in the latest Eocene or earliest Oligocene of the John Day Basin in eastern Oregon (MANCHESTER pers. comm.). Whether these fossils exhibit the floral characteristics of the extant genus is currently unknown.

Platanoid-*Debeya-Dewalquea* complex. Palmately compound leaves with a distinctive pedate arrangement of leaflets are common in Late Cretaceous and Early

Tertiary floras from the Northern hemisphere and have been assigned to the genera *Debeya* and *Dewalquea* (Fig. 10 M – P) (SAPORTA & MARION 1873, KNOWLTON 1917, STOCKMANS 1932, KRASSILOV 1973, NĚMEJC & KVAČEK 1975). *Debeya-Dewalquea* foliage has been compared to leaves of extant *Araliaceae, Sterculiaceae*, and certain *Ranunculaceae* (*Helleborus*) but their systematic position has generally been considered uncertain (SAPORTA & MARION 1873, KNOWLTON 1917). JOHNSON & GILMORE (1921) compared *Dewalquea* to leaves of extant *Juglandaceae* but more recently it has been recognized that *Debeya-Dewalquea* leaves exhibit similarities in morphology, venation and cuticular structure to extant and fossil *Fagaceae* and *Platanaceae*. Some taxa (*Debeya pentaphylla* VELENOVSKÝ, *Debeya coricea* (VELEN.) KNOBL., KVAČEK 1983) are particularly similar to the "*Aralia*" and "*Platanus*" components of the Mid Cretaceous platanoid complex (NĚMEJC & KVAČEK 1975, KVAČEK 1983) while others [*Debeya insignis* (HOS. & MARCK.) KNOBL., *Dewalquea hibernica* JOHNSON & GILMORE 1921, NĚMEJC & KVAČEK 1975, KVAČEK 1983] have been compared with leaves of extant *Castanopsis* (*Fagaceae*, RÜFFLE 1978) or the simple leaved *Platanaceae*. *Dicotylophyllum flexosum* (NEWBERRY) WOLFE (Paleocene, CRANE & al. 1988 b) *Platanus kerrii* GAGNEPAIN (extant), and *P. neptuni* BŮŽEK, HOLÝ & KVAČEK (1967, Miocene). *Dewalquea fraxinifolia* JOHNSON & GILMORE (1921) has been reassigned to *Platanus* largely on the basis of cuticular similarities (WALTHER 1985).

Leaves of *Dewalquea pulchella* KNOWLTON (1917) (Fig. 10 O) are abundant in the Frontier Formation (Cenomanian-Turonian, NICHOLS & JACOBSON 1982) near Cumberland Gap in southwestern Wyoming. The leaves are five-lobed and exhibit considerable variation in the extent to which the lamina is divided (KNOWLTON 1917, CRANE unpubl. results). In many specimens the lamina of the central lobe is free from that of the lateral lobes and may have a well developed petiolule (cf. Fig. 10 N, P). Isolated leaflets of this material closely resemble leaves of *Dryophyllum* (Fig. 10 K) and leaflets of other *Dewalquea* species have occasionally been assigned to this genus (RÜFFLE 1978). *Dryophyllum* is known to be related to extant *Fagaceae* based on studies of leaf venation and cuticles in Early Tertiary material (DILCHER 1971, RÜFFLE & al. 1976, JONES 1986).

The *Debeya-Dewalquea* complex has therefore been regarded as "intermediate" between Mid Cretaceous platanoids and Late Cretaceous-Early Tertiary *Fagaceae* (RÜFFLE & KNAPPE 1977, RÜFFLE 1980, JONES 1986). In certain features of venation, details of marginal teeth and cuticular structure these leaves overlap with the foliar variability seen in the platanoid-*Sapindopsis* complex (Figs. 10 and 13). Information on the reproductive structures of these plants will be necessary to clarify further their phylogenetic position.

Discussion and conclusions

The paleobotanical record indicates that the lineages leading to extant *Chloranthaceae*, *Platanaceae* and *Trochodendrales* were already differentiated in the Mid Cretaceous, and that the Aptian-Cenomanian interval was the critical phase during which nonmagnoliid dicotyledons underwent their initial major radiation. Although the *Magnoliidae* were also present and diversifying at this time (*Lactoridaceae*, ZAVADA & TAYLOR 1987, *Magnoliaceae*, DILCHER & CRANE 1984, ?*Myristicaceae*, WALKER & WALKER 1984, *Winteraceae*, WALKER & al. 1983, *Magnoliidae* incertae

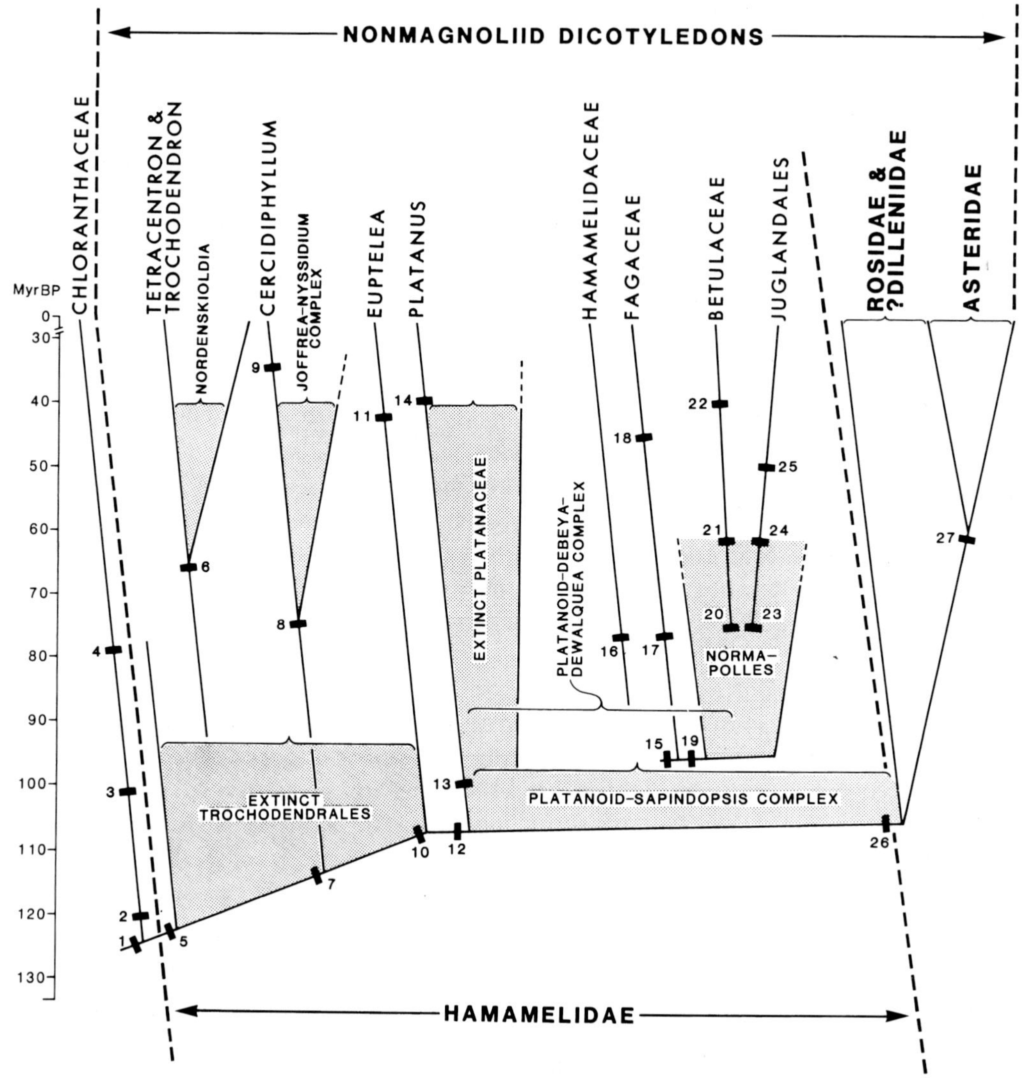

Fig. 13. Diagram summarizing hypothesized relationships between selected dicotyledons showing approximate times of appearance in the fossil record and possible position of critical fossil taxa. The *Caryophyllidae* and *Ranunculidae* are omitted. Numbers refer to possible autapomorphies and synapomorphies and are summarized below (see also text). Cited literature is the source for systematic and stratigraphic data. Specific reference to time of appearance in the fossil record is only made if this conflicts with the position on the diagram dictated by patterns of relationships. 1 Valvate anther dehiscence, lost in *Cercidiphyllum* and nodes 15 and 26, rudimentary in *Euptelea* (ENDRESS 1986 a, 1987 b); not known in the fossil record until Mid Albian (FRIIS & al. 1986): chloranthoid teeth, modified at node 10 (HICKEY & WOLFE 1975, ENDRESS 1986 a); not known in fossil record until the Aptian (UPCHURCH 1984 b): reduction of tepals, ? replaced by "staminodal petals" at node 12 (WALKER & WALKER 1984, ENDRESS 1986 a); earliest occurrence in fossil record uncertain. 2 *Clavatipollenites hughesii* (HUGHES & al. 1979) indicating possible presence of *Ascarina-Hedyosmum* type *Chloranthaceae* (WALKER & WALKER 1984), similar grains may extend into the Hauterivian (HUGHES & MCDOUGALL 1987). 3 Trilobed androecium as in *Chloranthus* but each stamen with two pairs of pollen sacs as in *Sarcandra* (FRIIS & al. 1986, CRANE & al. in prep.). 4 Trilobed androecium with one pair of pollen sacs on each

sedis, e.g., *Lesqueria*, "?*Williamsonia*" *recentior*, Crane & Dilcher 1984, *Caloda*, Dilcher & Kovach 1986, *Prisca* Retallack & Dilcher 1981) the radiation of "higher" dicotyledons may have been a major factor in the marked reduction in diversity, and eventual elimination of several groups of fossil gymnosperms. In association with this radiation, several features including triaperturate pollen grains, vessels, palmate and pinnate foliage, and pentamerous floral organization appear in the fossil record for the first time. In addition, the stabilization of whorled, pentamerous floral anthotaxy may have been a significant factor in facilitating further modification of floral structure and the elaboration of associated pollination mechanisms. Although these features were evidently acquired independently in several angiosperm lineages (Bailey 1944, Kubitzki 1987), it was at the level of

---

lateral lobe as in *Chloranthus* (Friis & al. 1986, Crane & al. in prep.). 5 Triaperturate pollen (Doyle 1978) not known in fossil record until Barremian-Aptian boundary: actinodromous leaf venation; not known in fossil record until Albian (Hickey & Doyle 1977). 6 *Nordenskioldia, Tetracentron, Trochodendron* clade defined by presence of a sessile syncarpous semi-inferior gynoecium (Endress 1986a, Crane & al. 1986a, 1988b). 7 Loss of "tepalar" petals (see Walker & Walker 1984, Endress 1986a); earliest occurrence in fossil record uncertain: vessels (Bailey 1944, Stopes 1912, Hughes 1976). 8 Inflorescences diagnostic of *Cercidiphyllaceae* including the *Joffrea-Nyssidium* complex (Endress 1986a, Dorf 1942, Crane 1984a, Crane & Stockey 1985). 9 Inflorescences diagnostic of extant *Cercidiphyllum* (Jähnichen & al. 1980, Manchester pers. comm.). 10 Reduction of lateral veins of chloranthoid teeth (Hickey & Wolfe 1975). 11 Pollen diagnostic of extant *Euptelea* (Muller 1981). 12 Flowers with five stamens and/or carpels (Crane & al. 1986b, Friis & al. 1988, Kubitzki 1987), modified at node 15, extant *Platanus* (see text) and many *Rosidae:* origin of "staminodal petals" (see Walker & Walker 1984). 13 Unisexual platanoid flowers (Crane & al. 1986b, Friis & al. 1988). 14 Obovoid achenes with basal hairs diagnostic of extant *Platanus* (Manchester, pers. comm.). 15 Modification of apocarpous pentamerous gynoecium to syncarpous tricarpellate ovary (Kubitzki 1987, Friis & Crepet 1987). 16 Flowers of *Hamamelidoideae* with a pair of anther thecae opening by a single valve (Friis 1985), diagnostic of extant *Dicoryphe, Ostrearia, Neostrearia, Noahdendron, Trichocladus,* and *Hamamelis* (also *Exbucklandia, Symingtonioideae* Endress 1989). 17 Bisexual flowers of probable *Fagales* (Tiffney & Friis in Tiffney 1986). 18 Cupulate fruits ? diagnostic of extant *Quercus* or *Lithocarpus* (Bones 1979) and ? *Castanea* (Crepet & Daghlian 1980); infructescences and inflorescences of trigonobalanoid and *Castanopsis/Chrysolepis* type first appear slightly earlier around the Paleocene/Eocene boundary (Crepet & Nixon 1986). 19 Triporate pollen (Doyle 1978). 20 Pollen with arci diagnostic of extant *Alnus* (*Betulea*) (Miki 1977, Muller 1981). 21 Fruits diagnostic of extant *Corylus* (*Coryleae*), *Palaeocarpinus* also present (Crane 1981, 1984b). 22 Fruits diagnostic of extant *Carpinus, Ostrya* and extinct *Asterocarpinus* (*Coryleae,* Crane 1984, Manchester & Crane 1987). 23 Bisexual flowers with in situ Normapolles pollen diagnostic of extant *Juglandales* (Friis 1983). 24 Fruits diagnostic of extant *Cyclocarya* (Manchester & Dilcher 1982, Crane & al. 1988b, Manchester 1987). 25 Infructescences, fruits, inflorescences, pollen and leaves diagnostic of extant *Platycarya* (Reid & Chandler 1933, Wing & Hickey 1984, Manchester 1987). 26 Pinnatifid and pinnately compound leaves (Hickey & Wolfe 1975, Doyle & Hickey 1976, Hickey & Doyle 1977). 27 Sympetalous flowers "in which the stamens are isomerous and alternate with the corolla-lobes, or fewer than the corolla lobes" (Cronquist 1981: 852): earliest occurrence in fossil record uncertain: pollen diagnostic of *Apocynaceae* and *Gentianaceae* (Muller 1984)

the *Hamamelidae* that they apparently became fixed as part of the "Bauplan" of the majority of living dicotyledons.

Although the *Chloranthaceae*, *Trochodendrales* and *Platanaceae* may have been involved in the transition from the magnoliid grade to "higher" dicotyledons the details of how these taxa are related remain uncertain. One hypothesis of relationships among these groups is given in Fig. 13, based largely on the views of others (HICKEY & WOLFE 1975, DOYLE & HICKEY 1976, ENDRESS 1977, CRONQUIST 1981, WALKER & WALKER 1984) as well as the recently obtained neobotanical and paleobotanical information reviewed here. The suggestions embodied in this hypothesis need to be tested by a more comprehensive analysis of extant taxa, but it is also clear that the considerable extinct diversity in the *Hamamelidae* will require a more refined understanding of fossil taxa for detailed resolution of phylogenetic relationships (see CRANE 1985, DOYLE & DONOGHUE 1987). Important paleobotanical priorities will be: recognition of the pistillate reproductive structures of Mid Cretaceous *Chloranthaceae*, additional information on early *Trochodendrales*, and clarification of reproductive structures in the *Sapindopsis* and *Debeya-Dewalquea* groups. Much of this work requires emphasis on the Mid Cretaceous fossil record, but experience with extinct *Platanaceae* (*Macginitiea*), *Trochodendrales* (*Joffrea-Nyssidium, Nordenskioldia*) and *Fagaceae* (*Fagopsis,* MANCHESTER & CRANE 1983) suggests that useful information can also be obtained from Early Tertiary material. With respect to the "higher *Hamamelidae*" (e.g., *Betulaceae, Juglandaceae, Myricaceae*) the major outstanding question concerns the position of plants that produced the abundant and diverse triporate *Normapolles* pollen in eastern North America and Europe during the Late Cretaceous. While it is now clear that at least some *Normapolles* taxa were closely related to the extant *Juglandales* (FRIIS 1983, 1984, 1985), their position with respect to the *Fagales* and other *Hamamelidae* remains unresolved and will be critical to further clarification of relationships among extant *Hamamelidae*.

Note added in proof. The demonstration of vessels in the root secondary xylem of *Sarcandra glabra* (THUNB.) NAKAI (CARLQUIST, S., 1987: Presence of vessels in wood of *Sarcandra* (*Chloranthaceae*): comments on vessel origins in angiosperms. – Amer. J. Bot. **74**: 1765 – 1771) is consistent with the pattern of relationships among extant *Chloranthaceae* summarized in Fig. 3 B. If vessels are general within the genus then there is no necessity to invoke either loss of vessels in *Sarcandra,* or parallel acquisition of vessels in *Chloranthus* and the *Ascarina-Hedyosmum* clade.

I am grateful to Prof. H. MEUSEL, Prof. F. EHRENDORFER and the Deutsche Akademie der Naturforscher Leopoldina for the opportunity to participitate in this symposium. I am indebted to many colleagues, especially D. L. DILCHER, J. A. DOYLE, L. RÜFFLE, G. R. UPCHURCH, J. A. WALKER, and J. A. WOLFE for helpful discussions of some of the ideas presented in this paper. I am particularly grateful for the advice and encouragement of P. K. ENDRESS, E. M. FRIIS, S. R. MANCHESTER, and K. RAUNSGAARD PEDERSEN as well as their participation in much of the work discussed in this paper. I also thank D. W. KRAUSE, P. & C. RECHTEN, P. & D. HERENDEEN, and A. ROURKE for assistance in the field. J. KISSINGER, D. BETINIS, and C. RICHARDSON helped to prepare the illustrations. This work was supported in part by NSF grants BSR 8314592 and BSR 8708460.

References

BAILEY, I. W., 1944: The development of vessels in angiosperms and its significance in morphological research. – Amer. J. Bot. **31**: 421 – 428.

Becker, H. F., 1973: The York Ranch flora of the Upper Ruby River Basin, southwestern Montana. – Palaeontographica Abt. B, Paläophytol. **143**: 18–93

Bell, W. A., 1956: Lower Cretaceous floras of Western Canada. – Geol. Surv. Canad. Mem. **285**: 1–331

Berry, E. W., 1921 a: The flora of the Woodbine Sand at Arthurs Bluff, Texas. – Profess. Pap. U.S. Geol. Surv. **129 G**: 153–180.

– 1921 b: The flora of the Cheyenne Sandstone of Kansas. – Profess. Pap. U.S. Geol. Surv. **129 I**: 199–226.

– 1931: A flora of Green River age in the Wind River Basin of Wyoming. – Profess. Pap. U. S. Geol. Surv. **165**: 55–80.

Bones, T. J., 1979: Atlas of fossil fruits and seeds from north-central Oregon. – Oregon Mus. Sci. Ind. Occ. Pap. Nat. Sci. **1**: 1–23.

Brenner, G. J., 1963: The spores and pollen of the Potomac Group of Maryland. – Bull. State of Maryland Dept. Geol., Mines & Water Res. **27**: 1–215.

– 1984: Late Hauterivian angiosperm pollen from the Helez Formation, Israel. – Abstr. 6th Int. Palynol. Conf., Calgary **1984**: 15.

Brown, R. W., 1933: Fossil plants from the Aspen Shale of southwestern Wyoming. – Proc. U.S. Natl. Mus. **82**: 1–10.

– 1936: Additions to some fossil floras of the western United States. – Profess. Pap. U.S. Geol. Surv. **186**: 163–206

– 1939: Fossil leaves, fruits and seeds of *Cercidiphyllum*. – J. Paleontol. **13**: 485–499.

– 1962: Palaeocene flora of the Rocky Mountains and Great Plains. – Profess. Pap. U.S. Geol. Surv. **375**: 1 119

Bůžek, C., Holý, F., Kvaček, Z., 1967: Eine bemerkenswerte Art der Familie *Platanaceae* Lindl. (1836) im nordböhmischen Tertiär. – Monatsber. Deutsch. Akad. Wiss. Berlin **9**: 203–215.

Chandler, M. E. J., 1961: The Lower Tertiary floras of Southern England, I, Paleocene floras; London Clay Flora (Supplement): & Atlas. – London: British Museum Natural History.

Chandrasekharam, A., 1974: Megafossil flora from the Genesee locality, Alberta, Canada. – Palaeontographica Abt. B, Paläophytol. **147**: 1–41

Cornet, B., 1979: Angiosperm-like pollen with tectate-columellate wall structure from the Upper Triassic (and Jurassic) of the Newark Supergroup, U.S.A. – Palynology **3**: 281–282.

Couper, R. A., 1958: British Mesozoic microspores and pollen grains. – Palaeontographica Abt. B, Paläophytol. **103**: 75–119.

Crane, P. R., 1981: Betulaceous leaves and fruits from the British Upper Palaeocene. – Bot. J. Linn. Soc. **83**: 103–136.

– 1984 a: A re-evaluation of *Cercidiphyllum*-like plant fossils from the British Early Tertiary. – Bot. J. Linn. Soc. **89**: 199–230.

– 1984 b: Early fossil history of the *Betulaceae*: a preliminary report. Amer. J. Bot **71(5,2)**: 109.

– 1985: Phylogenetic analysis of seed plants and the origin of angiosperms. – Ann. Missouri Bot. Gard. **72**: 716–793.

– 1987: Vegetational consequences of the angiosperm diversification. – In Friis, E. M., Chaloner, W. G., Crane, P. R., (Eds.): The origins of angiosperms and their biological consequences, pp. 107–144. – Cambridge: Cambridge University Press.

– Dilcher, D. L., 1984: *Lesqueria*: an early angiosperm fruiting axis from the mid-Cretaceous. – Ann. Missouri Bot. Gard. **71**: 384–402.

– Stockey, R. A., 1985: Growth and reproductive biology of *Joffrea speirsii* gen. et sp. nov., a *Cercidiphyllum*-like plant from the Late Paleocene of Alberta, Canada. – Canad. J. Bot. **63**: 340–364.

CRANE, P. R., STOCKEY, R. A., 1986: Morphology and development of pistillate inflorescences in extant and fossil *Cercidiphyllaceae*. – Ann. Missouri Bot. Gard. **73**: 382–393.

– DILCHER, D. L., MANCHESTER, S. R., 1986 a: *Nordenskioldia:* a vesselless angiosperm from the Paleocene of North America. – Amer. J. Bot. **73** (5): 697

– FRIIS, E. M., PEDERSEN, K., 1986 b: Lower Cretaceous angiosperm flowers: fossil evidence on early radiation of Dicotyledons. – Science **232**: 852–854.

– MANCHESTER, S. R., DILCHER, D. L., 1988 a: The morphology and relationships of *Platanites hebridicus* FORBES from the Palaeocene of Scotland. – Palaeontology, (in press).

– – – 1988 b: A preliminary survey of fossil leaves and well preserved reproductive structures from the Sentinel Butte Formation (Paleocene) near Almont, North Dakota, U.S.A. – Fieldiana, Geol., (in press).

CREPET, W. L., DAGHLIAN, C. P., 1980: Castaneoid inflorescences from the Middle Eocene of Tennessee and the diagnostic value of pollen (at the subfamily level) in the *Fagaceae*. – Amer. J. Bot. **67**: 739–757.

– NIXON, K. C., 1986: The earliest megafossil evidence of the *Fagaceae* and its implications. – Amer. J. Bot. **73** (5): 758 (Abstract).

CRONQUIST, A., 1981: An integrated system of classification of flowering plants. – New York: Columbia University Press.

DILCHER, D. L., 1971: A revision of the Eocene flora of southeastern North America. – Palaeobotanist **20**: 7–18.

– 1979: Early angiosperm reproduction: an introductory report. – Rev. Palaeobot. Palynol. **27**: 291–328.

– CRANE, P. R., 1984: *Archaeanthus:* an early angiosperm from the Cenomanian of the western interior of North America. – Ann. Missouri Bot. Gard. **71**: 351–383.

– KOVACH, W. L., 1986: Early angiosperm reproduction: *Caloda delevoryana* gen. et sp. nov., a new fructification from the Dakota Formation (Cenomanian) of Kansas. – Amer. J. Bot. **73**: 1230–1237.

DORF, E. R., 1942: Upper Cretaceous floras of the Rocky Mountains region. – Publ. Carnegie Inst. Wash., Contr. Paleontol. **508**: 1–168.

DOYLE, J. A., 1969: Cretaceous angiosperm pollen of the Atlantic coastal plain and its evolutionary significance. – J. Arnold Arbor. **50**: 1–35.

– 1978: Origin of angiosperms. – Ann. Rev. Ecol. Syst. **9**: 365–392.

– DONOGHUE, M. J., 1986: Seed plant phylogeny and the origin of angiosperms: An experimental cladistic approach. – Bot. Rev. **52**: 321–431.

– – 1987: The importance of fossils in elucidating seed plant phylogeny and macroevolution. – Rev. Paleobot. Palynol. **50**: 63–96.

– HICKEY, L. J., 1976: Pollen and leaves from the Mid Cretaceous Potomac Group and their bearing on early angiosperm evolution. – In BECK, C. B., (Ed.): Origin and early evolution of angiosperms, pp. 139–206. – New York: Columbia University Press.

ENDRESS, P. K., 1977: Evolutionary trends in the *Hamamelidales-Fagales* group. – Pl. Syst. Evol., Suppl. **1**: 321–347.

– 1986 a: Floral structure, systematics, and phylogeny in *Trochodendrales*. – Ann. Missouri Bot. Gard. **73**: 297–324.

– 1986 b: Reproductive structures and phylogenetic significance of extant primitive angiosperms. – Pl. Syst. Evol. **152**: 1–28.

– 1987 a: Floral phyllotaxis and floral evolution. – Bot. Jahrb. Syst. **108**: 417–438.

– 1987 b: The *Chloranthaceae*: reproductive structures and phylogenetic position. – Bot. Jahrb. Syst. **109**: 153–226.

– 1989: Aspects of evolutionary differentiation of the *Hamamelidaceae* and the Lower *Hamamelididae*. – Pl. Syst. Evol. **162**: 193–211.

FRIIS, E. M., 1983: Upper Cretaceous (Senonian) floral structures of juglandalean affinity containing Normapolles pollen. – Rev. Palaeobot. Palynol. **39**: 161–188.

FRIIS, E. M., 1984: Preliminary report of Upper Cretaceous angiosperm reproductive organs from Sweden and their level of organization. – Ann. Missouri Bot. Gard. **71**: 403–418.
– 1985: Structure and function in Late Cretaceous angiosperm flowers. – Biol. Skr. **25**: 1–37.
– CREPET, W. L., 1987: Time of appearance of floral features. – In FRIIS, E. M., CHALONER, W. G., CRANE, P. R., (Eds.): The origins of angiosperms and their biological consequences, pp. 145–179. – Cambridge: Cambridge University Press.
– CRANE, P. R., RAUNSGAARD PEDERSEN, K., 1986: Floral evidence for Lower Cretaceous chloranthoid angiosperms. – Nature **320**: 163–164.
– – – 1988: Reproductive structures of Cretaceous *Platanaceae*. – Biol. Skr. **31**: 1.
HEER, O., 1870: Die miozäne Flora und Fauna Spitzbergens. – Kongl. Svenska Vetenskapsakad. Handl. **8** (7): 1–98.
HICKEY, L. J., 1977: Stratigraphy and paleobotany of the Golden Valley Formation Early Tertiary of western North Dakota. – Mem. Geol. Soc. Amer. **150**: 1–181.
– DOYLE, J. A., 1977: Early Cretaceous fossil evidence for angiosperm evolution. – Bot. Rev. **43**: 3–104.
– WOLFE, J. A., 1975: The bases of angiosperm phylogeny: vegetative morphology. – Ann. Missouri Bot. Gard. **62**: 538–589.
HUGHES, N. F., 1976: Palaeobiology of angiosperm origins. – Cambridge: Cambridge University Press.
– MCDOUGALL, A. B., 1987: Records of angiospermid pollen entry into the English Early Cretaceous succession. – Rev. Palaeobot. Palynol. **50**: 255–272.
– DREWRY, G. E., LAING, J. F., 1979: Barremian earliest angiosperm pollen. – Palaeontology **22**: 513–535.
ILJINSKAJA, J., 1974: *Trochodendroides* BERRY, *Nyssidium* HEER. – In TAKHTAJAN, A. L., (Ed.): *Magnoliophyta* Fossilia U.R.S.S., I, *Magnoliaceae*–*Eucommiaceae*, pp. 115–124. – Leningrad: Nauka.
JÄHNICHEN, H., MAI, D. H., WALTHER, H., 1980: Blätter und Früchte von *Cercidiphyllum* SIEBOLD and ZUCCARINI im mitteleuropäischen Tertiär. – Schriftenreihe Geol. Wiss. Berlin **16**: 357–399.
JOHNSON, T., GILMORE, J. G., 1921: The occurrence of *Dewalquea* in the coal-bore at Washing Bay. – Scient. Proc. Roy. Dublin Soc. **16**: 323–333.
JONES, J. H., 1986: Evolution of the *Fagaceae:* the implications of foliar features. – Ann. Missouri Bot. Gard. **73**: 228–275.
KNAPPE, H., RÜFFLE, L., 1975: Beiträge zu den Platanaceen-Funden und einigen *Hamamelidales* der Oberkreide. – Wiss. Z. Humboldt-Univ. Berlin, Math.-Naturwiss. Reihe **24**: 487–492.
KNOWLTON, F. H., 1917: A fossil flora from the Frontier Formation of southwestern Wyoming. – Profess. Pap. U.S. Geol. Surv. **108 F**: 73–107.
– 1926: Flora of the Latah Formation of Spokane, Washington and Coeur d'Alene, Idaho – Profess. Pap. U.S. Geol. Surv. **140**: 17–79.
KRASSILOV, V. A., 1973: Cuticular structure of Cretaceous angiosperms from the Far East of the USSR. – Palaeontographica, Abt. B, Paläophytol. **142**: 105–116.
– 1976: The Tsagayan Flora of Amur Region. – Moscow: Nauka, (in Russian).
KRYSHTOFOVICH, A. N., 1956: Oligocene flora of Mount Ashutus in Kazakstan – In KOMAROV, L., (Ed.): Paleobotanica Fasciculus 1. Institutum Botanicum nomine 5. Academiae Scientarum URSS. ACTA Series 8. Moscow.
– 1958: The Cretaceous flora of Anadyrland. – In TAKHATAJAN, A. L., (Ed.): Paleobotanica Fasciculus 3, pp. 9–70. – ACTA Series 8. Moscow.
KUBITZKI, K., 1987: Origin and significance of trimerous flowers. – Taxon **36**: 21–28.
KUPRIANOVA, L. A., 1967: Palynological data for the history of the *Chloranthaceae*. – Pollen & Spores **9**: 95–100.

KVAČEK, Z., 1983: Cuticular studies in angiosperms of the Bohemian Cenomanian. – Acta Palaeontol. Polon. **28**: 159–170.

LEROY, J. F., 1982: Origine et évolution du genre *Platanus* (*Platanaceae*). – Compt. Rend. Hebd. Séances Acad. Sci. **295**: 251–254.

LESQUEREUX, L., 1892: The flora of the Dakota Group. – Monogr. U.S. Geol. Surv. **18**: 1–400.

MANCHESTER, S. R., 1986: Vegetative and reproductive morphology of an extinct plane tree (*Platanaceae*) from the Eocene of western North America. – Bot. Gaz. **147**: 200–226.

– 1987: The fossil history of the *Juglandaceae*. – Ann. Missouri Bot. Gard. Monog. **21**: 1–137.

– CRANE, P. R., 1983: Attached leaves, inflorescences, and fruits of *Fagopsis*, an extinct genus of fagaceous affinity from the Florissant Flora of Colorado, U.S.A. – Amer. J. Bot. **70**: 1147–1164.

– – 1987: A new genus of *Betulaceae* from the Oligocene of western North America. – Bot. Gaz. (Crawfordsville) **148**: 263–273.

– DILCHER, D. L., 1982: Pterocaryoid fruits *Juglandaceae* in the Paleogene of North America and their evolutionary and biogeographic significance. – Amer. J. Bot. **69**: 275–286.

MIKI, A., 1977: Late Cretaceous pollen and spore floras of northern Japan: composition and interpretation. – J. Fac. Sci. Hokkaido Imp. Univ., Ser. 4, Geol. **17**: 399–436.

MULLER, J., 1970: Palynological evidence on the early differentiation of the angiosperms. – Biol. Rev. Cambridge Philos. Soc. **45**: 417–450.

– 1981: Fossil pollen records of extant angiosperms. – Bot. Rev. **47**: 1–142.

– 1984: Significance of fossil pollen for angiosperm history. – Ann. Missouri Bot. Gard. **71**: 419–443.

NEE, M., 1981: *Platanaceae*. – In GÓMEZ-POMPA, A., (Ed.): Flora de Veracruz **19**, pp. 1–9. – Xalapa: INIREB.

NĚMEJC, F., KVAČEK, Z., 1975: Senonian plant macrofossils from the region of Zliv and Hloboká (near Cěské Budějovice) in South Bohemia. Prague: Universita Karlova Praha.

NICHOLS, D. J., JACOBSON, S. R., 1982: Palynostratigraphic framework for the Cretaceous (Albian-Maestrichtian) of the overthrust belt of Utah and Wyoming. – Palynology **6**: 119–147.

REID, E. M., CHANDLER, M. E. J., 1933: The London Clay flora. – London: British Museum Natural History.

RETALLACK, G., DILCHER, D. L., 1981: Early angiosperm reproduction: *Prisca reynoldsii* gen. et sp. nov. from mid-Cretaceous coastal deposits, Kansas, U.S.A. – Palaeontographica, Abt. B, Paläophytol. **179**: 103–137.

RÜFFLE, L., 1978: Evolutionary and ecological trends in Cretaceous floras particularly in some *Fagaceae*. – Cour. Forsch.-Inst. Senckenberg **30**: 77–83.

– 1980: Wachstums-Modus und Blatt-Morphologie bei altertümlichen *Fagales* und *Hamamelidales* der Kreide und der Gegenwart. – 100 Jahre Arboretum (1980), 329–341.

– KNAPPE, H., 1977: Entwicklungsgeschichtliche und ökologische Aspekte zur Oberkreide Flora, besonders einiger *Fagaceae* (*Hamamelididae*). – Zeit. Geol. Wiss., Berlin **5**: 269–303.

– MÜLLER-STOLL, W. R., LITKE, R., 1976: Weitere *Ranales, Fagaceae, Loranthaceae, Apocynaceae*. – Abh. Zentralen Geol. Inst. **26**: 199–282.

SAPORTA, G., MARION, A. F., 1873: Essai sur l'état de la végétation à l'epoque des marnes heersiennes de Gelinden. – Mém. Couronnés Mém. Savants Étrangers Acad. Roy. Sci. Belgique **37**: 1–95.

SCHLOEMER-JÄGER, A., 1958: Alttertiäre Pflanzen aus Flözen der Brögger-Halbinsel Spitzbergen. – Palaeontographica, Abt. B, Paläophytol. **104**: 39–103.

SCHWARZWALDER, R. N., DILCHER, D. L., 1988: Systematics and early evolution of the *Platanaceae*. – Ann. Missouri Bot. Gard. (in press).
STOCKEY, R. A., CRANE, P. R., 1983: In situ *Cercidiphyllum*-like seedlings from the Paleocene of Alberta, Canada. – Amer. J. Bot. **70**: 1564–1568.
STOCKMANS, F., 1932: Sur des épidermes de dicotylédonées (*Dewalquea gelindenensis* Saporta et Marion et *Litsea elatinervis* SAPORTA et MARION) des marnes de Gelinden (Paléocène). – Bull. Mus. Roy. Hist. Nat. Belgique **8** (9): 1–10.
STOPES, M. C., 1912: Petrifactions of the earliest European angiosperms. – Phil. Trans. Roy. Soc., ser. B **203**: 75–100.
SWAMY, B. G. L., 1953: The morphology and relationships of the *Chloranthaceae*. – J. Arnold Arbor. **34**: 375–411.
– BAILEY, I. W., 1950: *Sarcandra*, a vesselless genus of *Chloranthaceae*. – J. Arnold Arbor. **31**: 117–129.
TANAI, T., 1981: The revision of the so-called "*Cercidiphyllum*" leaves from the Paleogene of Japan. – J. Fac. Sci. Hokkaido Imp. Univ., Ser. 4, Geol., **19**: 451–484.
TIFFNEY, B. H., 1986: Fruit and seed dispersal and the evolution of the *Hamamelidae*. – Ann. Missouri Bot. Gard. **73**: 394–416.
UPCHURCH, G. R. JR., 1984 a: Cuticle evolution in Early Cretaceous angiosperms from the Potomac Group of Virginia and Maryland. – Ann. Missouri Bot. Gard. **71**: 522–550.
– 1984 b: Cuticular anatomy of angiosperm leaves from the Lower Cretaceous Potomac Group. I. Zone I leaves. – Amer. J. Bot. **71**: 192–202.
– WOLFE, J. A., 1987: Mid-Cretaceous to Early Tertiary vegetation and climate: evidence from fossil leaves and woods. – In FRIIS, E. M., CHALONER, W. G., CRANE, P. R., (Eds.): The origins of angiosperms and their biological consequences, pp. 75–105. – Cambridge: Cambridge University Press.
WALKER, J. W., WALKER, A. G., 1984: Ultrastructure of Lower Cretaceous angiosperm pollen and the origin and early evolution of flowering plants. – Ann. Missouri Bot. Gard. **71**: 464–521.
– BRENNER, G. J., WALKER, A. G., 1983: Winteraceous pollen in the Lower Cretaceous of Israel: early evidence of a magnolialean angiosperm family. – Science **220**: 1273–1275.
WALTHER, H., 1985: Das Vorkommen der Gattung *Platanus* L. im Tertiär des Weisselster-Beckens (Bezirk Leipzig, DDR). – Hallische Jahrb. Geowiss. **10**: 9–19.
WING, S. L., HICKEY, L. J., 1984: The *Platycarya* perplex and the evolution of the *Juglandaceae*. – Amer. J. Bot. **71**: 388–411.
WOLFE, J. A., 1966: Tertiary plants from the Cook Inlet Region, Alaska. – Profess. Pap. U.S. Geol. Surv. **398 B**: 1–32.
– DOYLE, J. A., PAGE, V. M., 1975: The bases of angiosperm phylogeny: paleobotany. – Ann. Missouri Bot. Gard. **62**: 801–824.
YOUNG, D. A., 1981: Are the angiosperms primitively vesselless? – Syst. Bot. **6**: 313–330.
ZAVADA, M. S., TAYLOR, T. N., 1987: Pollen morphology of *Lactoridaceae*. – Pl. Syst. Evol. **154**: 31–39.

Address of the author: PETER R. CRANE, Department of Geology, Field Museum of Natural History, Roosevelt Road at Lake Shore Drive, Chicago, IL 60605, U.S.A.

Pl. Syst. Evol. 162, 193–211

—Plant—
Systematics
and
Evolution

Aspects of evolutionary differentiation of the *Hamamelidaceae* and the Lower *Hamamelididae*

PETER K. ENDRESS

Received February 2, 1987

Key words: Angiosperms, *Hamamelididae, Trochodendrales, Hamamelidales, Hamamelidaceae, Hamamelideae,* fossil *Hamamelidaceae.* – Floral structure, fruit, seed, valvate anther dehiscence. – Floral evolution.

Abstract: New investigations on the flower and fruit structure of extant *Hamamelidaceae* and other Lower *Hamamelididae* together with new finds of fossil flowers and seeds from the Upper and Lower Cretaceous provide the outline of an increasingly more differentiated picture of the early evolution of the subclass. Three patterns of valvate anther dehiscence are recognized in the subfamily *Hamamelidoideae* (and the subclass *Hamamelididae*). The basic (plesiomorphic) type within the *Hamamelididae* has 2 valves per theca. The type with 1 valve but 2 pollen sacs per theca is both consistent and exclusive for the 5 southern genera of the *Hamamelidaceae*. They seem to be the remnants of a homogeneous group that originated before the Upper Cretaceous. This is supported by fossil hamamelidaceous flowers from the Upper Cretaceous that have thecae with 1 valve. Since several-seeded *Hamamelidaceae* predate one-seeded forms in the fossil seed record (in Europe) and the systematic structure of the one-seeded group is relatively more homogeneous, several-seeded groups are considered to be more ancient. Several parallel evolutionary trends are recognized within the *Hamamelidaceae* as well as within the Lower *Hamamelididae*: anther dehiscence with 2 valves per theca → 1 slit or 1 valve; pollen sacs per theca 2 → 1; pollen tricolpate → polyforate; exine coarsely reticulate → finely reticulate; loss of perianth (tepals or petals and sepals) and concomitant loss of fixed number of floral organs; differentiation of exposed nectaries.

That the Lower *Hamamelididae* (*Trochodendrales* and *Hamamelidales*) are systematically a closely knit group has become increasingly established (e.g., TAKHTAJAN 1969, CRONQUIST 1981, ENDRESS 1986). The focus of this study is to show how the *Hamamelidaceae* are embedded in the Lower *Hamamelididae* and the parallel evolutionary trends that are repeated in different groups. It is even unclear at present how to delimit the *Hamamelidales* from the *Trochodendrales*, mainly with regard to the *Platanaceae* and *Myrothamnaceae*. Another question is whether such isolated genera as *Eucommia* and *Daphniphyllum* belong to the subclass *Hamamelididae*. Within the *Hamamelidaceae* the tribal delimitation has not yet been sufficiently resolved. But these problems will not be discussed here.

At the present it is fascinating tracing the early evolution of the *Hamamelididae* because spectacular finds of well-preserved fossil flowers from the Cretaceous and

Table 1. Selected reproductive features of the Lower *Hamamelididae*. Special features *Eupteleaceae, Cercidiphyllaceae* (Endress 1986); *Myrothamnaceae* (Jäger-Zürn 1966); 1978, unpubl. obs.; Bogle 1970, 1984, 1986; Bogle & Philbrick 1980; Wisniewski &

Taxon		Sepals	Petals	Stamens/ Staminodes	Locules per theca	Valves per theca
Trochodendraceae						
Trochodendron	1	0–5	–	34–70	2	2
Tetracentron	1	4	–	4	2	2
Eupteleaceae						
Euptelea	2	–	–	6–19	2	–(2)
Cercidiphyllaceae						
Cercidiphyllum	2	–	–	1–7?	2	–
Myrothamnaceae						
Myrothamnus	2	–	–	3–8	2	–
Platanaceae						
Platanus	6–7	0–6	?	3–6	2	2
Hamamelidaceae						
Hamamelidoideae						
Loropetalum	1–3	4–5	4–5	4–5	2	2
Tetrathyrium	1	5	5	5	2	2
Embolanthera	2	5	5	5	2	2
Maingaya	1	5	5	5/5	2	2
Ostrearia	1	5	5	5/5	2	1
Neostrearia	1	5	5	5	2	1
Noahdendron	1	5	5	5	2	1
Trichocladus	5	5	5	5	2	1
Dicoryphe	c. 13	4–5	4–5	4–5/ 4–5	2	1
Hamamelis	c. 10	4	4	4/4	1	1
Corylopsis	7–36	5	5	5/5	2	2
Eustigma	2	5	?	5/?	2	2
Fortunearia	1	5	5	5	2	2
Sinowilsonia	1	5	0–5	5	2	2
Molinadendron	3	±5	–	5–9	2	2
Fothergilla	2	±7	–	17–24	2	2
Parrotiopsis	1	~	–	15–20	2	2
Parrotia	1	7–8	–	9–14	2	–
Sycopsis	c. 9	5–6	–	6–10	2	–
Distyliopsis	c. 6	–	–	1–8	2	–
Distylium	c. 18	–	–	1–8	2	–
Matudaea	2	–	–	12–22	2	2
Rhodoleioideae						
Rhodoleia	1–10	~	0–4	7–11	2	2
Disanthoideae						
Disanthus	1	5	5	5	2	–
Exbucklandioideae						
Exbucklandia	1–4	–	0–5	10–15	1	1
Mytilaria	1	5	5	10–13?	2	2
Chunia	1	–	–	8?	2	2
Liquidambaroideae						
Liquidambar	3–4	–	–	4–10	2	–(2)
Altingia (incl. *Semiliquidambar*)	1–13	–	–	?	2	2

characterizing certain groups of genera are in frames. Main sources: *Trochodendraceae, Platanaceae* (SCHOENLAND 1883); *Hamamelidaceae* (HARMS 1930; ENDRESS 1967, 1970, BOGLE 1982; CHANG 1973)

Pollen apertures	Carpels	Ovules per carpel	Site of nectary
3	4–17	15–30	Gynoecium
3	4	5–6	Gynoecium
3–6	8–31	1–4	–
3	1	17–24	–
3	3–4	20–30?	–
3	5–10	1	–
3	2	1	Disc
3	2	1	Disc
3	2	1	–?
3	2	1	Disc
3	2	1	–
3	2	1 (+2)	–
3	2	1 (+2)	–
3	2	1	–
3	2	1	–?
3	2	1	Staminodes
3	2	1 (+2)	Disc
3	2	1	–?
3	2	1	Disc?
3	2	1	–
3	2	1	–
3–6	2	1	–
3	2	1	–
3	2	1	–
±10	2	1	–
3	2	1	–
4–10	2	1	–
±10	2	1	–
3	2	10–20	Disc
3	2	5–6	Petals
3	2	6–8	–?
3	2	6	–?
3–6	2	6	–?
±12	2	20–30	–
±15	2	28–47	–

early Tertiary help establish a framework for the phylogenetic reconstruction, which, thus, no longer has to rely solely on extrapolations from extant material.

The most important fossil finds in this respect are hamamelidaceous flowers from the Upper Cretaceous of Sweden (FRIIS 1985 b) and and platanoid flowers from the Lower Cretaceous of North America (CRANE & al. 1986), Upper Cretaceous of Sweden (FRIIS 1984, 1985 b), and Eocene of North America (MANCHESTER 1986). Platanoid leaves have also been found from the Lower Cretaceous of North America (UPCHURCH 1985) and have been known from the Upper Cretaceous of North America and Eurasia for a long time (e.g., RÜFFLE 1968, KNAPPE & RÜFFLE 1975, BUDANTSJEV 1975, HERMAN 1984). Hamamelidaceous seeds from the Upper Cretaceous of Europe are also of great interest (KNOBLOCH & MAI 1986).

Some guiding features of extant groups for a systematic evaluation

Gynoecium, fruit, seeds. A completely uniform character in the family *Hamamelidaceae* is the gynoecium consisting of two carpels (Table 1). This is true for all 29 genera of the family. Within many species occasional deviations of this number occur (e.g., 3 or 1 carpel) but in each species the main number is two. Outside the *Hamamelidaceae*, on the other hand, the number is highly variable (Table 1). However, in *Cercidiphyllum*-like fossils dimerous gynoecia often occurred as well (CRANE & STOCKEY 1986).

Also uniform within the *Hamamelidaceae* is the stigma with multicellular, often shallow papillae (Fig. 3 e) or ridges (Fig. 3 d, f, h), while in the five other families (of Table 1) stigmatic papillae are unicellular (Fig. 3 a, b). In some *Hamamelidaceae* (e.g., *Liquidambar*) a unicellular pattern is superimposd on the multicellular basal pattern (Fig. 3 c).

On the next lower level, in the subfamilies, in the *Hamamelidoideae* the ovule number is highly constant: 1 per carpel. In all other groups (subfamilies of *Hamamelidaceae* and families of Lower *Hamamelididae*) it again varies considerably. In *Altingia* it reaches about 50 (cf. BOGLE 1986).

This fixation on a single ovule is functionally correlated with the explosive seed discharge in the *Hamamelidoideae*. Each of the two carpels of a fruit has an extremely hard endocarp. At desiccation it changes shape and moves like tongs to forcibly eject the hard, smooth, spindle-shaped seed out of the locule (Fig. 2 d – i). The anatomical-histological differentiation of the endocarp of this fruit type that causes the hygroscopical movement has been analyzed by EICHHOLZ (1886) in *Hamamelis virginiana*. This mechanism only works if a single seed is present in a locule. It further constrains the range of variability in shape of fruits and seeds. The structural range where the mechanism functions seems to be very narrow. A systematic consequence is that fruit and seed structure is extremely uniform throughout the subfamily *Hamamelidoideae*.

It is for this reason that fossil endocarps and seeds of the *Hamamelidoideae* are so difficult to determine down to the generic level. Such remains are quite common in Tertiary deposits of Europe (e.g., surveys: TAKHTAJAN 1974, GREGOR 1982; Eocene: COLLINSON 1982, 1983, 1984, MAI & WALTHER 1985; Oligocene: COLLINSON 1983, MAI & WALTHER 1978; Miocene: GREGOR 1975, 1977, 1978, 1986, FRIIS 1985 a; Pliocene: VAN DER BURGH 1983). They have usually been ascribed to a variety of extant genera. But their identity is mostly uncertain. For a better

Fig. 1. Flowers of Lower *Hamamelididae*. *a Trochodendron aralioides* Sieb. & Zucc., flower in female stage (E 6684), × 6; *b Corylopsis spicata* Sieb. & Zucc., two flowers, left in female, right in male stage (E 7150), × 16; *c Disanthus cercidifolius* Maxim., inflorescence with two flowers in female stage (E 7118), × 6; *d Hamamelis japonica* Sieb. & Zucc. var. *flavopurpurascens* Rehd., flower in male stage, pollen packets on open anther valves (E 7176), × 16; *e Trichocladus crinitus* Pers., flower in male stage, two pollen packets on each open anther valve (E 4908), × 15; *f Matudaea trinervia* Lundell, inflorescence, flowers in female stage (E 7141), × 6; *g Rhodoleia championii* Hook., f., opening inflorescence (Floyd s. n.), × 0.8; *h Trichocladus crinitus* Pers., opening flower in female stage (E 4908), × 12

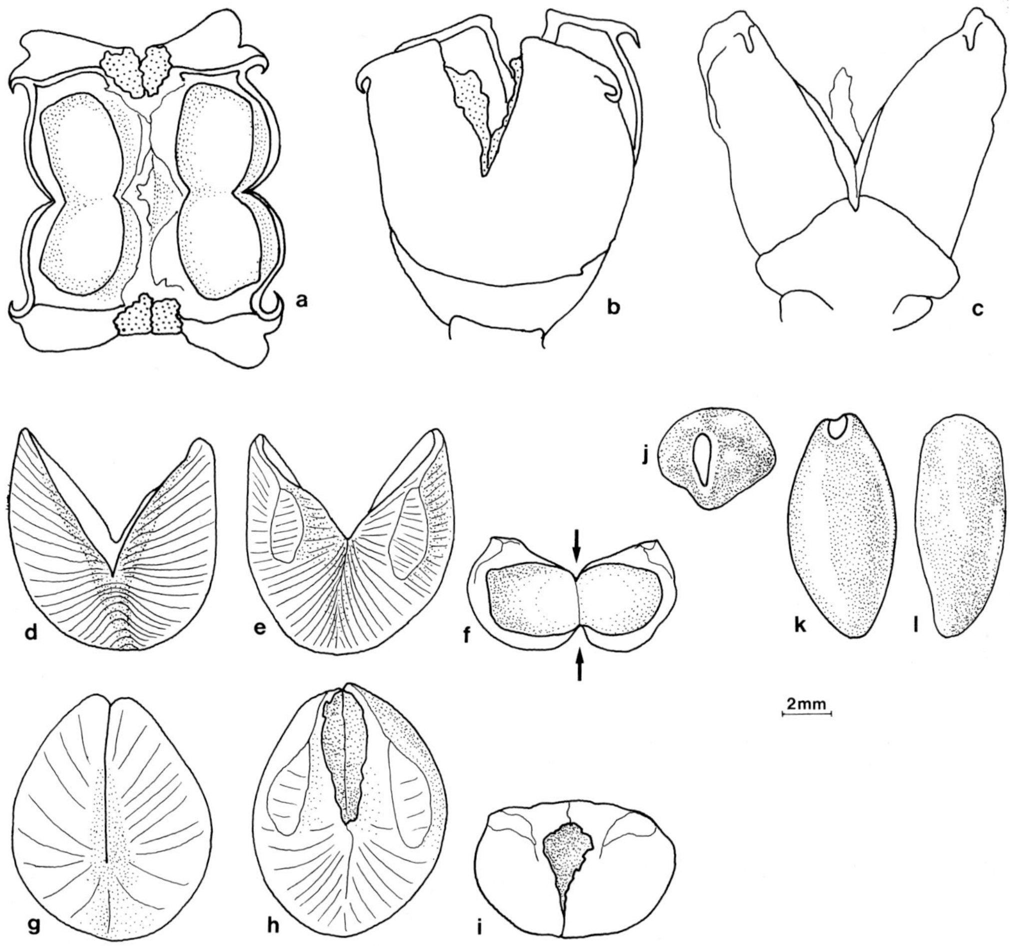

Fig. 2. Typical fruit and seed of a representative of the *Hamamelidoideae*: *Hamamelis mollis* OLIV. (E 7135). *a–c* Open fruit, seeds ejected (thick dots: septicidally dehisced faces of exocarp); *a* from above; *b* the two carpels in lateral view (with amount of septicidal dehiscence); *c* one carpel in dorsal view (with amount of loculicidal dehiscence). *d–f* Isolated endocarp of a carpel in dry condition (direction of fibres on the surface indicated); *d* from dorsal side; *e* from ventral side; *f* from above, ventral side up (arrows: tongs). *g–i* Isolated endocarp in wet condition (same views as *d–f*). *j–l* Seed; *j* from above, ventral side (with regard to carpel) up; *k* from ventral side (with regard to carpel); *l* from lateral

evaluation of the potentials and limitations of interpretations of fossil material it would be extremely helpful to have a comparative treatment of extant fruits of the *Hamamelidoideae*. The part that is probably the most useful for fruit identification, the exocarp, is often absent or only fragmentary in fossil material.

How are the fruits differentiated within the subfamilies of the *Hamamelidaceae*? In the *Hamamelidoideae* the capsule is (1) loculicidal down to at least the upper rim of the floral cup (if present) (Fig. 2 c), (2) ventricidal along the free parts of the carpel (Fig. 2 a), and (3) somewhat septicidal (Fig. 2 b). The two bony endocarps of the fruit are separate from each other by the more woody or leathery gynoecial

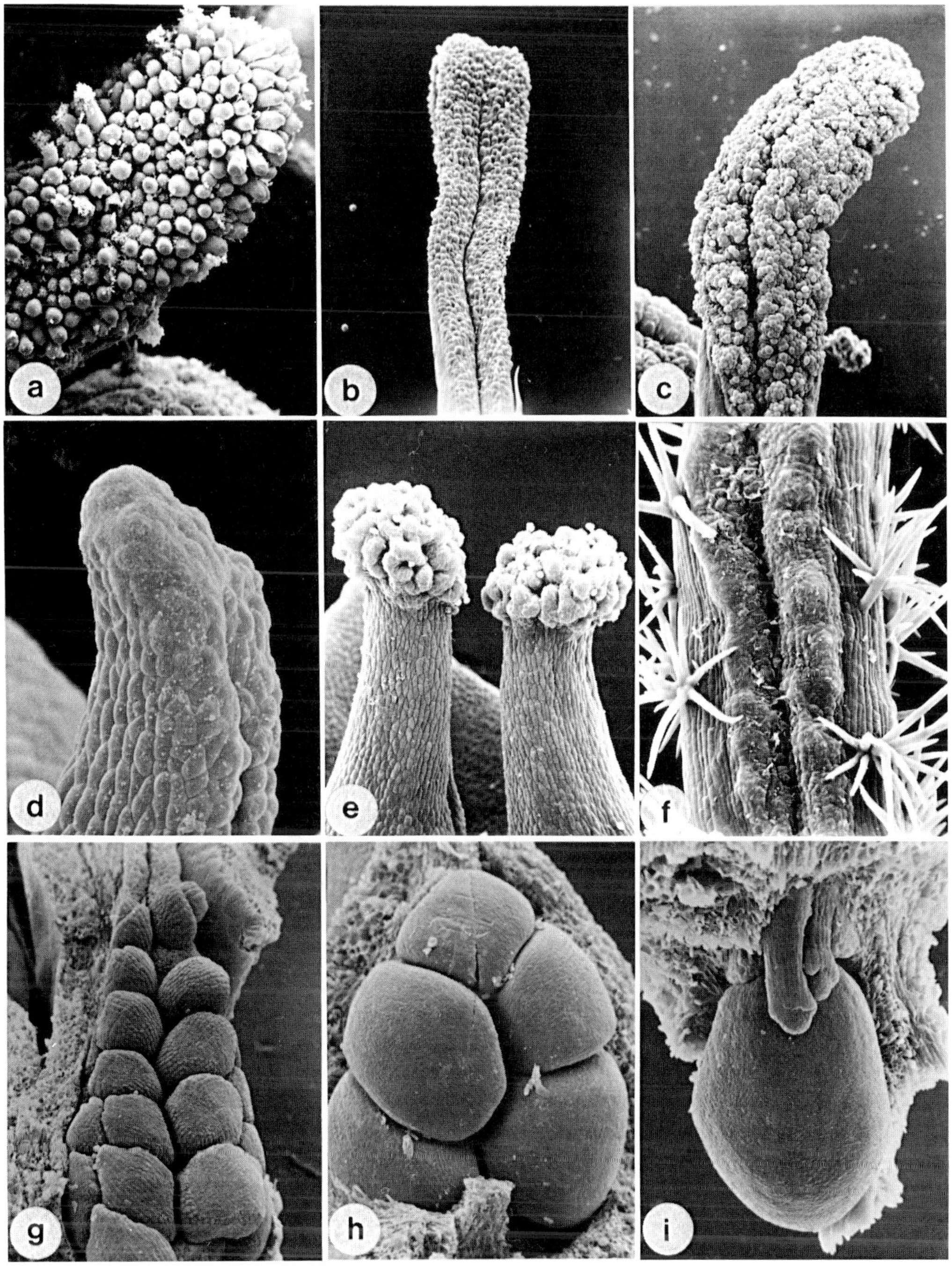

Fig. 3. *a–f* Stigmas in flowers of Lower *Hamamelididae; a Tetracentron sinense* Oliv. (Bot. Gard. Edinburgh s. n.), × 170; *b Platanus orientalis* L. (E 2032), × 50; *c Liquidambar orientalis* Mill. (E 2033), × 30; *d Disanthus cercidifolius* Maxim. (E 7118), × 320; *e Corylopsis platypetala* Rehd. & Wils. (E 3211), × 50; *f Parrotia persica* C. A. Mey. (E 465), × 110; *g–i* Ovules in a carpel at anthesis, as seen from the dorsal side; *g Liquidambar orientalis* Mill. (E 2221), × 45; *h Disanthus cercidifolius* Maxim. (E 7118), × 95; *i Corylopsis willmottiae* Rehd. & Wils. (E 3577), × 90

centre, and they become separated from the exocarp by the hygroscopic deformation at maturity. They do not extend to the top of the capsule.

The spindle-shaped seeds have a very smooth, and hard seed coat. The elongate hilum is on the broader (upper) end of the seed (Fig. 2j–l).

In the other subfamilies of the *Hamamelidaceae* fruits and seeds are more diverse. In the pluriovulate *Exbucklandia* (*Exbucklandioideae*) and in the multiovulate *Liquidambaroideae* and *Rhodoleioideae* seeds are flat, winged, and have their hilum about in the middle. In the pluriovulate *Disanthoideae* the seeds resemble more those of the *Hamamelidoideae* but they are flattened where they touch each other. It is interesting that in *Exbucklandia* (*Exbucklandioideae*) and *Rhodoleioideae* dehiscence zones and extensions are the same as described for *Hamamelidoideae*, although there is no ejecting mechanism. Although in *Exbucklandia* and *Rhodoleia* an endocarp is present, it does not separate from the exocarp nor is it bony and hard, but has about the same woody, brittle consistency as the remaining pericarp. But unlike the *Hamamelidoideae* it extends to the top of the capsule. In contrast, in *Liquidambar* (*Liquidambaroideae*), with similar seeds, the capsule is smaller, thinner-walled, more simplified, the endocarp being thin and papery. Dehiscence is not loculicidal, but only ventricidal (follicular) and septicidal. However, in *Altingia*, the other genus of the *Liquidambaroideae*, loculicidal dehiscence is common (cf., MAI 1968). This is also true for the related Tertiary genus *Steinhauera* (= *Protaltingia*) (MAI 1968, VAUDOIS-MIÉJA 1979).

A puzzling case is *Mytilaria*, related to *Exbucklandia* and placed in *Exbucklandioideae* (or in a subfamily of its own: *Mytilarioideae*, CHANG 1973). The fruits are much larger (about the largest in the family!) and thicker than in *Exbucklandia* and resemble more those of *Hamamelidoideae*. Also in *Disanthus* (*Disanthoideae*) the fruit walls are thick and have a massive bony endocarp as in *Hamamelidoideae*. But the shape of the endocarp is slightly different. It does not form tongs. An ejection mechanism is probably lacking, as would be expected since usually several seeds per carpel are developed.

A consistent difference of all non-ejecting fruits (*Disanthoideae, Rhodoleioideae, Liquidambaroideae, Exbucklandioideae*) from the ejecting fruits (*Hamamelidoideae*) is that the endocarp in the ripe, open fruit is almost undeveloped in the vertical direction on the ventral side of each carpel. This ventral extension of the fruit of the *Hamamelidoideae* is part of the tongs constituting the ejection mechanism.

In view of the evolution of one-seeded carpels with an ejecting mechanism it is interesting to consider the development of ovules and seeds in the entire group. Three observations are noteworthy: (1) Also in the subfamilies with pluri- or multiovulate carpels usually only the lowermost one or two seeds reach maturity; the others remain much smaller and are sterile (*Rhodoleioideae, Exbucklandioideae, Liquidambaroideae*; cf., VINK 1957, KAUL & KAPIL 1975, BOGLE 1986). The only exception is *Disanthoideae* with more fertile seeds per carpel. (2) On the other hand, there are at least three genera in the *Hamamelidoideae*, where regularly two reduced ovules are formed above the fertile one per carpel: *Corylopsis* (BAILLON 1894, ENDRESS 1967), *Neostrearia* (SMITH 1958), and *Noahdendron* (ENDRESS & al. 1985) (cf., Table 1). Supernumerary reduced or aborting ovules have been reported for various other genera of the *Hamamelidoideae* (e.g., HARMS 1930) but they represent abnormal cases or are based on inaccurate observations. (3) In all *Hamamelidaceae* with more than one ovule in a carpel there is an acropetal gradient of developmental

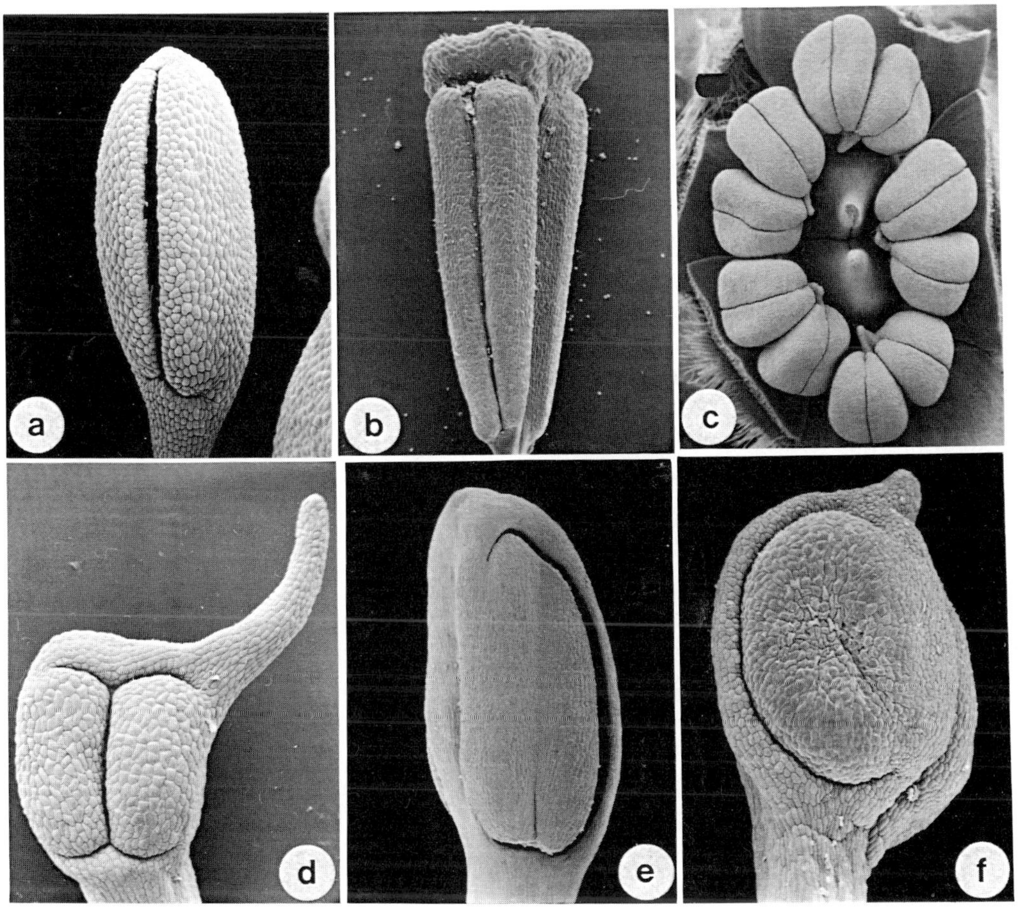

Fig. 4. Anther dehiscence patterns in Lower *Hamamelididae*, anthers from the side (except *c*); *a Trochodendron aralioides* SIEB. & ZUCC. (type 1) (E 6684), × 25; *b Platanus orientalis* L. (type 1) (E 2032); *c Disanthus cercidifolius* MAXIM., androecium from above, anthers with longitudinal slits (E 7118), × 25; *d Loropetalum chinense* R. BR. (type 1) (E 791), × 65; *e Trichocladus crinitus* PERS. (type 2) (E 4908), × 35; *f Hamamelis virginiana* L. (type 3) (E 3170), × 80

retardation mostly in correlation with eventual abortion of the retarded ovules (Fig. 3 g – i) (FLINT 1959, BOGLE 1986).

This is in contrast to the pluri- or multiovulate *Trochodendrales*, where numerous seeds per carpel reach maturity and there is no obvious gradient of retardation from base to apex. According to SMITH (1945) *Tetracentron* has on the average 6 mature seeds per carpel (only 10% sterility) and *Trochodendron* has 7 – 12 mature seeds (65% sterility).

In *Trochodendrales* capsules (*Trochodendron, Tetracentron*) and follicles (*Cercidiphyllum*) the fruit walls are thinner and the dust like or winged seeds smaller than in *Hamamelidaceae*. Fruit dehiscence is ventricidal and septicidal, and in *Cercidiphyllum* and *Trochodendron* also dorsicidal (loculicidal) for a short distance (*Tetracentron* has not been studied in this respect).

In the capsular *Trochodendrales* and many *Hamamelidaceae* the ventral region of the (apocarpous) ovary top elongates more than the dorsal region so that the styles are eventually much divergent or even recurved (extremely so in *Tetracentron*, cf., NAST & BAILEY 1945).

Androecium, anther dehiscence patterns, pollen. Anther structure and the modes of anther opening are salient features in the systematics and evolution of the subclass *Hamamelididae*. A constitutive feature of the Lower *Hamamelididae* is the presence of a connective tip and the basifixed anthers. It has long been known that in many *Hamamelidaceae* the anthers open by valves, and not by simple longitudinal slits. As it has turned out this is another constitutive feature for the entire group of the Lower *Hamamelididae* (*Hamamelidales* and *Trochodendrales*) (ENDRESS 1977, 1986). Thus, valvate anthers are not an extravagant elaboration of the *Hamamelidaceae* but a plesiomorphic feature for the subclass *Hamamelididae*. In the Lower *Hamamelididae* the feature is absent only in few genera, most of which are anemophilous (Table 1; Fig. 6). on the other hand, among the *Hamamelidaceae* those genera with the most bizarre valves are strictly entomophilous (mainly myophilous?). Valves are also lacking in the Higher *Hamamelididae*, which are predominantly wind-pollinated.

Within the *Hamamelididae* there are three types of anthers with valvate dehiscence, which are systematically also intriguing (Fig. 6). This aspect is all the more important since even in the oldest described fossil hamamelidaceous flowers from the Upper Cretaceous (FRIIS 1985 b) the anther type can be recognized (see section on phylogenetic considerations). Evidently, the most primitive anther type (type 1) in the subclass *Hamamelididae* is such that each theca opens like a window with two wings (Figs. 4 a, b, d and 5 a). The dehiscence area is a longitudinal slit that extends horizontally in both directions at its upper and lower end. This type is by far the most common in the *Hamamelidaceae*. But it also occurs in *Platanaceae, Trochodendraceae* (incl. *Tetracentron*), while in *Eupteleaceae* dehiscence is almost by simple longitudinal slit, though with rudiments of horizontal extensions, which suggests derivation from a valvate pattern as well (Fig. 6).

In another anther type (type 3) each theca has but one pollen sac. Accordingly, there is only one valve.This valve bends toward the floral centre as it opens. The dorsal pollen sac is lacking. This pattern occurs in only two genera: *Hamamelis* and *Exbucklandia* (Figs. 4 f and 5 c).

The third anther type (type 2) is the most unusual, extravagant and systematically most interesting. It is in some way between type 1 and 3. Although each theca has

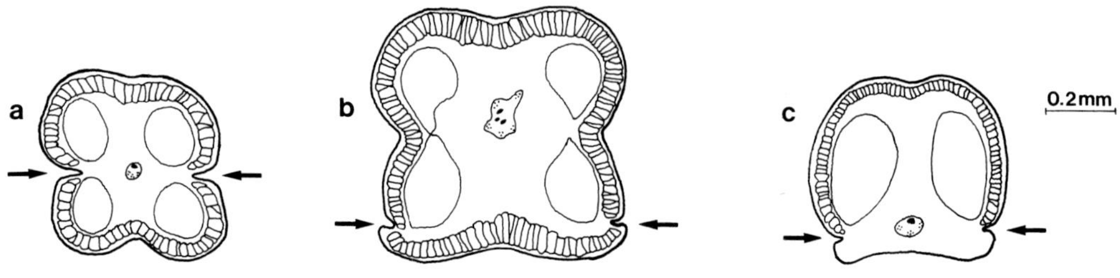

Fig. 5. Anther dehiscence patterns in *Hamamelidaceae*, anthers in transverse section, ventral side up (arrows: site of dehiscence; locules, endothecium and vascular bundle indicated); *a Loropetalum chinense* R. BR. (type 1) (E 791); *b Trichocladus crinitus* PERS. (type 2) (E 4908); *c Hamamelis virginiana* L. (type 3) (E 3170)

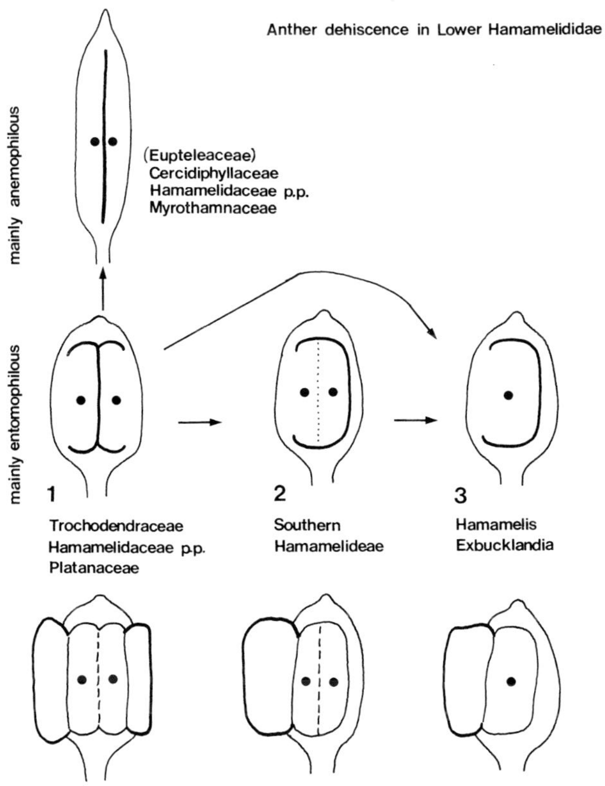

Fig. 6. Anther dehiscence patterns in Lower *Hamamelididae*, anthers from the side, ventral side on the left (arrows: possible evolutionary directions; thick dots: centre of a pollen sac); Upper row: dehiscence with longitudinal slit; Middle and lower row: the 3 patterns of valvate dehiscence; Middle row: closed anthers (thick line: dehiscence line; dotted: border line between 2 pollen sacs); Lower row: open anthers (thick line: outline of dehisced valves; interrupted line: septum between 2 pollen sacs)

two pollen sacs, it has but one valve, which serves both pollen sacs. This valve also opens toward the floral centre as in type 3 (Figs. 4 e and 5 b). Strangely enough, in some of the groups, the longitudinal part of the dehiscence line follows neither the septum between the two loculi nor the dorsal rim of the dorsal theca but extends amidst over the dorsal pollen sac (Endress & al. 1985). Type 2 is only known from the tribe *Hamamelideae* (sensu Harms 1930). Here it occurs in all five genera that are confined to the southern hemisphere, and only in these: *Trichocladus* (Southeastern Africa), *Dicoryphe* (Madagascar), *Ostrearia, Neostrearia,* and *Noahdendron* (all Northeastern Australia) (cf., Table 1). Type 3, on the other hand, is confined, within the tribe *Hamamelideae,* to the northernmost and only deciduous genus *Hamamelis*.

Either type 2 is an evolutionarily intermediate pattern between type 1 and type 3, or type 2 and 3 are parallel modifications of type 1.

Since type 3 (as probably the most derived type) is already known from the Upper Cretaceous of Sweden (cf. Figs. 2–4 in Friis 1985 b) it can be concluded that the evolutionary differentiation into the different anther types had already taken place at that time. Therefore, differentiation within the *Hamamelidaceae* into the subfamilies, tribes and certain infratribal groups could also have taken place.

As mentioned above, almost all *Hamamelididae* with valvate anther opening are zoophilous. What is the functional significance of the valves? Pollen of valvate anthers is generally very sticky (cf. Hesse 1978) and forms entire packets. When the valve opens the locule(s) often the entire pollen content is attached to the inner surface of the valve and is now exposed by the recurving of the valve (Fig. 1 b, d).

In those genera with type 2 anthers, where one valves serves two pollen sacs, each valve presents then two pollen packets (Fig. 1 e). It seems that, at least in some members of this floral type, dipters play an important role in pollination (e.g., *Hamamelis;* DE STEVEN 1983). Dipters move on the flat, widely open flowers and the pollen packets become attached to their body. An exception among fly flowers is *Disanthus* (cf. MIZUSHIMA 1968) in having anthers without valves. But here, the thecae are so much convex by enlargement of the dorsal anther surface that the longitudinal dehiscence lines are prominently curved and cause an especially wide opening of the thecae and open presentation of the pollen (Fig. 4 c). However, observations on pollination in the field are lacking for most genera.

Pollen structure is relatively uniform in the Lower *Hamamelididae*. The basic pattern is tricolpate with a reticulate exine and often conspicuous gemmate structures in the apertures (e.g., PRAGLOWSKI 1974, BOGLE & PHILBRICK 1980, ZAVADA & DILCHER 1986). This pollen type is also present in the Upper Cretaceous hamamelidaceous flowers described by FRIIS (1985 b, Fig. 4 c) and even in the Lower Cretaceous platanoid flowers (CRANE & al. 1986). In *Trochodendron* and *Tetracentron* the exine is striate rather than reticulate (ENDRESS 1986). *Myrothamnus*, although exceptional in forming permanent tetrads, resembles *Cercidiphyllum* in its exine structure (ZAVADA & DILCHER 1986).

Within the entire group of the Lower *Hamamelididae* there is a trend of parallel evolution towards more and smaller apertures and finer exine reticulum in various predominantly anemophilous groups (Table 1). This is the case in the *Eupteleaceae* among the *Trochodendrales* (PRAGLOWSKI 1974) and among the *Hamamelidales* in the *Liquidambaroideae*, in some apetalous and perianthless genera of the *Hamamelidoideae* (*Sycopsis, Fothergilla, Distylium, Matudaea*) and in the genus *Chunia* among the *Exbucklandioideae* (ENDRESS 1977, BOGLE & PHILBRICK 1980). A similar trend is perhaps initiated in *Platanus* (cf., Fig. 9 in ZAVADA & DILCHER 1986).

Perianth. One of the most salient features is another parallel evolutionary trend, which has evidently occurred in several groups of the Lower *Hamamelididae*: the differentiation of completely perianthless flowers. This is true for the *Eupteleaceae, Cercidiphyllaceae* (ENDRESS 1986), *Myrothamnaceae* (JÄGER-ZÜRN 1966), and among the *Hamamelidaceae* in the *Liquidambaroideae* (WISNIEWSKI & BOGLE 1982), in *Chunia* of the *Exbucklandioideae*, and in *Distylium, Distyliopsis,* and *Matudaea* of the *Hamamelidoideae* (ENDRESS 1978). In *Matudaea* fused bracts on the floral axis form a calyx-like floral envelope (Fig. 1 f). In *Trochodendron,* which has also been claimed to be perianthless, floral ontogenetic studies have shown that rudimentary organs that may be interpreted as tepals can be detected in early developmental stages (ENDRESS 1986).

Petals among the subclass *Hamamelididae* are known to occur only in the *Hamamelidaceae*. Beginning with the studies by SCHOENLAND (1883), the question whether petals are present in *Platanaceae* has still not yet been resolved. In most genera of the *Hamamelidaceae* with petaliferous flowers, the petals are unusual in shape as compared to other angiosperms. They are narrow, linear, ribbon-like (Fig. 1 c). This is well-known for *Hamamelis,* but characterizes the entire tribe *Hamamelideae*, the *Disanthoideae*, and part of the *Exbucklandioideae* and *Rhodoleioideae* (HARMS 1930, VINK 1957). Only in *Corylopsis* and in part of *Rhodoleia* are the petals spathulate or circular and resemble more conventional angiosperm

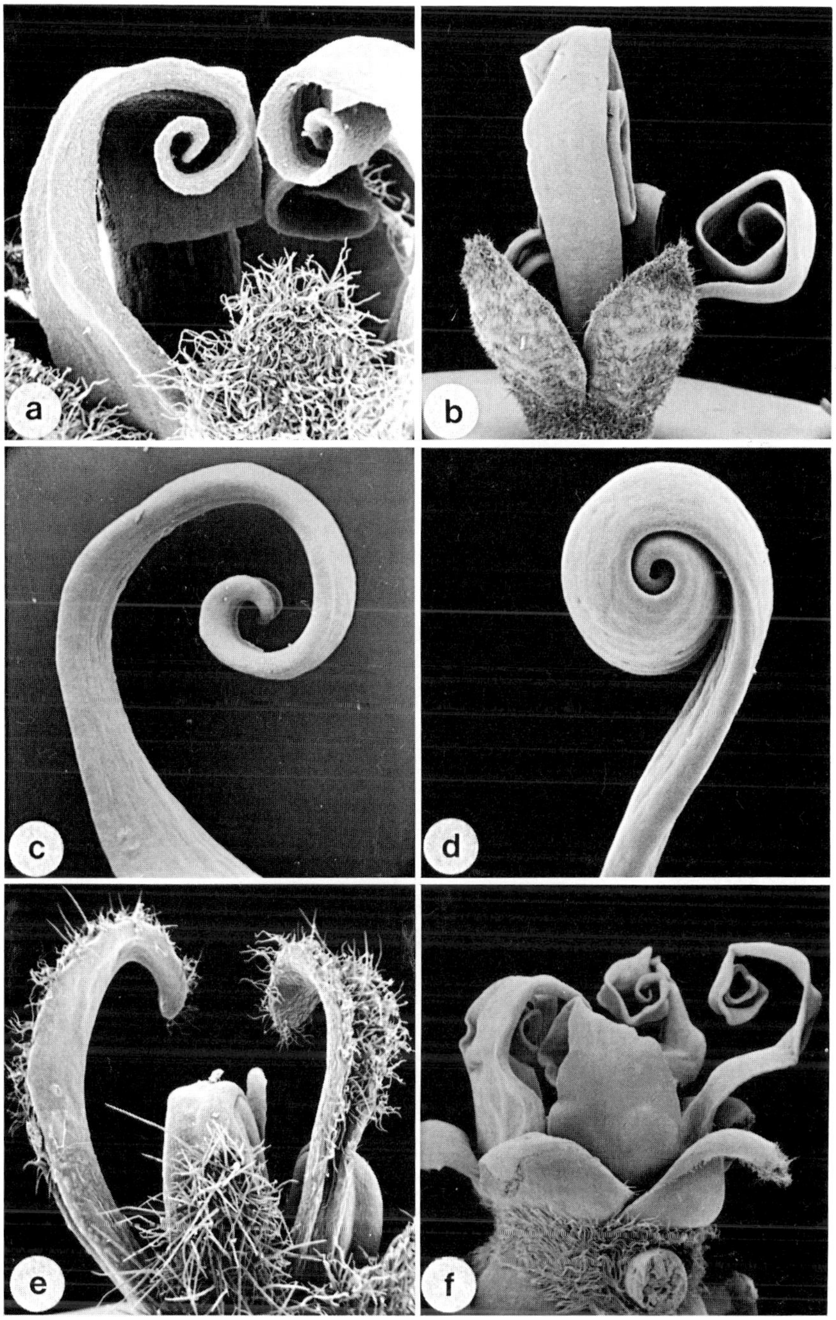

Fig. 7. Circinate petals in young flowers of *Hamamelidaceae*; *a Hamamelis virginiana* L. (E 3177), × 35; *b Loropetalum chinense* R. Br. (E 791), × 15; *c Ostrearia australiana* Baill. (E 7106), × 35; *d Noahdendron nicholasii* Endress & al. (Tracey 14945), × 25; *e Trichocladus crinitus* Pers. (E 4908), × 25; *f Disanthus cercidifolius* Maxim. (E 7118), × 15

petals (Fig. 1 b, g). Perhaps petals have evolved on the level of the *Hamamelidales* (cf., EHRENDORFER 1977) but have been lost again in various groups of the *Hamamelidaceae* (and other *Hamamelidales*?) (ENDRESS 1977, 1986).

The most striking feature in the ribbon-like morphs is that most of them are circinate in bud (Fig. 7 a – f). This is the case in the tribe *Hamamelideae* and in the *Disanthoideae*, while they are ± straight in bud in the *Rhodoleioideae* (Fig. 1 g) and probably the *Exbucklandioideae*. The circinate shape seems to arise by differential elongation predominantly of the apical portion in bud. Before or at early anthesis the petals unfold to be more or less expanded with a flat or wrinkled blade. In contrast, in *Corylopsis* with petals of a conventional angiospermous shape, they are much retarded in development until shortly before anthesis (ENDRESS 1967).

There are, however, modifications in the degree of circination and in the unfolding pattern from genus to genus. One of the most unusual cases is the genus *Noahdendron* (Fig. 7 d) (ENDRESS & al. 1985). Here the relatively thick and fleshy petals are tightly circinate. They remain in this position for a longer time, when they are already exposed. It is not yet known whether they completely expand at all. In older stages the circinate apical portion is often lacking (damaged by animals?). The reverse is the case in certain species of *Trichocladus* (e.g., *T. crinitus*). Here the petals are not circinate in bud, only slightly involute (Fig. 7 e). But at the same time the lateral margins are revolute. While expanding the petals form a lantern-like construction (Fig. 1 h). It is possible that small flies are attracted through the five lateral openings of each flower formed by the petals. However, observations on the pollination biology of the genus have not been recorded so far. Furthermore, the petals are unusual in producing stellate hairs (like the sepals), while they are glabrous in the other genera. Ribbon-like petals with dull red or yellowish colours are often part of a myophily syndrome (Fig. 1 c, d) (KUGLER 1956). Records on fly pollination in *Disanthus* and *Hamamelis* are mentioned in the preceding section.

In the Lower *Hamamelididae* there is an obvious correlation between the degree of differentiation of the perianth and the plasticity in floral organ number. It seems that a well differentiated perianth with calyx and corolla canalizes the organ number. Such flowers are uniformly 5- or 4-merous. As soon as the corolla is no longer present and the calyx is no longer an enveloping, protective organ, this fixation of organ number is suspended. Where even the calyx disappears, the fluctuation in stamen number becomes still more pronounced. This is true for both *Hamamelidales* and *Trochodendrales* (cf., Table 1, and ENDRESS 1978, 1987). Also in *Platanaceae* loss of floral symmetry and of fixed organ numbers seems to have occurred. In the Lower Cretaceous (CRANE & al. 1986) and Eocene platanoid floral fossils (MANCHESTER 1986) 5-mery prevails, while in extant *Platanus* organ number is more variable.

Nectaries. In the subclass *Hamamelididae* floral nectaries are known from only two families: *Trochodendraceae* (*Trochodendron* and *Tetracentron*) (Fig. 1 a, and ENDRESS 1986) and *Hamamelidaceae* (several genera, cf., Table 1). But they exhibit an unusual variety in the position of the nectaries (ENDRESS 1977).

In the *Trochodendraceae* the nectaries occupy the dorsal side of the ovary providing an open platform for visiting insects (predominantly dipters?). The secreted nectar reaches the surface through stomata. In the *Hamamelidaceae* the nectariferous region is either on the base of the petals (*Disanthus*; Fig. 1 c, and MIZUSHIMA

1968), on staminodes (*Hamamelis*; Fig. 1 d, and ENDRESS 1967), or on ontogenetically late-formed scales between androecium and gynoecium (*Loropetalum, Tetrathyrium, Maingaya, Corylopsis, Fortunearia, Rhodoleia;* DOCTERS VAN LEEUWEN 1927, 1938, ENDRESS 1967). In contrast to *Trochodendraceae* stomata have not been found in nectaries of the *Hamamelidaceae* (but *Fortunearia*; ENDRESS 1967). In some of the nectariferous representatives the nectaries are freely exposed as part of a myophily syndrome (*Trochodendron, Hamamelis, Disanthus*).

However, the flowers of most genera of the *Hamamelidaceae* do not produce any nectar (and do not have nectary-like structures). This is true not only for the anemophilous taxa but also for many of the apparently entomophilous genera, even for such genera as *Ostrearia* and *Trichocladus*, as observed in the greenhouse.

Breeding systems. There is little known on the breeding systems of the Lower *Hamamelididae*. As in many woody *Magnoliidae* proterogyny is common in taxa with bisexual flowers (Fig. 1 a – c, f, h; see also MORLEY & CHAO 1977). Within the *Trochodendrales* and *Hamamelidaceae* bisexual flowers prevail. In anemophilous *Hamamelidaceae* often andromonoecy occurs. Monoecy is rare in the family (*Liquidambar, Altingia, Sinowilsonia, Fortunearia*). Dioecy probably occurs only outside the *Hamamelidaceae*, i.e., in *Cercidiphyllum* and *Myrothamnus*. Incidental observations in botanical gardens suggest that self-incompatibility is common in the group, but self-compatibility seems to occur in some genera (e.g., *Trochodendron; Parrotiopsis; Hamamelis:* DE STEVEN 1983).

Phylogenetic considerations

The southern *Hamamelidaceae* as a marker group. The five genera of the *Hamamelidaceae* that are restricted to the southern hemisphere, *Ostrearia, Neostrearia, Noahdendron, Trichocladus,* and *Dicoryphe*, have been considered to be offshoots of various groups within the tribe *Hamamelideae*, and not as a coherent group (e.g., HARMS 1930, SKVORTSOVA 1975, CHANG 1979). They have some odd features as compared to other genera of the family, such as congenitally united staminal filaments and syncarpous style in *Dicoryphe*, opposite leaves and revolute petal margins in some *Trichocladus* species, but tightly circinate petals in *Noahdendron*.

However, it has now become evident that the unusual anther type 2 containing thecae with two pollen sacs but a single valve occurs exclusively and at the same time consistently in the southern genera of the *Hamamelidaceae* (Table 1; and ENDRESS & al. 1985). This is strong evidence for the assumption that the five genera form a coherent group within the tribe *Hamamelideae*. This is further emphasized by the presence of persistent, large, stalked stipules showing pinnate, reticulate venation, which may not only protect their own leaf lamina but (after leaf expansion) also the subsequent terminal bud. This feature is well developed in the *Exbucklandioideae* and *Disanthoideae* but not in the *Hamamelidoideae*, with the exception of three of the southern genera: *Noahdendron, Dicoryphe,* and *Trichocladus* (Fig. 8).

These five southern genera can tentatively been used as a framework of evolutionary reconstruction for the *Hamamelidaceae*. The southern group probably goes back at least to the Upper Cretaceous, which is corroborated by the fact that fossil flowers of *Hamamelidaceae* with univalvate thecae are known from the Upper Cretaceous (Upper Santonian or Lower Campanian) (of Sweden) (FRIIS 1985 b) as already mentioned. The five extant genera are highly endemic relics (some almost

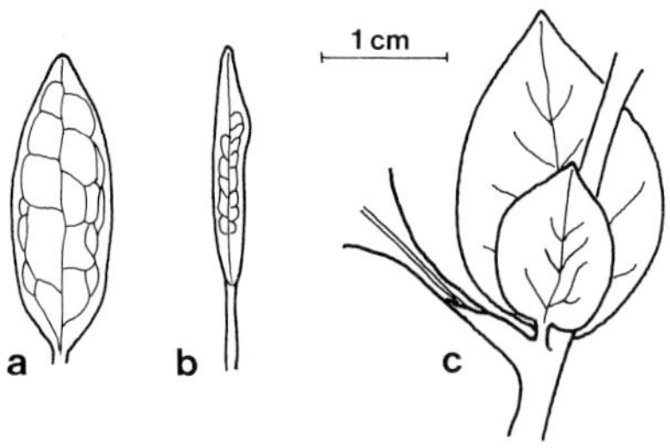

Fig. 8. Stipules; *a Noahdendron nicholasii* ENDRESS & al. (HYLAND 25058), from below; *b Trichocladus ellipticus* ECKL. & ZEYH. (E 7175), stipule peltate, from above; *c Dicoryphe stipulacea* JAUME ST. HIL., leaf base with 2 stipules of unequal size (from DU PETIT-THOUARS 1805, pl. VII)

extinct) of a primitive group of the tribe *Hamamelideae*, which was already present in the Upper Cretaceous or earlier in Gondwanaland.

Fossil hamamelidaceous seeds as guiding organs. The fruit type with one-seeded carpels where the seed is ejected ballistically, which is so characteristic for the subfamily *Hamamelidoideae*, appears first in the fossil record of the Eocene (TIFFNEY 1986) and there is a rich record of fossil seeds from the Eocene onwards in Europe (but this is not so in North America) (cf. references in the section on fruits). In contrast, fruits of the *Liquidambaroideae* alliance are already known from the Paleocene (TIFFNEY 1986). Seeds ascribed to *Disanthus, Rhodoleia*, and *Klikovispermum* (*Rhodoleia*-like?) have been found in some variety from the Upper Cretaceous (Maastrichtian) of Europe (KNOBLOCH & MAI 1984, 1986). The finds of *Klikovispermum* have been ascribed to not less than 17 species (KNOBLOCH & MAI 1986). This makes it probable (at least for Europe) that multiovulate *Hamamelidaceae* were diverse already in the Upper Cretaceous, while the uniovulate subfamily *Hamamelidoideae* became prominent only from the Eocene onwards in certain regions of Europe. However, it does not exclude the possibility that uniovulate groups were present as early as the Upper or Mid Cretaceous, especially with regard to the southern genera. In this respect, it would be of interest to know whether the Upper Cretaceous flower fossils from Sweden (FRIIS 1985 b) have uni- or pluriovulate carpels. At any rate, the overall fossil seed record in Europe suggests an evolutionarily advanced state of the uniovulate carpels as compared to the pluriovulate carpels. This is also supported by the functional specialization of the one-seeded fruits and the comparatively higher systematic homogeneity of the one-seeded group.

Other families. The presence of putative platanoids in the Lower Cretaceous (Late Albian) (CRANE & al. 1986, UPCHURCH 1985) suggests that the evolutionary roots of the subclass *Hamamelididae* have to be sought for here at the latest. Unfortunately the fossil record of the *Trochodendrales* is not yet traceable that far back (cf., TIFFNEY 1986, TAO & XIONG 1986, CRANE 1989).

I am greatly indebted to numerous colleagues for their help in the field or for providing material, especially A. FLOYD, Coffs Harbour, Dr J. L. GRESSITT, Wau, Dr B. P. M. HYLAND, Atherton, Prof. A. MOLINA, Tegucigalpa, Dr F. S. P. NG, Kepong, Dr R. SCHMID, Berkeley, J. G. TRACEY, Atherton, Prof. Dr A. WEBER, Vienna, Dr L. O. WILLIAMS,

Chicago. For valuable discussions I thank Dr P. R. CRANE, Chicago, Dr E. M. FRIIS, Åarhus, Dr E. KNOBLOCH, Prague, and Dr D. H. MAI, Berlin. I am grateful to U. JAUCH, Institute of Plant Biology, University of Zurich, for use of the SEM. I thank the "Georges-und-Antoine-Claraz-Schenkung" for financial support of my trips to Central America, Eastern Australia, and Papua-New Guinea.

References

BAILLON, H., 1894: Les ovules des *Corylopsis*. – Bull. Soc. Linn. Paris **2** (142): 1122–1123.

BOGLE, A. L., 1970: Floral morphology and vascular anatomy of the *Hamamelidaceae*: The apetalous genera of *Hamamelidoideae*. – J. Arnold Arbor. **51**: 310–366.

– 1984: Floral morphology and vascular anatomy of *Maingaya* OLIV. (*Hamamelidaceae, Hamamelidoideae, Hamamelideae*). – Amer. J. Bot. **71** (5, 2): 19.

– 1986: The floral morphology and vascular anatomy of the *Hamamelidaceae*: subfamily *Liquidambaroideae*. – Ann. Missouri Bot. Gard. **73**: 325–347.

– PHILBRICK, C. T., 1980: A generic atlas of hamamelidaceous pollens. – Contr. Gray Herb. **210**: 29–103.

BUDANTSJEV, L. J., 1975: Morphology of the leaves and taxonomy of the genera *Protophyllum* LESQ. and *Pseudoprotophyllum* HOLLICK. – In BUDANTSJEV, L. J., (Ed.): Problems of comparative morphology of the seed plants, pp. 90–107. – Leningrad: Nauka (in Russian).

BURGH, J. VAN DER, 1983: Allochthonous seed and fruit floras from the Pliocene of the Lower Rhine basin. – Rev. Palaeobot. Palynol. **40**: 33–90.

CHANG, H. T., 1973: A revision of the hamamelidaceous flora of China. – Sunyatsen Univ. Bull. **1**: 54–71 (in Chinese).

CHANG, K.-T., 1979: The spore and pollen morphology in relation to the taxonomy and phylogeny of some plant groups. – Acta Phytotax. Sin. **17** (2): 1–8 (in Chinese).

COLLINSON, M. E., 1982: A preliminary report on the Senckenberg-Museum collection of fruits and seeds from Messel bei Darmstadt. – Cour. Forsch.-Inst. Senckenberg **56**: 49–57.

– 1983: Fossil plants of the London Clay. – London: The Palaeontological Association.

– 1984: Early Tertiary *Hamamelidae:* a contribution from isolated fossil organs. – Amer. J. Bot. **71** (5, 2): 109.

CRANE, P. R., 1989: Paleobotanical evidence on the early radiation of nonmagnoliid dicotyledons. – Pl. Syst. Evol. **162**: 165–191.

– STOCKEY, R. A., 1986: Morphology and development of pistillate inflorescences in extant and fossil *Cercidiphyllaceae*. – Ann. Missouri Bot. Gard. **73**: 382–393.

– FRIIS, E. M., PEDERSEN, K. R., 1986: Lower Cretaceous angiosperm flowers: fossil evidence on early radiation of dicotyledons. – Science **232**: 852–854.

CRONQUIST, A., 1981: An integrated system of classification of flowering plants. – New York: Columbia University Press.

DOCTERS VAN LEEUWEN, W. M., 1927: Vogelbloemen. 3. *Rhodoleia Teysmanni* MIQ. – Trop. Natuur **16**: 2–6.

– 1938: Observations about the biology of tropical flowers. – Ann. Jard. Bot. Buitenzorg **48**: 27–68.

EHRENDORFER, F., 1977: New ideas about the early differentiation of angiosperms. – Pl. Syst. Evol. Suppl. **1**: 227–234.

EICHHOLZ, G., 1886: Untersuchungen über den Mechanismus einiger zur Verbreitung von Samen und Früchten dienender Bewegungserscheinungen. – Jahrb. Wiss. Bot. **17**: 543–590.

ENDRESS, P. K., 1967: Systematische Studie über die verwandtschaftlichen Beziehungen zwischen den Hamamelidaceen und Betulaceen. – Bot. Jahrb. Syst. **87**: 431–525.

Endress, P. K., 1970: Die Infloreszenzen der apetalen Hamamelidaceen, ihre grundsätzliche morphologische und systematische Bedeutung. – Bot. Jahrb. Syst. **90**: 1–54.

– 1977: Evolutionary trends in the *Hamamelidales-Fagales*-group. – Pl. Syst. Evol. Suppl. **1**: 321–347.

– 1978: Blütenontogenese, Blütenabgrenzung und systematische Stellung der perianthlosen *Hamamelidoideae*. – Bot. Jahrb. Syst. **100**: 249–317.

– 1986: Floral structure, systematics, and phylogeny in *Trochodendrales*. – Ann. Missouri Bot. Gard. **73**: 297–324.

– 1987: Floral phyllotaxis and floral evolution. – Bot. Jahrb. Syst. **108**: 417–438.

– Hyland, B. P. M., Tracey, J. G., 1985: *Noahdendron*, a new Australian genus of the *Hamamelidaceae*. – Bot. Jahrb. Syst. **107**: 369–378.

Flint, F. F., 1959: Development of the megagametophyte in *Liquidambar styraciflua* L. – Madroño **15**: 25–29.

Friis, E. M., 1984: Organisation og bestøvningsformer hos blomster fra Øvre Kridt. – Dansk Geol. Foren, Årsskr. **1983**: 1–8.

– 1985a: Angiosperm fruits and seeds from the Middle Miocene of Jutland (Denmark). – Biol. Skr. **24** (3): 1–165.

– 1985b: Structure and function in Late Cretaceous angiosperm flowers. – Biol. Skr. **25**: 1–37.

Gregor, H.-J., 1975: Die mittelmiozäne Mastixioideen-Flora aus dem Braunkohlen-Tagebau Oder II bei Wackersdorf (Oberpfalz). – Diss., Univ. München.

– 1977: Subtropische Elemente im europäischen Tertiär (Fruktifikationen). – Paläont. Z. **51**: 199–226.

– 1978: Die miozänen Frucht- und Samen-Floren der Oberpfälzer Braunkohle. 1. Funde aus den sandigen Zwischenmitteln. – Palaeontographica, Abt. B, **167**: 8–103.

– 1982: Die jungtertiären Floren Süddeutschlands. – Stuttgart: Enke.

– 1986: Zur Flora des Randecker Maares (Miozän, Baden-Württemberg). – Stuttgarter Beitr. Naturk., Ser. B, **122**: 1–29.

Harms, H., 1930: *Hamamelidaceae*. – In Engler, A., Prantl, K., (Eds.): Die natürlichen Pflanzenfamilien **18a**, pp. 303–345, 487, 2 ed. – Leipzig: Engelmann.

Herman, A. B., 1984: A new genus of platan-tree-like angiosperms. – Paleont. Ž. **1984** (1): 71–79.

Hesse, M., 1978: Entwicklungsgeschichte und Ultrastruktur von Pollenkitt und Exine bei nahe verwandten entomophilen und anemophilen Angiospermensippen: *Ranunculaceae, Hamamelidaceae, Platanaceae* und *Fagaceae*. – Pl. Syst. Evol. **130**: 13–42.

Jäger-Zürn, I., 1966: Infloreszenz- und blütenmorphologische sowie embryologische Untersuchungen an *Myrothamnus* Welw. – Beitr. Biol. Pfl. **42**: 241–271.

Kaul, U., Kapil, R. N., 1975: *Exbucklandia populnea* – from flower to fruit. – Phytomorphol. **24**: 217–228.

Knappe, H., Rüffle, L., 1975: Beiträge zu den Platanaceen-Funden und einigen *Hamamelidales* der Oberkreide. – Wiss. Z. Humboldt-Univ. Berlin, math.-nat. R., **24**: 487–492.

Knobloch, E., Mai, D. H., 1984: Neue Gattungen nach Früchten und Samen aus dem Cenoman bis Maastricht (Kreide) von Mitteleuropa. – Feddes Repert. **95**: 3–41.

– – 1986: Monographie der Früchte und Samen in der Kreide von Mitteleuropa. – Rozpravy Ústř. Úst. Geol. Praha **47**: 1–219.

Kugler, H., 1956: Über die optische Wirkung von Fliegenblumen auf Fliegen. – Ber. Deutsch. Bot. Ges. **69**: 387–398.

Mai, D. H., 1968: Zwei ausgestorbene Gattungen im Tertiär Europas und ihre florengeschichtliche Bedeutung. – Palaeontographica, Abt. B, **123**: 184–199.

– Walther, H., 1978: Die Floren der Haselbacher Serie im Weisselster-Becken (Bezirk Leipzig, DDR). – Abh. Staatl. Mus. Mineral. Geol. Dresden **28**: 1–200.

MAI, D. H., WALTHER, H., 1985: Die obereozänen Floren des Weisselster-Beckens und seiner Randgebiete. – Abh. Staatl. Mus. Mineral. Geol. Dresden **33**: 5–176.

MANCHESTER, S. R., 1986: Vegetative and reproductive morphology of an extinct plane tree (*Platanaceae*) from the Eocene of Western North America. – Bot. Gaz. **147**: 200–226.

MIZUSHIMA, M., 1968: On the flower of *Disanthus cercidifolius* MAXIM. – J. Japan. Bot. **43**: 522–524 (in Japanese).

MORLEY, B., CHAO, J.-M., 1977: A review of *Corylopsis* (*Hamamelidaceae*). – J. Arnold Arbor. **58**: 382–415.

NAST, CH. G., BAILEY, I. W., 1945: Morphology and relationships of *Trochodendron* and *Tetracentron*. II. Inflorescence, flower, and fruit. – J. Arnold Arbor. **26**: 267–276.

PETIT-THOUARS, A. A. DU, 1805: Histoire des végétaux recueillis dans l'isles australes d'Afrique. 1. – Paris: Levrault, Schoell.

PRAGLOWSKI, J., 1974: The pollen morphology of the *Trochodendraceae, Tetracentraceae, Cercidiphyllaceae*, and *Eupteleaceae*, with reference to taxonomy. – Pollen & Spores **16**: 449–467.

RÜFFLE, L., 1968: Merkmalskomplexe bei älteren Angiospermen-Blättern und die Kutikula von *Credneria* ZENKER (*Menispermaceae*). – Palaeontographica, Abt. B, **123**: 132–143.

SCHOENLAND, S., 1883: Über die Entwicklung der Blüten und Frucht bei den Platanen. – Bot. Jahrb. Syst. **4**: 308–327.

SKVORTSOVA, N. T., 1975: Comparative morphological studies on representatives of the family *Hamamelidaceae* and their phylogenetic relationships. – In BUDANTSJEV, L. J., (Ed.): Problems of comparative morphology of the seed plants, pp. 7–24. – Leningrad: Nauka (in Russian).

SMITH, A. C., 1945: A taxonomic review of *Trochodendron* and *Tetracentron*. – J. Arnold Arbor. **26**: 123–142.

SMITH, L. S., 1958: New species of and notes on Queensland plants 3. – Proc. Roy. Soc. Queensland **69**: 43–51.

STEVEN, D. DE, 1983: Floral ecology of witch-hazel (*Hamamelis virginiana*). – Michigan Bot. **22**: 163–171.

TAKHTAJAN, A. L., 1969: Flowering plants. Origin and dispersal. – Edinburgh: Oliver & Boyd.

– (Ed.) 1974: Fossil flowering plants of the USSR. 1. *Magnoliaceae–Eucommiaceae*. – Leningrad: Nauka (in Russian).

TAO, J.-R., XIONG, X.-Z., 1986: The latest Late Cretaceous flora of Heilongjiang Province and the floristic relationship between East Asia and North America. – Acta Phytotax. Sin. **24**: 1–15, 121–135.

TIFFNEY, B. H., 1986: Fruit and seed dispersal and the evolution of the *Hamamelidae*. – Ann. Missouri Bot. Gard. **73**: 394–416.

UPCHURCH, G. R. JR., 1985: Cuticle evolution in early Cretaceous angiosperms from the Potomac Group of Virginia and Maryland. – Ann. Missouri Bot. Gard. **71**: 522–550.

VAUDOIS-MIÉJA, N., 1979: Sur un fruit fossile d'Hamamélidacées dans les Grès à Sabals de l'ouest de la France. – C. R. Congr. Nat. Soc. Sav., I, **104**: 263–273.

VINK, W., 1957: *Hamamelidaceae*. – In STEENIS, C. G. G. J. VAN, (Ed.): Flora Malesiana, Ser. I, **5**, pp. 363–379. – Dordrecht: Nijhoff.

WISNIEWSKI, M., BOGLE, A. L., 1982: The ontogeny of the inflorescence and flower of *Liquidambar styraciflua* L. (*Hamamelidaceae*). – Amer. J. Bot. **69**: 1612–1624.

ZAVADA, M. S., DILCHER, D. L., 1986: Comparative pollen morphology and its relationship to phylogeny of pollen in the *Hamamelidae*. – Ann. Missouri Bot. Gard. **73**: 348–381.

Address of the author: Prof. Dr PETER K. ENDRESS, Institut für Systematische Botanik der Universität, Zollikerstrasse 107, CH-8008 Zürich, Switzerland.

Pl. Syst. Evol. 162, 213 – 229

—Plant—
Systematics
and
Evolution

Paleobotanical studies in *Fagaceae* of the European Tertiary

Z. Kvaček and H. Walther

Received February 2, 1987

Key words: Angiosperms, *Fagaceae.* – Cuticles, Recent, Tertiary, evolution.

Abstract: A systematic reassessment of megafossil records of *Fagaceae* in Central Europe has been undertaken on the basis of leaf cuticular characters. The oldest representatives date back to the Eocene: *Quercus subhercynica* spec. nova, *Dryophyllum furcinerve* (Rossm.) Schmalh., *Trigonobalanopsis rhamnoides* (Rossm.) gen. & comb. nov. In the Oligocene other members of extant genera appear: *Quercus rhenana* (Weyl. & Kilpp.) Knobloch & Kvaček, *Fagus attenuata* Goepp., *Lithocarpus saxonicus* spec. nova. In the Neogene these ancient taxa (except in *Fagus* lineage), are gradually replaced by deciduous species of *Quercus* and *Castanea. Trigonobalanus* and *Castanopsis* are recorded by fruits (or wood) only.

The following review includes recent advances in cuticular analyses of *Fagaceae* from various Tertiary deposits in Europe. In order to evaluate the diagnostic importance of leaf epidermal features, many extant species of all *Fagaceae* genera (except *Nothofagus* Blume) have been investigated. Our results agree with those obtained by an independent study of extant taxa by Jones (1986). For details of our research see Kvaček & Walther (1987).

Materials and methods

The examined fossils were collected for the most part in the lignite basins of the German Democratic Republic (Weisselster, Geiseltal, Saxony, Lausitz) and Czechoslovakia (Cheb, Sokolov, Bílina, Hrádek, Nováky). Additional material was obtained from Federal Republic of Germany (Achldorf, Rhine Valley), Great Britain (Bovey Tracey) and Greece (Aliveri, Lava). Comparative studies of extant taxa included leaf samples of the majority of the *Chrysolepsis, Trigonobalanus, Fagus,* and *Castanea* species, of about half of the *Castanopsis* and *Lithocarpus* species, and of less than one third of the *Quercus* species. Cuticular features in fossils were observed in macerated samples (Schulze solution), both under LM and SEM. Leaves of extant taxa were macerated in eau de Javelle and stained in safranin. Additional observation of unmacerated samples by SEM revealed the arrangement of surface features in space. Documentation is deposited at the Staatliches Museum f. Mineralogie u. Geologie, Dresden (MMG), the Institute of Geology and Geotechnics, Czechoslovak Academy of Sciences (UGG), the National Museum in Prague (PRNM), and elsewhere (as stated).

Fagus L.

On the basis of distinct leaf morphological traits it has been possible to divide the genus roughly into form groups. Convergent leaf forms are exceptional, e.g. in the extinct *Pseudofagus* SMILEY & HUGGINS of the N. American Tertiary. When studied more extensively, the foliage of extant species exhibit similarities and transitions. Overlaps in the specific variation of leaf size and marginal features apparently is typical in this genus. Detailed epidermal studies offer several characters which can aid in the separation of extant species. Narrow bases with uniseriate remnants of glandular filamentous hairs occur throughout the genus. Solitary thick-walled hairs are variously distributed among different species. Specific differential features are the types of stomata (anomocytic versus cyclocytic) and subsidiary cells (mostly narrow versus isodiametric), the papillate adaxial leaf surface, and to a certain extent the degree of anticline undulation. The form and size of cupules as well as the length of the peduncles are further characters which help to elucidate the relationships among beeches.

Evidence for fossil beech remains has been presented already in the last century (e.g., UNGER 1847). It mostly consists of leaf imprints. Unless more complete analyses based on several complementary methods are available, all the formal species described must be considered with caution. First attempts to reconstruct the evolutionary history of *Fagus* were misleading (ETTINGSHAUSEN 1880).

The earliest reliable evidence of fossil *Fagus* in Central Europe is known from the Middle and Upper Oligocene. All leaves so far studied with cuticles preserved, e.g., from Borna-Ost (Upper Oligocene), Hrádek (Lower Miocene), Wischgrund (Middle Miocene), Sośnica, Ruszów (Upper Miocene), are uniform in epidermal characters (Fig. 1 a): cyclocytic stomata with mostly narrow subsidiary cells, anticlines mostly slightly wavy, solitary thickened hairs only on the midvein, serial bases of glandular hairs evenly dispersed over the adaxial side. Various associated *Fagus* cupules were mostly assigned to two fossil species: *F. deucalionis* UNGER (Oligocene) (Fig. 1 d) and *F. decurrens* REID & REID (Pliocene). In our opinion, the cupules from the Miocene are referrable to transitional forms (Fig. 1 b, c). These fossil leaf and fruit remains are therefore considered as one evolutionary lineage which can be traced from the Oligocene to the Pliocene. The foliage then should bear the name *Fagus attenuata* GOEPPERT (Fig. 1 e – g). Such a form complex exhibit trends from large but slender and distinctly dentate leaf forms to smaller and ovate ones with partly reduced marginal teeth.

The similarity in leaf morphology of the *F. attenuata* complex to the extant *F. grandifolia* EHRH. and *F. longipetiolata* SEEM. appears to be due to parallel evolution because their leaf epidermal characters (anomocytic stomatal type, papillate abaxial side in *F. longipetiolata*) differ distinctly from the fossils studied. Cyclocytic stomata are confined to the extant *F. orientalis* LIPSKY, *F. crenata* BL., and *F. lucida* REHD. & WILS. The latter species matches well (also in cupules) with the fossil remains. Our studies support the hypothesis that the European *Fagus sylvatica* L. might be of post-Tertiary origin. This idea corresponds with the postglacial migrations of

---

Fig. 1. Fossile leaf and cupule structures of *Fagus*. *a F. attenuata*, abaxial cuticle, Espenhain, Upper Oligocene, microslide No. 18/85 (MMG), × 500. *b, c F. decurrens*, cupules, Hrádek n. N., Lower Miocene (PRNM), × 2. *d F. deucalionis*, cupule, Podlesí, Oligocene (PRNM), × 2. *e F. attenuata*, leaf impression, Oligocene, No. BOB II, 1649: 2 (MMG), 1 : 1. *f F. attenuata*, leaf compression, Hrádek n. N., Lower Miocene (UGG), × 2. *g F. attenuata*, leaf impression, Borna-Ost, Oligocene, No. BOB III 1885 a (MMG), 1 : 1

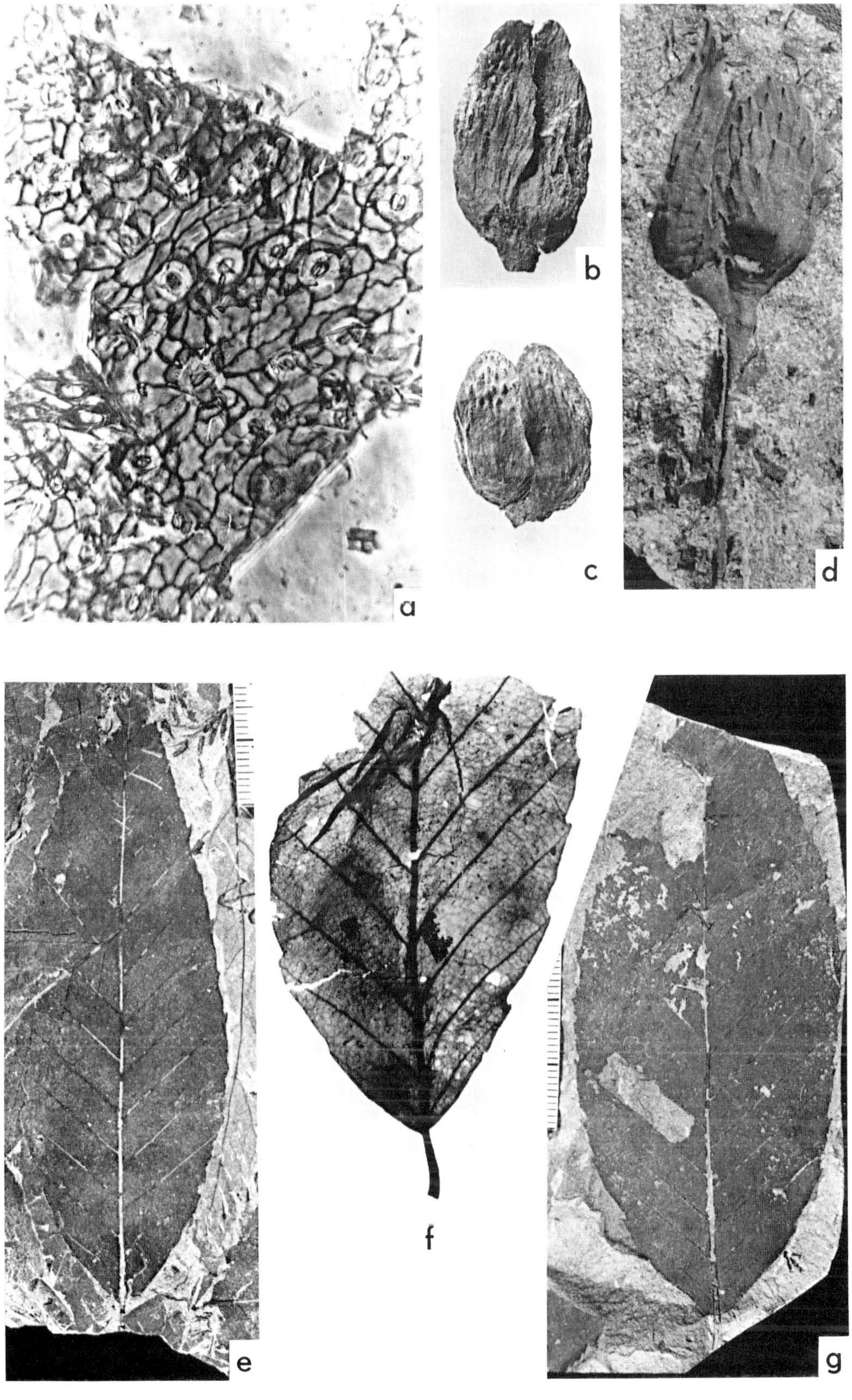

European beech as reflected in palynological data (cf. Środoń 1985). Beside the lineage of *F. attenuata*, typical for the late Tertiary of Central Europe, other species can be traced, mainly from the Tertiary of Central Asia and South Europe, i.e. large-leaved forms like *F. antipofii* Heer in the Oligocene and the closely allied *F. juliae* Jakubovskaja, as well as *F. gussonii* Massalongo sensu Knobloch & Velitzelos (1986) in the Miocene. So far, no anatomical and only scarce carpological evidence on this group is available; in its leaf form it recalls *F. orientalis*.

Lithocarpus Blume

In *Lithocarpus* the stomata are cyclocytic except for the N. American *L. densiflorus* (Hook. & Arn.) Rehd. which tends towards an anomocytic arrangement. The guard cells bulge outwards, mostly by their outer periclinal walls. The leaves of many species on their adaxial side bear typical appressed finger-like tufted (to stellate) hairs, not met with in other *Fagaceae* (Fig. 2 e). This indumentum obviously serves as a protection for the stomata. In *L. densiflorus* the hairs are multiradiate with heavily cutinized bases, and similar to those occurring in *Quercus*. Glandular hairs, mostly cylindrical, capitate or peltate, or their bases, are present in all studied species; short papillae are met with occasionally.

Although *Lithocarpus* has been reported several times from the Tertiary (e.g., Andreánszky & Kovács 1964, Kolakovskij in Takhtajan 1982), only the leaf fossils described below can be safely referred to this genus.

L. saxonicus **Walther & Kvaček, spec. nova,** Fig. 2 a – c.

Holotype: No. KS 61 a/1 (MMG) Fig. 2 a, microslides Nos. KS 121 – 127/72 (MMG) Fig. 2 c.

Paratypes: Nos. KS 57, microslides KS 64 – 88/72 (MMG), KS 61/1, Fig. 2 b, microslides 466 – 470/72 (MMG).

Locus typicus: Core Kleinsaubernitz near Bautzen, German Democratic Republic.

Stratum typicum: Diatomite, Volcanogenic Series, Oligocene.

Tree (?). Leaves incomplete, lanceolate, 8.8 – 8.9 cm long, 1.6 to 3.1 cm wide, entire margined, firm to coriaceous, margin slightly wavy and thickened, apex attenuate, base cuneate, petiolate. Venation pinnate, brochiodrome, midvein moderately thick, secondaries alternate, at angles of 27 – 40° and 31° to 43°, looping very near the margin, intersecondaries rare, tertiaries oblique or at right angles, often forked, c. 9 per 1 cm.

Adaxial cuticle showing outlines of straight-walled to slightly curved cells (10 – 25 µm across) and scattered simple hair bases, abaxial cuticle slightly papillate, cells mostly straight-walled. Stomata cyclocytic, 22 – 27 µm long and 15 – 23 µm

---

Fig. 2. Fossile leaf structures of *Lithocarpus* and *Dryophyllum*. *a, b L. saxonicus* spec. nova holotype and paratype, Core Kleinsaubernitz, Oligocene, Nos. KS 61 a/1, KS 61/1 (MMG), 1 : 1. *c L. saxonicus*, abaxial cuticle of the holotype, Core Kleinsaubernitz, Oligocene, microslide No. KS 127/72 (MMG), × 1000. *d Dryophyllum furcinerve* f. *haselbachense* forma nova, Core Kleinsaubernitz, Oligocene, No. 27 d/3 (MMG), × 2. *e Lithocarpus submonticolus*, abaxial leaf side, Elmer 11894 (UGG) × 660. *f Dryophyllum furcinerve* f. *lyellii* (Heer) comb. et stat. nov., abaxial cuticle, Bovey Tracey, Oligocene, microslide No. GSM 77271 (MMG), × 1000

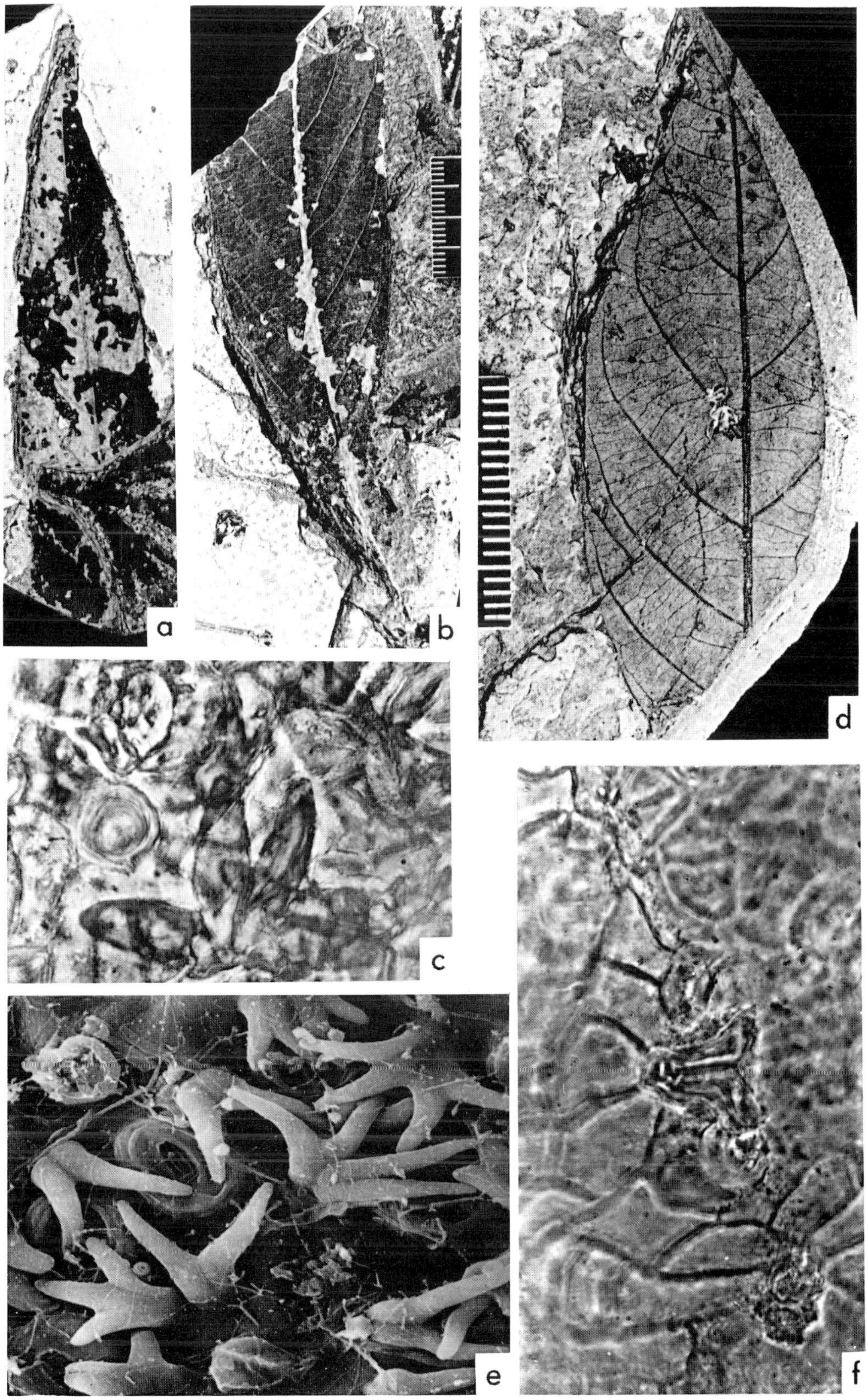

wide, guard cells (14 – 22 µm long) sunken below the level of up to 4 subsidiary cells, ornamented with thick cutin structure. Simple glandular hair bases (10 – 15 µm across) frequent, capitate glands rarely preserved, finger-like tufted appressed hairs with 2 – 4 arms 35 – 55 µm long, with rounded tips, thickly covering stomatas.

Castanopsis Spach

The leaf morphology varies in a similar way as in *Lithocarpus* so that a generic distinction on that basis cannot be suggested. *Castanopsis,* contrary to most *Lithocarpus* species, has cyclocytic stomata in dense groups with more or less thickened narrow subsidiary cells. The stomata are protected by a fine cover of peltate-stellate thin-walled hairs; they can persist, but in several species are shed at maturity, leaving broad bases. Typical are solitary thick-walled hairs sparsely distributed on the midrib, leaving polygonal bases. Tufted hairs are limited to a few species. *Castanopsis chrysophylla* Hjelmq. and *C. sempervirens* (Kell.) Dudl., very similar to each other, differ from E. Asiatic species by their indistinct grouping of stomata with non-thickened subsidiary cells and persistent shallow-lobed peltate hairs. They are often placed into an independent genus, *Chrysolepsis* Hjelmq.

In Europe several species of *Castanopsis* were reported on the basis of cuticular studies (e.g., Kräusel & Weyland 1950, 1954), from the Upper Eocene to the Middle Miocene. Leaves of *C. toscana* (Band.) Kr. & Weyl. and *C. furcinervis* (Rossm.) Kr. & Weyl., regularly associated with fruits identified with *Trigonobalanus* by Mai (1970), deviate in certain respects from extant members of *Castanopsis,* as shown below, and are removed to *Dryophyllum* and *Trigonobalanopsis.* The affinities of *C. glandulosa* Rüffle (1963), *C. bavarica* Knobloch & Kvaček (1976), and *C. atavia* (Ung.) Iljinskaja (in Takhtajan 1982) still have to be verified. Thus, the only megafossil evidence of *Castanopsis* in the European Tertiary seems to consist of the fruits of *C. salinarum* (Ung.) Kirchh. and *C. pyramidata* Kirchh. and some species of *Castanoxylon* Navale (Selmeier 1970 a, b, 1972; v. d. Burgh 1973).

Trigonobalanus Forman

Trigonobalanus shows little uniformity in its leaf morphology and anatomy (except for, e.g., cyclocytic type of stomata – Baas 1982, Jones 1986). It recalls *Castanopsis* by several epidermal features. Peltate lobed hairs occur in *T. doichangensis* (Camus) Forman, but this species has isodiametric and thinly cutinized subsidiary cells, and a coarsely papillate abaxial epidermis. In *T. excelsa* Lozano & al. and *T. verticillata* Forman, on the other hand, the subsidiary cells are thickened, while the glandular hairs are cylindric or capitate and partly caducous. In addition thick-walled solitary and tufted hairs (in *T. excelsa* mostly in pairs) are well developed. An important generic character are the trigonal fruits with persistent remains of perianth and capitate stigmas, enclosed from 1 to 27 in cupules which are lobed and widely open at maturity and exhibit distinct fruit scars.

Forman (1964) drew attention to the close affinity of the fossil fruits known as *Fagus succinea* Goepp. & Menge from the Baltic amber to *Trigonobalanus.* The combination was made by Mai (1970). This unique specimen can be linked with *Dryophyllum berentianum* (Goepp.) Kirchh. from the same deposits, of which unfortunately only fragmentary information on epidermal structure is available

(KIRCHHEIMER 1937). However, the morphology of the leaves, strongly reminiscent of other Upper Eocene sharply toothed forms of *Dryophyllum*, is noteworthy. MAI (1970) has suggested three more species of *Trigonobalanus*. Carpological remains of Central European Upper Eocene to early Middle Oligocene were referred to *T. andreanszkyi* MAI. Althouth they do not correspond in all details to those of modern species (see also MANCHESTER & CRANE 1983), they suggest close affinities to *Trigonobalanus* by their fruit form and the cupules being 5-lobed, open and with rows of lamellae on the outside. These fruits are associated with undoubtedly fagalean leaves known as *Dryophyllum furcinerve*, which, however, correspond little with extant *Trigonobalanus* species and are left in a separate genus. *T. minima* (CHANDLER) MAI is known as fruits only from the type locality at Bovey Tracey. Associated leaves, described as *Quercus lyellii* HEER, also match by their cuticles with *Dryophyllum furcinerve* group. Thus, the occurrence of *Trigonobalanus* in the European Tertiary relies upon carpological evidence only.

Dryophyllum DEBEY ex SAPORTA

This name is attached to *Fagaceae*-like leaf fossils known mainly from the Palaeogene. ILJINSKAJA (in TAKHTAJAN 1982) was the first to select *D. palaeocastanea* SAPORTA as a legitime type for this genus and to preserve the customary circumscription of this genus. She was perfectly right (in respect to Art. 8 of ICBN) and must be given priority over the procedure of JONES & al. (in press) who restore the illegitime and mechanical selection of ANDREWS (1970) and wish to transfer *Dryophyllum* to the *Juglandaceae*. Still by far not all fossil remains assigned to *Dryophyllum* can be attributed to the *Fagaceae* on leaf epidermal evidence. In the Middle and Upper Eocene *Dryophyllum* fossils appear as an important forest element, documented by an abundant record in N. America and W. Eurasia.

The form complex of *Dryophyllum furcinerve* includes foliage, highly variable, particularly in general form, size and margin. The leaf epidermal characters, however, are typical of the whole complex: stomata in groups, cyclocytic, with slightly thickened subsidiary cells, bases of glandular hairs simple, partly with asymmetrical hair attachment, upper epidermis composed of minute cells. Beside the typical form of *D. furcinerve* with appressed tufted hairs, some differentiation can be recognized, which do not hold as independent species and thus are given infraspecific rank (as forms). They partly occur together throughout their range. The differences are in the absence of *Lithocarpus*-like hairs, weak thickening on guard cells and leaf form. In epidermal characters the *Dryophyllum furcinerve*-complex matches well with *Castanopsis* (see also KRÄUSEL & WEYLAND 1950), but in hairs it recalls *Lithocarpus*. In view of the rare occurrence of undulated anticlines affinities to *Trigonobalanus* (*T. excelsa*) also cannot be excluded. The opinion, already expressed e.g. by ILJINSKAJA (1980) that *Dryophyllum* might represent an ancestral line to both *Castanopsis* and *Lithocarpus*, and even *Trigonobalanus*, is thus supported by the present study. In any case, *Dryophyllum furcinerve* including all its forms must be interpreted as a primitive member of *Fagaceae*, typical for the Palaeogene of Central Europe.

Dryophyllum furcinerve (ROSSMÄSSLER) SCHMALHAUSEN

Lectotype: ROSSMÄSSLER (1840), pl. 7, f. 36, Staré Sedlo (ČSSR), Upper Eocene – (Mus. Nat. Berlin, see RÜFFLE & al. 1976: 241, pl. 28, f. 5.)

— — **forma *furcinerve,*** Fig. 3 b.

Leaves ovate-lanceolate, dentate to entire-margined, venation semicraspedodrome to camptodrome. Abaxial leaf side with simple rounded hair-bases and small tufted hairs, mostly four-armed, often preserved as mere bases.

Occurrence: Middle Eocene, Mine Königsaue; Upper Eocene, Complex Zeitz, Mine Profen, Haselbach (German Democratic Republic); Staré Sedlo Formation, Sokolov Basin, Staré Sedlo, Nové Sedlo, Jehličná, Cheb Basin, Nový Kostel (Č.S.S.R.); Svenigorodka (U.S.S.R.).

— — **forma *haselbachense* forma nova,** Figs. 2 d and 3 a, c.

Holotype: No. HbO 214, microslides HbO 237/77-HbO 240/77 (MMG), MAI & WALTHER (1978, pl. 30, f. 1), Haselbach Mine, early Middle Oligocene.

Leaves broadly ovate to lanceolate, coarsely dentate to entire-margined, venation semicraspedodrome to camptodrome. Abaxial leaf side only with simple hair-bases of two kinds: simple rounded and larger or irregularly elliptic in outline; stomata outlines thickened.

Occurrence: Oligocene, Complex Haselbach, Göhren, Mines Schleenhaim, Schnauderhainichen; Core Kleinsaubernitz; Complex Witznitz, Witznitz Mine (German Democratic Republic); Altenrath, Ludwigshütte (Federal Republic of Germany); Sokolov Basin, Erika Mine, Lomnice (Č.S.S.R.); Kiscel Clays, Budapest (Hungary).

— — **forma *lyellii* (HEER), comb. et stat. nov.**, Fig. 2 f.

Basionym: *Quercus lyellii* HEER (1863, Phil. Trans. Roy. Soc. London **152**: 40, pl. 12, f. 2–9, pl. 13, f. 1–4, pl. 14,f,12 b, pl. 15, f. 1–2, pl. 17, f. 4–5).

Syntypes: Geol. Surv. Mus. London, Bovey Tracey.

Leaves ovate-lanceolate, entire-margined, venation camptodrome. On abaxial leaf side beside simple rounded hair-bases also sparse appressed two-armed tufted hairs with very short arms, spreading nearly at right angle.

Occurrence: Oligocene, Bovey Tracey (G.B.).

— — **forma *robustum* (RÜFFLE & al.), stat. nov.**, Fig. 3 d and e.

Basionym: *Dryophyllum robustum* RÜFFLE, MÜLLER-STOLL & LITKE (1976, Abh. Zentr. Geol. Inst. **26**: 248, pl. 19, f. 19, 20, pl. 29, textf. 28, 29).

Holotype: RÜFFLE & al. (1976, pl. 19, f. 19, textf. 28 microslide + 224 +, Mus. Nat. Berlin), Neumark-Süd, Mine Geiseltal.

Leaves small, lanceolate-ovate, entire-margined. Anticlines straight, stomata without thickened elements, thick walled. Abaxial leaf side with four-armed tufted hairs.

Occurrence: Middle Eocene, Geiseltal, Mines Neumark-Süd, Kanya-Süd, Upper Eocene, Gaumnitz (German Democratic Republic).

Fig. 3. Fossile leaf structures of *Dryophyllum furcinerve* f. *haselbachense* forma nova, *a* holotype, Haselbach Mine, Oligocene, No. HbO 215 (MMG), 1 : 1. *b D. f.* f. *furcinerve*, abaxial cuticle, Profen Mine, Upper Eocene, microslide N. Pf. 69/82 (MMG), × 500. *c D. f. haselbachense* forma nova, abaxial cuticle, Witznitz, microslide No. Wz. 23/85 (MMG), × 500. *d D. f.* f. *robustum* (RÜFFLE & al.) stat. nov., leaf, Gaumnitz, Upper Eocene, No. Gz 1 (Mus. Halle), 1 : 1. *e D. f.* f. *robustum*, abaxial cuticle of *d*, Gaumnitz, Upper Eocene, microslide No. Gz 6/82 (MMG), × 400

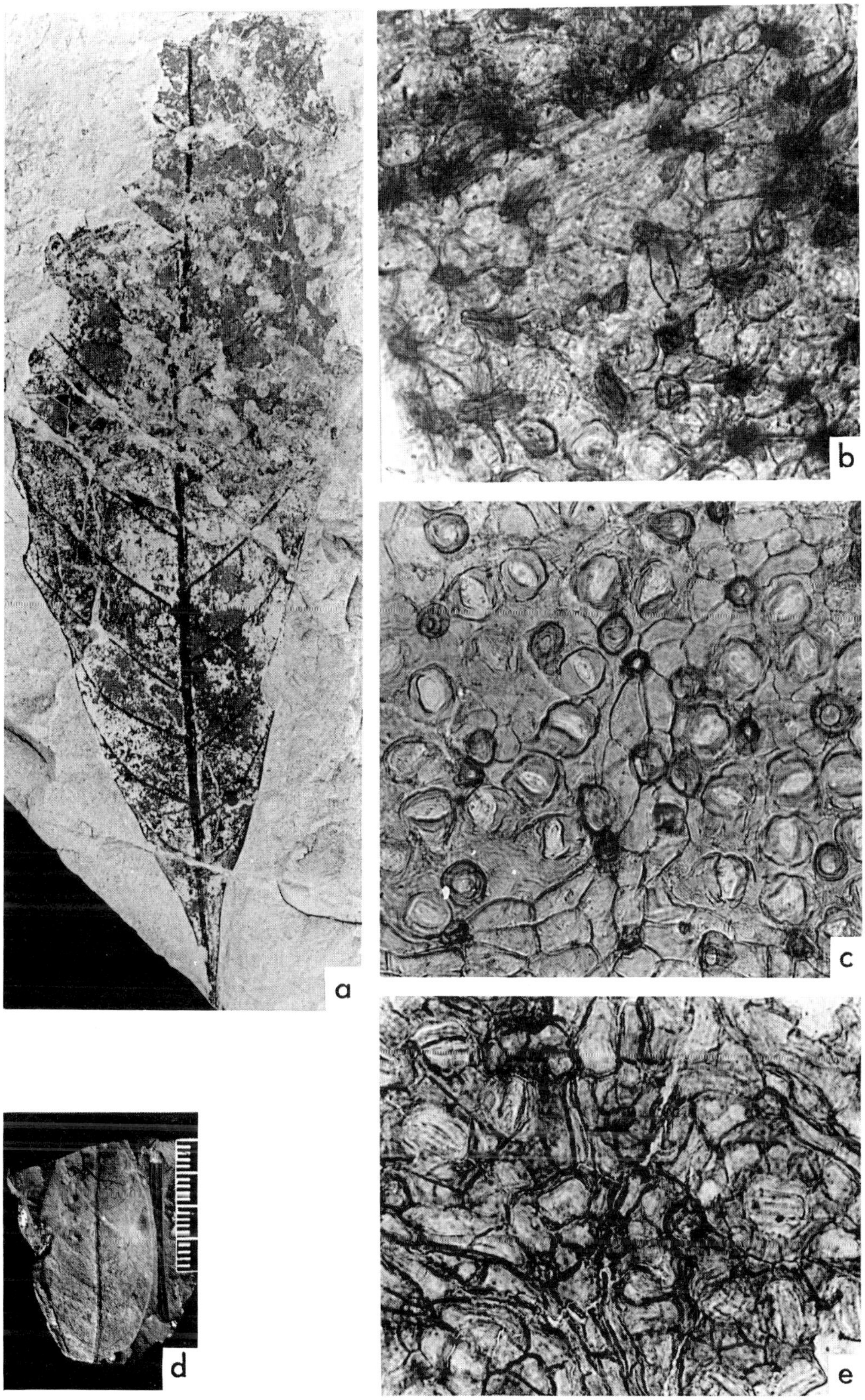
a
b
c
d
e

Excluded species:

***Carya knauensis* (MAI & WALTHER), comb. nova**

Basionym: *Dryophyllum knauense* MAI & WALTHER (1985, Abh. Staatl. Mus. Min. Geol. Dresden **33**: 67, pl. 16, f. 4–8, textf. 9/11, 12/1).

Holotype: Kn 15 a, microslides Kn 113/82, Kn 183-186/82, Knau near Altenburg (MMG).

The affinity to *Juglandaceae* is suggested by the peltate hairs with a simple stalk. The size of the hairs (45–75 µm across) and the craspedodrome venation are characters of *Carya*.

Trigonobalanopsis

The cupules and fruits, on which *Trigonobalanus exacantha* MAI has been based, occur chiefly in the Europoean Lower Miocene. However, they little resemble the true *Trigonobalanus* cupules. They contain 1–2 fruits, wedged in between the 3–4 closely adhering valves, of which the admedial one is not reduced, the pericarp, contrary to *Trigonobalanus,* shows radially elongated sclereids (see also MANCHESTER & CRANE 1983, MELVILLE 1982). MAI (1970) links these fruit remains with the dentate *Castanea*-like leaves from Dłuzyna Dolna (Poland); but this has to be rejected. The leaves belong to the *Quercus kubinyi* form complex, differ decidedly from *Dryophyllum furcinerve* in their cuticular characters and are associated with oak fruits (HUMMEL 1983).

Together with cupules of *T. exacantha* there do occur fagalean leaves, which, however, have been so far referred to as *Castanopsis toscana* or *C. dechenii*. Their connection is supported by identical epidermal structures of leaf petioles and peduncles. These leaves share some features, e.g. minute undulation of abaxial anticlines, thickened subsidiary cells, cylindric glandular hairs, with *T. excelsa,* but at the same time also with *Castanopsis* (solitary hairs on the midrib). From either genus they differ by a tendency towards strong but minute undulations of the adaxial anticlines. Leaves and cupules with fruits, therefore, are considered to belong to the same extinct and so far undescribed genus. The earliest name for the leaves was published by ROSSMÄSSLER (1840) as *Phyllites rhamnoides* from the Upper Eocene of the Staré Sedlo formation. The topotypical specimens show the same cuticular patterns as those from the Oligocene and Miocene.

***Trigonobalanopsis* KVAČEK & WALTHER, genus novum**

Type: ***Trigonobalanopsis exacantha* (MAI), comb. nova**

Trees (?). Leaves evergreen, entire-margined, elliptic, acuminate, venation brochiodrome, stomata cyclocytic with thickened subsidiary cells, trichomes unicellular,

---

Fig. 4. Leaf structures of *Trigonobalanopsis* and *Quercus* (m). *a, b T. exacantha* (MAI) comb. nova, cupules, Mine Kristina, Hrádek n. N., Lower Miocene (PRNM), × 5. *c–f T. exacantha*, cupules and fruits, Wiesa, Lower Miocene (PRNM), × 5. *g, h T. rhamnoides* (ROSSM.) comb. nova, leaf compression, Jehličná, Core 36 H (UGG), × 4, × 1.5. *i T. rhamnoides*, leaf compression, Hrádek n. N., Lower Miocene, No. Kr 196 (UGG), × 2. *j T. rhamnoides*, abaxial cuticle showing glandular hairs, Adendorf, Miocene, microslide NMo. Ad-1/1 (UGG), × 500. *k T. exacantha*, fruit surface showing simple hairs, Wiesa, Lower Miocene (UGG), × 300. *l T. rhamnoides*, abaxial cuticle, Hrádek n. N., Lower Miocene, microslide No. Kr 306/1 (UGG), × 200. *m Quercus rhenana*, abaxial cuticle showing stellate hair. Cheb Basin, Core V 3 a, microslide No. V 3 a-16/2 (UGG), × 500

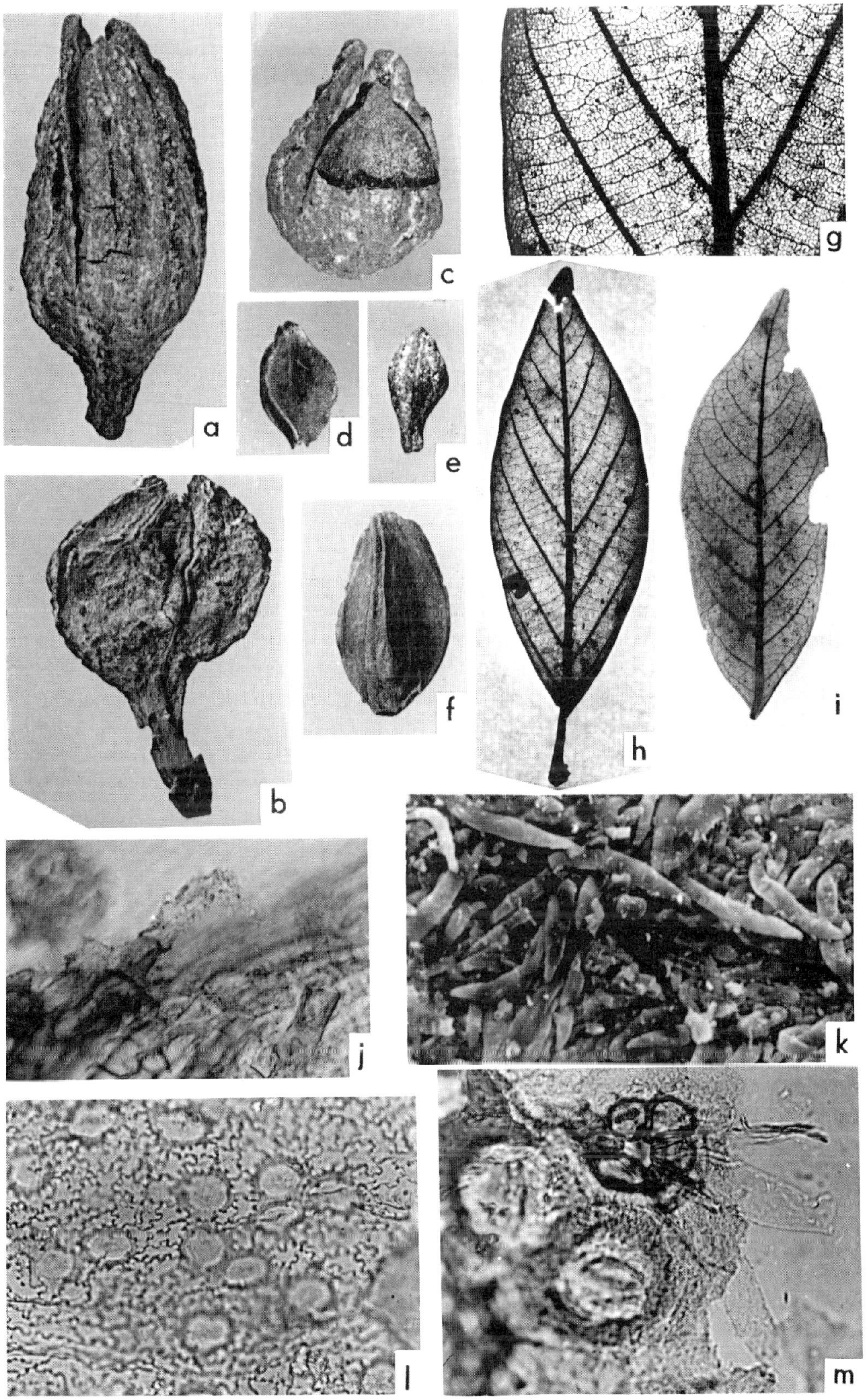
a
b
c
d
e
f
g
h
i
j
k
l
m

solitary, thick walled, confined to abaxial costal areas, and glandular, serial, mostly sheding terminal cylindric part. Fruits achenes, trigonous, sharply edged to slightly winged, hairy, terminally with remnants of 3 capitate (?) styles, one or rarely two in closed or slightly open cupules with 3 to 4 equivalent valves, rugulose outside, covered by trichome bases, hairy inside.

T. exacantha **(MAI), comb. nova**

Fruit remains (Fig. 4a–f, k), syn. *Trigonobalanus exacantha* MAI (1970, Jahrb. Geol. **3**: 385, pl. 1, f. 19–26, pl. 2, f. 1–22, pl. 3, f. 1–6, 14–19), Lower Miocene of Central Europe.

T. rhamnoides **(ROSSMÄSSLER), comb. nova**

Leaves (Fig. 4g–j, l), syn. *Phyllites rhamnoides* ROSSMÄSSLER (1840, Verst. Braunk. Geg. Altsattel. **35**, pl. 8, f. 30, 31), *Rhamnus dechenii* WEBER (1852, Palaeontographica **2**: 204, pl. 23, f. 2a–f), *Tristania toscana* BANDULSKA (1931, J. Linn. Soc. Bot. **48**: 668, pl. 40, f. 9–11), *Castanopsis toscana* (BANDULSKA) KRÄUSEL & WEYLAND (1954, Palaeontographica B **96**: 135, pl. 29, f. 3–9, textf. 14), *Castanopsis dechenii* (WEBER) KRÄUSEL & WEYLAND (1954, Palaeontographica B **96**: 137, pl. 17, f. 6–8, pl. 30, f. 1, textf. 14), Upper Eocene to Pliocene of Europe.

Castanea MILL.

The seemingly characteristic leaves of *Castanea* have morphological parallels in many *Quercus* species, notably in sect. *Cerris* SPACH. In *Castanea*, as in many deciduous *Fagaceae*, the cyclocytic stomatal type is not clearly differentiated and transitions to an anomocytic arrangement are encountered. Leaves of most species on their abaxial side bear a thick indumentum of stellate to tufted hairs, beside sparsely distributed cylindric or capitate glandular hairs with large basal cell and asymmetrically oriented hair attachments. Similar epidermal patterns make it extremely difficult to distinguish the leaves of *Castanea* from those of *Quercus*. HUMMEL (1983) has shown that *Castanea* lacks the fimbrial veins normally developed in oak leaves. According to this character, only few fossils have turned out to belong to *Castanea*, which seems to be confined to the latest Tertiary strata in Europe, as already suggested by ILJINSKAJA (in TAKHTAJAN 1982). Carpological evidence for *Castanea* in the European Tertiary so far is missing (except for an unripe fruit from the Pliocene of the Rhine Valley, V. D. BURGH 1978).

Quercus L.

This strongly diversified genus, rich in species, still presents many taxonomical problems. Various approaches in the delimitation of infrageneric taxa are apparent from several monographic treatments (CAMUS 1934–1954, SCHWARZ 1936, SOEPADMO 1968), and judgements on the diagnostic value of different organs have

---

Fig. 5. Leaf structures of *Quercus*. *a Q. mediterranea*, leaf, Lava (Greece), Miocene, No. La 7 (UGG), × 1.5. *b Q. mediterranea*, abaxial cuticle of the above specimen, microslide No. LAČ 1/86 (MMG), × 400. *c Q. pseudocastanea*, leaf, Achldorf, Upper Miocene (private coll.), × 1.5. *d Q. pseudocastanea*, abaxial cuticle of the above specimen (UGG), × 500. *e Q. subhercynica* spec. nova, abaxial cuticle of the holotype, microslide No. MMG Kö 72/86 (Harz Mus.), × 500. *f Q. subhercynica*, holotype, Königsaue Mine, Middle Eocene, No. 2136/2 (Harz Mus.), 1 : 1

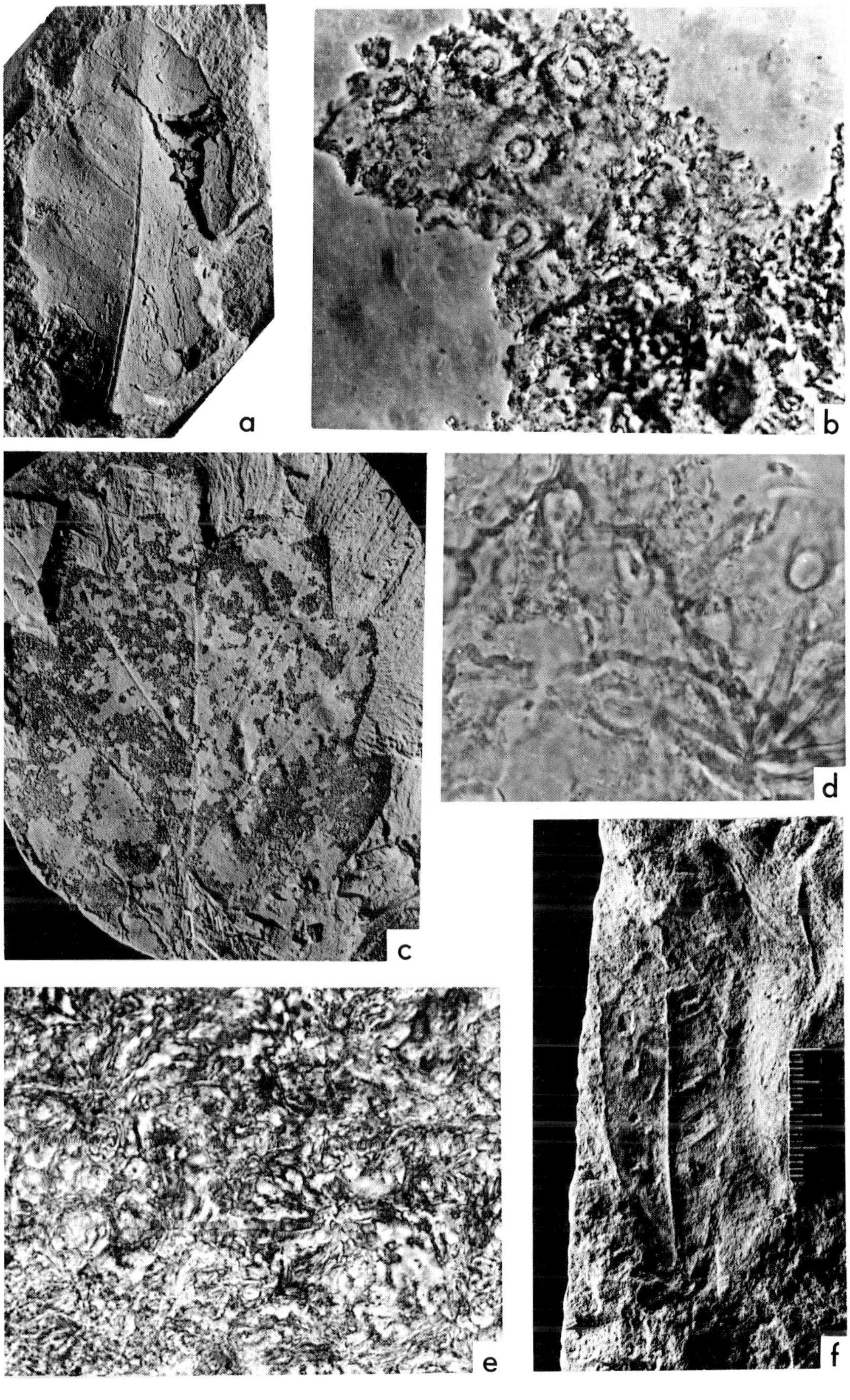

been inconsistent. Leaf epidermal studies have lately been employed as an aid in oak systematics species (Olsson 1975; Hardin 1976, 1979: Thompson & Mohlenbrock 1979; Jain & Singh 1974), but our knowledge in this regard still remains rather fragmentary. The occurrence of cyclocytic versus anomocytic stomata seems to be of significance for the differentiation of infrageneric groups. The indumentum mostly consists of glandular simple seriate and capitate and of non-glandular simple, tufted, stellate and multiradiate trichomes. The latter type with a massive base was found only in this genus (and in *Lithocarpus densiflorus*). Papillae, typical for subgen. *Cyclobalanopsis* (Oerst.) Schneider, are of the same type as in *Lithocarpus*.

In studies of fossil oaks taxonomic difficulties are still more pronounced. Most of the fossil species have been based on leaf morphological characters alone. This has been caused not so much by the limited experience of investigators in cuticular analysis but rather by difficulties in obtaining cuticles, particularly from deciduous late Tertiary oaks. The carpological evidence is largely limited by the scarcity of well preserved acorns and cupules.

Cuticular studies so far have revealed a unique leaf specimen belonging to the true oaks in the Middle Eocene of Central Europe (see below), while no reliable record of *Quercus* is known from the Upper Eocene strata.

Qu. subhercynica **Walther & Kvaček, spec. nova**, Fig. 5 e and f

Holotype: No. F 2136-1, microslides No. MMG Kö 72/86 (Harzmuseum Wernigerode, German Democratic Republic).

Locus typicus: Open cast mine Königsaue near Aschersleben (German Democratic Republic).

Stratum typicum: silty clay, Middle Eocene.

Tree (?). Leaf incomplete, petiolate, longly ovate, cuneate at the base, 5.5 cm long and 1.8 cm wide, entire-margined, petiole 1.6 cm long, margin thickened, venation camptodrome, midvein thick, secondaries moderately thick, alternate to opposite, with one or two intersecondaries, at angles of 51° to 63°, leaf consistence thick, cuticles poorly preserved. Abaxial leaf side covered by stellate hairs (30 – 60 µm in diam.) with 6 and more arms, stomatas of the cyclocytic (?) type, more or less rounded, 19 – 27 µm across, elliptic to rounded hair-bases (15 – 20 µm across) with small rounded remnants of their terminal parts.

In the Oligocene an evergreen oak *Qu. rhenana* (Kräusel & Weyland) Knobloch & Kvaček (syn. *Qu. lusatica* Jähnichen) has spread into Europe. By its epidermal features (cyclocytic stomata, massive hair-bases) and associated cupules, it belongs most probably to subgen. *Erythrobalanus*, but deviates from comparable extant species by strong wax layers around stomata (Fig. 4 m). Mai (1964) has compared its deformed acorns with *Castanopsis*. Rüffle & Palamarev (1979) unjustly introduced for this species the old name *Qu. apocynophyllum* Ett. which is based on a leaf with a different form (Ettingshausen 1870: pl. 2, f. 15).

Of younger origin are oaks with toothed or lobed margin: *Qu. kubinyi* (Kovats ex Ettingshausen) Berger form complex and *Qu. pseudocastanea* Goepp. (Fig. 5 c and d). Both have been assigned to sect. *Cerris* and their epidermal features, i.e. stellate to tufted hairs, anomocytic stomata, and short appressed asymmetric glandular hairs bases, support such a relationship. The same can be said about associated cupules. A number of leaf forms are attached to the *Qu. kubinyi* complex (Knobloch 1986), and are currently described as independent taxa. At present their status is difficult to establish more precisely, but studies on their fine venation and anatomy might possibly help to find acceptable modern counterpart.

The occurrence of true roburoid oaks in Europe only dates back to the Pliocene.

The past evolution and dispersal of oaks during the Tertiary was obviously controlled by climate and environment. An example of the adaptive evolution of an oak lineage from the originally humid-mesic to the present Mediterranean climate is afforded by *Qu. mediterranea* UNG. (Fig. 5 a and b) from the Miocene of southern Europe. Its morphology closely resembles *Q. coccifera* L., whereas its leaf consistence was chartaceous, judging by the extremely thin cuticles (in comparison to willow leaves from the same deposits). This undoubtedly deciduous oak must have changed only recently into the present sclerophyll and evergreen type, but we lack Quaternary material to follow this evolution more precisely.

We are grateful for support to keepers of herbaria, particularly to L. L. FORMAN (K), L. A. LAUENER, I. C. HEDGE (E), C. G. LOZANO (COL), E. EBEL, K. WERNER (HAL), and W. SPANOWSKI (DR), to our colleagues C. HILL, H. JÄHNICHEN, D. H. MAI, L. RÜFFLE, A. KNAPPE, Č. BŮŽEK, E. KNOBLOCH, and the late F. HOLÝ for providing fossil material, and to D. L. DILCHER for pre-print typescripts.

References

ANDREÁNSZKY, G., KOVÁCS, É., 1964: Der Verwandtschaftskreis der Eiche in der Unteroligozänen Flora von Kiseged bei Eger (Oberungarn). – Egri Muz. Évk. **2**: 7–42.

ANDREWS, H. N., 1970: Index of generic names of fossil plants, 1820–1965. – U.S. Geol. Surv. Bull. **1300**: 1–354.

BAAS, P., 1982: Comparative leaf anatomy of *Trigonobalanus* FORMAN *Fagaceae*. – Blumea **28**: 171–175.

BURGH, J. V. D., 1973: Hölzer der niederrheinischen Braunkohlenformation, 2. – Rev. Palaeobot. Palyn. **15**: 73–275.

– 1978: The Pliocene flora of Fortuna-Garsdorf 1. – Rev. Palaeobot. Palyn. **26**: 173–211.

CAMUS, A., 1934–1954: Les Chênes. Monographie du genre *Quercus* (et *Lithocarpus*). – Paris: Lechevalier.

ETTINGSHAUSEN, C. V., 1870: Beiträge zur Kenntnis der Tertiärflora Steiermarks. – Sitz.-Ber. Akad. Wiss. Math.-Nat. Kl. **60**: 17–100.

– 1880: Beiträge zur Erforschung der Phylogenie der Pflanzenarten 3–7. – Denkschr. K. Akad. Wiss. Math.-Nat. Kl. **43**: 93–102.

FORMAN, L. L., 1964: *Trigonobalanus*, a new genus of *Fagaceae*, with notes on the classification of the family. – Kew Bull. **17**: 381–396.

HARDIN, J. W., 1976: Terminology and classification of *Quercus* trichomes. – J. Elisha Mitchell Sci. Soc. **92**: 151–161.

– 1979: Patterns of variation in foliar trichomes of eastern North American *Quercus*. – Amer. J. Bot. **66**: 576–585.

HUMMEL, A., 1983: The Pliocene leaf flora from Ruszów near Żary in Lower Silesia, SW Poland. – Prace Muz. Ziemi **36**: 9–104.

ILJINSKAJA, I. A., 1980: K sistematike iskopaemych *Fagaceae* SSSR (ustanovlennych po ostatkam list'ev i cvetkov i otpečatkam plodov). – In ŽILIN, S. G., (Ed.): Sistematika i evoljucija vysšich rastenij, pp. 20–29. – Leningrad: Nauka.

JAIN, D. K., SINGH, V., 1974: Epidermal studies in some Himalayan species of oaks and their taxonomic significance. – Proc. Ind. Acad. Sci. **1380**: 188–196.

JONES, J. H., 1986: Evolution of the *Fagaceae*: the implications of foliar features. – Ann. Missouri Bot. Gard. **73**: 228–275.

– MANCHESTER, S. R., DILCHER, D. L., in press: *Dryophyllum* DEBEY ex SAPORTA, juglandaceous not fagaceous. – Rev. Palaeobot. Palyn.

KIRCHHEIMER, F., 1937: Beiträge zur Kenntnis der Flora des baltischen Bernsteins 1. – Beih. Bot. Centralbl. B, **57**: 441–484.

KNOBLOCH, E., 1986: Die Flora aus der Oberen Süßwassermolasse von Achldorf bei Vilsbiburg (Niederbayern). – Docum. Nat. **30**: 14–48.

– KVAČEK, Z., 1976: Miozäne Blätterfloren vom Westrand der Böhmischen Masse. – Rozpr. Ústř. úst. geol. **42**: 1–131.

– VELITZELOS, E., 1986: Die obermiozäne Flora von Likudi bei Elassona/Thessalien, Griechenland. – Docum. Nat. **29**: 5–20.

KRÄUSEL, R., WEYLAND, H., 1950: Kritische Untersuchungen zur Kutikularanalyse tertiärer Blätter 1. – Palaeontographica B **91**: 7–92.

– – 1954: Kritische Untersuchungen zur Kutikularanalyse tertiärer Blätter 2. – Palaeontographica B **96**: 106–163.

KVAČEK, Z., WALTHER, H., 1987: Revision der mitteleuropäischen tertiären Fagaceen nach blattepidermalen Charakteristiken. 1. Teil – *Lithocarpus* BLUME. – Feddes Repert. **98**: 637–652.

MAI, D. H., 1964: Die Mastixioideen-Floren im Tertiär der Oberlausitz. – Paläont. Abh. B, **2**: 1, 1–192.

– 1970: Die tertiären Arten von *Trigonobalanus* FORMAN (*Fagaceae*) in Europa. – Jahrb. Geol. **3 (1967)**: 281–409.

– WALTHER, H., 1978: Die Floren der Haselbacher Serie im Weisselster-Becken (Bezirk Leipzig, DDR). – Abh. Staatl. Mus. Min. Geol. Dresden **28**: 1–200.

– – 1985: Die obereozänen Floren des Weisselster-Beckens und seiner Randgebiete. – Abh. Staatl. Mus. Min. Geol. Dresden **33**: 1–260.

MANCHESTER, S. R., CRANE, P. R., 1983: Attached leaves, inflorescences, and fruits of *Fagopsis*, an extinct genus of fagaceous affinity from the Oligocene Florissant flora of Colorado, U.S.A. – Amer. J. Bot. **70**: 1147–1164.

MELVILLE, R., 1982: The biogeography of *Nothofagus* and *Trigonobalanus* and the origin of the *Fagaceae*. – Bot. J. Linn. Soc. **85**: 75–88.

OLSSON, V., 1975: The structure of stellate trichomes and their taxonomic implication in some *Quercus* species (*Fagaceae*). – Bot. Not. **128**: 412–424.

ROSSMÄSSLER, E. A., 1840: Die Versteinerungen des Braunkohlensandsteines aus der Gegend von Altsattel in Böhmen (Elnbogener Kreises). – Dresden, Leipzig: Arnoldische Buchhandlung.

RÜFFLE, L., 1963: Die obermiozäne (sarmatische) Flora vom Randecker Maar. – Paläont. Abh. B, **1**: 139–295.

– PALAMAREV, E., 1979: Verwandtschaftskreis und die Zönogenese von *Quercus apocynophyllum* ETT. in der Tertiärflora Mittel- und Südosteuropa. – Phytology, Bulg. Acad. Sci. **13**: 31–58.

– MÜLLER-STOLL, W. R., LITKE, R., 1976: Weitere *Ranales, Fagaceae, Loranthaceae, Apocynaceae*. – In Eozäne Floren des Geiseltales, Abh. Zentr. Geol. Inst. **26**: 199–282.

SCHWARZ, O., 1936: Entwurf zu einem natürlichen System der Cupuliferen und der Gattung *Quercus* L. – Notizbl. Bot. Gart. Berlin-Dahlem **8**: 1–22.

SELMEIER, A., 1970 a: Ein *Castanopsis*-Holz aus jungtertiären Schichten Südbayerns (Schrobenhausen). – N. Jahrb. Geol. Paläont. Mh. **1970**: 235–250.

– 1970 b: *Castanopsis*-Hölzer aus obermiozänen Glimmersanden der südlichen Franckenalb. – Mitt. Bayer. Staatssamml. Paläont. Hist. Geol. **10**: 309–320.

– 1972: Ein *Castanopsis*-Holz aus oberchattischen Steigbach-Schichten des Allgaus. – Mitt. Bayer. Staatssamml. Paläont. Hist. Geol. **12**: 97–104.

SOEPADMO, E., 1968: A revision of *Quercus* L. subgen. *Cyclobalanopsis* (OERSTED.) SCHNEIDER in Malesia. – Gardens Bull. Singapore **22**: 355–427.
ŚRODOŃ, A., 1985: *Fagus* in the forest history of Poland. – Acta Palaeobot. **25**: 119–137.
TAKHTAJAN, A. L., (Ed.), 1982: *Magnoliophyta* fossilia USSR 2. – Leningrad: Nauka.
THOMPSON, P. M., MOHLENBROCK, R. H., 1979: Foliar trichomes of *Quercus* subgenus *Quercus* in the eastern United States. – J. Arnold Arbor. **60**: 350–366.
UNGER, F., 1847: Chloris protogaea. – Leipzig: Engelmann.

Addresses of the authors: Dr sc. ZLATKO KVAČEK, Institute of Geology and Geotechnics, Okořská 11/340 CS-18100 Praha 8, Czechoslovakia. – Dr sc. HARALD WALTHER, Staatliches Museum für Mineralogie und Geologie zu Dresden, Augustusstrasse 2, Dresden, German Democratic Republic.

Pl. Syst. Evol. 162, 231 – 250

—Plant—
Systematics
and
Evolution

Early history of the *Juglandaceae*

STEVEN R. MANCHESTER

Received August 3, 1987

Key words: Angiosperms, *Juglandaceae*. – Paleobotany, pollen, fruits, evolution, Cretaceous, Tertiary.

Abstract: The major radiation of the *Juglandaceae* occurred during the early Tertiary as recorded by the proliferation of juglandaceous pollen and the appearance of fruits representing extinct and extant genera of the family. Juglandaceous pollen types of the Paleocene were predominantly triporate and exhibited a greater diversity in patterns of exinous thinning than occurs in the family today. Analyses of in situ pollen from early Tertiary juglandaceous inflorescences confirms the taxonomic value of certain patterns of exinous thinning. Data from co-occurring fruits and pollen indicate that relatively unspecialized, isopolar triporate pollen of the type presently confined to the tribe *Engelhardieae* also occurred in other tribes of the family during the Paleocene. Pollination has been mostly anemophilous throughout the Tertiary. Both wind and animal fruit-dispersal syndromes were established early in the radiation of the family but a greater diversity of wind-dispersed genera has prevailed.

The *Juglandaceae* is widespread in the Northern hemisphere today with eight genera and about 60 species. The fossil record of the family, which includes pollen, leaves, wood, inflorescences and fruits, is extensive and provides insight into the evolutionary and biogeographic history of the family through the Tertiary in North America, Europe and Asia (reviewed MANCHESTER 1987). The present paper considers the early history of the family, including probable Cretaceous forerunners, the early Tertiary diversification, and the appearance of extant genera. This review focuses on the fossil record of reproductive structures, chiefly pollen and fruits; other papers have reviewed the record of vegetative organs including leaves (e.g., JÄHNICHEN & al. 1977, WING & HICKEY 1984, DILCHER & MANCHESTER 1986, MANCHESTER 1987) and wood (MANCHESTER 1983, 1987).

Specimens illustrated in this paper are from several sources, including the following institutions (abbreviations as indicated): Indiana University (IU), United States National Museum, Washington, D.C. (USNM), Museum National d'Histoire Naturelle de Paris (MNHNP), Field Museum of Natural History, Chicago (FM), and British Museum (Natural History) (BM).

Within the *Hamamelidae,* the *Juglandaceae* is distinguished by the combination of pinnately compound leaves, unisexual flowers, bicarpellate, inferior ovaries, unilocular fruits with one or more incomplete septa and porate pollen with regularly spaced, fine spinules. The basic floral envelope within the catkins typically includes a bract, two bracteoles, and four tepals although variations occur (MANNING 1938, 1940, 1948, 1978). Inflorescences and pollen are, in general, well adapted for wind pollination, although *Platycarya strobilacea* has a syndrome of features adaptive to pollination by syrphid flies (ENDRESS 1986). Characters of the porate pollen including the footlayer, granular interstidium, tectum traversed by microchannels, and microspinulose ornamentation (STONE & BROOME 1975) are shared with the *Rhoipteleaceae, Myricaceae, Betulaceae,* and *Casurinaceae* (type III pollen of ZAVADA & DILCHER 1986) and help to differentiate the family from other families of the *Hamamelidae.*

The monotypic family *Rhoipteleaceae* is usually classified together with the *Juglandaceae* in the order *Juglandales. Rhoiptelea* shares with the *Juglandaceae* pinnately compound foliage, bicarpellate pistils and porate pollen (STONE & BROOME 1971), although it is distinguished by the completely bilocular ovary, presence of stipules, and bisexual flowers. WOLFE (1973) attributed fossil pollen from the Upper Cretaceous (Maastrichtian) of Maryland, U.S.A., to extant *Rhoiptelea* and KNOBLOCH & MAI (1986) assigned fruits to the genus from the Upper Cretaceous of Hergenrather near Aachen, Federal Republic of Germany. In contrast, extant genera of the *Juglandaceae* have not been confirmed prior to the Tertiary.

The eight extant genera of the *Juglandaceae* are divided into four tribes (MANNING 1978); the *Juglandeae* (*Juglans, Pterocarya, Cyclocarya*), the *Hicoreae* (*Carya*), the *Platycaryeae* (*Platycarya*), and the *Engelhardieae* (*Engelhardia, Oreomunnea, Alfaroa*). Each of the tribes has a fossil record that can be traced back to the Eocene or Paleocene and includes extinct as well as extant genera (MANCHESTER 1987). Characters separating the tribes include pith (septate in *Juglandeae*), bract shape (trilobed in *Engelhardieae*), participation of the bract in fruit formation (free from dispersed fruit and persisting on axis in *Platycaryeae*) and trends of pollen specialisation (pseudocolpi in *Platycaryeae*; more than three pores in *Juglandeae*; subisopolarity and exinous thinning at the proximal pole in *Hicoreae*) (MANNING 1978).

Pollen ornamentation and wall structure are remarkably uniform in extant *Juglandaceae*, and these characters are of limited value in the distinction of genera and tribes within the family (WHITEHEAD 1963, 1965, STONE & BROOME 1975). This uniformity supports the concept of the *Juglandaceae* as a very natural group despite its original circumscription based on characters of other organs. In exine stratification and ornamentation consisting of fine spinules that are invisible or only barely resolvable with light microscopy, pollen of the *Juglandaceae* is difficult to distinguish from that of *Betulaceae, Casurinaceae, Myricaceae*, and *Rhoipteleaceae*. However, in the *Juglandaceae*, the spinules are very evenly distributed over an otherwise smooth surface (Fig. 2 D), whereas in the other families the distribution of spinules is less uniform and the surface may be slightly rugulate or verrucate. These differences are readily observed with SEM (e.g., LIEUX 1980, ZAVADA & DILCHER 1986) and are useful in assessing the similarity of fossil pollen with extant *Juglandaceae*.

More variable features of the pollen that are of greater taxonomic utility within extant *Juglandaceae* include the number, distribution and shape of pores, grain

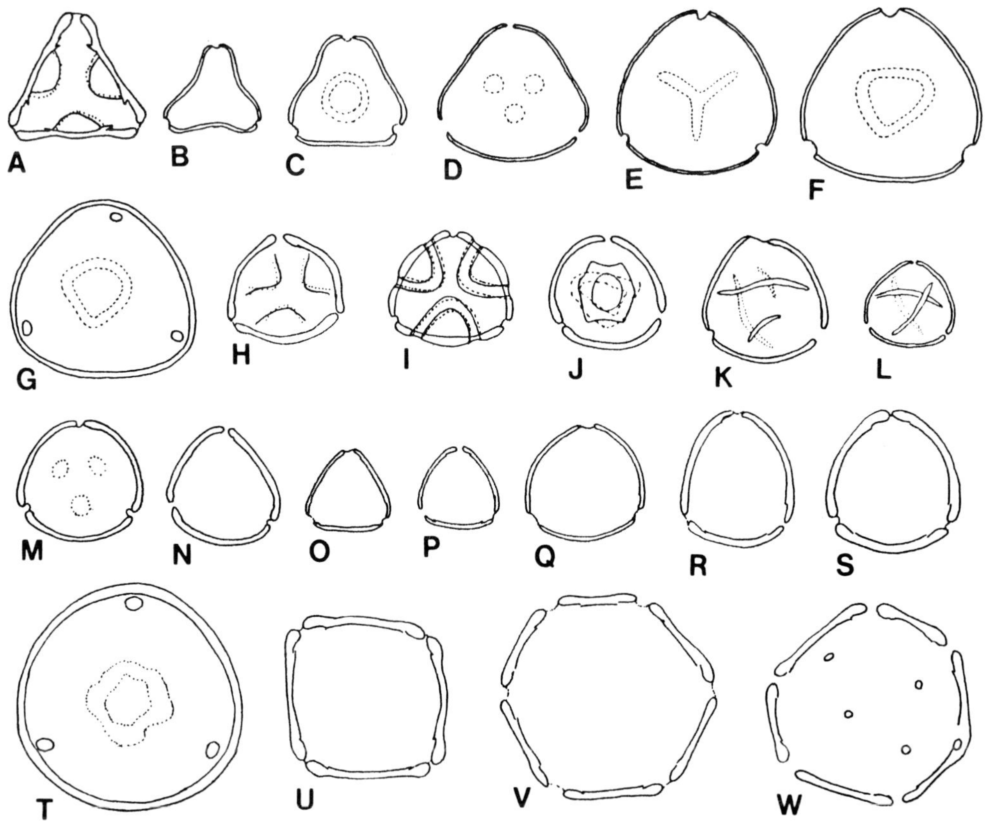

Fig. 1. Pollen grain diagrams of *Plicapollis* and of fossil and extant *Juglandaceae*. Thin areas of exine outlined with dotted (proximal surface) and/or solid (distal surface) lines, all × 800. *A Plicapollis retusus* TSCHUDY, showing thin areas of exine near the equator (Upper Cretaceous of Tennessee, U.S.A.), after TSCHUDY (1975: Pl. 9, Figs. 15–17); *B Momipites fragilis* FREDERIKSEN & CHRISTOPHER (Upper Cretaceous of S. Carolina, U.S.A.), after FREDERIKSEN & CHRISTOPHER (1978: Pl. 1, Fig. 11). *C Maceopolipollenites leffingwelli* (NICHOLS & OTT) comb. nov., showing ring of thin exine at one pole (Paleocene, Wyoming, U.S.A.), after NICHOLS & OTT (1978: Pl. 1, Fig. 29); *D Maceopolipollenites triorbicularis* LEFFINGWELL, with three symmetrically placed thin spots in the exine (Paleocene of Wyoming), after LEFFINGWELL (1971: Pl. 7, Fig. 5); *E Maceopolipollenites leboensis* LEFFINGWELL showing triradiate thin area, after LEFFINGWELL (1971: Pl. 7, Fig. 6); *F Maceopolipollenites amplus* LEFFINGWELL showing thin ring in exine at one pole (Paleocene of Wyoming), after LEFFINGWELL (1971: Pl. 6, Fig. 2); *G Caryapollenites veripites* (WILSON & WEBSTER) NICHOLS & OTT showing thin ring, but distinguished from *Maceopolipollenites* by the subequatorial pores, cf. *Carya* (Paleocene of Wyoming), after NICHOLS & OTT (1978: Pl. 2, Fig. 13); *H Plicatopollis plicatus* (middle Eocene Geiseltal, German Democratic Republic), after FREDERIKSEN (1979: Pl. 1, Fig. 34); *I P. plicatus* (middle Eocene Geiseltal), after KRUTZSCH (1962: Fig. 6); *J "Momipites tenuipolus* group" (lower Eocene of Virginia, U.S.A.), after FREDERIKSEN (1979: Pl. 1, Fig. 25); *K Platycaryapollenites* spec. (Eocene of Wyoming), after NICHOLS & OTT (1978: Pl. 2, Fig. 14); *L* Extant *Platycarya strobilacea*, after STRONE & BROOME (1975: Fig. 1 a); *M Maceopolipollenites triorbicularis* subsp. *menatensis* KEDVES (Paleocene of Menat, France), after KEDVES (1982: Pl. 10, Fig. 22); *N Momipites* spec. from catkin of *Eokachyra* (Eocene of Tennessee, U.S.A.), after CREPET & al. (1980: Fig. 7); *O Momipites* species from catkin of *Eoengelhardia* (Eocene of Tennessee), after CREPET & al. (1980: Fig. 6); *P Engelhardia roxburghiana*, after STONE & BROOME (1975: Fig. 1 b); *Q E. spicata*, after STONE & BROOME (1975: Fig. 1 c); *R Oreomunnea pterocarpa*, after STONE & BROOME (1975: Fig. 1 d); *S Alfaroa costaricensis*, after STONE & BROOME (1975: Fig. 1 e); *T Carya tonkinensis*, after STONE & BROOME (1975: Fig. 1 d); *U Cyclocarya paliurus*, after STONE & BROOME (1975: Fig. 1 f); *V Pterocarya delavayi*, after STONE & BROOME (1975: Fig. 1 g); *W Juglans cinerea*, after STONE & BROOME (1975: Fig. 1 j)

size, and the presence or absence of, and various patterns of, thin areas in the exine (WHITEHEAD 1963, 1965, STONE & BROOME 1975). Thus, *Engelhardia, Alfaroa* and *Oreomunnea* (*Engelhardieae*) are characterized by small to medium isopolar triporate pollen without prominent exine specialization (Fig. 1 P – S); *Platycarya* (*Platycaryeae*) is distinguished by small isopolar triporate grains with thin creases (pseudocolpi) developed on both hemispheres (Fig. 1 L); *Carya* has medium to large triporate grains that are subisopolar, with one or more of the pores offset toward the distal pole and with a thin area in the exine at the proximal pole (Fig. 1 T); *Cyclocarya, Pterocarya* and *Juglans (Juglandeae)* typically have larger grains with four to many pores and range from stephanoporate, with all of the pores equatorial (the usual case in *Cyclocarya, Pterocarya*; Fig. 1 U – V) to heteropolar, with one or more of the pores confined to one of the polar hemispheres (as is typical for *Juglans*; Fig. 1 W). However, there is a large amount of variation in the number pores even within a single anther, and it is best to base determinations on populations of grains from the same species (WHITEHEAD 1965). In practice, however, it is difficult to delimit conspecific populations of dispersed fossil grains, and the data obtained from in situ pollen of fossil inflorescences are invaluable.

Fruits are particularly useful in the generic taxonomy of the *Juglandaceae*. Extant genera are readily distinguished on the basis of differences in fruit morphology, such as septation, configuration of husk or wings, size, vasculature, and stylar orientation, etc. (MANNING 1978), and it is fortunate that fruits of this family are frequently preserved as fossils. Fruits are commonly small nutlets with attached wings adapted for wind dispersal or large husk-enclosed nuts adapted for rodent dispersal (STONE 1973).

Cretaceous precursors

Evidence of pollen (WOLFE 1973, 1976), flower and fruit (FRIIS 1983) similarity indicates that the *Juglandaceae* evolved from a portion of the Normapolles complex during the late Cretaceous or earliest Tertiary. The term Normapolles (PFLUG 1953) is applied to dispersed fossil pollen characterized by typically brevaxonate, triaperturate grains with complex, typically thick-walled, protruding pore regions. This kind of pollen appears in the middle part of the Cretaceous (Middle Cenomanian), becomes diverse and common in the Upper Cretaceous of Europe and Eastern North America and declines in diversity and abundance during the Early Tertiary through extinction and intergradation with extant taxa (review by BATTEN 1981). The Normapolles complex includes more than 100 form genera (BATTEN & CHRISTOPHER 1981) and is considered to be a heterogeneous assemblage of taxa including genera with diverse affinities to modern taxa (BATTEN 1981). Some of the genera have tectate-columellate grains, but many have a granular interstidium and thick imperforate tectum like that observed in the *Juglandaceae* and other extant *Hamamelidae* with type III pollen (ZAVADA & DILCHER 1986).

FRIIS (1983) described three genera of flowers that produced Normapolles pollen from the Upper Cretaceous (Upper Santonian or Lower Campanian) of Sweden and the Federal Republic of Germany. All three are small and bisexual, with inferior ovaries producing unilocular nuts with a basal orthotropous seed, and, as indicated by FRIIS, are suggestive of myricalean and juglandalean affinities. The most similar in many respects to extant *Juglandaceae* is the genus described as *Caryanthus*. This genus was widespread and apparently diverse in Europe during the Late Cretaceous, with 3 species recognized in Sweden (FRIIS 1985) and eight in Central Europe (KNOBLOCH & MAI 1986). *Caryanthus* flowers are bisexual and bisymmetrical with a single small bract and two lateral bracteoles united with the base of the ovary, an epigynous perianth of four tepals, six to eight stamens opposite the lateral tepals and a gynoecium of two median carpels. The fruit is a small laterally compressed nut about 0.5 to 1.5 mm in diameter that is apically flattened into a short wing and is unilocular with an orthotropous seed arising from a raised placental area. *Plicapollis* pollen was observed within the perianth of many specimens from Scania, Sweden (FRIIS 1983) and in situ within attached anthers of specimens from Aachen, Federal Republic of Germany (FRIIS 1985).

Caryanthus resembles extant *Juglandaceae* in several important characters including inferior ovary, basal orthotropous seed, bract and tepal numbers, 2 carpels, and pollen ornamentation. Characters that distinguish *Caryanthus* from all extant *Juglandaceae* include the regular occurrence of bisexual flowers, the lack of a prominent septum within the fruit, the very small size of the fruit and features of the *Plicapollis* pollen.

Plicapollis differs from pollen of extant *Juglandaceae* in the pronounced thickening of the exine near the apertures and in the triradiate polar plication on each hemisphere (Fig. 1 A). The presence of fine spinules on the surface of the grains as seen in SEMs (STANLEY & KEDVES 1975, FRIIS 1983, 1985) resembles that in extant families including the *Betulaceae, Casurinaceae, Myricaceae, Rhoipteleaceae*, and *Juglandaceae*. However, the spacing of spinules is not as regular as in pollen of extant *Juglandaceae* (see FRIIS 1985: Fig. 4 D). In addition, the pronounced thickening of wall layers near the apertures, a feature of many Normapolles taxa and of *Rhoiptelea* (STONE & BROOME 1971), is not characteristic of the *Juglandaceae* (STONE & BROOME 1975). Although the triradiate plication of the exine in *Plicapollis* is not found in extant genera of the *Juglandaceae,* the extinct juglandaceous pollen genus *Plicatopollis* KRUTZSCH, common in the Early Tertiary of Europe and North America, has a similar plication (Fig. 1 H, I). Similarly well developed plicae occur in the pollen of *Rhoiptelea* (GOCZAN & al. 1967, WOLFE 1973). Numerous species of *Plicapollis* pollen have been described and it remains to be determined whether the *Plicapollis* pollen type was produced only by flowers of *Caryanthus*. If *Plicapollis* is accepted as a monophyletic taxon, then its stratigraphic and geographic range may be relevant to the origin of the *Juglandaceae*. TSCHUDY (1981) reported that *Plicapollis* ranges from the Cenomanian to the Early Eocene in North America and from the Turonian to at least the Late Eocene in Europe.

The overlap in pollen morphological characters between certain types of Normapolles taxa and extant *Juglandaceae,* and reports of triporate pollen very similar to that of the *Juglandaceae* from the Upper Cretaceous, for example *Momipites tenuipolus* ANDERSON from the Maastrichtian of California (DRUGG 1967, CHMURA 1973), *Momipites fragilis* FREDERIKSEN & CHRISTOPHER from the Lower Campanian of southeastern United States (FREDERIKSEN & CHRISTOPHER 1978), and *Platycaryapollenites,* "*microcoryphaeus* group" from the Maastrichtian of middle Europe (KRUTZSCH 1970) indicates that the *Juglandaceae* may have originated in the Upper Cretaceous. Unfortunately, none of the purported juglandaceous pollen from the Cretaceous has been documented with EM studies, and the characteristic even distribution of spinules remains to be demonstrated. More work, particularly on

Upper Cretaceous megafossils, and including electron microscopy and high resolution light microscopy of small triporate pollen may help to clarify the transition between Normapolles taxa and extant families such as the *Juglandaceae, Rhoipteleaceae, Myricaceae,* and *Betulaceae*. Although it is possible that the *Juglandaceae* originated in the Late Cretaceous, it is clear from both pollen and megafossil data that the major radiation of the family took place in the Early Tertiary both in North America and Europe.

Paleocene diversification of juglandaceous pollen

Pollen with characteristic juglandaceous exine ornamentation diversified during the Paleocene in Europe (Krutzsch 1970, Kedves 1982) and North America (Nichols 1973, Nichols & Ott 1978, Frederiksen & Christopher 1978, Frederiksen 1979, 1980). Most of the observed Paleocene diversity is in triporate grains, although some 4 – 6 porate stephanoporate grains similar to those of extant *Pterocarya* occur in the late Paleocene (for review, see Manchester 1987). Numerous triporate juglandaceous taxa have been described from the Early Tertiary based upon differences in curvature of the sides (concave, straight or convex), position of the pores (equatorial, subequatorial, or mixed), occurrence and patterns of thin spots and folds in the exine and size (Fig. 1). The taxonomy of dispersed triporate juglandaceous grains is rather confused; different workers have applied different concepts of generic and species classification. At some Paleocene localities, the range of morphological variation in triporate juglandaceous pollen appears to be almost

Fig. 2. Paleogene triporate juglandaceous pollen grains from North America and England. *A Momipites* spec., dispersed grain from same shale as fruits of *Cyclocarya brownii*, Paleocene Fort Union Formation, Almont, North Dakota, × 2 000; *B Maceopolipollenites* cf. *annelus* from anther of an inflorescence recovered from the same deposit as *Polyptera manningii* fruits, Paleocene Mexican Flats locality, Fort Union Formation, southern Wyoming, × 2 000; *C* Light micrograph of pollen grains of *Maceopolipollenites* from the same inflorescence as *B*, showing the presence of a thin ring in the exine delimiting an island of normally thickened exine at the pole, × 500; *D* Detail of the spinulate ornamentation characteristic of fossil and modern *Juglandaceae*: *Maceopolipollenites annelus* pollen grain from an inflorescence associated with *Polyptera* fruits, Paleocene Earnest Butte locality, southwestern Wyoming, × 4 000; *E Momipites* pollen grain from dispersed anthers in sediments including fruits of *Casholdia* from the Paleocene Reading Beds, England, courtesy M. Collinson, × 2 000; *F Maceopolipollenites* cf. *triorbicularis,* from stamen of an inflorescence from the Oligocene of Huntsville, Texas, courtesy C. Daghlian, × 2 000; *G Caryapollenites veripites,* dispersed grain from shale of Upper Paleocene of Signor Ridge, Wyoming, showing the pores offset from the equator of the grain as in extant *Carya,* but relatively small size, × 2 000; *H* Larger *Caryapollenites* pollen grain, from a catkin (figured, Manchester 1987) from Lower Oligocene, Florissant beds, Colorado, × 2 000; *I Platycaryapollenites* pollen grain from the stamens of a platycaryoid catkin (figured, Manchester 1987), showing prominent pseudocolpi, Lower Eocene Camels Butte Member, Golden Valley Formation, North Dakota, × 2 000; *J* Mass of *Platycaryapollenites* pollen grains from the same staminate inflorescence as *I*, showing typical extent of variation in shape and position of pseudocolpi, which occur on both sides of each grain, × 500; *K* Mass of pollen grains from an anther of the same inflorescence as *B* and *C* showing depressions on the surfaces of about half of the grains due to the thin area that is present only on one hemisphere of each grain, × 500

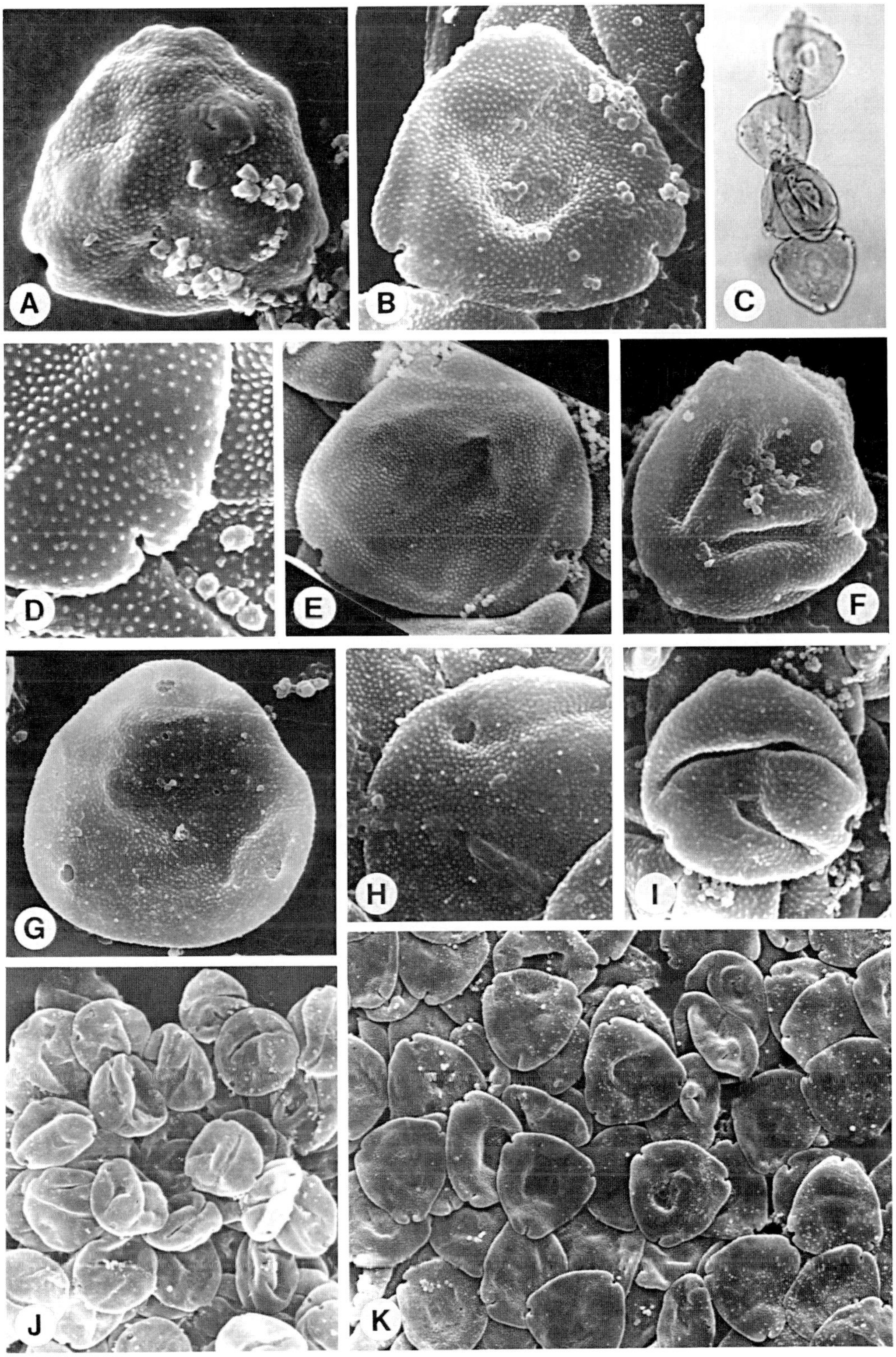
A
B
C
D
E
F
G
H
I
J
K

continuous, and discrimination of individual taxa may become somewhat arbitrary (FREDERIKSEN 1979). Studies of in situ pollen from fossil catkins of *Platycaryeae, Juglandeae* and *Hicoreae* reviewed here indicate that many of the characters traditionally used in distinguishing fossil juglandaceous pollen genera are indeed taxonomically useful, although some overlap does occur, particularly between species of the same pollen genus. There was a greater extent of morphological diversity in triporate grains in the Early Tertiary than there is today in the *Juglandaceae*; the following review of fossil juglandaceous pollen genera is presented to provide a sense of the Early Tertiary morphological diversity.

The genus *Momipites* accommodates relatively unspecialized isopolar triporate grains (Figs. 1 B and 2 A, E) of the type found today in *Engelhardia* (Fig. 1 P, Q), *Oreomunnea* (Fig. 1 R) and *Alfaroa* (Fig. 1 S). Pollen of this kind has also been documented in Eocene catkins with trilobate bracts diagnostic of the *Engelhardieae* (Fig. 1 N, O) (CREPET & al. 1975, 1980). *Momipites* pollen is recorded from the Upper Cretaceous and is abundant in the Paleocene, where it co-occurs with fruits of *Cyclocarya* and *Casholdia*. Although some authors have equated *Momipites* with *Engelhardia* or the *Engelhardieae* (NICHOLS 1973, MULLER 1981, FREDERIKSEN & al. 1983), it is likely that this pollen type evolved prior to the *Engelhardieae* (MANCHESTER 1987). The relatively complete record of fruits discussed later in this paper suggests that the *Engelhardieae*, as recognized by the trilobed inflorescence bract, did not evolve until the Early Eocene. NICHOLS (1973) emended the genus *Momipites,* to make it a broader taxon including pollen with various patterns of polar exinous thickening or thinning that do not occur in extant *Engelhardieae*. However, I think the genus is more useful in its more strict usage, which excludes the distinctive exinous thickening and thinning patterns here attributed to *Maceopolipollenites* or *Plicatopollis*.

Maceopolipollenites accommodates triporate isopolar grains in which the exine in the polar area of one hemisphere is thinned in various regular patterns, such as circles (Fig. 1 C, F), triangles, y-shapes (Fig. 1 E), or three small thin spots arranged in a triangular pattern (Fig. 1 D) (LEFFINGWELL 1971). This kind of pollen does not typically occur in extant *Juglandaceae* but is common in the Tertiary of North America (LEFFINGWELL 1971, NICHOLS & OTT 1978, FREDERIKSEN & CHRISTOPHER 1978) and Europe (KEDVES 1982). Pollen similar to *Maceopolipollenites amplus*, having a ring of thin exine delimiting an island of normally thickened exine at the pole (Fig. 2 B, C) has been recovered from catkins associated with fruits of the extinct genus *Polyptera* (Fig. 3 G, H) (MANCHESTER & DILCHER 1985). Pollen of the *M. triorbicularis* type, having three thin spots arranged in a triangular pattern, becomes common during the Paleocene (NICHOLS & OTT 1978) and persists into the Oligocene in some areas of North America. The recovery of this pollen type from the anthers of a juglandaceous catkin from the Oligocene of Huntsville, Texas (Fig. 1 F) (DAGHLIAN, pers. comm., MANCHESTER 1987), indicates that such grains were not just isolated aberrant grains of *Momipites*. Although there is some morphological intergradation between the patterns of exinous thinning included within *Maceopolipollenites* (NICHOLS & OTT 1978), the consistency of thinning patterns of the grains recovered from fossil catkins indicates that the taxon may include more than one natural genus. The type species, *M. triorbicularis,* and other species with the three thin spots may represent a genus distinct from those having the circular ring of thin exine.

Caryapollenites includes heteropolar (subisopolar) grains having one or more pores offset from the equator and having a thin spot or thin ring in the exine at one pole (Krutzsch 1961, Nichols & Ott 1978) (Figs. 1 G and 2 G, H). Among extant genera, this kind of pollen is diagnostic only of *Carya* (Fig. 1 T) and many authors have assigned fossil pollen of this morphology directly to the modern genus. However, the occurrence of this pollen type in the Paleocene record of Europe and North America predates the Late Eocene appearance of fruits attributable to *Carya*, and it is likely that such pollen was produced by extinct genera of the *Hicoreae*.

Nichols & Ott (1978) observed intergradation between isopolar pollen having a circular thin ring in the exine at one pole (e.g., *Maceopolipollenites amplus*, Fig. 1 F, referred to by them as *Momipites*) and subisopolar pollen of *Caryapollenites* (Fig. 1 G) through a Paleocene stratigraphic sequence in central Wyoming and suggested that an evolutionary trend from *Momipites* to *Caryapollenites* occurred during that time interval. Within *Caryapollenites*, several authors, including Tschudy (1973), Nichols & Ott (1978), and Frederiksen & Christopher (1978) have observed a trend of size increase through the Early Tertiary. In comparison with the pollen of extant *Carya* species which usually range from 35 – 45 μm, the earliest pollen of *Caryapollenites*, which became abundant in the Upper Paleocene of North America (e.g., Fig. 2 G), was small, mostly within the range of 22 – 28 μm. Larger grains, mostly in the range of 29 – 39 μm become common in the Eocene. In situ pollen from catkins associated with unequivocal *Carya* fruits from the Lower Oligocene of Colorado (Manchester 1987), is 30 – 38 μm (Fig. 2 H) and thus comparable in size to extant *Carya* pollen. Some of the Early Tertiary European dispersed pollen formerly attributed to *Carya* was transferred to *Subtriporopollenites* by Krutzsch (1961). Although similar to *Carya* in the triporate subisopolar morphology, this pollen does not possess the characteristic exinous thinning, and detailed microscopy reveals a "double-structure" ornamentation in which the spinules are distributed on a verrucate-rugulate surface (Kedves & Stanley 1976) unlike that of *Juglandaceae*.

Plicatopollis (Krutzsch 1962) includes triporate isopolar grains with three thin areas or pseudocolpi symmetrically arranged about the pole on both sides of the grain, sometimes with a correlated triradiate fold or thickened area at each pole (Fig. 1 H, I). There is intergradation from grains with three thin spots and no plications to grains with a thick plication and no thin spots (Frederiksen 1979). The basic difference between this kind of pollen and that of *Maceopolipollenites* and *Caryapollenites* is that the exine modifications occur on both polar hemispheres, rather than just one. In this respect, the pollen is similar to that of the Normapolles genus Plicapollis and to extant *Platycarya*. Although *Plicatopollis* pollen is not identical to that of any extant juglandaceous genus, and has not been recovered from fossil catkins, high resolution light microscopy (Krutzsch 1962) and electron microscopy (Kedves & Stanley 1976) show characters, such as the ornamentation of evenly spaced spinules, that support assignment to the family. Frederiksen (1973, 1980) observed that it is sometimes difficult to determine whether the tri radiate structure and/or thin spots are on both sides or just one, so that in practice it may be difficult at some localities to distinguish some *Plicatopollis* grains from *Maceopolipollenites triorbicularis* type.

Platycaryapollenites (Nagy 1969) applies to isopolar grains with one or more asymmetrically distributed thin areas, or pseudocolpi, on each polar hemisphere

(Fig. 1 K). Such pollen is produced today only by *Platycarya* (Fig. 1 L) and many investigators have attributed dispersed fossil pollen of this kind to the extant genus. However, association with extinct platycaryoid fruit types such as *Hooleya* (Fig. 5 F) and *Paleoplatycarya* (Fig. 5 G) and with extinct platycaryoid foliage types suggests that very similar, possibly identical, pollen was produced by extinct genera of the *Platycaryeae* (MANCHESTER 1987). In examining pollen from the anthers of *Platycarya americana* catkins from the Early Eocene of North Dakota (Fig. 2 I, J), WING & HICKEY (1984) observed variation in pseudocolpi distribution and shape, from one to two per hemisphere and from straight to arcuate, circular, or Y-shaped. By implication, the various form species of *Platycaryapollenites* (KEDVES 1982 recognized eight species from the Paleocene of Menat, France) may represent variability within species rather than species diversity. Nevertheless, most of the pollen in the Eocene *Platycarya* catkins is easily distinguishable from that of *Momipites* and *Maceopolipollenites*. Although KRUTZSCH (1970) reported *Platycaryapollenites* back to the Maastrichtian, he noted that grains similar to those of extant *Platycarya* first occur in the Late Paleocene. FREDERIKSEN (1979) recorded a *Momipites-Plicatopollis-Platycaryapollenites*-complex from the Lower Eocene of northeastern Virginia in which there is so much overlap that individual generic assignments become impractical.

Although triporates dominate the juglandaceous pollen scene through much of the Early Tertiary, multiporate isopolar (stephanoporate) pollen similar to that produced by extant *Pterocarya* and *Juglans* (sect. *Cardiocaryon*) appears in the late Paleocene. The correct name for dispersed pollen of this type is *Polyatriopollenites* PF. (KRUTZSCH 1970). Although the earliest records are Paleocene, such pollen does not become abundant until the Oligocene. Pollen of the multiporate, heteropolar type typical of extant *Juglans* (especially sectt. *Rhysocaryon* and *Dioscaryon*) is rare in the Early Tertiary, both in Europe and North America. This type of dispersed pollen, often attributed to the fossil taxon *Multiporopollenites maculosus* (R. POT.) TH. & PF. (KRUTZSCH 1966) is not confirmed from the Paleocene and is known only from a few Eocene specimens before becoming common in the Oligocene (references cited, MANCHESTER 1987).

Early Tertiary fruit record

Fossil fruits of the *Juglandaceae* are well represented in the Early Tertiary of Europe and North America, and are important in the systematic resolution that they provide through comparison with extant genera. Presently, at least four genea of juglandaceous fruits are known in the Paleocene of the Northern hemisphere: *Cyclocarya, Polyptera, Juglandicarya, Casholdia.* These will be discussed here, followed by a consideration of other genera that first appear in the Eocene.

Cyclocarya. Among extant genera of the *Juglandaceae, Cyclocarya* has the oldest fossil record, extending well into the Paleocene, and apparently predating the appearance of the tribes *Platycaryeae* and *Engelhardieae* (MANCHESTER 1987). *Cyclocarya* has a single modern species in eastern Asia and is easily recognized by its fruits which consist of a small nut in the center of a circular disk-like wing with radiating, subparallel, dichotomous venation (Fig. 3 A). Although only one species exists today, three are known from the Paleocene to lowermost Eocene of the Rocky Mountain region in North America based upon differences in nutlet morphology

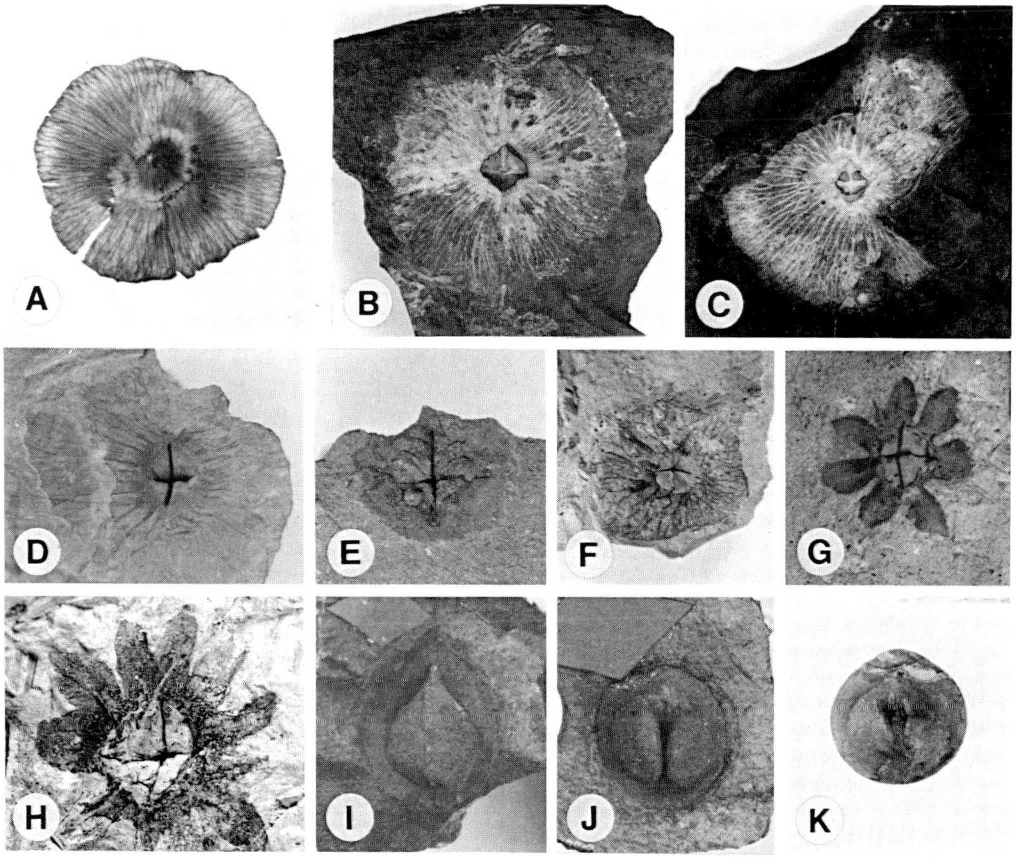

Fig. 3. Extant and fossil fruits of *Juglandaceae*. *A* Extant *Cyclocarya paliurus* (Batalin) Iljinsk., showing round central nutlet surrounded by a circular wing, IU mod. ref. coll. 3386, × 1; *B* Holotype of *Cyclocarya brownii* Manchester & Dilcher showing quadrangular nutlet and circular wing, from the Paleocene Fort Union Formation of Almont, North Dakota (IU 4031), × 1; *C C. brownii* from the same locality showing basal side of the nutlet with locule cast exposed showing that the seed was basally four-lobed (IU 4032), × 1; *D C. brownii* from the Paleocene Fort Union Formation near Broadus, Montana, showing prominent grooves in the position of the primary and secondary septa of the nutlet (USNM 298887), × 1; *E* Holotype of *C. coalmontensis* showing relatively large quadrangular nutlet with prominent septa and small wing, from the Coalmont Formation of Colorado (USNM 298886), × 1.5; *F Cyclocarya minuta* Manchester showing rounded nutlet and smaller wing than that of *C. brownii*, from the Paleocene Fort Union Formation of Central Wyoming (USNM 364531), × 1.5; *G Polyptera manningii* Manchester & Dilcher, extinct fruit type consisting of a quadrangular nutlet with prominent primary and secondary septa surrounded by a wing of 8 lobes from Paleocene Fort Union Formation of southern Wyoming (IU 4864), × 1.5; *H* Larger fruit of *P. manningii* with a wing of 10 lobes (IU 7272), × 1.5; *I Juglandicarya simplicarpa* Manchester showing smooth surface of locule cast, from the Paleocene Fort Union Formation of Wyoming, × 1.5; *J J. simplicarpa* from the same locality, lateral view of locule cast showing position of primary septum; *K Juglandicarya cantia* Ried & Chandler fruit from the Lower Eocene London Clay, England (BM-v 22121) × 1.5

and relative size of the wing and nutlet (MANCHESTER & DILCHER 1982, MANCHESTER 1987).

Fruits of *Cyclocarya brownii* from the Paleocene of Wyoming, Montana and North Dakota (Fig. 3 B – D) closely resemble those of extant *C. paliurus* in venation, shape, size and orientation of the wing and in the thick nutshell lacking lacunae, smooth locule surface and stylar orientation (MANCHESTER & DILCHER 1982). Unlike extant *C. paliurus,* which has fruits on short pedicels and globose nutlets, the fruits of this fossil species are borne on long pedicels (MANCHESTER 1987) and the nutlets are pyramid-shaped, with a quadrangular cross section and triangular longitudinal section. In addition, the wing is attached to the base of the nut in the fossil species but is approximately equatorial in the living species (MANCHESTER & DILCHER 1982). *C. coalmontensis* from the uppermost Paleocene or lowermost Eocene Coalmont Formation of northern Colorado (Fig. 3 E) resembles *C. brownii* in having a pyramidal nutlet and basal wing attachment, but is unique in having a nutlet that is relatively large in proportion to the wing (MANCHESTER & DILCHER 1982). *C. minuta* from the Paleocene of Central Wyoming (Fig. 3 F) resembles *C. coalmontensis* in small overall wing size, but has a small nutlet that is circular rather than square in cross section (MANCHESTER 1987).

Tetraporate pollen characteristic of extant *Cyclocarya* has not been recovered from the sediments in which these fossil *Cyclocarya* fruits are preserved. At the Almont locality of North Dakota, where *C. brownii* fruits are very numerous, the only kind of juglandaceous pollen recovered from the matrix of the megafossils is *Momipites* (Fig. 2 A). Other kinds of juglandaceous fruits are not known from this locality, and the apparent absence of engelhardioid fruits from this and all other deposits of the Fort Union Formation that have been sampled suggests that this kind of isopolar triporate pollen, resembling that of extant *Engelhardieae*, was produced by the *Cyclocarya brownii* plant.

Cyclocarya vanished from the North American fossil record during the Eocene, but has good records in the Oligocene to Pliocene of Europe and Asia. Fruits of the other extant genera of the tribe *Juglandeae, Juglans* and *Pterocarya,* make their first appearances in the middle Eocene of North America and become common in the Oligocene throughout the Northern hemisphere (MANCHESTER 1987).

Polyptera. Fruits of the extinct genus *Polyptera* from the Paleocene of Wyoming resemble those of *Cyclocarya* in consisting of a nutlet in the center of a prominent disk-like wing with radiating subparallel venation (Fig. 3 G, H). However, the fruit wing of *Polyptera* is deeply dissected into distinct radially arranged lobes. The type species, *P. manningii* from southern Wyoming has 8 to 12 lobes, although newly recovered material from northern Wyoming has only 4 to 5 lobes (SCOTT WING, pers. comm. 1986). Like *Cyclocarya brownii* and *C. coalmontensis, Polyptera manningii* has a pyramidal nutlet and basal wing attachment. The primary and secondary septa are both well developed, as in the fossil *Cyclocarya* species. Based upon the similarities with *Cyclocarya,* this genus has been attributed to the *Juglandeae* tribe (MANCHESTER & DILCHER 1982). Laterally compressed specimens often show filamentous strands ensheathing the nut impression (MANCHESTER & DILCHER 1982) that might represent resistent vascular bundles in the outer covering of the nut. Newly recovered specimens from the Earnest Butte locality show that the fruits were borne on short (1.5 mm), stout pedicels unlike the long (20 – 30 mm) thin pedicels of *C. brownii.*

Juglandaceous catkins have been recovered from two of the localities in southern Wyoming where *Polyptera* fruits are found. Anthers from these inflorescences contain triporate isopolar pollen of the *Maceopolipollenites amplus* type (Fig. 2 B, C), having a thin ring in the exine at one pole. In view of the occurrence of a similar exinous thin ring in *Caryapollenites* and in pollen of extant *Carya*, it is possible that *Polyptera* (if correctly associated with these catkins) shares a relationship with the *Hicoreae* as well as with the *Juglandeae* (MANCHESTER & DILCHER 1985).

Juglandicarya. The genus *Juglandicarya* is diagnosed to include "fruits which, although clearly referable to the *Juglandaceae*, are of doubtful relationship both to living genera and to one another" (REID & CHANDLER 1933: 140). Thus, it is an artificial genus that may include species with diverse affinities within the family. Of the four species originally described from the Lower Eocene London Clay in England (REID & CHANDLER 1933), *Juglandicarya cantia* (Fig. 3 K) is of interest here because its large size and apparent lack of wings suggests that it may have been adapted for animal-dispersal. Fruits similar to *J. cantia* occur somewhat earlier, in the Paleocene Fort Union Formation in Wyoming (Fig. 3 I, J) (BROWN 1962: Pl. 19, Figs. 5, 8 – 11) and are described as *Juglandicarya simplicarpa* (MANCHESTER 1987). The American specimens are preserved only as siltstone molds and casts, but, like the single British specimen, show the impression of a smooth, thick nutshell and a locule cast with two prominent basal lobes indicating the development of a primary septum at the base. A slight emargination at the base of each lobe of the locule cast indicates only weak development of a secondary septum. Although similar in its relatively large size and lack of dispersal wings to fruits of extant *Carya, Juglans* and *Alfaroa*, the simplicity of the locule rules out placement in any of these modern genera.

Obvious pollen of *Juglans* type is absent from the sediments associated with *Juglandicarya cantia* and *J. simplicarpa*, both in England (KEDVES 1967, GRUAS-CAVAGNETTO 1976) and North America (MANCHESTER, unpubl. data). However, *Caryapollenites* pollen similar to, but smaller than that of most extant *Carya* species, occurs at both locations, indicating the possibility that these fruits may belong to the *Hicoreae*. The first well documented records of *Carya* fruits are Late Eocene and a major radiation of this genus is evident in the Oligocene to Miocene of Europe (MAI 1981).

Casholdia. Fruits of *Casholdia* (Fig. 4 A) from the Late Paleocene of England and France appear to be relevant to the evolution of the *Engelhardieae* tribe (CRANE & MANCHESTER 1982). They consists of a small nutlet with two elongate styles at the base of two wings, one of them large and elongated, interpreted as a bract, and the other smaller and rounded, interpreted as a prophyllum originating from the fusion of two bracteoles. Fruits of extant *Engelhardia* and *Oreomunnea* (Fig. 4 D; *Engelhardieae*) have the same construction and in observable characters, *Casholdia* appears to be more similar to the *Engelhardieae* than to other extant tribes. Moreover, the venation of the wings is very similar to that in the wings of *Oreomunnea*. However, the trilobed inflorescence bract, considered diagnostic for the *Engelhardieae* (MANNING 1978), does not occur in *Casholdia*. Although the bract of *Casholdia* is easily observed because it forms the major wing of the fruit, it shows no evidence of lobing.

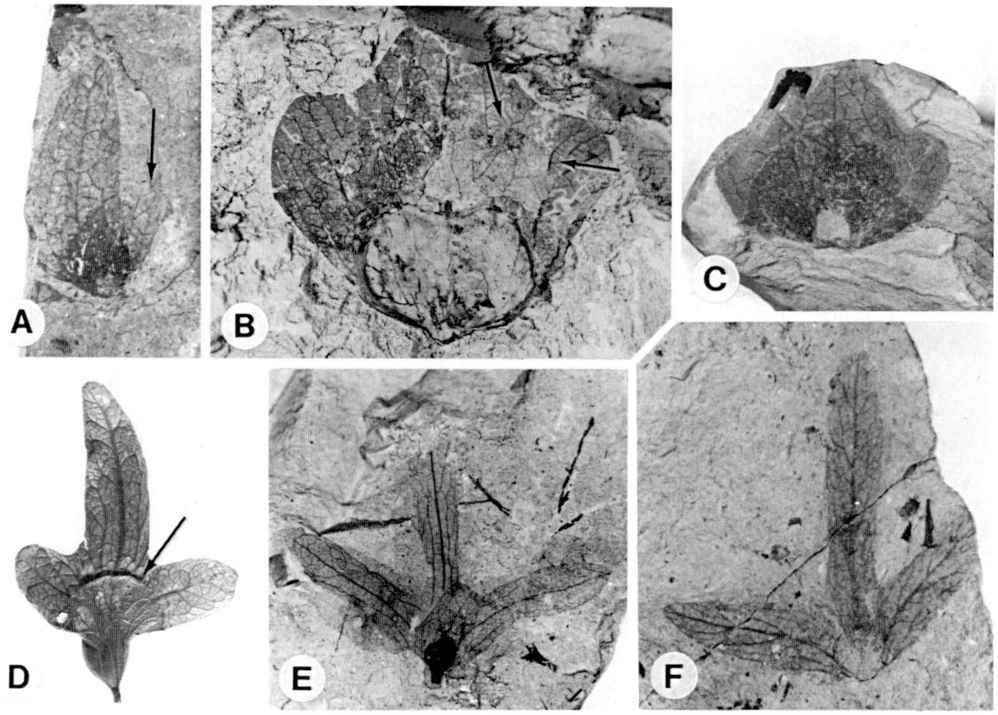

Fig. 4. Fruits of *Casholdia* and *Engelhardieae*. *A Casholdia microptera* CRANE & MANCHESTER showing nutlet at the base of an elongate, unlobed bract and remains of a prophyllum (arrow), from the Paleocene of Menat, France (MNHNP-Men 31), × 3; *B* Holotype of *Paleooreomunnea stoneana* DILCHER, POTTER & CREPET, showing trilobed bract with shallow sinuses, large nutlet and faint, rounded outline of the prophyllum (arrows) from the Middle Eocene Claiborne Formation, Warman Clay Pit, Tennessee (IU 1879), × 1.5; *C Paraengelhardtia eocenica* BERRY showing trilobed bract with shallow sinuses, rounded margin of the prophyllum, and small basal nutlet from the Middle Eocene Claiborne Formation, Puryear, Tennessee (FM-PP 7966), × 1.5; *D* Extant *Oreomunnea mexicana* (STANDL.) LEROY showing typical trilobed bract with well developed sinuses, and large rounded prophyllum (arrows), central mountains, Nicaragua (IU mod. ref. coll. 4034), × 1; *E Palaeocarya nevadensis* (MACGINITIE) MANCHESTER, showing similarity to *Oreomunnea* in venation of the trilobed bract and well developed prophyllum from the Lower Eocene of the Sierra Nevadas, California (UCMP 2157), × 1.5; *F Engelhardia oxyptera* SAPORTA from the Upper Oligocene of Armissan, France (MNHNP 12742), × 1.4

Dispersed anthers with intact juglandaceous pollen occur in the Upper Paleocene Reading Beds at the same locality as fruits of *Casholdia* (Fig. 2 E). The triporate, isopolar pollen is similar to that of extant *Engelhardieae*, and conforms to the dispersed pollen genus *Momipites*.

Radiation of engelhardioid fruits. Evidence from the fossil fruit record suggests that the *Engelhardieae* diversified during the Early and Middle Eocene (DILCHER & al. 1976, CRANE & MANCHESTER 1982, MANCHESTER 1987). Although conspicuously absent from Paleocene sediments so far investigated, trilobate engelhardioid fruits become abundant and diverse during the Early to Middle Eocene. The earliest record, *Palaeocarya nevadensis* (MACGINITIE) MANCHESTER is a fruit closely re-

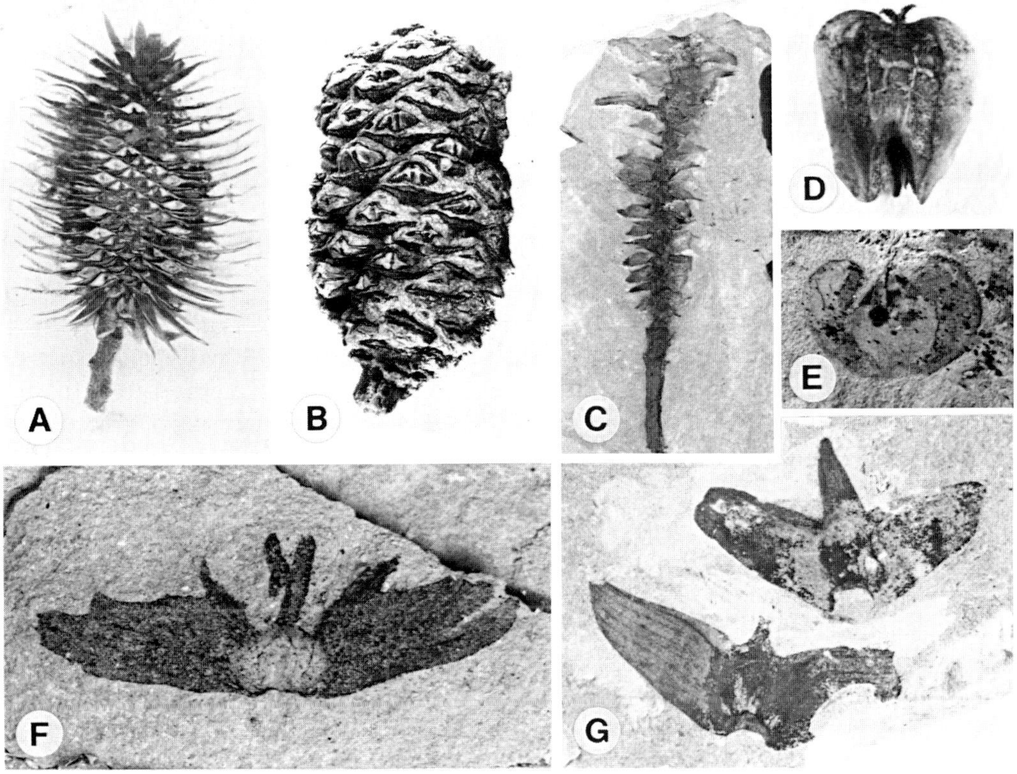

Fig. 5. Fruits of *Platycaryeae*. *A* Infructescence of extant *Platycarya strobilacea*, longitudinally sectioned to show position of winged fruits subtended by bracts (IU mod. ref. coll. 4018), × 1.5; *B* Infructescence of fossil *Platycarya richardsoni* (BOWERBANK) CHANDLER with surface abraded showing spirally arranged fruits and subtending bracts from the Lower Eocene London Clay of Herne Bay, Kent (BM-v. 29803), × 1.5; *C* Extinct playtcaryoid infructescence, *Paleoplatycarya wingii* MANCHESTER showing short rounded bracts and a single remaining fruit (arrow) of the kind shown in *G*, from the early Eocene Wind River Formation, Wyoming, × 1.5; *D* Fruit of *Platycarya strobilacea* from the infructescence in *A*, showing striated nut, apical stigmas, prominent basal attachment scar and two lateral wings, × 5; *E* Fruit of *Platycarya americana* HICKEY from the early Eocene Golden Valley Formation, North Dakota (IU 4766), × 5; *F* Fruit of *Hooleya lata* WING & HICKEY from the Eocene Clarno Formation, Oregon showing two large lateral wings with subparallel venation and four perianth parts (IU 5298), × 4; *G* Fruits of *Paleoplatycarya wingii* MANCHESTER nutlets with prominent basal attachment scars and a pair lateral wings; lateral sepals are united with the wings whereas the median sepals are free (USNM 387391), × 4

sembling that of extant *Oreomunnea* from the Early Eocene Chalk Bluffs of California (Fig. 4 E, MACGINITIE 1941). Several genera are recognized in the middle to Late Eocene of the Mississippi Embayment region DILCHER & al. 1976, MANCHESTER 1987) including species similar to *Oreomunnea* and *Engelhardia*. The extinct morphotypes include *Paleoreomunnea* (Fig. 4 B) and *Paraengelhardtia* (Fig. 4 C) in which the bract-wing is only faintly tri-lobed and thus conceptually intermediate between *Casholdia* and extant *Engelhardieae*. Although presently confined to Central America and eastern Asia, the *Engelhardieae* is well represented in the fruit and foliage record of Europe, with *Palaeocarya* (very similar to *Oreomunnea*) extending from the Eocene to the Miocene (JÄHNICHEN & al. 1977). In Asia, the tribe is first recorded on the basis of fruits from the Oligocene of Korea and Japan (TANAI & UEMURA 1983).

Staminate catkins yielding *Momipites* type pollen (Fig. 1 N, O) co-occur with the engelhardioid fruits in Tennessee (CREPET & al. 1975, 1980). Fossil fruits and foliage similar to *Oreomunnea* appear in the Eocene to Miocene of Europe (JÄHNICHEN & al. 1977). Fruits of the *Engelhardia* type occur in the Oligocene of Korea and Japan (Fig. 4 F).

Radiation of platycaryoid fruits. Unequivocal fruits of *Platycaryeae* first occur in the earliest Eocene of Europe (REID & CHANDLER 1933) and North America (WING & HICKEY 1984), somewhat prior to the appearance of the first fruits of *Engelhardieae*. Fruits of *Platycarya richardsoni* (REID & CHANDLER) CHANDLER from the London Clay flora (Fig. 5 B) and *Platyacarya americana* from the Golden Valley Formation of North Dakota, U.S.A. (Fig. 5 E) are virtually indistinguishable from those of the extant east Asian species *P. strobilacea* (Fig. 5 A, D), although the fruit wings are slightly larger. Foliage associated with *Platycarya americana* differs in many characters from that of the single extant species of *Platycarya*, and WING & HICKEY (1984) consider some of these to be simply more primitive features and some to be characters indicative of a close relationship to *Engelhardieae*. Extinct platycaryoid fruit morphotypes with large wings and prominent venation were also present during the Eocene, including *Paleoplatycarya* (Fig. 5 C, G) from the Lower Eocene of Wyoming (MANCHESTER 1987), *Hooleya* from the Eocene of Oregon (Fig. 5 F) and Oligocene of England and Czechoslovakia (REID & CHANDLER 1926, WING & HICKEY 1984) indicate that the tribe was more diverse and widespread during the Early Tertiary than it is today.

The oldest fossil fruits that have been attributed to *Platycarya* are those described as *P. cordiformis* from the Lower Paleocene of Gonna near Sangerhausen, German Democratic Republic (MAI 1987). This species is based only upon locule casts with fragmentary remains of "exocarp" without preservation of wings. Although the locule configuration is clearly juglandaceous and agrees with *Platycarya,* more characters, especially those of the wings, would be necessary to confirm its placement within the *Platycaryeae*. Although a position within *Platycarya* is possible, I suggest that this species is better placed in the noncommital genus *Juglandicarya* until the presence and nature of the wings can be determined. It would also be useful to know if typical pseudocolpate pollen of the *Platycaryeae* is present in the same sediment as the fruit because *Platycaryapollenites* pollen has not been clearly documented prior to the uppermost Paleocene in other areas.

Discussion

The fossil records of pollen and fruits corroboratively indicate a Paleocene radiation of the *Juglandaceae*, during which extant tribes became established. However, the question of when the family first evolved remains a matter of interpretation, and depends in part on what characters are required in order for an extinct species to be included within the modern family. It is likely that pollen characters consistent with the *Juglandaceae* evolved prior to some of the floral and fruit features diagnostic of the family (MANCHESTER 1987). We do not know whether the leaves of

Caryanthus and related Normapolles genera were compound, as in the *Juglandales*, or simple, as in other extant *Hamamelidae*; however, similarities in reproductive morphology documented by FRIIS (1983) suggest the *Caryanthus* or a similar Upper Cretaceous taxon was ancestral to the *Juglandaceae*. Placement of *Caryanthus* within the *Juglandaceae*, suggested by TIFFNEY (1986), would require a broadened concept of the family that would, in my opinion, mask some significant differences of fruit and pollen morphology.

The transition from bisexual flowers like those of *Caryanthus* to unisexual flowers and catkins was an important step in the evolution of *Juglandaceae* that apparently occurred near the Cretaceous-Tertiary boundary. It is possible that the Paleocene explosion in abundance and diversity of juglandaceous pollen represents an adaptive radiation linked with improved adaptation for wind pollination. Although FRIIS (1985) pointed out a suite of features in *Caryanthus* that suggest anemophily, including simple undifferentiated perianth, small, dry pollen grains and a single ovule per ovary, other features such as the bisexual flowers and limited number of stamens per flower suggest that they were not as highly adapted to wind-pollination as most fossil and extant *Juglandaceae*. The pollen isolated from the *Caryanthus* flowers (about 14 μm) is smaller than the optimum for wind-pollination. Among extant *Juglandaceae* similarly small pollen occurs in *Platycarya strobilacea* which is now known to be entomophilous (interpreted as secondarily evolved entomophily, ENDRESS 1986). Typical juglandaceous staminate catkins are not known from the Cretaceous; the earliest are those associated with *Polyptera* in the Middle to Late Paleocene, with pollen 20 – 25 μm (MANCHESTER & DILCHER 1985).

Fruit size is another criterion by which the *Juglandaceae* is distinguished from its precursors and it is apparent that the evolution of the family was linked in part with an increase in seed reserves. The fruits of *Caryanthus* and related Normapolles-producing genera are an order of magnitude smaller than those of extant and fossil *Juglandaceae* and have been recovered only as a result of sediment sieving. Whereas *Caryanthus* fruits range from 0.4 to 1.7 mm (KNOBLOCH & MAI 1986), the fruits of extant and fossil *Juglandaceae* range from about 5 mm (*Platycarya*) to more than 60 mm (e.g., *Juglans, Engelhardia*). Even the smallest juglandaceous fruits (*Platycarya* and *Casholdia*) have seed sizes considerably greater than those of *Caryanthus*. The timing the appearance of *Juglandaceae* in the fossil record coincides with that of a marked increase in mean seed size in widespread angiosperm assemblages and may reflect changing ecological conditions and dispersal agents (TIFFNEY 1985, 1986).

The fossil fruit record indicates that most of the extant genera, as well as several extinct genera, evolved between the Mid Paleocene and Late Eocene. The Paleocene diversification in fruit morphology resulted in large wingless nuts evidently adapted for animal dispersal (e.g., *Juglandicarya simplicarpa*) as well as a variety of winged nutlets adapted for wind dispersal (e.g., *Cyclocarya, Polyptera, Casholdia*). Although both animal and wind dispersal strategies were established early, the greatest generic diversity in the Tertiary, as well as today, appears to have been among fruits adapted for wind dispersal.

I thank Prof. H. MEUSEL, Prof. F. EHRENDORFER and the Deutsche Akademie der Naturforscher Leopoldina for the invitation to participate in this symposium.Helpful discussion was provided by P. R. CRANE, D. L. DILCHER, H. JÄHNICHEN, W. KRUTZSCH, D.

H. MAI, T. TANAI, and K. UEMURA. I thank C. BLANC, M. COLLINSON, P. R. CRANE, C. P. DAGHLIAN, J. FERIGNO, H. SCHORN, S. WING for making specimens or photographs available for illustration in this paper. This work was supported in part by NSF grant BSR 84-07841 to the author and by NSF facilities grant PCM 82-12660 to the Indiana University Biology Department for the purchase and maintenance of SEM equipment used in this research.

References

BATTEN, D. J., 1981: Stratigraphic, paleogeographic, and evolutionary significance of Late Cretaceous and Early Tertiary Normapolles pollen. – Rev. Palaeobot. Palyn. **35**: 125–137.

– CHRISTOPHER, R. A., 1981: Key to the recognition of Normapolles and some morphologically similar pollen genera. – Rev. Palaeobot. Palyn. **35**: 359–383.

BROWN, R. W., 1962: Paleocene floras of the Rocky Mountains and Great Plains. – U.S. Geol. Surv. Prof. Pap. **375**: 1–119, pl. 1–69.

CHMURA, C. A., 1973: Upper Cretaceous (Campanian-Maastrichtian) angiosperm pollen from the western San Joachin Valley, California, USA. – Palaeontographica, Abt. B. Paläophytol. **141**: 89–171.

CRANE, P. R., MANCHESTER, S. R., 1982: An extinct juglandaceous fruit from the Upper Palaeocene of southern England. – Bot. J. Linn. Soc. **85**: 89–101.

CREPET, W. L., DILCHER, D. L., POTTER, F. W., 1975: Investigations of angiosperms from the Eocene of North America: a catkin with juglandaceous affinities. – Amer. J. Bot. **62**: 813–823.

– DAGHLIAN, C. P., ZAVADA, M. S., 1980: Investigations of angiosperms from the Eocene of North America: a new juglandaceous catkin. – Rev. Palaeobot. Palyn. **30**: 361–370.

DILCHER, D. L., POTTER, F. W., JR., CREPET, W. L., 1976: Investigations of angiosperms from the Eocene of North America: juglandaceous fruits. – Amer. J. Bot. **63**: 532–544.

– MANCHESTER, S. R., 1986: Investigations of angiosperms from the Eocene of North America: leaves of the *Engelhardieae* (*Juglandaceae*). – Bot. Gaz. **147**: 189–199.

DRUGG, W. S., 1967: Palynology of the Upper Moreno Formation (Late Cretaceous-Paleocene), Escarpado Canyon, California. – Palaeontographica, Abt. B, Paläophytol. **120**: 1–71.

ENDRESS, P. K., 1986: An entomophily syndrome in *Juglandaceae*: *Platycarya strobilacea*. – Veröff. Geobot. Inst. ETH, Stiftung Rübel, Zürich, **87**: 100–111.

FREDERIKSEN, N. O., 1979: Paleogene sporomorph biostratigraphy, northwestern Virginia. – Palynology **3**: 129–167.

– 1980: Paleogene sporomorphs from South Carolina and quantitative correlations with the Gulf Coast. – Palynology **4**: 125–179.

– CHRISTOPHER, R. W., 1978: Taxonomy and biostratigraphy of late Cretaceous and Paleogene triatriate pollen from South Carolina. – Palynology **2**: 113–145.

– CARR, D. R., LOWE, G. D., WOSIKA, E. P., 1983: Middle Eocene Palynomorphs from San Diego, California. – Am. Ass. Strat. Palyn. Contrib. Ser. **12**: 1–110.

FRIIS, E. M., 1983: Upper Cretaceous (Senonian) floral structures of juglandalean affinity containing Normapolles pollen. – Rev. Palaeobot. Palyn. **39**: 161–188.

– 1985: Structure and function in late Cretaceous angiosperm flowers. – Biol. Skr. **25**: 1–37.

GOCZAN, F., GROOT, J. J., KRUTZSCH, W., PACLTOVA, B., 1967: Die Gattungen des „Stemma Normapolles PFLUG 1953 b" (*Angiospermae*). – Paläontol. Abh., Abt. B, Paläobot. **2**: 429–539.

GRUAS-CAVAGNETTO, C., 1976: Etude palynologique de Paleogene du Sud de l'Angleterre. – Cah. Micropal. **1**: 1–49.

JÄHNICHEN, H., MAI, D. H., WALTHER, H., 1977: Blätter und Früchte von *Engelhardia* LESCH. ex BL. (*Juglandaceae*) aus dem europäischen Tertiär. – Feddes Repert. **88**: 323 – 363.

KEDVES, M., 1967: Spore pollen data from the London Clay. – Acta Biologica, n. s. **8**: 25 – 30.

– 1982: Palynology of the Thanetian layers of Menat. – Palaeontographica, Abt. B, Paläophytol. **182**: 87 – 150.

– STANLEY, E. A., 1976: Electronmicroscopical investigations of the Normapolles group and some other selected European and North American angiosperm pollen 2. – Pollen & Spores **18**: 105 – 127.

KNOBLOCH, E., MAI, D. H., 1986: Monographie der Früchte und Samen in der Kreide von Mitteleuropa. – Rozpr. Ústr. Úst. Geol. **47**: 3 – 219.

KRUTZSCH, W., 1961: Beitrag zur Sporenpaläontologie der präoberoligozänen kontinentalen und marinen Tertiärablagerungen Brandenburgs. – Ber. Geol. Ges. **4**: 290 – 343.

– 1962: Stratigraphisch bzw. botanisch wichtige neue Sporen- und Pollenformen aus dem deutschen Tertiär. – Geologie **11**: 265 – 308.

– 1966: Zur Kenntnis der präquartären periporaten Pollenformen. – Geologie **15**: 16 – 72.

– 1970: Die stratigraphisch verwertbaren Sporen- und Pollenformen des mitteleuropäischen Alttertiärs. – Jahrb. Geol. **3**: 309 – 379.

LEFFINGWELL, H. A., 1971: Palynology of the Lance (Late Cretaceous) and Fort Union (Paleocene) Formations of the type Lance area, Wyoming. – Geol. Soc. Amer. Spec. Pap. **127**: 1 – 64.

LIEUX, M. H., 1980: An atlas of pollen of trees, shrubs, and woody vines of Louisiana and other southeastern states, 2. *Platanaceae* to *Betulaceae*. – Pollen & Spores **22**: 191 – 243.

MACGINITIE, H. D., 1941: A Middle Eocene flora from the central Sierra Nevada. – Carnegie Inst. Wash. Pub. **534**: 1 – 198.

MAI, D. H., 1981: Der Formenkreis der Vietnam-Nuss [*Carya poilanei* (CHEV.) LEROY] in Europa. – Feddes Repert. **92**: 339 – 385.

– 1987: Neue Früchte und Samen aus paläozänen Ablagerungen Mitteleuropas. – Feddes Repert. **98**: 197 – 229.

MANCHESTER, S. R., 1983: Fossil wood of the *Engelhardieae* (*Juglandaceae*) from the Eocene of North America: *Engelhardioxylon* gen. nov. – Bot. Gaz. **144**: 157 – 163.

– 1987: The fossil history of the *Juglandaceae*. – Monogr. Syst. Bot. Missouri Bot. Gard. **21**: 1 – 137.

– DILCHER, D. L., 1982: Pterocaryoid fruits (*Juglandaceae*) in the Paleogene of North America and their evolutionary and biogeographic significance. – Amer. J. Bot. **69**: 275 – 286.

– – 1985: Multiple organ reconstruction of an extinct juglandaceous genus from the Paleocene of the Fort Union Formation in Wyoming. – Amer. J. Bot. (Abstract) **72**: 896.

MANNING, W. E., 1938: The morphology of the flowers of the *Juglandaceae* I. The inflorescence. – Amer. J. Bot. **25**: 407 – 419.

– 1940: The morphology of the flowers of the *Juglandaceae* 2. The pistillate flowers and fruit. – Amer. J. Bot. **27**: 839 – 852.

– 1948: The morphology of the flowers of the *Juglandaceae* 3. The staminate flowers. – Amer. J. Bot. **35**: 606 – 621.

– 1978: The classification within the *Juglandaceae*. – Ann. Missouri Bot. Gard. **65**: 1058 – 1087.

MULLER, J., 1981: Fossil pollen records of extant angiosperms. – Bot. Rev. **47**: 1 – 142.

NAGY, E., 1969: Palynological investigations of the Miocene in Macsek Mountains. – Magyar Állami Földt. Intéz. Évk **52**: 235 – 649.

NICHOLS, D., 1973: North American and European species of *Momipites* ("*Engelhardtia*") and related genera. – Geoscience & Man **7**: 103–117.
– OTT, H. L., 1978: Biostratigraphy and evolution of the *Momipites-Caryapollenites* lineage in the Early Tertiary in the Wind River Basin, Wyoming. – Palynology **2**: 93–112.
PFLUG, H. D., 1953: Zur Entstehung und Entwicklung des angiospermiden Pollens in der Erdgeschichte. – Palaeontographica, Abt. B, Paläophytol. **95**: 60–171.
REID, E. M., CHANDLER, M. E. J., 1926: The Bembridge flora. Catalogue of Cainozoic plants in the Department of Geology, 1. – London: British Museum (Natural History).
– – 1933: The London Clay flora. – London: British Museum (Natural History).
STANLEY, E. A., KEDVES, M., 1975: Electronmicroscopical investigations of the Normapolles group and some other selected European and North American angiosperm pollen, 1. – Pollen & Spores **17**: 233–271.
STONE, D. E., 1973: Patterns in the evolution of amentiferous fruits. – Brittonia **25**: 371–384.
– BROOME, C. R., 1971: Pollen ultrastructure: evidence for relationship of the *Juglandaceae* and *Rhoipteleaceae*. – Pollen & Spores **13**: 5–14.
– – 1975: *Juglandaceae*. – World Pollen & Spore Flora **4**: 1–31.
TANAI, T., UEMURA, K., 1983: *Engelhardia* fruits from the Tertiary of Japan. – J. Fac. Sci., Hokkaido Univ., Ser. IV. **20**: 249–260.
TIFFNEY, B. H., 1985: Seed size, dispersal syndromes, and the rise of the angiosperms: evidence and hypothesis. – Ann. Missouri Bot. Gard. **71**: 551–576.
– 1986: Fruit and seed dispersal and the evolution of the *Hamamelidae*. – Ann. Missouri Bot. Gard. **73**: 394–416.
TSCHUDY, R. H., 1973: Stratigraphic distribution of significant Eocene palynomorphs of the Mississippi embayment. – U.S. Geol. Surv. Prof. Pap. **743 B**: 1–24.
– 1975: Normapolles pollen from the Mississippi embayment. – U.S. Geol. Surv. Prof. Pap. **865**: 1–40.
– 1981: Geographic distribution and dispersal of Normapolles genera in North America. – Rev. Palaeobot. Palyn. **35**: 284–318.
WHITEHEAD, D. R., 1963: Pollen morphology in the *Juglandaceae*, 1: pollen size and pore number variation. – J. Arnold Arbor. **44**: 101–110.
– 1965: Pollen morphology in the *Juglandaceae*, 2: survey of the family. – J. Arnold Arbor. **46**: 369–410.
WING, S. L., HICKEY, L. J., 1984: The *Platycarya* perplex and the evolution of the *Juglandaceae*. – Amer. J. Bot. **71**: 388–411.
WOLFE, J. A., 1973: Fossil forms of Amentiferae. – Brittonia **25**: 334–355.
– 1976: Stratigraphic distribution of some pollen types from the Campanian and Lower Maastrichtian rocks (Upper Cretaceous) of the Middle Atlantic states. – U.S. Geol. Surv. Prof. Pap. **997**: 1–108.
ZAVADA, M. S., DILCHER, D. L., 1986: Comparative pollen morphology and its relationship to phylogeny of pollen in the *Hamamelidae*. – Ann. Missouri Bot. Gard. **73**: 348–381.

Address of the author: STEVEN R. MANCHESTER, Department of Geology, Indiana University, Bloomington, IN 47405, U.S.A.

Pl. Syst. Evol. 162, 251–265

—Plant— Systematics and Evolution

Aspects of the fossil record and phylogeny of the family *Rutaceae* (*Zanthoxyleae, Toddalioideae*)

H. J. Gregor

Received January 26, 1987; in revised form July 25, 1988

Key words: Angiosperms, *Rutaceae*. – Phylogeny, fructifications, morphology, anatomy, paleogeography, Cretaceous, Tertiary.

Abstract: A comparative analysis of the seed morphology and anatomy of fossil and extant *Rutaceae* (mainly *Zanthoxyleae* and *Toddalioideae*) is presented. This allows to place the most important fossil taxa in a time-table and on paleogeographical maps. A phylogenetic scheme demonstrates the postulated historical relationships of *Evodia, Zanthoxylum, Fagara, Rutaspermum, Acronychia, Toddalia, Fagaropsis*, and *Phellodendron*.

Surveying fossil seeds and fruits, and whole floras in the Tertiary, one notices that there are some families, which have been wide-spread since the beginning of the Tertiary, and have an abundant fossil record, e.g., the *Magnoliaceae, Hamamelidaceae, Juglandaceae, Mastixiaceae*, and *Rutaceae*. Concentrating on the latter, I soon was impressed by the clear morphological characters in their seeds and by the good possibilities for separating genera and species, particularly in *Rutaceae-Zanthoxyleae* and -*Toddalioideae*. My first attempt at putting fossil *Rutaceae* (*Zanthoxyleae*) together into a phylogenetic scheme was made 1975. This was followed by special studies on fossil species of *Zanthoxylum, Toddalia, Fagaropsis* and *Rutaspermum* (Gregor 1977, 1978 a, 1979, 1984, Gregor & Hantke 1983, Collinson & Gregor 1987).

The abundance of fossil *Rutaceae-Zanthoxyleae* and -*Toddalioideae* in the European and American Tertiary is evident from several papers (Gregor 1978 b, 1980, 1982 a, b, 1983, 1985, 1986, Schmid & Gregor 1983, Holy 1975, Reid & Reid 1915, Reid 1920, 1923, Mai & Walther 1978, 1985, Kirchheimer 1957, Tiffney 1980). *Rutaceae* have also been found in Cretaceous sediments (Knobloch & Mai 1986, Velenovsky & Viniklar 1926), and there is much literature on the English and Eastern Paleogene (Chandler 1925, 1957, 1961 a, b, 1962, 1963 a, b, 1964, Reid & Chandler 1926, 1933, Mai 1976) and the Russian and Japanese Neogene (Negru 1969, 1972, Miki 1937, 1941). Not only fossil fructifications were found in the Tertiary, but also leaves and woods (*Toddalia hofmanni* Walther in Mai & Walther 1985, *Skimmia tortonica* Palamarev & Usunova 1970, cf. *Toddalia aculeata* Persoon, Menzel 1920). New and yet unpublished material has been obtained from the Lower Miocene of the Mainz area (Knoerzer & Meller, in prep.), the Messinian of the Stirone river, in the Ottnangian of N-Evia (in prep. GREGOR), and the Miocene/Pliocene of Alsace (GEISSERT & al. 1988, in prep.).

Fossil fructifications of *Rutaceae*

From available literature sources (e.g., ENGLER 1931, BRIZICKY 1962) one can place the *Rutaceae* genera for which fossil material is available into the following subfamilies and tribes:

Rutoideae	
Zanthoxyleae	*Zanthoxylum, Fagara, Evodia, Rutaspermum, Orixa*?
Ruteae	*Dictamnus, Ruta*?
Boronieae	–
Diosmeae	–
Cusparieae	*Pilocarpus*? *Erythrochiton*?
Dictyolomatoidea	–
Flindersioideae	–
Spathelioideae	–
Toddalioideae	*Phellodendron, Ptelea* (*Pteleaecarpum*), *Vepris* vel *Fagaropsis, Acronychia*? *Toddalia, Skimmia, "Toddaliospermum"*
Citroideae	–
Rhabdodendroideae	–

In the following, mainly the *Zanthoxyleae* and *Toddalioideae* will be discussed, two groups with certain affinities to one another.

Fossil capsular fruits are hardly documented for *Rutaceae*, as their seeds nearly always occur isolated. In one case (Geiseltal, MAI 1976) a fruit could be reconstructed, as its seeds where found in the original crosswise position in the sediment.

Fossil seeds and closed fruits of the different taxa can be differentiated by the morphology of their testa, inner anatomy and surface. Note that we find no integuments, subdermal tissue or anything else important for recent systematics (see BOESEWINKEL 1977: 202). Anatomical or chemical research would be necessary for further studies, especially for separating the oldest taxa definitively (also see CORNER 1976: 233, Key). – Some important genera are distinguished by the following characters (see Fig. 1).

***Zanthoxylum* L.** Seeds with hard testa, oval to ellipsoidal and boat-like; exotesta with palisade cells; mesotesta with pitted surface, mostly ornamented; long hilar scar; prominent raphe region with basal chalaza; apical micropyle, symmetrical dehiscing line; resin cells (starch grains?).

***Fagara* L.** Same as *Zanthoxylum*, perhaps with a more shortened or rounded hilar scar with rim around it.

***Evodia* J. R. & G. FORSTER.** Seeds with thin testa; ellipsoidal, boat-shaped; with prominent apical knob; elongate, straight ventral hilum with margins; dorsal margin convex; outer integument double-layered, inner integument thin; shining black outer testa.

***Dictamnus* L.** No fossil seeds in Europe; droplike seeds with S-curved hilum (triangular), apical micropyle on incised knob; basal chalaza; shining black testa surface; thin and brittle testa wall.

***Orixa* THUNB.** No certain fossils in Europe; seeds rounded oval; thin testa with apical micropyle on small wart; long narrow hilum with basal chalaza.

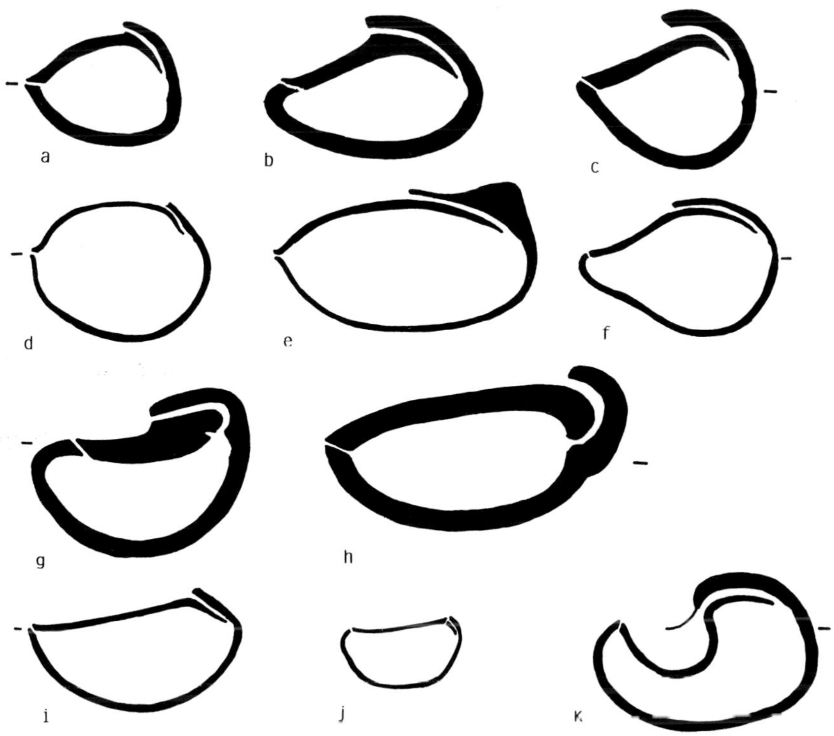

Fig. 1. Longitudinal sections through seeds of *Rutaceae-Zanthoxyleae* (*a*–*e*), *Ruteae* (*f*) and *Toddalioideae* (*g*–*k*): *a Zanthoxylum, b Rutaspermum, c Fagara, d Orixa, e Evodia, f Dictamnus, g Toddalia, h Acronychia, i Phellodendron, j Skimmia, k Vepris* vel *Fagaropsis*. Left and right bars: orientation line micropyle–chalaza

Phellodendron **Rupr.** Seeds with brittle testa, with the shape of a laterally compressed hemisphere; convex dorsal side; linear sharp ventral face with raphe (not prominent); long hilar scar with basal chalaza, apical micropyle, sometimes on protrusion; reticulate surface pattern; two layers in outer integument, three in inner; symmetrical dehiscing line.

Pteleaecarpum **Weyland.** Stalked and winged samaras, oval to ovate in outline; wing prominently reticulate-veined anastomosing from the middle line; seed cavity central, two-loculed, with a symmetrical dehiscing line; seeds unknown.

Fagaropsis **Mildbr. ex Siebenlist** (= *Vepris* Comm.). Drupe with relatively brittle surface; seeds broadly boat-shaped with deeply incised triangular hilar scar; prominent raphe region with basal chalaza; micropyle on small tuberculum showing backwards; surface of testa finely pitted; no clear dehiscence.

Acronychia **Forster.** Fruit drupaceous, hard, with 4 carpels and 1 seed per carpel; seed with a very thick testa; asymmetrical halfmoon-shaped, with straight ventral side and long hilum; micropyle in curved knoblike area; surface of testa ornamented with longitudinal ribs, which anostomose and partly have cross ribs.

Toddalia **Juss.** Drupe; seeds with hard testa, slightly curved, boat-shaped to elongate-reniform; mesotesta with finely pitted surface without ornamentation; medially long and sunken hilar scar; prominent raphe region, apical micropyle; asymmetrical dehiscing line; exocarp unknown.

Skimmia **THUNBG.** Only fossil leaves. No fructifications (seeds with a thin testa only; no possibility for fossilization).

Rutaspermum **CHANDLER.** Form genus of *Rutaceae* with boat-shaped seeds and ornamented testa surface; without close relationship to extant genera.

Toddaliospermum **CHANDLER.** Form genus with halfmoon-shaped seeds and less ornamented testa compared to *Rutaspermum* (mostly non-rutacean).

Anatomical aspects

There is not too much literature about the anatomy of rutacean seeds or fruits (NETOLITZKY 1926, CORNER 1976, BOESEWINKEL 1977), and even less about the fossil ones. The following gives an impression of the possibilities of future work and compares some important features of histology and anatomy.

Surface of seed. The typical pitted surface of many fossils is due to the cell character of the mesotesta with wide lumina. It is comparable in *Toddalia asiatica, Rutaspermum ornatum, R. chandlerae* and *Zanthoxylum* spp. (see Fig. 2 a – d).

Inner tegmen. Four species are illustrated in their tegmal structure: *Toddalia asiatica, Rutaspermum chandlerae, R. ornatum,* and *Zanthoxylum simulans* (Fig. 2 e – h). The typical fibres are comparable particularly in the two latter taxa.

Testa anatomy. The sclerotesta (mesotesta) of the seeds consists of a mass of cells (often compressed in the fossil state) with "starch grains", "resin cells", "oil bodies" or whatever. A chemical analysis would help to solve this question. We find it in *Zanthoxylum simulans* and also in *Rutaspermum chandlerae,* but not in *Toddalia* (Fig. 2 i – l).

Compression of seeds. As in *Zanthoxylum*, some seeds show flat faces, because mostly two seeds develop per carpel and are thus compressed. The same occurs in *Rutaspermum chandlerae* and in fossil *Toddalia* species, but also in *Zanthoxylum* (Fig. 3 ü, y).

Fossil *Rutaceae* in Europe

A short survey of the most important fossil species of the *Zanthoxylum* and *Toddalia* type from the Molasse area (Southern West-Germany) and adjacent areas demonstrates that they mostly resemble extant species very much and gives an impression of the historical variation of the two genera (GREGOR 1978 a, 1979). Two species have to be emended: *Z. negrui* is declared not valid (GREGOR 1987: 18), and should now be called *Z. pararugosum* NEGRU (1982) (formerly *Z. rugosum* NEGRU 1972); the other is *Rutaspermum kristinae* HOLY, now being combined to *Z. kristinae*. Of the Paleogene species only one is considered here (*Rutaspermum chandlerae*) – most of the others should be re-studied with new techniques.

Two species, *Toddaliospermum leonhardii* and *T. neomarcense* MAI (1976: 116, 117), according to my personal opinion, do not belong to the *Rutaceae*. In addition, *Shrubseola trebecensis* MAI (KNOBLOCH & MAI 1985: 99) and *Sh. jenkinsi* REID & CHANDLER must be studied again, as their rutacean character is not quite clear. The same applies to the species of *Canticarya, Eozanthoxylon, Clausenispermum* and *Caxtonia* (REID & CHANDLER 1933). *Orixa reticulata* (REID & REID 1915: 104, Plate 10, Figs. 3 – 5) is a hamamelidoidean type. Furthermore, the somewhat doubtful *Acronychia ornata* (CHANDLER) MAI should be studied again; *Melicope* and *Evodia* also have crossed carpels, which split up in the same way, when they fall down unripe.

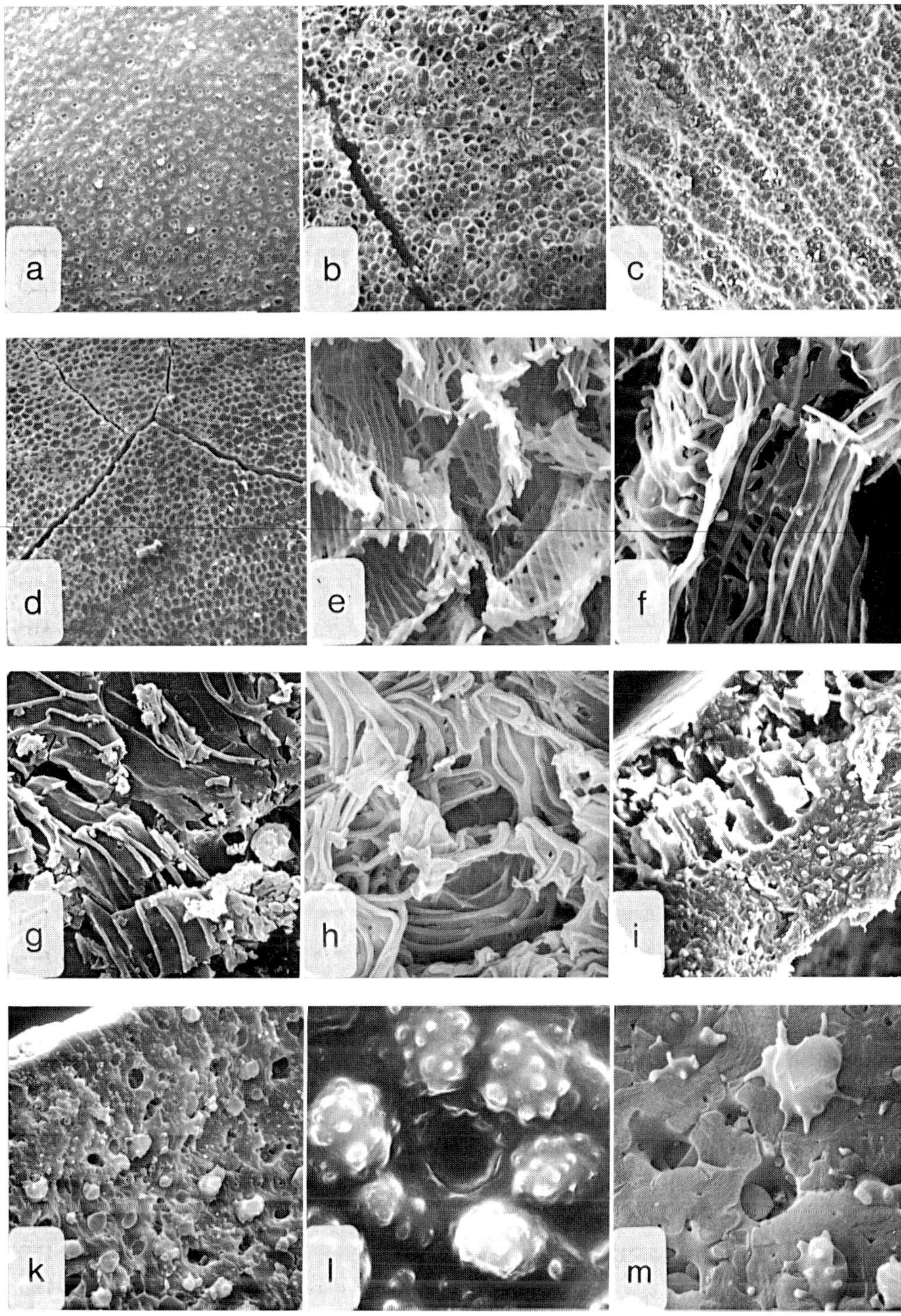

Fig. 2. REM-fotos of anatomical characters from different *Rutaceae*. *a – d* Surface of testa with pits (cell-lamina sunken), *a Zanthoxylum americanum* Mill., × 55; *b Zanthoxylum kristinae* (Holý) Gregor, × 4; *c Rutaspermum ornatum* Chandler [= *Acronychia ornata* (Chandler) Mai], × 50; *d Toddalia asiatica* (L.) Lam., × 55. *e – h* Criss-crossed spiral lignifications of tegmen, *e Toddalia asiatica*, × 1 000; *f Zanthoxylum simulans* Hance, × 1 000; *g Rutaspermum ornatum* [= *Acronychia ornata* (Chandler) Mai], × 1 000; *h Rutaspermum chandlerae*, × 1 000. *i – l* Testa wall with resin (starch) grains (?), *i Zanthoxylum simulans*, × 100; *k Rutaspermum chandlerae*, × 270; *l Zanthoxylum americanum*, × 1 000; *m Rutaspermum chandlerae*, × 1 025

Fig. 3. Morphological comparison among seeds of *Zanthoxylum, Toddalia, Fagaropsis* and *Rutaspermum*. Material stored in the Bayerische Staatssammlung für Paläontologie und historische Geologie München (BSPG); British Museum of Natural History London (BMNH); Instytut Botaniki PAN Krakow (IBPAN); Staatliches Museum für Naturkunde Stuttgart (SMNS); Geologisches Institut der Eidgenössischen Technischen Hochschule Zürich (GIETH); Geologisch-Paläontologisches Institut Würzburg (GPIW); Geologisch-Paläontologisches Institut Köln (GPIK); Museum für Naturkunde – Paläontologisches Museum Berlin (MNPM); Palaeontological Dept., National Museum Prague, Nat. Hist. Mus. (NMP). *a – i Zanthoxylum* type, *a – d* small-seeded group, *e* smooth-surfaced type, *f – i* big-sized group (Middle America!). *j Vepris* (*Fagaropsis, Clausenopsis*) type, linking the *Zanthoxylum Toddalia* types. *k – s Toddalia* type, *k – o* small-sized group, *ö – p* medium-sized types, *q – s* big-sized types. *t* Shows anatomy of *Toddalia* seed, comparable also with *Acronychia, Zanthoxylum* (with large seeds) and *Rutaspermum*. *t, u, z Toddalia* and *Rutaspermum* seeds where the raphe excrescence has been lost because of abrasion (arrows). *v, x Toddalia* seed (*v*) with typical asymmetrical dehiscing lines (arrows) similar to *Rutaspermum* (*x*). *ü, y Toddalia* and *Rutaspermum* seeds with flattened sides (two seeds per carpel!). *w Rutaspermum* type, combining *Toddalia* and *Zanthoxylum* characters (compare *e* and *u*). Magnification × 5, if not stated otherwise. – *a Zanthoxylum ailanthiforme* (Gregor) Gregor; BSPG 1970 X 47; Lower Miocene, open mine Oder II, sand (Oberpfalz); *b Z. kristinae* (Holy) Gregor; NMP 64 317; Lower Miocene, open mine Kristina; Hradek basin (C.S.S.R.); *c Z. schoetzii* Gregor; SMNS P 1244/1; Upper Miocene, clay pit Schandl, Achldorf (Lower Bavaria); *d Z. tertiaria* (Heer) Gregor & Hantke; Gieth LT 1983/1 Lower Miocene (MN 2a, b), Tunnel de Lausanne (Switzerland), *e Z. tiffneyi* Gregor; BSPG 1970 X 236; Lower Miocene, open pit Oder-Süd, sand (Oberpfalz); *f Z. müller-stollii* Gregor; GPIW 1603; Lower Miocene, Kaltennordheim browncoal (Rhön); *g Z. holyi* Gregor; BSPG 1979 V 58; Lower Miocene, Autobahn parking Passau-Nord (Bavaria); *h Z. giganteum* (Gregor) Gregor; BSPG 1970 X 46; Lower Miocene open pit Oder II, sand (Oberpfalz); *i Z. wemdingense* Gregor; BSPG 1966 XLII 8; Middle Miocene, Wemding/Ries (Bavaria); *j Fagaropsis huardii* Gregor; BSPG 1976 VII 1; Lower – Middle Miocene, open mine EDF Morcenx-Arjuzanx (France); *k Toddalia maii* Gregor; BSPG 1970 X 132; Lower Miocene, open mine Oder II, sand (Oberpfalz); *l T. rhenana* Gregor; BSPG 1976 XVII 1; Lower – Middle Miocene, open mine Zukunft-West/Eschweiler (Rheinish area); *m T. thieleae* Gregor; BSPG 1970 X 133; Lower Miocene, open mine Oder II, sand (Oberpfalz) *n T. maerkeri* Gregor; BSPG 1972 XXI 17; Middle Oligocene, Märker quarry, Harburg/Ries (Bavaria); *o T. maii* IBPAN 37/41; Middle Miocene, Wieliczka salt mine (Poland); *ö T.* cf. *latisiliquata* (Ludwig) Gregor var. *gracilis* Gregor; GPIK F 213 (Coll. Weyland); Upper Oligocene to Lower Miocene, clay pit Niederpleis (Rheinish area); *p T. latisiliquata*; MNPM 3473; Lower – Middle Miocene, Laubach browncoal (Wetterau); *q T.* cf. *latisiliquata* IBPAN 37/66; Middle Miocene, Wieliczka salt mine (Poland); *r T. turovensis* (Cz. & Sk.) Gregor; BSPG 1970 X 457; Lower Miocene, open mine Brükkelholz, upper seam/sand (Oberpfalz); *s T. naviculaeformis* (Reid) Gregor; BSPG 1970 X 129; Lower Miocene, open mine Brückelholz, upper seam/sand (Oberpfalz); *t T. naviculaeformis*; × 10, BMNH V 25667; Upper Miocene, Pont-de-Gail/Cantal (France) (M micropyle, R raphe, CH chalaza); *u T. naviculaeformis*; BMNH V 25585: Upper Miocene, Pont-de-Gail/Cantal (France); *ü T. maii* Gregor × 10; Coll. Geissert G-1976-1; Lower Pliocene, open pit Grandes Carrières, Soufflenheim (Alsace); *v T. maii* var. *abnormis* Gregor; × 10; Coll. Geissert G-1976-4; Lower Pliocene, boring Schwighouse/Haguenau (Alsace); *w Rutaspermum chandlerae* Collinson & Gregor, × 10, Coll. Gregor G-1983-MES-1; Middle Eocene, open pit Messel site SMF 7 (Darmstadt); *x R. chandlerae*, × 10, Coll. Gregor G-1983-MES-36; Middle Eocene, open pit Messel site SMF 7 (Darmstadt); *y R. chandlerae*, × 10; Coll. Gregor G-1983-MES-30: Middle Eocene, open pit Messel site SMF 7 (Darmstadt); *z R. chandlerae*, × 10. Coll. Gregor G-1983-MES-40; Middle Eocene, open pit Messel site SMF 7 (Darmstadt); notice the abraded raphe region

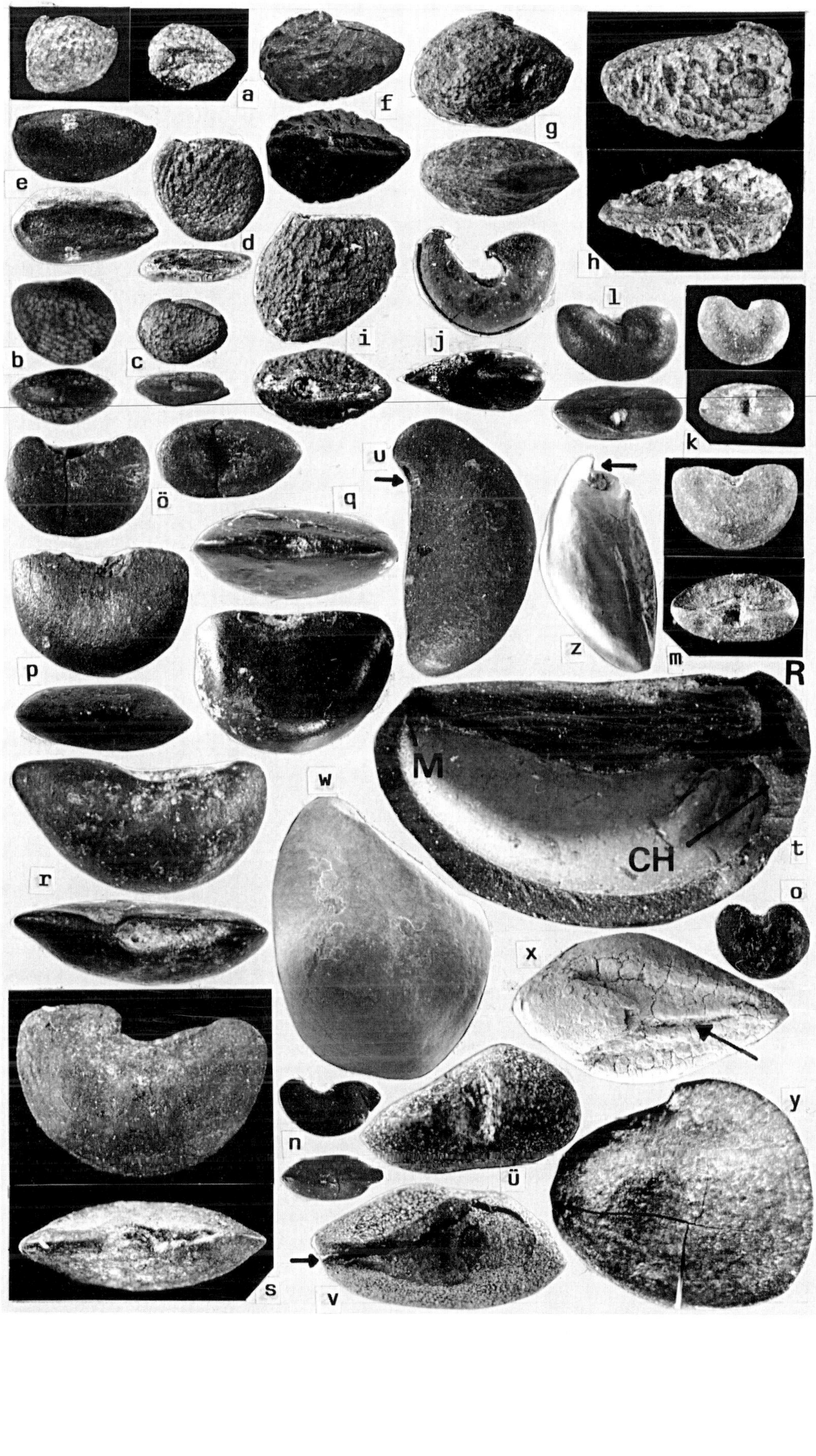
a
f
g
h
e
d
l
b
c
i
j
k
u
ö
q
z
m
p
R
M
w
CH
t
r
o
x
y
n
ü
s
v

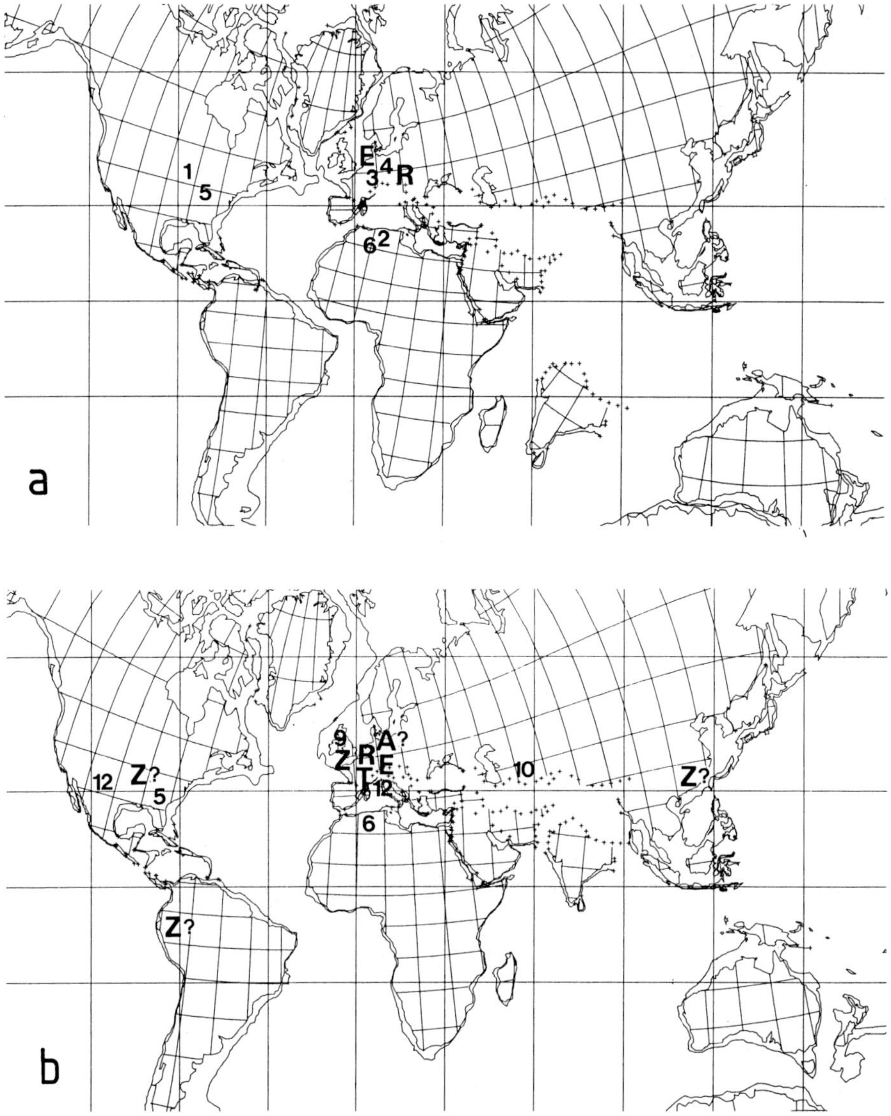

Fig. 4. Paleogeographic distribution of *Rutaceae* (*Zanthoxyleae, Toddalioideae*) in time and space (Cretaceous to Recent, after ENGLER 1896, 1931, etc.). *a* Cenomanian, Santonian, (Cretaceous); Paleocene to Lower Eocene, *b* Middle Eocene to Lower Oligocene, *c* Middle Oligocene to Late Miocene/Early Pliocene, *d* Late Pliocene/Pleistocene to Recent. Extant genera: A *Acronychia,* E *Evodia,* F *Fagara,* P *Phellodendron,* R *Rutaspermum*, T *Toddalia*, V *Vepris* vel *Fagaropsis*, Z *Zanthoxylum*. Extinct genera and doubtful fossils: 1 *Citrophyllum*, 2 *Evodioxylon,* 3 *Rutaecarpus*, 4 *Shrubseola*, 5 *Rutaphyllum*, 6 *Rutoxylon*, 7 *Dictamnus*, 8 *Pteleaecarpum*, 9 *Canticarya*, 10 *Rutaceites*, 11 *Phellodendron* pollen grains, 12 *Fagaroxylon*, 13 *Ptelea*. Solid line in *d* for recent distribution of *Zanthoxylum* and *Fagara*, broken line for *Toddalia*

***Rutaceae* distribution in space and time**

Beginning with the Cretaceous, a short review occurrence of fossil *Rutaceae* in time and space is given in Fig. 4, even if some of the data mentioned in the older literature are doubtful. This is a preliminary attempt on the paleogeographic maps from

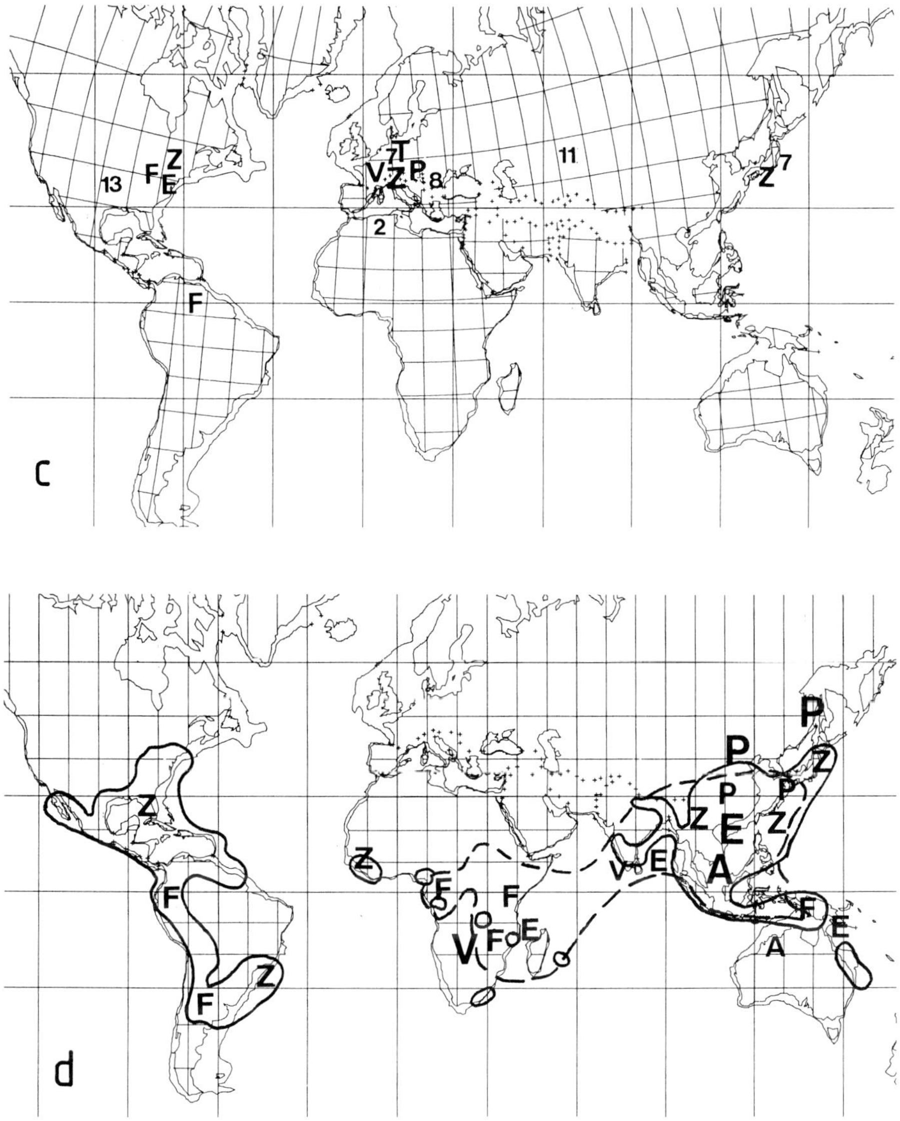

SMITH & BRIDEN (1979), as those from KRUTZSCH (see this volume) are not yet available.

Mid Cretaceous (100 million years, Fig. 4a): only doubtful fossils like *Citrophyllum* in the U.S.A. (leaves!) and *Evodioxylon*-wood in N. Africa.

Late Cretaceous (80 million years, Fig. 4a): *Rutaecarpus* in the Czechoslovakian area (doubtful), *Shrubseola* (Rutaceae?) in the G.D.R. and a significant first *Rutaspermum biornatum* KNOBLOCH & MAI (1985: Taf. 25, Figs. 16–18) in the G.D.R.

Palaeocene – Middle Eocene (60–50 million years, Fig. 4b): a doubtful leaf-type (*Rutaphyllum*) in the United States, a fossil wood of *Rutoxylon* in Northern Africa (age?), *Acronychia ornata* (CHANDLER) MAI (?) and different *Rutaspermum* and *Zanthoxylum* species in England, Bulgaria and W. Germany. *Toddaliospermum leonhardii* MAI and *T. neomarcense* MAI (1976) are doubtful.

Late Eocene – Early Oligocene (40 – 35 million years, Fig. 4 b): *Rutaphyllum* and *Rutoxylon*, including *Rutaceites*, are doubtful, also the *Fagara*-woods in N. America and Europe. The same is the case with *Zanthoxylum* in America (North and South) and SE. Asia. In Europe we have certain fossils of *Zanthoxylum, Evodia, Rutaspermum* and *Toddalia. Phellodendron* definitely in N. America.

Middle Oligocene – Early Miocene (30 – 18 million years, Fig. 4 c): *Evodioxylon* remains doubtful, *Zanthoxylum* surely existed in Europe, Asia and N. America; *Fagara, Evodia* and *Ptelea* in southeastern N. America; *Phellodendron* pollen grains in Asia, seeds in England; *Toddalia* in Europe, also *Fagaropsis* and *Pteleaecarpum*. A fossil wood of *Fagara* was found in W. Germany.

Late Miocene – Early Pliocene (10 – 5 million years, Fig. 4 c): There are scarce findings during this period: *Fagara* (S. America), species of *Toddalia, Zanthoxylum, Acronychia* and *Phellodendron* in Europe (incl. *Dictamnus*-leaf). *Fagaroxylon* is described from W. Germany.

Late Pliocene, Pleistocene and Holocene (1 – 0 million years, Fig. 4 d). *Toddalia* is mentioned for Africa from the Pleistocene, but not from the Holocene. Today we find *Zanthoxylum* widely distributed in eastern N. America, in N., SE.

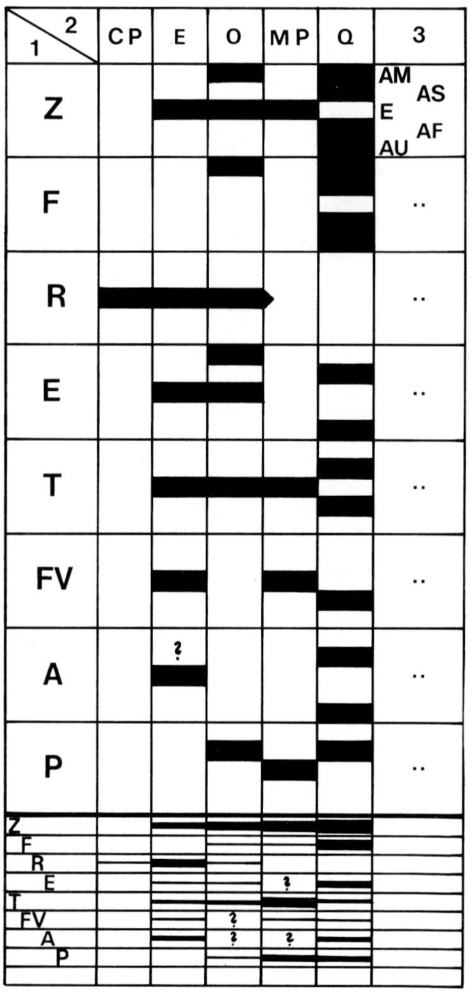

Fig. 5. Appearance of *Zanthoxyleae* and *Toddalioideae* taxa in the time span from the Cretaceous to the Teritary, and their recent distribution (upper part), and frequency of fossil types (from number of specimens and taxa: thin line = few, thick line = many (lower part). 1 Genera: Z *Zanthoxylum*, F *Fagara*, R *Rutaspermum*, E *Evodia*, T *Toddalia*, FV *Fagaropsis* vel *Vepris*, A *Acronychia*, P *Phellodendron*. 2 Time table: C Cretaceous, P Paleocene, E Eocene, O Oligocene, MP Mio-Pliocene, Q Quaternary. 3 Paleogeography: AM America, AS Asia, E Europe, AF Africa, AU Australia

and E. Asia, *Fagara* in C. and S. America, in the southern part of SE. Asia and in E. and W. Africa. *Toddalia* today is restricted to W. and E. Africa and SE. Asia, *Phellodendron* to E. Asia (northern part), *Fagaropsis* to tropical Africa and India, *Evodia* to Australasia and Madagascar.

If we take a look at the following table (Fig. 5), we find the distribution in time and space of eight genera of *Rutaceae,* which are not too doubtful in their determination (it is clear that some problems remain, for example the proof for *Acronychia*). *Rutaspermum* clearly shows the oldest type, followed by *Zanthoxylum, Evodia, Acronychia, Toddalia* and *Fagaropsis.*

The Oligocene record is meagre, but we have first records of *Fagara* and *Phellodendron; Rutaspermum* dies out in the Upper Oligocene (to Miocene), *Zanthoxylum* is recently wide-spread but more in subtropical areas, *Fagara* in tropical ones.

In ASA-GRAY-distribution is found in *Zanthoxyleae* (*Zanthoxylum*), an Asian-African distribution in *Toddalioideae* (*Toddalia, Fagaropsis* p.p.) and *Fagara*, an Australian-Asian one in *Evodia, Acronychia* and *Zanthoxyleae.*

Interpretation and discussion

In conclusion we can try to sketch a phylogenetic scheme, extending from the Cretaceous to the Holocene, for some important genera of the *Zanthoxyleae: Zanthoxylum* s.l. (incl. *Fagara*), *Evodia, Rutaspermum* and *Toddalioideae: Phellodendron, Fagaropsis* (vel *Vepris*), *Acronychia, Toddalia, Toddaliospermum* (Fig. 6).

The first certain *Rutaceae* fossil in the Cretaceous is *Rutaspermum biornatum,* followed from the Paleocene onward (through the Eocene/Oligocene) by other well comparable *Rutaspermum* species. Subsequently we find *Zanthoxylum* types (incl. *Fagara*) up to the present. If *Fagara* can be separated from *Zanthoxylum* by its seed characters (*Fagara* seeds have a more curved sharpe and a rim running around the hilum), *Rutaspermum ceciliense* MAI and *Zanthoxylum rhabdosperma* (LESQU.) TIFFNEY would fit very well into this *Fagara* line.

A similar evolution becomes apparent in *Toddalia* with a precursor in *Rutaspermum chandlerae* or a still theoretical "*protochandlerae*" (Fig. 6, 3). In the Eocene we already have a small- and a big-seeded *Toddalia* species. *Fagaropsis* also seems to be an old type – connected loosely with *Toddalia.*

In my opinion and on the basis of seed characters, *Zanthoxylum* – *Fagara* and *Evodia* have closer phylogenetic connections to *Toddalia* – than to *Phellodendron* – *Acronychia.* This is an approach different from the "chemical" one (FISH & WATERMAN 1973: 192), but also shows the clear connections between *Zanthoxyleae* and *Toddalioideae.* An important question in this respect concerns the phylogenetic divergence of *Zanthoxylum* and *Toddalia,* and the position of *Rutaspermum.* It is relevant in this context that there are "resin cells" in *Zanthoxylum* and *Rutaspermum,* but not in *Toddalia.*

There is neither time nor space here to consider further aspects of *Rutaceae* evolution, but I am prepared to do so in the future. There is the unsolved question of the systematic separation or fusion of *Zanthoxylum* and *Fagara,* for which one should consider not only flower morphology and chemistry (see FISH & WATERMAN 1973), but also seed characters.

Also, an examination of the whole systematic order will be useful in the light of paleobotanical research, including the solution of the question, whether the

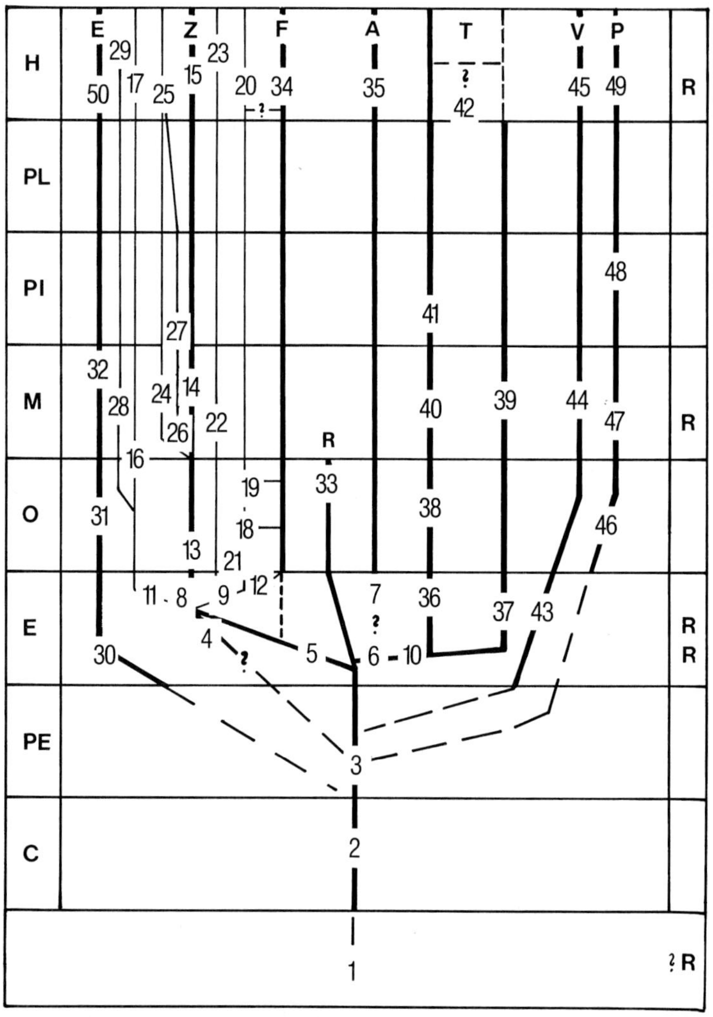

Fig. 6. Phylogenetic scheme of *Rutaceae* with the two recent tribes *Zanthoxyleae* and *Toddalioideae*, depending on the morphology of the seeds in comparison to one another. Left row (time table): H Holocene, PL Pleistocene, PI Pliocene, M Miocene, O Oligocene, E Eocene, PE Paleocene, C Cretaceous. Upper row (recent genera): E *Evodia*, Z *Zanthoxylum*, F *Fagara*, A *Acronychia*, T *Toddalia*, V *Fagaropsis* vel *Vepris*, P *Phellodendron*, R *Rutaspermum*. Numbers: species of *Rutaceae:* 1 hypothetical proto-rutacean, 2 *Rutaspermum biornatum*, 3 *R. "protochandlerae"*, 4 *R. ornatum*, 5 *R. excavatum*, 6 *R. chandlerae*, 7 "*Acronychia ornata*", 8 *Rutaspermum rugosum*, 9 *R. magnificum*, 10 *R. minimum*, 11 *R. compressum* and *bognorense*, 12 *R. messeliense* and *ceciliense*, 13 *Zanthoxylum hordwellense*, 14 *Z. ailanthiforme*, 15 *Z. ailanthoides*, 16 *Z. tiffneyi*, 17 *Z. americanum*, 18 *Z. echinospermum*, 19 *Z. rhabdospermum*, 20 *Z. monophylla*, 21 *Rutaspermum striatum* and *glabrum*, 22 *Zanthoxylum giganteum*, *Z. holyi* and *müller-stollii*, 23 *Z. hawaiense* and *kauaiense*, 24 *Z. wemdingense*, 25 *Z. alatum* and *piperitum*, 26 *Z. kristinae*, 27 *Z. schoetzii* and *pararugosum*, 28 *Z. tertiaria*, 29 *Z. hamiltoniana*, 30 *Evodia costata*, 31 *E. lignita*, 32 *E. nitida*, 33 *Rutaspermum "ornatum"* (hypothetical), 34 *Fagara ovatifolia* and *macrophylla*, 35 *Acronychia laurina*, 36 *Toddalia excavata*, 37 *T.* cf. *naviculaeformis*, 38 *T. maerkeri*, 39 *T. turovensis*, *T. naviculaeformis* and *latisiliquata*, 40 *T. maii* and *thieleae*, 41 *T. rhenana*, 42 *T. asiatica* (vel *T. aculeata*?), 43 *Fagaropsis ornata*, 44 *F. huardii*, 45 *F. angolensis*, 46 *Phellodendron novo-anglicae*, 47 *Ph. lusatica*, 48 *Ph. elegens*, 49 *Ph. amurensis*, 50 *Evodia hupehensis*. Right row: R Radiation time

Rutaceae really are primitive and if they had "oil glands" already in their early evolution.

I wish to thank my colleagues D. H. Mai and M. E. Collinson, for discussions and help for this work.

References

Boesewinkel, F. D., 1977: Development of ovule and testa in *Rutaceae*, 1. *Ruta, Zanthoxylum*, and *Skimmia*. – Acta Bot. Neerl. **26**: 193–211.

Brizicky, G. K., 1962: The genera of *Rutaceae* in the southeastern United States. – J. Arnold Arbor. **43**: 1–2.

Chandler, M. E. J., 1925: The Upper Eocene flora of Hordle, Hants. – Monogr. Palaeont. Soc. London **1**: 1–32.

– 1957: The Oligocene flora of the Bovey Tracey Lake Basin, Devonshire. – Bull. Brit. Mus. (Nat. Hist.) Geol. **3**: 71–123.

– 1961 a: Flora of the Lower Headon Beds of Hampshire and the Isle of Wight. – Bull. Brit. Mus. (Nat. Hist.) Geol **5**: 91–158.

– 1961 b: The Lower Tertiary floras of Southern England. 1. Palaeocene floras. London Clay flora. – Bull. Brit. Mus. (Nat. Hist.) Suppl. 1960: 1–354.

– 1962: The Lower Tertiary floras from southern England. 2. Flora of the Pipe-clay Series of Dorset (Lower Bagshot). – Bull. Brit. Mus. (Nat. Hist.) Suppl.: 1–176.

– 1963 a: Revision of the Oligocene floras of the Isle of Wight. – Bull. Brit. Mus. (Nat. Hist.) Geol. **6**: 321–384.

– 1963 b: The Lower Tertiary floras of Southern England. 3. Flora of the Bornemouth Beds, the Boscombe, and the Highcliff Sands. – Bull. Brit. Mus. (Nat. Hist.) Suppl.: 1–169.

– 1964: The Lower Tertiary floras of southern England. 4. A summary and survey of findings in the light of recent botanical observations. – Bull. Brit. Mus. (Nat. Hist.) Suppl: 1–151.

Collinson, M. E., Gregor, H.-J., 1987: *Rutaceae* from the Eocene of Messel, West Germany. – Tertiary Res. (in print).

Corner, E. J. H., 1976: The seeds of Dicotyledons, **1** and **2**. – Cambridge: Cambridge Univ. Press.

Engler, A 1986: Über die geographische Verbreitung der Rutaceen im Verhältnis zu ihrer systematischen Gliederung. – Abh. Akad. Wiss. Berlin: 1–27.

– 1931: *Rutaceae*. – In Engler, A., Prantl, K., (Eds.): Die natürlichen Pflanzenfamilien **19 a.** 2nd edn. – Leipzig: Engelmann.

Fish, F., Waterman, P. G., 1973: Chemosystematics in the *Rutaceae*. 2. The chemosystematics of the *Zanthoxylum/Fagara* complex. – Taxon **22**: 177–203.

Geissert, F., Gregor, H.-J., Mai, D. H., 1988: Die „Saugbaggerflora" – eine Frucht- und Samenflora aus dem Mio-Pliozän des Elsaß. – Stuttgart: Enke.

Gregor, H.-J., 1975: Die Rutaceen aus dem Mittel-Miozän der Oberpfälzer Braunkohle. – Cour. Forsch.-Inst. Senckenberg **13**: 119–128.

– 1977: *Zanthoxylum wemdingense* nov. spec. aus untersarmatischen Riessee-Ablagerungen. – Mitt. Bayer. Staatssamml. Paläont. hist. Geol. **17**: 249–256.

– 1978 a: Subtropische Elemente im europäischen Tertiär III – Rutaceae. – Acta Palaeobotanica **19**: 21–40.

– 1978 b: Die miozänen Frucht- und Samenfloren der Oberpfälzer Braunkohle 1. Funde aus den sandigen Zwischenmitteln. – Palaeontographica B **167**: 8–103.

– 1979: Systematics, biostratigraphy and palaeoecology of the genus *Toddalia* Jussieu (*Rutaceae*) in the European Tertiary. – Rev. Palaeobot. Palyn. **28**: 311–363.

– 1980: Die miozänen Frucht- und Samen-Floren der Oberpfälzer Braunkohle. 2. Funde aus den Kohlen und tonigen Zwischenmitteln. – Palaeontographica B **174**: 7–94.

GREGOR, H.-J., 1982a: Ein Samenfund aus der Kiesgrube Maßendorf (Kurzberichte IV). – Documenta naturae **4**: 28.

– 1982b: Die jungtertiären Floren Süddeutschlands – Paläokarpologie, Phytostratigraphie, Paläoökologie, Paläoklimatologie. – Stuttgart: Enke.

– 1983: A miocene fruit- and seed flora from the browncoal of Aliveri (Island of Evia, greece). – Documenta naturae **6**: 1–26.

– 1984: Subtropische Elemente im europäischen Tertiär 4 (*Onagraceae, Rutaceae, Vitaceae, Theaceae, Elaeagnaceae*). – Documenta naturae **16**: 1–37.

– 1985: Vorläufiger Bericht über neue Pflanzenfossilien aus tertiären Sedimenten Süd-Frankreichs. – Documenta naturae **25**: 12–14.

– 1986: Die Früchte und Samen aus der Oberen Süßwassermolasse von Achldorf (Vilsbiburg, Niederbayern). – Documenta naturae **30**: 49–59.

– 1987: Revision of two Rutacean taxa (*Zanthoxylum negrui* GREGOR and *Zanthoxylum kristinae* (HOLY) nov. comb.) Kurzmitteilungen 1, Documenta naturae **38**: 18–19.

GREGOR, H.-J., HANTKE, R., 1983: Eine untermiozäne *Zanthoxylum*-Art aus der Schweizer Molasse. – Documenta naturae **13**: 19–25.

HOLY, F., 1975: On some new species from the *Mastixiaceae*-Flora Taphocenose from the Miocene near Hradek nad Nisou (Zittau Basin, North Bohemia). – Sborn. Narodniho Muzea, B **31**: 109–122.

KIRCHHEIMER, F., 1957: Die Laubgewächse der Braunkohlenzeit. – Halle/Saale: VEB Wilh. Knapp.

KNOBLOCH, E., MAI, D. H., 1985: Monographie der Früchte und Samen in der Kreide von Mitteleuropa. – Rozpr. Ustr. Ust. Geol. **47**: 1–219.

MAI, D. H., 1976: Fossile Früchte und Samen aus dem Mitteleozän des Geiseltales. – Abh. Zentr. Geol. Inst. Paläont. Abh. **26**: 93–149.

– WALTHER, H., 1978: Die Floren des Haselbacher Serie im Weißelster-Becken (Bezirk Leipzig, DDR). – Abh. Staatl. Mus. Min. Geol. Dresden **28**: 1–200.

– – 1985: Die obereozänen Floren des Weißelster-Beckens und seiner Randgebiete. – Abh. Staatl. Mus. Min. Geol. Dresden **33**: 1–220.

MENZEL, F., 1920: Über Pflanzenreste aus Basalt-Tuffen des Kamerungebietes. – Beitr. Geol. Erforsch. Deutsch. Schutzgebiete **18**: 17–30.

MIKI, S, 1937: Plant fossils from the *Stegodon* beds and the Elephant beds near Akashi. – Japan. J. Bot. **8**: 303–341.

– 1941: On the change of Flora in Eastern Asia since Tertiary Period. (1). – Japan. J. Bot. **11**: 237–303.

NEGRU, A. G., 1969: Die miozäne Flora von Bursuk in Moldavien. – ANSSSR, Bot. Ž. **44**: 1727–1738.

– 1972: Die frühsarmatische Flora des südöstlichen Moldaviens. – ANSSSR, Bot. Garten, Kishinev: Shtiintsa.

– 1982: Über die Namensänderung von *Zanthoxylum rugosum* N. – Paläokarpol. Untersuch. im Känozoikum. – Inst. f. Geochemie u. Geophys., Akad. Beloruss. S.S.R.

NETOLITZKY, D., 1926: Anatomie der Angiospermen-Samen. – Handb. Pfl. Anat. **10**. – Berlin.

PALAMAREV, E., 1973: Die Eozäne Flora des Burgas-Beckens. – Bull. Inst. Bot. **24**: 75–124.

PALAMAREV, D., USUNOVA, K., 1970: Morphologisch-anatomischer Nachweis der Gattung *Skimmia* in der Tertiärflora Bulgariens. – Compt. Rend. l'Acad. Bulg. Sci. **23**: 835–838.

REID, C., REID, E. M., 1915: La Flore Pliocene de Bidart et Cenitz. – Bull. Soc. Géol. France **4**: 420–427.

REID, E. M., 1920: Recherches sur quelques graines pliocenes du Pont-de-Gail (Cantal). – Bull. Soc. Geol. France **4**: 48–87.

– 1923: Nouvelles recherches sur les graines du Pliocènc infćricur du Pont-de-Gail (Cantal). – Bull. Soc. Géol. France **23**: 308 – 355.

– Chandler, M. E. J., 1926: The Bembridge flora. Catalogue of Cainozoic plants in the Department of Geology. 1. – Bull. Brit. Mus. (Nat. Hist.) Suppl.: 1 – 206.

– – 1933: The flora of the London Clay. – Bull. Brit. Mus. (Nat. Hist.), Suppl.: 1 – 561.

Schmid, W., Gregor, H.-J., 1983: Gallenbach — eine neue mittelmiozäne Fossilfundstelle in der westlichen Oberen Süßwassermolasse Bayerns. – Ber. Naturwiss. Ver. Schwaben e. V. **87**: 51 – 63.

Smith, A. G., Briden, J. C., 1979: Mesocoic and Cenozoic paleocontinental maps. – p. 63, 52 maps. – London: Cambridge.

Tiffney, B. H., 1980: Fruits and seeds of the Brandon Lignite, 5. *Rutaceae*. – J. Arnold Arbor. **61**: 1 – 36.

Velenovsky, J., Viniklar, L., 1926: Flora Cretacea Bohemia, Rozpravy Statniho Geol. Ust. I.

Address of the author: Dr Hans-Joachim Gregor, Prähistorische Staatssammlung München, Lerchenfeldstrasse 2, D-8000 München 2, Federal Republic of Germany.

Pl. Syst. Evol. 162, 267–283

—Plant—
Systematics
and
Evolution

The genus *Buxus* (*Buxaceae*): aspects of its differentiation in space and time

EGON KÖHLER and PAUL BRÜCKNER

Received January 23, 1987; in revised form March 14, 1988

Key words: Angiosperms, *Buxaceae, Buxus.* – Pollen morphology, leaf venation, fossil records, systematics, evolution, chorogenesis.

Abstract: Trends of pollen grain aperture evolution and exine characters as well as characters of leaf venation, petiole and axial vascularization are briefly described and related to geographical distribution and classification of *Buxus*. A review of fossil records is given. Three major taxonomic groups can be delimitated within the genus, and aspects of their relationships and chorogenesis are presented. The level of differentiation, the pattern of distribution and the fossil record speak in favour of an ancient origin of the genus.

Our interest in the family *Buxaceae* was focussed initially on the phylogenetic position and the delimitation of the family (KÖHLER 1980, KÖHLER & BRÜCKNER 1983). After preliminary investigations by BRÜCKNER (1971) and MUNSCHKE (1974), and after taking over the revision of the family for the new Flora of Cuba and the Flora Neotropica, we have concentrated on the genus *Buxus*.

There is no modern review of the genus, except the anatomical investigations by MATHOU (1940) and the work by HATUSIMA (1942) on the Asiatic taxa. Thus, in addition to work on pollen morphology of the American (KÖHLER 1981), African (KÖHLER & BRÜCKNER 1982), and Eurasiatic species (KÖHLER 1979, BRÜCKNER 1985, BRÜCKNER, in prep.) and the leaf venation of these taxa (KÖHLER 1982, 1984, BRÜCKNER 1985 and in prep.), investigations of leaf anatomy (SCHIRAREND 1981, HARTUNG 1984, GRAJCAREK 1986) were undertaken to verify and update the results of MATHOU (1940) and finally studies on gynoeceum morphology (BURKHARDT 1986) were carried out. These results have led to a thorough knowledge of the structure of extant taxa and their systematic relationships. Together with palaeobotanical evidence (KEDVES 1962, KRUTZSCH 1966, KVACEK & al. 1982, BESSEDIK 1983, and others), these findings now allow insights into the world-wide differentiation and possible chorogenesis of this undoubtedly quite old genus.

Distribution and classification of the genus

The genus *Buxus* has a centre of diversity in the Caribbean-Latin American area (Fig. 6) with about 50 species, originally described as the genus *Tricera* (SCHREBER

1791). Subsequently, they were classified by BAILLON (1859) and finally by MATHOU (1940) as sect. *Tricera* of *Buxus*. With more than 30 species, Cuba is a main centre of diversity; followed by Jamaica with only 4 – 5, Puerto Rico with 2, and Martinique and the Bahamas, each with one species. The genus is represented in C. America and Mexico by 4 species and reaches S. America with one species (*B. citrifolia*) in Venezuela, which recently was shown to occur in Panama also (GENTRY 1978). Meanwhile, further *Buxus* samples have been collected in Columbia and Guyana (GENTRY 1985, pers. comm.).

A second centre of diversity with sect. *Buxus* is located in E. Asia, where approximately 40 species occur from China, Korea, and Japan to Malaysia, the Philippines and Indonesia. As western representatives, this section includes the Mediterranean and North African taxa. Six of the tropical SE. Asian species were separated as sect. *Eugeniobuxus* (HATUSIMA 1942).

Furthermore, a comparatively small number of 7 – 8, definitely relictual species occurs in Africa. They can be divided into 2 groups, for which VAN TIEGHEM (1897) originally described the E. African genus *Buxanthus* and the S. African-Malagasy genus *Buxella*, which were last classified by MATHOU (1940) as subsections of the African sect. *Probuxus*.

The closely related genus *Notobuxus* is represented in Africa with 4 species, whose systematic position is the subject of controversial interpretations (PHILLIPS 1943, HUTCHINSON 1967, and others).

Pollen morphology

The morphological diversity of the pollen wall, whose basic pattern is genetically determined by the male gametophyte, is – due to its high specificity and its remarkable evolutionary stability – a strong indicator for systematic and phylogenetic affinities.

Aperture characteristics. The basic pollen form of *Buxus* is a sphaeroid, zono-3-colporate grain with long colpi, each of which contains 3 – 4 clearly defined, roundish ora, surrounded by slight nexine thickenings (Fig. 2). This is a specific aperture configuration, which occurs in a comparable form only in a few families within the angiosperms (KÖHLER 1981).

From this basic structure, a remarkable pattern of aperture evolution within the American species can be demonstrated. It starts with an increase of colpi number from 6-9-12, accompanied by a transition to global arrangement. Then the number of ora decreases from 3-2-1 per colpus, finally leading to pantoporate pollen (Fig. 1). In some side lines one can observe an increase in the number of ora in the colpi up to 6 and an increase of the number of colpi in equatorial position to 7 – 8. Occasionally, there is a tendency towards syncolpy.

This pattern of pollen evolution becomes manifest in differing extent within 4 pollen types (*B. macrophylla, B. glomerata, B. vahlii,* and *B. bahamensis* types) of the American species and reappears with identical successions in the African taxa. Whereas the *B. hildebrandtii* type shows only occasional increases in aperture number (4 – 6) within one species, the complete series of aperture evolution is observed in the *B. macowani* type (Fig. 4).

The Eurasiatic species, however, all have pantoporate pollen, also characteristic for the genera *Sarcococca, Pachysandra*, and *Styloceras*.

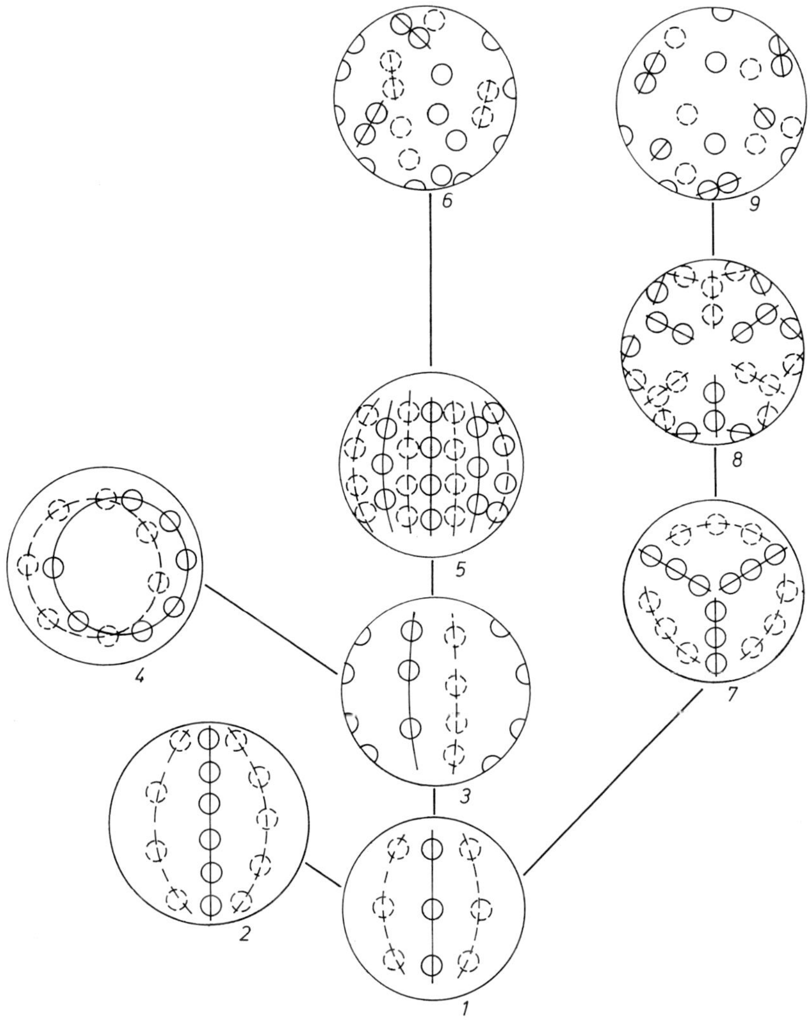

Fig. 1. Aperture evolution of the pollen of the genus *Buxus*

This pattern of aperture evolution, common to different pollen types in the New and Old World obviously represents a basic genetic pattern of pollen variation reflecting common genetic information and similar evolutionary strategies. With respect to the concept of homologous series of variation (VAVILOV 1922) this validates the common descent of taxa. This pattern of aperture evolution can be seen as a variation of the "successiformous pattern" (VAN CAMPO 1976).

With fossil pollen from the Early Eocene to the Middle Miocene, covering a time span of 35 million years, BESSEDIK (1983) was able to demonstrate a chronological sequence of the described morphological series and evolutionary tendencies. In the Early and Middle Eocene an increase in the number of endoapertures takes place, clearly preceding the "successiformy" phenomenon. In the Middle and Late Eocene, the transition to pantocolporate forms occurs, and in the Late Eocene and Early Oligocene, finally, the pantoporate forms appear.

Exine sculpture. The characters of exine sculpturing, often regarded as being variable and less important for the determination of phylogenetic interrelationships, form the basis for the description of pollen types in *Buxus*.

The most primitive pattern (*B. macrophylla* type) in the sect. *Tricera* consists of relatively free, interwoven, and anastomizing strands (Fig. 2). Increasing condensation and fusion of the strands (*B. aneura* type) has obviously led to the formation of a vertically undulating exine sculpture (*B. bahamensis* type). Thereby, the outer sculpture elements are reinforced compared with the inner ones, which leads to a differentiation into thicker and thinner muri. This principle occurs in different variations in almost all groups of the genus and can be regarded as a basic pattern of phylogenetic importance.

The tendency towards stronger development of the outer sculpture elements results, in the case of *B. vahlii,* in a reticulum with wide lumina and more strongly thickened muri. In the case of *B. retusa,* it leads to cross-ribbed large verrucoid muri, leaving space for smaller lumina only. In *B. glomerata,* the surface of the vertically undulating, almost equally strong muri is studded with small spinulae or verrucae.

Among the Eurasiatic taxa of sect. *Buxus*, all being pantoporate, a striking similarity of the most important sculpture types is obvious (Fig. 3). In the *B. rupicola* type, there appear smooth, yet spatially branched muri, reminiscent of the New World *B. macrophylla* type. The *B. rolfei* type, with its relatively coarse reticulum and cross-ribbed muri, is similar to *B. vahlii*; the extremely verrucoid development of the muri in *B. liukiuensis* shows an astonishing resemblance to *B. retusa*.

The *B. balearica* type, with its undulating muri studded with small spinulae or verrucae, resembles the *B. glomerata* type to the point of confusion. This similarity has caused the senior author (1981) to search for a conceivable plant geographical connection between the Mediterranean-NW. African (*B. balearica*) and the Caribbean taxa (*B. glomerata*). This connection, however, can be clearly excluded on the basis of differences in leaf venation and vascularization.

The pilate-verrucoid sculpture with elements of varying size, as observed in the *B. loheri* type from the Philippines, shows a strong similarity to the S. African-Malagasy species, especially to *B. madagascarica*. The pollen grains of *B. loheri* differ in their relatively large pores with richly sculptured membranes, whereas the African taxa exhibit the evolutionary series from 3-colporate to pantoporate.

In the S. African-Malagasy taxa of the subsect. *Buxella*, we observe a trend from reticulate-pilate to verrucoid exine sculptures, with traces of a microreticulate to perforate tectum visible only at the base between the verrucae.

On the other hand, the NE. African taxa of the *B. hildebrandtii* type have exine patterns and aperture characters which strongly resemble those of the Caribbean

---

Fig. 2. Pollen morphology and exine details of the American *Buxus* species. *a B. marginalis* (BRITT.) URB. (CLEMENTE 3567), × 2 000; *b B. lancifolia* BRANDEGEE (PURPUS 5308), × 2 000; *c B. mexicana* BRANDEGEE (PURPUS 2972), × 3 000; *d B. macrophylla* (BRITT.) FAWC. & RENDLE (PROCTOR & STEARN 11819), × 5 000; *e B. aneura* URB. (EKMAN 15176), × 5 000; *f B. pilosula* URB. (EKMAN 6014), × 6 500; *g B. foliosa* (BRITT.) URB. (FIGUEIRAS 651), × 5 000; *h B. flaviramea* (BRITT.) MATHOU (EKMAN 18831), LM × 1 000; *i B. pilosula* (EKMAN 6014), LM × 1 000; *j B. vahlii* BAILL. (ALAIN 10802), × 7 000; *k B. retusa* (GRISEB.) MUELL. ARG. (WRIGHT 1921), × 6 500

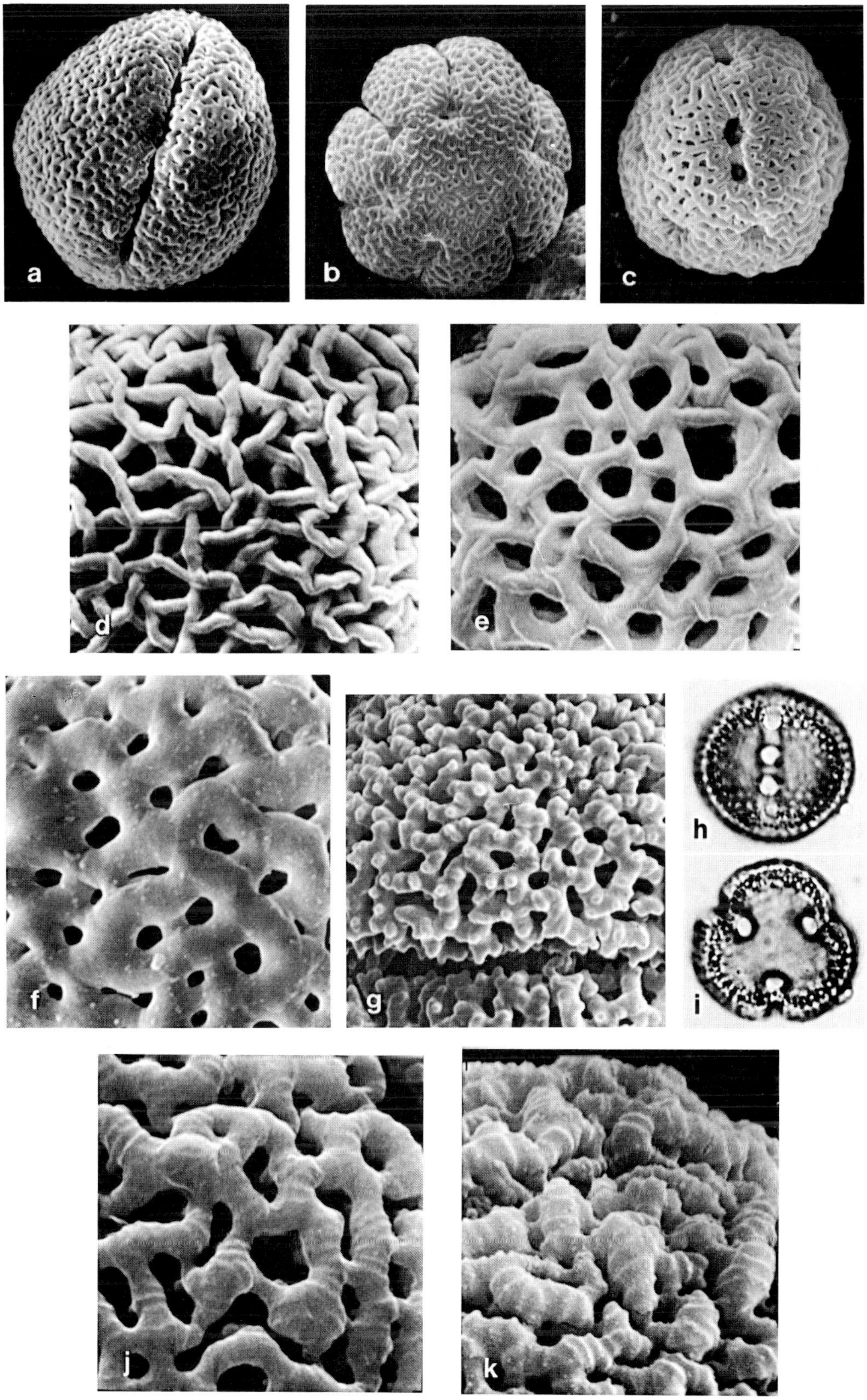
a
b
c
d
e
f
g
h
i
j
k

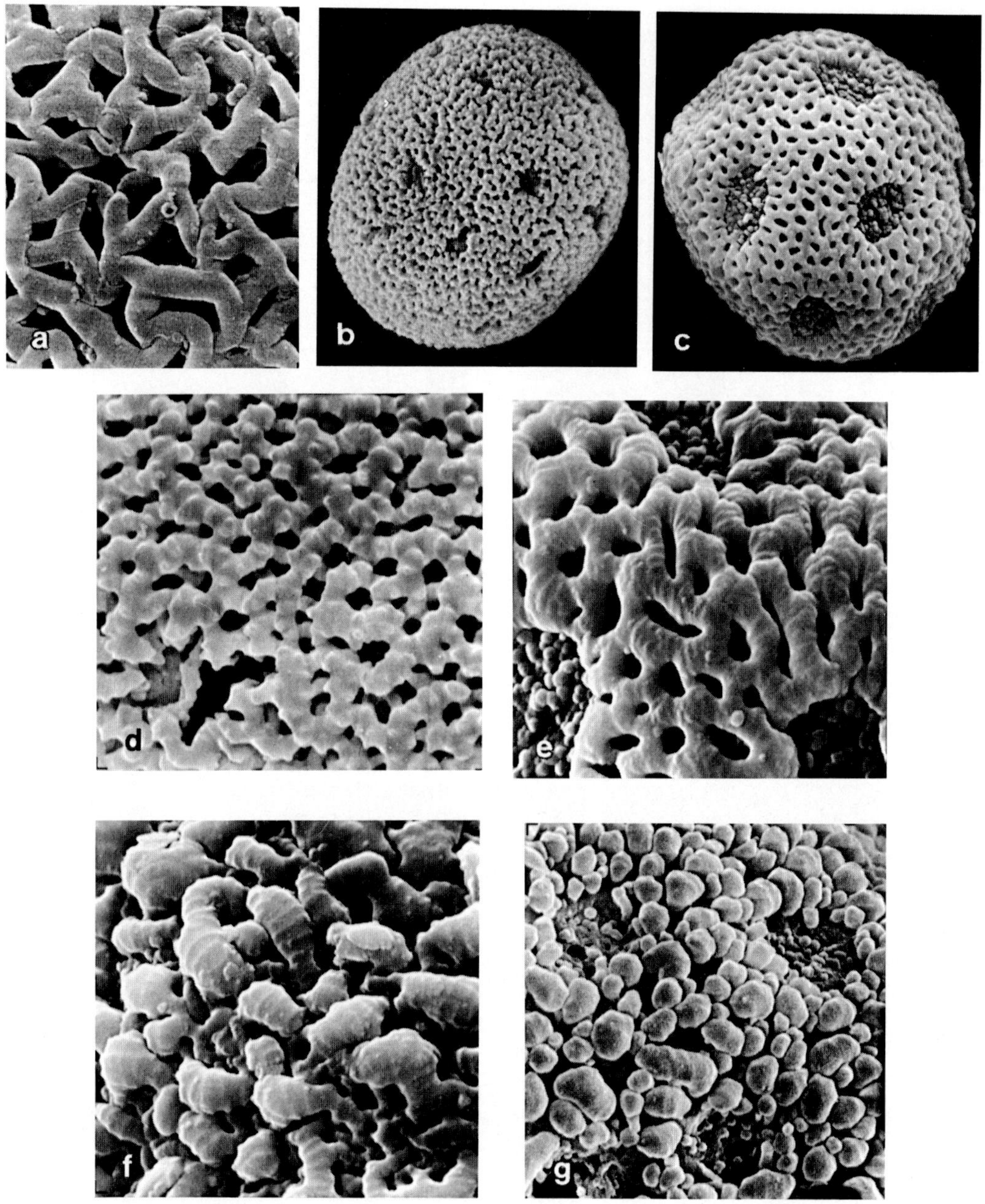

Fig. 3. Pollen morphology and exine details of the Eurasiatic *Buxus* species. *a B. rupicola* RIDL. (ROBINSON), × 5 000; *b, d B. balearica* LAM. (KENNEDY 83), *b* × 1 500, *d* × 5 000; *c, e B. cochinchinensis* PIERRE ex GAGNEP. (KERR 12405), *c* × 2 000, *e* × 5 000; *f B. ichangensis* HATUS. (WILSON 3399), × 5 000; *g B. loheri* MERR. (LOHER 13507), × 5 000

species. Interestingly, there is a tendency towards partial thickening of the muri in *B. hildebrandtii.*

Finally, the coarsely reticulate pattern of the genus *Notobuxus* remains. By virtue of its undulating, cross-ribbed, and spinulate muri, it can be traced back to the basic pattern described above.

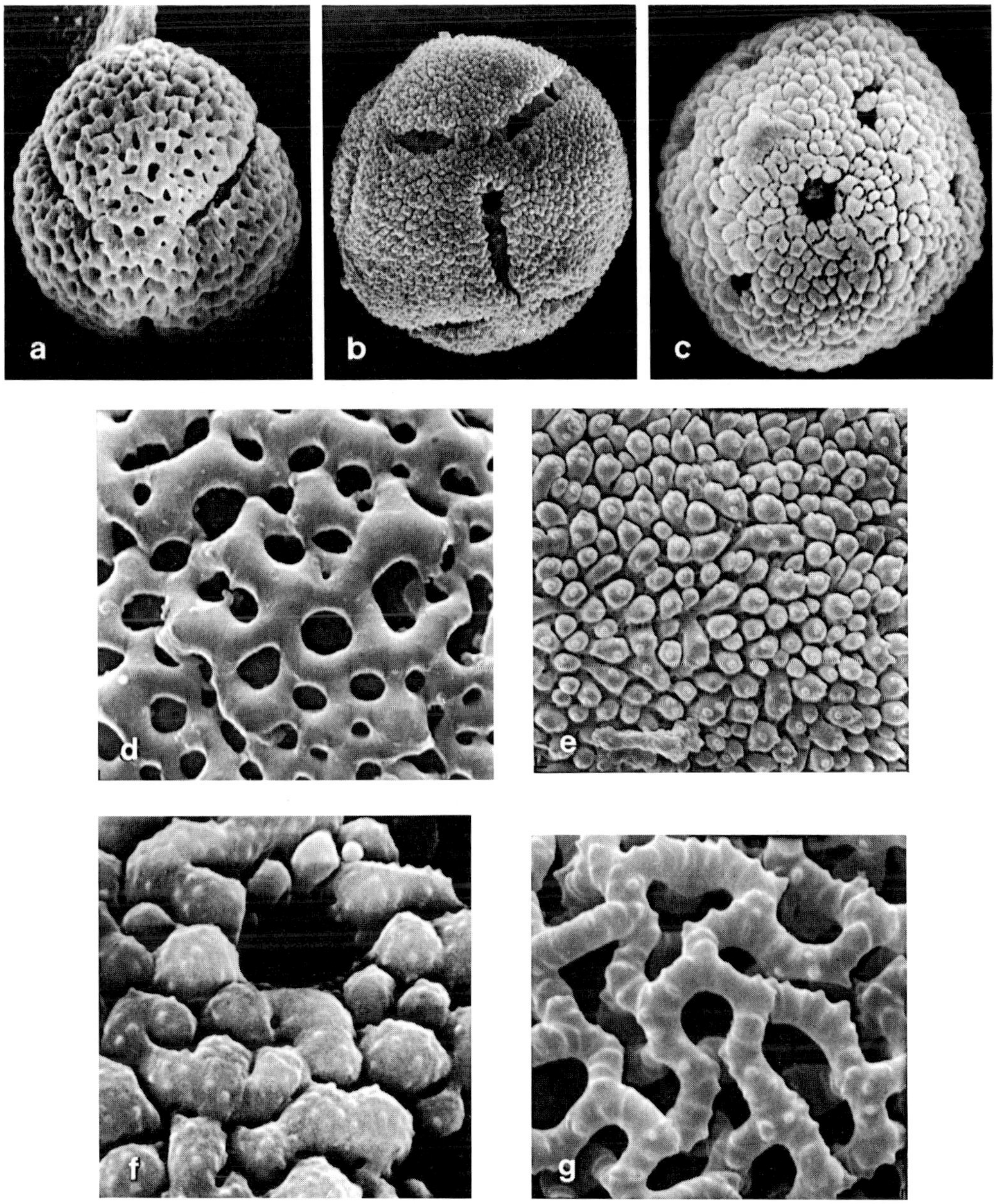

Fig. 4. Pollen morphology and exine details of the African species of *Buxus* and *Notobuxus*. *a, d B. hildebrandtii* BAILL. (HILDEBRANDT 891), *a* × 3 000, *d* × 5 000; *b, e B. benguellensis* GILG. var. *hirta* HUTCH. (BARBOSA & MORENO), *b* × 3 000, *e* × 5 000; *c, f B. macowani* OLIV. (LONG 698), *c* × 2 500, *f* × 5 000; *g Notobuxus natalensis* OLIV. (COOPER 129), × 5 000

Thus, the mosaic evolution of characteristics is apparent for the apertures as well as for the exine sculptures. This shows again the independent development of similar characters based on the common genetic background, uniting the separate lines of descent within *Buxus*.

Leaf venation

Although the Austrian Palaeobotanist Ettingshausen (1861) has presented a terminology of leaf venation, these attributes were first used for systematic purposes by Mouton (1970, 1972), Hickey (1971, 1973, 1979), and Melville (1976). These investigations have resulted in the concept, that the angiosperm leaf has a constant and recognizable pattern of architecture, which can be helpful in taxonomic studies and for the understanding of phylogenetic relationships.

Leaves of the genus *Buxus* show a considerable variation in form and size, ranging from up to 25 cm in *B. macrophylla* from Jamaica to the almost needle- or scale-like leaves of *B. leoni* and *B. vaccinioides* from Cuba. The leaves offer the most important distinction characters of these species.

Our studies (Köhler 1984, Brückner 1985) have shown that patterns of leaf venation reflect rather well the spatial differentiation of the genus, thereby indicating close correlations with palynological and leaf- and shoot-anatomical findings.

The species of the New World have a brochidodromous venation, in which the arch segments of the secondary veins are joined to form a more or less distinct intramarginal vein (Figs. 5 and 6). The higher order venation is still occasionally branched, or randomly to orthogonally reticulate. Based on the number and course of the secondary veins, the development of the intersecondary veins, and the formation of the higher order venation net, four separate nervature patterns can be defined (Köhler 1984).

The basic pattern occurs in Africa in the *B. hildebrandtii* group, extending from Ethiopia to Somalia. With respect to its intramarginal vein, details of the marginal venation, and the incompletely developed areolation, the group is closely related to the *B. obovata* group from the mountains of E. Cuba.

This basic pattern occurs again in the *B. cochinchinensis* group, distributed from Thailand to Vietnam and Malaysia. In venation details it shows a strong similarity to the New World taxa, e.g., to *B. acuminata* from the *B. macrophylla* group.

Most of the Asiatic species can be included in the *B. rolfei* and *B. megistophylla* types (Brückner 1985). The *B. rolfei* type possesses a brochidodromous venation with a more or less pronounced intramarginal vein, which enters the primary vein basally in contrast to the New World taxa. Noteworthy is the formation of admedially branched intersecondary veins, whose further branchings are often strongly thickened by sclereides.

An increasing reduction in leaf size is accompanied by a tendency towards reduction of the marginal venation; in the *B. megistophylla* type this leads to the displacement of the intramarginal vein to the outer leaf edge. The fine venation is reduced more strongly here, whereas exmedially branched secondary veins are dominating.

Fig. 5. Leaf venation pattern of the genus *Buxus*. *a–e* Sect. *Tricera*, *a B. citrifolia* (Willd.) Spreng. (Steyermark 72595); *b B. vaccinioides* (Britt.) Urb. (Shafer 4122); *c B. leoni* (Britt.) Mathou (Leon & Charles s. n.); *d B. rheedioides* Urb. (Ekman 15175); *e B. crassifolia* (Britt.) Urb. (Alain & Figueiras 7357); *f* Sect. *Eugeniobuxus*, *B. rivularis* Merr. (Sulit, Conklin 5035); *g–h* Sect. *Buxus*, *g B. austro-yunnanensis* Hatus. (How 72326), *h B. rupicola* (Curtis 2662); *i–j* Sect. *Probuxus*, *i B. macowani* (Long 698); *j B. madagascarica* Baill. (Bernardi 11400). Bars: 1 cm

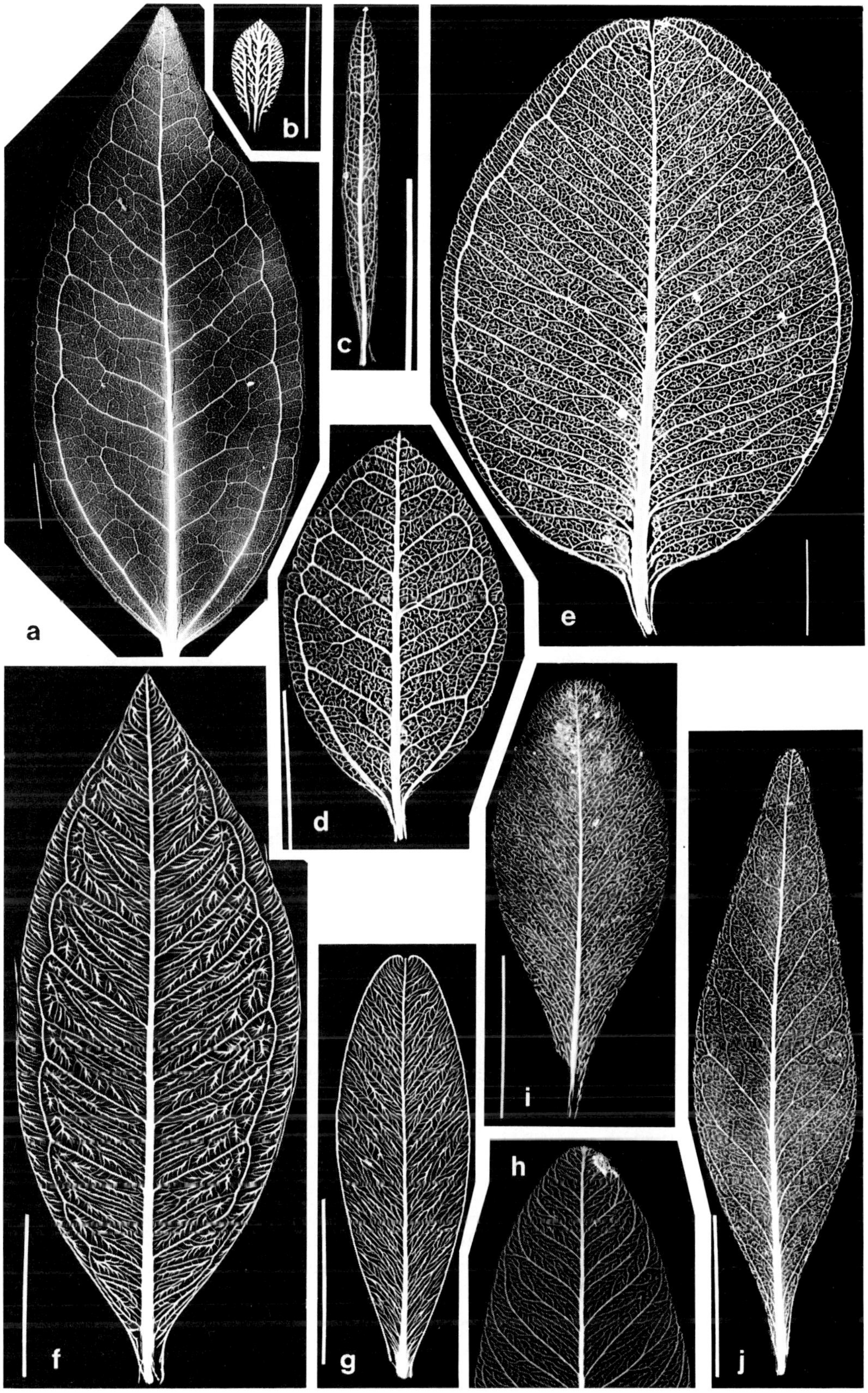
a
b
c
d
e
f
g
h
i
j

This pattern is characteristic for a group of small-leaved, extra-tropical species, to which also *B. sempervirens* belongs. *Buxus* leaves with similar venation have been demonstrated in the European Tertiary by KVAČEK & al. (1982). In *B. wallichiana*, free vein endings are almost exclusively formed and areolation is largely missing.

The *B. rupicola* type, occurring in 2 species in Thailand and Lankawi Island, occupies a special position among the Eurasiatic species. It lacks a closed intramarginal vein and often has a branched higher-order venation with poorly developed areolation, in some aspects analogous to *B. madagascarica.* A similar marginal pattern with disintegrating arches of the secondary veins among the New World species occurs only in pronounced xeromorphic taxa. For the Asiatic taxa with their 5 – 10 cm large, thin to leathery leaves, this aspect is of no importance.

Of the African-Malagasy species, *B. madagascarica,* with its loose brochidodromous venation, occasional anastomizing tertiary nerves, and weakly pronounced areolation, has the most primitive characteristics. The secondary arches are completely missing in *B. macowani*, and its secondary and intersecondary veins branch once or more towards the edge to form a more or less cladodromous pattern. The areolation is weakly developed and irregular, with many free vein endings; this is a venation pattern which obviously reflects an adaptation to strongly xerophytic conditions.

Anatomy of the petiole and shoot

A further differentiation pattern emerges from a comparative analysis of the anatomy of the shoot axis and the petiole, whose systematic importance has already been pointed out by VAN TIEGHEM (1897) and MATHOU (1940). Whereas the African species of the *B. hildebrandtii* group have 3 vascular bundles in the petiole and 2×2 cortical meristeles in the axis, the latter are missing in the neotropical taxa. Most of the Eurasiatic representatives possess only 1 median vascular bundle and 2 lateral fibre strands in the petiole; and 2 – 4 fibre strands in the axis. This petiole anatomy correlates to the reduction tendencies of the leaf venation. Preliminary investigations have shown that in several groups, the lateral fibre strands are completely reduced.

In the S. African-Malagasy taxa, the lateral bundles on the base of the petiole join the median one, which corresponds to the leaf venation pattern. Thus, only the NE. African taxa close to *B. hildebrandtii* possess what is considered to be the most primitive configuration, from which separate developments towards the New World, Eurasiatic, and S. African taxa could have proceeded.

Note that the genus *Notobuxus* also possesses the complete and primitive pattern of the NE. African taxa.

Fossil evidence

Pollen with the specific *Buxus* aperture characteristics and verrucate exine were described by KEDVES (1962) as *Nagyipollis* from the Hungarian Lower Eocene; and reticulate forms by CHATEAUNEUF (1980) from the Eocene and Oligocene of the Parisian Basin. The latter were subsequently found in the Eocene of the Massif Armoricain by OLLIVIER-PIERRE (1980), in the Upper Oligocence of the Lisbon region by PAIS (1981), in the Upper Burdigalia of Catalonia and S. France by

BESSEDIK (1981), as well as repeatedly in the Eocene of the GDR (KRUTZSCH, pers. comm.). The accompanying flora indicates a warm and humid climate, partially with swamp or mangrove vegetation (BESSEDIK 1983, OLLIVIER-PIERRE 1980, and others).

After examinations of subfossil sediments of the Zambesi Delta and the Gulf of Aden, as well as aeropalynological investigations in the Red Sea region and in S. Arabia (DUZER & VAN CAMPO from BESSEDIK 1983), one can assume for the E. African-Malagasy taxa, that much pollen is formed and is widely spread. Consequently, if *Buxus* was not directly an element of this vegetation, the taxa must have had at least larger leaves, like the extant *B. macrocarpa*.

With *Buxapollis* (KRUTZSCH 1966), the Eurasiatic taxa are already present in the early Oligocene. Their presence in Miocene was proved by KVAČEK & al. (1982), based on the characteristic venation of leaf remnants.

The aperture characters of fossil pollen specimens with reticulate exine are reminiscent of sectt. *Tricera, Buxus*, and the African *B. hildebrandtii* group. Together with leaf remnants, they demonstrate that the genus was already considerably differentiated in the European Tertiary, and that its complete basic stock was already existent. The New World and Eurasiatic groups, already separated in the Oligocene, can probably be traced back to a common laurasian-tethyan origin (BESSEDIK 1983), which certainly existed long before the separation of N. America and W. Europe. *Nagyipollis* thus could represent the missing link in the pollen evolution of the Eurasiatic taxa.

While the Early and Middle Eocene was characterized by floras rich in mesophilous (sub)tropical elements (KRUTZSCH 1967, CHÂTEAUNEUF 1980, OLLIVIER-PIERRE 1980, and others), in the course of the Oligocene, they declined and were replaced by temperate elements. According to MULLER (1980), many taxa from warm and humid regions in Europe have retreated after the upper Eocene from the North to the South as a consequence of the increasing aridization. They appear to have disappeared from the Early and Middle Miocene of the Tethys Province (BESSEDIK 1981, 1982, 1983). This is certainly applicable to parts of the *Buxus* taxa, whereas others obviously adapted to the increasing dryness. After a critical revision of the available evidence, BESSEDIK (1983) was able to show that the pollen morphological diversity has increased rapidly during the Tertirary, obviously reflecting a progressive differentiation of the taxa. These findings indicate that the more primitive pollen forms, linked today with mesophilic taxa, appear much earlier than the pollen from more evolved, xerophilic or ecologically specialized taxa. Thus, an in situ genesis of *B. sempervirens, B. balearica*, and *B. longifolia* is conceivable.

In Japan, the genus first appears from the Oligocene to the Early Miocene (*B. protojaponica*, UEMURA 1979). In the Middle Miocene, it was wide-spread and became frequent in the Late Miocene and Pliocene, having adapted to the cooling and the changing physiography of the Japanese islands (UEMURA 1979).

Discussion

The results of pollen morphology, petiole and axis anatomy, leaf venation and their distribution as well as fossil evidence give the following picture of the genus' differentiation.

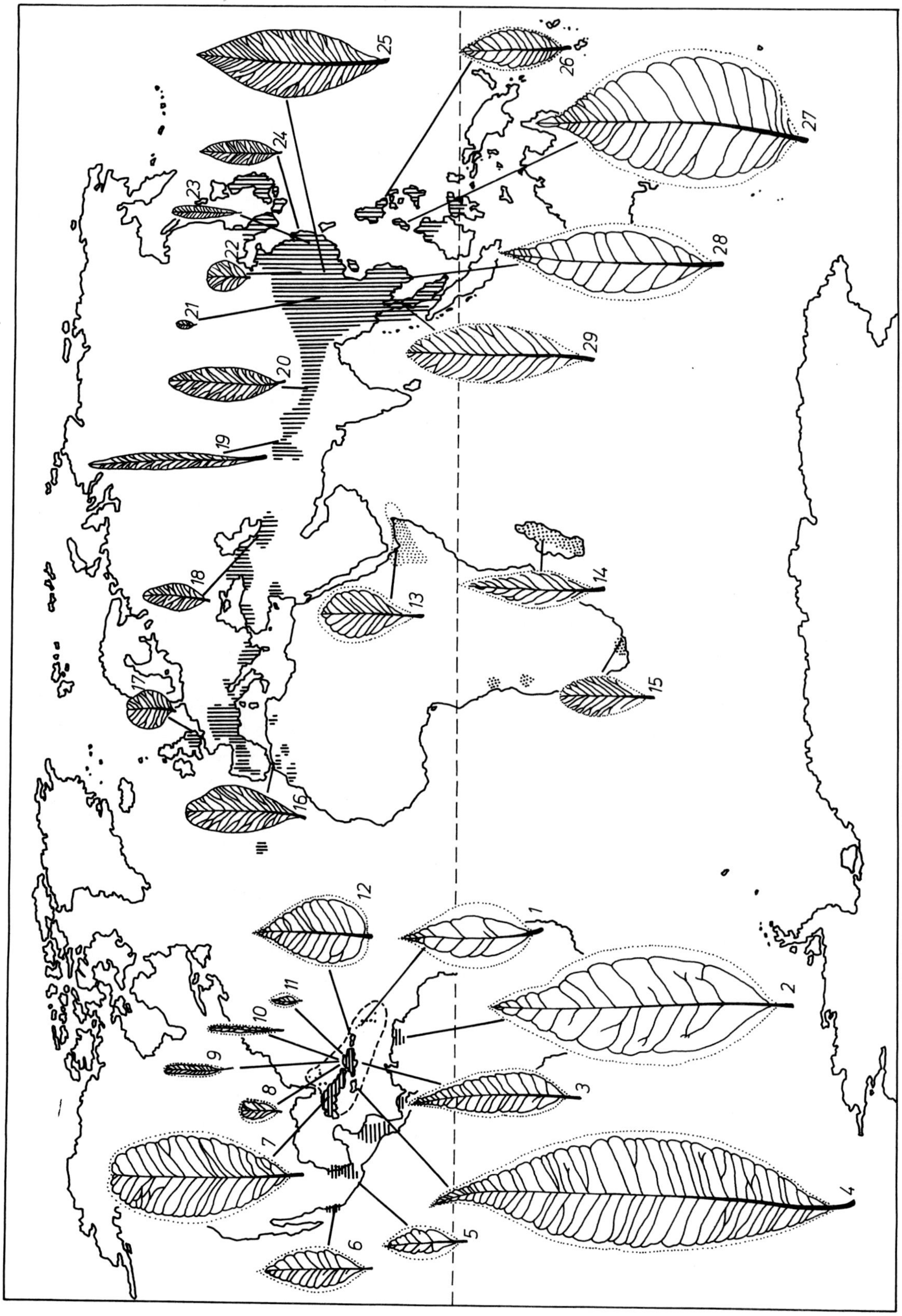
1
2
3
4
5
6
7
8
9
10
11
12
13
14
15
16
17
18
19
20
21
22
23
24
25
26
27
28
29

The brochidodromous venation type, with a more or less pronounced intramarginal vein, exhibits a geographical disjunction between the Caribbean, NE. Africa and SE. Asia and thus suggests an old Tethys connection (Fig. 6).

B. cochinchinensis is the only Asiatic taxon which corresponds in its fine venation to the New World taxa. It differs, however, in that the basal portion of the petiole possesses one median vascular bundle only. This represents a transition towards the more advanced venation pattern of the large-leaved species related to *B. rolfei*. In their coarse reticulate pollen exine, these species groups (except *B. loheri*) also correspond to each other.

The highly advanced *B. megistophylla* type arose through reduction of leaf size in conjunction with the occupation of arid-temperate areas. The pollen of the species belonging here is finely reticulate and the muri are partially studded with spinulae. The taxa of this type have gone through considerable adaptive radiation in E. Asia and form a centre of diversity, with *B. sempervirens* and *B. balearica* as its westermost representatives. Because of its leaf venation pattern, which is fully identical to that of the other taxa, the reticulate-verrucoid exine sculpture of the species with *B. liukiuensis* pollen type must be viewed as a parallel development.

In addition to the mentioned relationships to African taxa, *B. rupicola* exhibits a certain similarity of the more or less eucamptodromous venation with *B. madagascarica*. Because of qualitative differences between the spatially anastomizing muri of *B. rupicola* pollen and the pilate exine sculpture of African-Malagasy taxa, one has to assume parallel tendencies with respect to venation differentiation.

The relationships of the NE. African *B. hildebrandtii* group to the taxa of the European Tertiary were viewed by Bessedik (1983) as a disjunction resulting from the North-South floral retreat over the Tethys. They may even be older, since up to the early Palaeocene loose ties between Africa, Europe and Asia are assumed. We believe that such relationships could have existed only with the progenitors of the New World taxa, since they are very similar to the *B. hildebrandtii* group with respect to its leaf venation, pollen aperture and exine characters. Both groups conform in the trifascicular construction of their petioles and differ only in the absence of cortical meristeles in the neotropical taxa. These findings contradict the classification of these taxa into sect. *Buxus*, as proposed by Pax (1896) and Hutchinson (1912, 1967). However, the NW. African taxa are derivatives of this Eurasiatic group.

It is remarkable that among all the extant *Buxus* taxa, the *B. hildebrandtii* group exhibits the most primitive characteristics of petiole and axis anatomy. These characters also appear in the African genus *Notobuxus*, which, in addition, possesses

---

Fig. 6. Geographical distribution of leaf venation pattern of the genus *Buxus*. *1 B. portoricensis* Alain, *2 B. citrifolia*, *3 B. acuminata* (Griseb.) Muell. Arg., *4 B. macrophylla*, *5 B. mexicana*, *6 B. pubescens* Greenm., *7 B. crassifolia*, *8 B. rotundifolia* (Britt.) Mathou, *9 B. revoluta* (Britt.) Mathou, *10 B. foliosa*, *11 B. vaccinioides*, *12 B. cubana* (A. Rich.) Baill., *13 B. hildebrandtii*, *14 B. madagascarica*, *15 B. macowani*, *16 B. balearica*, *17 B. sempervirens* L., *18 B. hyrcana* Pojark., *19 B. papillosa* Schneid., *20 B. wallichiana* Baill., *21 B. rugulosa* Hatus. subsp. *prostrata* (Smith) Hatus., *22 B. microphylla* Sieb. & Zucc. subsp. *sinica* (Rehd. & Wils.) Hatus., *23 B. harlandi* Hance, *24 B. microphylla* subsp. *sinica* var. *aemulans* (Rehd. & Wils.) Hatus., *25 B. megistophylla* Lev., *26 B. rivularis*, *27 B. rolfei* Vid., *28 B. cochinchinensis*, *29 B. rupicola*

a very primitive brochidodromous venation and still more primitive features in higher stamen numbers. The remarkable combinations of primitive characteristics retained in these African relic groups suggest parallels with general consideration on the important role of Western Gondwanaland in angiosperm evolution (RAVEN & AXELROD 1975).

As to the question of the origin of the remaining African-Malagasy taxa (subsect. *Buxella*), the phylogenetic importance of the pilate-verrucate exine sculpture must be discussed. The occurrence of verrucoid next to reticulate patterns in American species and in two systematically well-defined Eurasiatic groups (sectt. *Buxus* and *Eugeniobuxus*) indicates a parallel development of verrucate exine sculpture. Transitions from reticulate to verrucate exines are not infrequent within genera, species, and even within individuals.

Fossil pollen from Europe strongly resembles that of representatives from the sect. *Buxus* near *B. loheri* (BESSEDIK 1983) with respect to size and arrangement of the verrucae, and, to a lesser degree, that of the African subsect. *Buxella* near *B. madagascarica*, particularly with respect to aperture characters.

On the basis of their pilate-verrucate pollen and specific leaf venation, the African-Malagasy species as delimitated by VAN TIEGHEM (1897) and MATHOU (1940), prove to be a clearly defined group of common descent. In contrast to the other African taxa, this group possesses the most highly advanced venation pattern of the genus. Its highly reduced or absent gynoeceum rudiments and interstylar nectaries give further proof to its derived position. The species of this relic taxon have survived under the increasing aridization only in favourable refugial areas.

On Madagascar, *B. madagascarica* represents a relatively primitive taxon, which still possesses 3-colporate pollen. With its brochidodromous venation, it could be related to forerunners of the *B. hildebrandtii* group and the neotropical taxa, relationships which may date to the early Tertiary.

From their remarkable similarities one has to conclude that the *B. hildebrandtii* group among extant Old World taxa stands closest to the progenitors of the New World group of *Buxus*. This relic group thus is a further example for the very old disjunction between Mid America and E. Africa, comparable to the situation in genera like *Dracaena, Savia, Ravenala*, and others.

With respect to the origin and distribution of the New World taxa, the genus *Buxus* often is interpreted as a representative of an old sclerophyll flora from the Northern coasts of the Tethys. AXELROD (1975) considers the genus, with the Mexican *B. lancifolia* and *B. mexicana* and with the Mediterranean *B. sempervirens*, as an example for his "Madrean-Tethyan links". Our investigations demonstrate that these types belong to lines of ancient separation, less closely related to each other than the E. African *B. hildebrandtii* is to the New World taxa.

In contrast to representatives of true sclerophyll communities, *Buxus* does not occur in N. American areas with a Mediterranean-type climate, like California, Arizona, New Mexico, etc., where it also is lacking in the fossil record. The populations of *Buxus* in S. Mexico and C. America are quite limited, have a relic character, and prefer more mesic habitats. This argues against an East-West migration along the northern Madrean-Tethyan connections.

The systematic relationships to the E. African *B. hildebrandtii* bring up questions as to the importance of the southern Tethyan coasts with their rich laurophyll vegetation existing up into the Miocene (AXELROD 1975). In this context, BORHIDI

(1982, 1985) talks of "peri-Afro-American" genera, which have settled in the broad coastal belts of these continents and have survived in the relic-areas of Macaronia, E. Africa, and Madagascar. The Caribbean plate and the Macaronesian plate obviously belong to this coastal belt. There is some evidence that the genus *Buxus* may have been an element of the southern wing of this Madrean-Tethyan flora that has not yet been well investigated.

Among the New World taxa of *Buxus* a centre of diversity has originated in the Caribbean region, where the genus has passed through a remarkable adaptive radiation, particularly in Cuba (Köhler 1984). There, we not only find those taxa most closely related to the African species, but also representatives of nearly all groups, which – in differing degrees – have extended to the other Antilles, and to C. and S. America (Köhler 1985). The latter mostly are derivatives of only one or two groups and later differentiations. In several of these New World groups, parallel evolution has occurred from mesophilic to more or less xeromorphic taxa.

The various character differentiations described for the genus *Buxus* allow a satisfactory delimitation of its different lines of descent and verify their relationships. The degree of their divergence and the pattern of their geographic distribution, together with their early fossil evidence, demonstrate the great age of the genus and its very early initial differentiation.

We wish to thank Dr Menniger, Mr Heinrich, and Miss Krause of the Central Institute for Electron Physics of the Academy of Sciences of the GDR for the opportunity to carry out scanning electron microscopical investigations. We also wish to thank Miss Chwalczyk and Mrs Gračner for very reliable technical assistance. To Dr Krutzsch we are grateful for invaluable advice concerning this manuscript.

References

Axelrod, D. I., 1975: Evolution and biogeography of Madrean-Tethyan sclerophyll vegetation. – Ann. Missouri Bot. Gard. **62**: 280 – 334.

Baillon, H., 1859: Monographie des Buxacées et des Stylocérées. – Paris.

Bessedik, M., 1981: Recherches palynologiques sur quelques sites du Burdigalien du Midi de la France. – Thèse de Spécialité. – Montpellier.

– 1982: Analyse pollinique des dépôts lacustres miocènes de Fabrèques (Aubs, Var, France): intérêt paléoclimatique. – Géobios **15**: 589 – 592.

– 1983: Le genre *Buxus* L. (*Nagyipollis* Kedves 1962) au Tertiaire en Europe occidentale: èvolution et implications paléogéographiques. – Pollen & Spores **25**: 461 – 486.

Borhidi, A., 1982: Die Entwicklung der Flora der Antillen im Lichte der Kontinentalverschiebungstheorie. – Stapfia **10**: 235 – 241.

– 1985: Phytogeographic survey of Cuba. 1. The phytogeographic characteristics and evolution of the flora of Cuba. – Acta Bot. Hungar. **31**: 3 – 34.

Brückner, P., 1971: Beiträge zur Pollenmorphologie der *Buxaceae*. – Master of Science Thesis, Berlin (unpubl.).

– 1985: Blattnervatur und Pollenmorphologie eurasiatischer Arten der Gattung *Buxus* L. (*Buxaceae* Dumort.) und ihre Bedeutung für die Systematik. – Ph.D., Berlin.

Burkhardt, D., 1986: Untersuchungen zur Morphologie des Gynoeceums der neotropischen Arten der Gattung *Buxus* L. (*Buxaceae*) und ihre Bedeutung für die Systematik. – Master of Science Thesis, Berlin (unpubl.).

Chateauneuf, J. J., 1980: Palynostratigraphie et paléoclimatologie de l'Eocène supérieur et de l'Oligocène du Bassin de Paris. – Mém. B.R.G.M. 116.

Ettingshausen, C. von, 1861: Die Blatt-Skelete der Dikotyledonen. – Wien.

Gentry, A. H., 1978: Flora of Panama: 6. Family 99 *Buxaceae*. — Ann. Missouri Bot. Gard. **65**: 5 — 8.

Grajcarek, C., 1986: Blattanatomische Untersuchungen an afrikanischen Arten der Gattungen *Buxus* L. und *Notobuxus* Oliv. (*Buxaceae*). — Master of Science Thesis, Berlin (unpubl.).

Hartung, M., 1984: Anatomische Untersuchungen von Blattstielen an neotropischen Arten der Gattung *Buxus* L. (*Buxaceae*). — Master of Science Thesis, Berlin (unpubl.).

Hatusima, S., 1942: A revision of the Asiatic *Buxus*. — J. Dept. Agric. Kyushyu Imp. Univ. **6**: 261 — 342.

Hickey, L. J., 1971: Evolutionary significance of leaf architectural features in the woody dicots. — Amer. J. Bot. **58**: 469.

— 1973: Classification of the architecture of dicotyledonous leaves. — Amer. J. Bot. **60**: 17 — 33.

— 1979: A revised classification of the architecture of dicotyledonous leaves. — In Metcalfe, C. R., Chalk, L., (Eds.): Anatomy of the dicotyledons. — Oxford: Clarendon Press.

Hutchinson, J., 1912: African *Buxeae*. — Kew Bull. 1912: 52 — 55.

— 1967: The genera of flowering plants **2**, pp. 105 — 109. — London: Oxford University Press.

Kedves, M., 1962: *Nagyipollis*, a new pollen fgen. from the Hungarian Lower Eocene. — Acta Biol., Szeged **8**: 83 — 84.

Köhler, E., 1979: Ergebnisse palynotaxonomischer Untersuchungen an cubanischen *Buxus*-Arten. — Wiss. Zeitschr. Friedr.-Schiller-Univ. Jena, math.-nat. R. **28**: 683 — 689.

— 1980: Zur Pollenmorphologie und systematischen Stellung der *Didymelaceae* Leandri. — Feddes Repert. **91**: 581 — 591.

— 1981: Pollenmorphology of the West Indian-Central American species of the genus *Buxus* L. (*Buxaceae*) with reference to taxonomy. — Pollen & Spores **23**: 38 — 91.

— 1982: Drei neue *Buxus*-Arten für die Flora von Cuba. — Wiss. Zeitschr. Friedr.-Schiller-Univ. Jena, math.-nat. R. **31**: 239 — 250.

— 1984: Zur Blattnervatur der neotropischen *Buxus*-Arten und ihre Bedeutung für die Systematik (*Buxaceae*). — Flora **175**: 345 — 374.

— 1985: Vorstellungen zur Evolution und Chorogenese der neotropischen *Buxus*-Arten. — Feddes Repert. **96**: 663 — 675.

— Brückner, P., 1982: Die Pollenmorphologie der afrikanischen *Buxus*- und *Notobuxus*-Arten (*Buxaceae*) und ihre systematische Bedeutung. — Grana **21**: 71 — 82.

— — 1983: Zur Pollenmorphologie und systematischen Stellung der Gattung *Simmondsia* Nutt. — Wiss. Zeitschr. Friedr.-Schiller-Univ. Jena, math.-nat. R. **32**: 945 — 955.

Krutzsch, W., 1966: Zur Kenntnis der präquartären periporaten Pollenformen. — Geologie **15**: 16 — 71.

— 1967: Der Florenwechsel im Alttertiär Mitteleuropas aufgrund von sporenpaläontologischer Untersuchungen. — Abh. Zentr. geol. Inst. **10**: 17 — 37.

Kvaček, Z., Bůzek, C., Holy, F., 1982: Review of *Buxus* fossils and a new large-leaved species from the Miocene of Central Europe. — Rev. Palaeobot. Palynol. **37**: 361 — 394.

Mathou, Th., 1940: Recherches sur la famille des Buxacées. — Toulouse.

Melville, R., 1976: The terminology of leaf architecture. — Taxon **25**: 549 — 561.

Mouton, J. A., 1970: Architecture de la nervation foliaire. — Compt. Rend. 92e Congr. Nat. Soc. Savantes, sect. sc. 3, 165 — 176.

— 1972: Contribution de la morphologie foliaire a la phylogénie des Angiospermes. — 93e Congr. Nat. Soc. Savantes, sect. sc. 3, 199 — 209.

Mueller Arg., J., 1869: *Buxaceae*. — In De Candolle: Prodromus **XVI**. — Paris.

Muller, J., 1980: Palynological evidence for Paleogene climatic changes. — Mem. Mus. Nat. Hist. Nat., Nouv. Sér., Sér. B, Botanique **27**: 211 — 218.

MUNSCHKE, M., 1974: Beiträge zur Pollenmorphologie der Gattungen *Buxus, Notobuxus, Styloceras, Sarcococca, Pachysandra* und *Simmondsia* (*Buxaceae*). – Master of Science Thesis, Berlin (unpubl.).

OLLIVIER-PIERRE, M. F., 1980: Etude palynologique (spores et pollens) de gisements paléogènes du Massif armoricain. Stratigraphie et paléogéographie. – Mém. Soc. geol. minéral. Bretagne **25**: 1–239.

PAIS, J., 1981: Contribuição para o conhecimento da vegetacão miocénica da parte ocidental dą Bacia to Tejo. – Thèse ès Sciences, Lisbonne.

PAX, F., 1896: *Buxaceae.* – In ENGLER, A., PRANTL, K., (Eds.): Die natürlichen Pflanzenfamilien **3**, pp. 130–135. – Leipzig: W. Engelmann.

PHILLIPS, E. P., 1943: Some changes in nomenclature IV., *Buxus* L. and *Notobuxus* OLIV. – J. South Afric. Bot. **9**: 138–140.

RAVEN, P. H., AXELROD, D. I., 1975: History of the flora and fauna of Latin America. – Amer. Sci. **63**: 420–429.

SCHIRAREND, C., 1981: Blattanatomische Untersuchungen an neotropischen Arten der Gattung *Buxus* L. (*Buxaceae*). – Master of Science Thesis, Berlin (unpubl.).

SCHREBER, J. C. D., 1791: *Tricera.* – In LINNAEUS, C. v., (Ed.): Genera plantarum, 8th ed. p. 630. – Francofurti a/M.: Varrentrapp.

UEMURA, K., 1979: Leaf compressions of *Buxus* from the Upper Miocene of Japan. – Bull. Nat. Sci. Mus., Tokyo, Ser. C, **5**: 1–8.

VAN CAMPO, M., 1976: Patterns of pollen morphological variation within taxa. – In FERGUSON, J. K., MULLER, J., (Eds.): The evolutionary significance of the exine. – Linn. Soc. Symp. Ser. **1**: 125–137.

VAN TIEGHEM, P., 1897: Sur les Buxacées. – Ann. Sci. Nat., Bot., Ser. 8, **5**: 289–338.

VAVILOV, N. I., 1922: The law of homologous series in variation. – J. Gen. **12**: 47–89.

Address of the authors: Prof. Dr E. KÖHLER and Dr P. BRÜCKNER, Bereich Botanik und Arboretum des Museums für Naturkunde der Humboldt-Universität Berlin, Späthstrasse 80/81, DDR-1195 Berlin, German Democratic Republic.

Pl. Syst. Evol. 162, 285–304

Plant
Systematics
and
Evolution

The ecogeographical differentiation of Amazonian inundation forests

K. Kubitzki

Received August 31, 1987

Key words: Amazon region, biogeography, biotic diversity, chorology, ecogeography, refuge hypothesis; tropical rivers, inundation forests, flood resistance; neotropical flora.

Abstract: Due to the considerable annual fluctuations of water level of the Amazonian rivers, their river banks are fringed with periodically flooded forests of vast extension. The biota of these communities are adapted to annual inundations that can last for more than half a year. Water chemistry is most important for the floristic differentiation of these flooded forests. White water rivers, which carry a rich load of suspended material originating from the erosion of the Andes, have a floristic composition related to that of the non-inundatable Amazonian forest. Clear water and black water rivers, which originate in the Amazon Basin or its adjacent crystalline shields, are nutrient-poor and more or less acidic; their flora is related to that of peculiar woodland and savannah vegetation on oligotrophic white sand. The distribution patterns of floodplain species of nutrient-poor waters point to a centre of diversity in the Upper Rio Negro region, and another one in the Guayana lowland. These coincide with diversity centres for species of non-flooded habitats. Hence it seems unlikely that species diversity is directly influenced by pluviosity. The flooded forests have developed biotic interactions with the fish fauna of the Amazon Basin, which are vital for their continued existence. It is assumed that the origin of these habitats, their biota and their interactions dates back long into the Tertiary.

There is hardly another region on earth that is influenced by water as strongly as Amazonia. Although the Amazon is not the Earth's longest river, it is by far the mightiest, and its discharge is 4–5 times that of the Congo, the only river that comes near to the Amazon. Its river banks and those of all its major and smaller affluents are bordered by forests subject to periodical inundation. Gourou (1950) and Pires (1973) made the estimate that 2% of the whole area of Brazilian Amazonia are covered by periodically flooded forests. If this estimate is applied to the whole of Amazonia, we have to count with an inundatable area of more than 120 000 km^2. Of this area, more than $^3/_4$ would be represented by várzea, which borders white water rivers, while a smaller fraction would be covered by igapó, which fills the floodplain of nutrient-poor clear and black water rivers.

Two factors are decisive for the existence and vast extension of periodically flooded forests in Amazonia. First, the unequal annual distribution of precipitations. In most parts of Amazonia, the rainy season is interrupted by a drier period

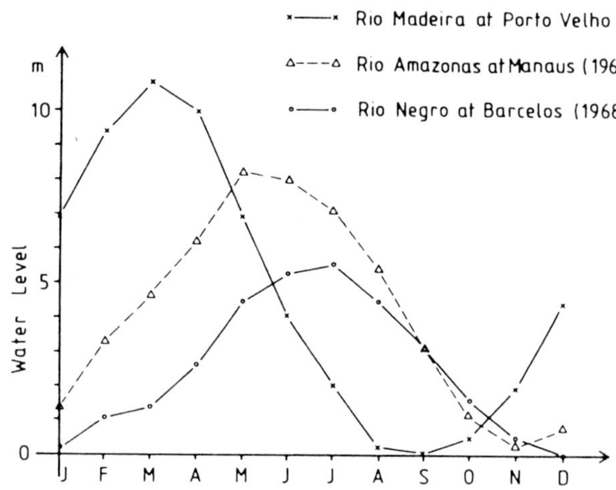

Fig. 1. Amplitude and timing of water level fluctuations of Amazonian rivers. (After JUNK 1984)

of several month's duration, which, however, is not synchronous in all parts of the Hylaea and is more or less absent in its northwestern part. Second, the Amazon Basin is a low-lying plain, as is evident from the fact that Iquitos at a distance of 3600 km from the mouth of the Amazon lies only 100 m above sea level. Therefore, the annual rise of the water level of the network of Amazonian rivers leads to inundations of enormous extension, in contrast, for instance, to the Orinoco, which never borders beyond its deeply incised bed. The amount of annual change in water level is highest in Central Amazonia; at the mouth of Rio Juruá the amplitude is 20 m; downstream the Amazon it decreases, reaching about 6 m at the mouth of the Rio Tapajoz (Fig. 1), and is so small in the region of the estuary that it is superimposed by the rise of the tide (SIOLI 1984).

Conditions for life in the seasonally flooded habitats are quite extreme. The length of the annual period of flooding can amount up to 10 months, and trees can be covered by water up to 15 m deep. Often, trees keep their entire foliage in the floeded stage, and after falling dry immediately seem to start with photosynthesis. Another stress factor is the poorness of oxygen in the water and the rhizosphere of flooded trees, which requires modifications of the respiratory pathway, preventing the accumulation of toxic end products of anaerobic respiration, such as ethanol. It is likely that this metabolic capacity is decisive in controlling the floristic composition of temporarily flooded forests.

Since the dry period of flooded forests can be very short, regeneration must imply special mechanisms of seedling establishment, apart from vegetative propagation (Fig. 2), although we do not know anything about this. The ecophysiological adaptations of periodically inundated forests have barely been touched upon (GESSNER 1968, SCHOLANDER & OLIVEIRA 1968, KEEL & PRANCE 1979).

The trees of the inundated forests have scarcely striking physiognomic traits, despite the heavy stress exerted upon them through the fluctuations of the water level. Morphologically differentiated respiratory roots, such as occurring in the mangrove, are clearly absent and would not be of any use during inundations lasting for months. In Amazonia such roots are only found in swamp forests subject

Fig. 2. *Couepia paraensis*, showing vegetative propagation. Rio Nhamundá, Pará, Brazil. (Phot. K. KUBITZKI, 1 Oct. 1984)

Fig. 3. *Eschweilera coriacea* in clear water *igapó* along the Rio Arapiuns, affl. to the Rio Tapajoz, Brazil. (Phot. H. SIOLI, 14 Nov. 1952)

to diurnal or other frequent changes of water level; *Symphonia globulifera* is a foremost example. A remarkable trait is that the crowns of the trees often look just as cut away on their lower surface, as has been illustrated by Gessner (1968). This is certainly due to the effect of the high flood, which prevents the growth of those branches that are submerged for the longest time. Being far from universal, this can be observed in *Simaba orinocensis, Eschweilera coriacea* (Fig. 3), and several members of the *Combretaceae*, such as *Buchenavia ochroprumna*.

A most characteristic adaptation of the trees of the inundated forests is the frequent occurrence of spongy aerenchymatous tissue in their diaspores, imparting buoyancy, as has been pointed out by Ducke (1949). The fruit production of the flooded forests has also great importance for the sustenance of the fish populations which, in turn, contribute towards the dissemination of their diaspores (Goulding 1980).

The floristic composition of the periodically inundated forests was described by pioneers such as Spruce (1908), Huber (1910), and Ducke (1913). Further contributions are due to Ducke & Black (1953), Rodrigues (1961), Takeuchi (1962), Hueck (1966), Keel & Prance (1979), Prance (1979), Worbes (1983), Junk (1984), and Adis (1984). A first attempt towards an ecophysiological study of the inundated forests has been made by Gessner (1968).

In view of the peculiar stress situation to which the floodplain forests are exposed, problems of ecologic, floristic, and evolutionary nature emerge. Here I want to deal with the differentiation of the flora of periodically inundated forests in dependence on water chemistry, and with the relationship of their elements to those of non-inundatable forests on terra firme. This implies the question of the age and origin of the flora of these forests.

Floristic differentiation according to water chemistry

The three main river types of Amazonia – white water, clear water, and black water rivers – are known since the time of A. R. Wallace and have carefully been studied and characterized by limnologists, such as H. Sioli and collaborators. However, the terminology and delimitation of Amazonian forest types subject to inundation is still controversial. Local people do not differentiate according to water quality; for them, all flooded forests is várzea, or, when it has more water, igapó (lit. "the place where the water stands") (J. M. Pires, pers. comm.). In the usage of biologists, the várzea comprises the alluvial floodplain including the forest growing on it, which originates from the annual flood of white-water rivers that are rich in suspended matter. The várzea fringes the Amazon itself, the Rio Madeira, Rio Purus and other white-water rivers rich in sediment load. Igapó, in contrast, is defined to comprise inundated river banks together with their vegetation that have originated by vertical erosion (Sioli 1954). In my usage, in which I follow Prance (1979), the river banks not only of black water rivers, but of clear water rivers as well are called by name igapó, even when their rivers have some insignificant sedimentation due to their sediment load stemming from the nutrient-poor Tertiary fresh water sediments of Central Amazonia. Due to their poverty in nutrients and relative low pH the conditions for life in clear water are much more alike to those in black water than in white water.

The floristic differentiation between várzea and igapó has often been emphasized, and characteristic elements are listed in Tables 1 and 2. Such a differentiation is also apparent with respect to the mycoflora, as has been discovered by SINGER (1984).

Biomass and local species diversity of the várzea are higher than of the igapó. However, the flora of the former is much more uniform than that of the latter which has much regional differentiation, so that in total the flora of the igapó is richer than that of the várzea. There is hardly an independent flora of clear-water igapós, although their flora may appear impoverished in comparison with that of black water regions, at least that of the Rio Negro basin. There is also no significant difference in the fish fauna between black and clear water (GOULDING 1980).

Despite the rather clear separation of the floras of várzea and igapó, there are some tree species in common to them; DUCKE (1913) mentioned *Campsiandra laurifolia, Macrolobium acaciaefolium,* and *Symmeria paniculata* for both river types; PRANCE (1979) listed *Virola elongata, Caryocar microcarpum,* and *Allantoma lineata* as common elements. In fact, the list could be considerably extended: *Caraipa densifolia, Pachira aquatica* and *P. insignis, Swartzia polyphylla, Vatairea guianensis,* and *Licaria armeniaca* are species that occur in all river types, the two last-mentioned also on terra firme. One has also to consider that the water quality of a river can change during the course of the year. This occurs regularly in the side branches of the Amazon river where, depending on the water level, either the white water of the main river, or the nutrient-poor water of an affluent may be dominant. Several rivers are known that are turbid in the rainy season but transparent in the dry one; the Rio Tocantins is the largest of them.

Altogether, the floristic differentiation between the domains of nutrient-poor and nutrient-rich water is clear although the reasons for this might be difficult to define, since the ecological optimum of a species under the influence of competition may be quite different from its physiological optimum. The species listed in Tables 1 and 2 have a high indicator value for trophic conditions, and in those places where nutrient-poor rivers cross bars of Palaeozoic limestone, várzea species are present. This can be observed in the lower course of the Rio Trombetas/Rio Cuminá de Oeste. A similar phenomenon is produced by the influx of white water of the Rio Branco into the lower Rio Negro.

Ecological relationships of tree species of flooded forests

The flora of Amazonian flooded forests represents no independent floristic stratum, but has close ties to the flora of non-flooded habitats. There is an especially close relationship between várzea and terra firme forest on latosol. In the várzea one can find wide-spread forest species of the neotropics, such as *Guazuma ulmifolia,* or species of even wider tropical distribution, such as *Spondias lutea,* which in Amazonia have developed flood-resistant ecotypes. The differentiation of várzea species proper is exemplified by the moraceous genus *Maquira,* in which one species (*M. coriacea*) is bound to the várzea, while another (*M. calophylla*) occurs there facultatively; the remaining species are restricted to terra firme (BERG 1972). In the western part of the Hylaea, the differentiation between várzea and terra firme fades away, and várzea species such a *Ceiba pentandra* or *Pseudobombax munguba* occur also in the non-inundated forest. This is because the soil of the terra firme

Fig. 4

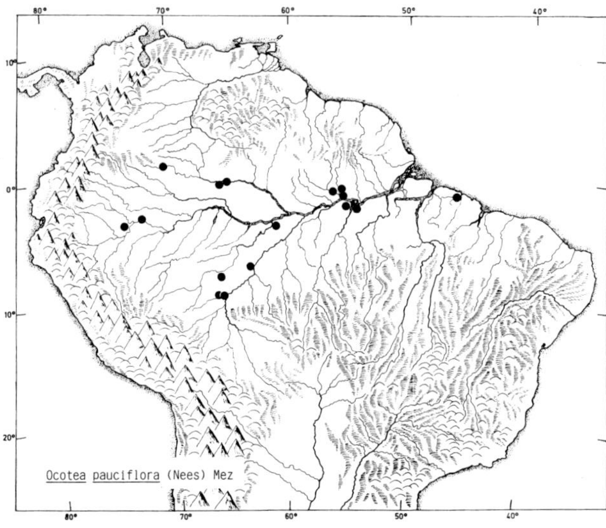

Fig. 5

Fig. 4. *Qualea retusa*, wide-spread in Amazonia on upper sandy river banks, tolerating short-termed inundation. (Orig.). – Fig. 5. *Ocotea pauciflora*, occurring on higher, periodically inundated river banks, on savannahs, and in terra firme forest. Its sister species, *O. cernua*, occurs equally in periodically flooded and non-flooded habitats. (After pers. comm. by J. ROHWER)

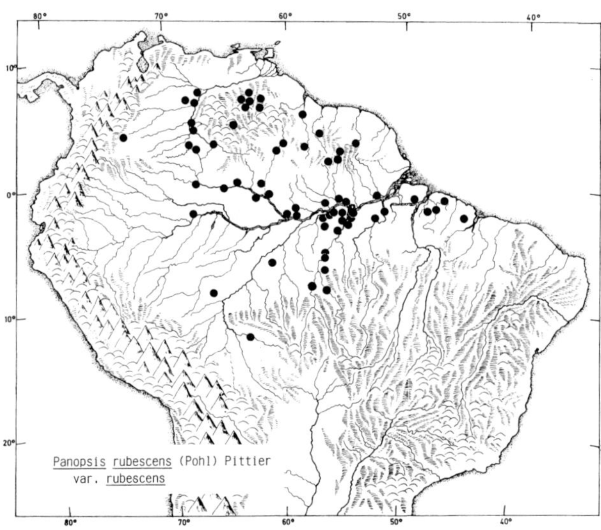

Fig. 6

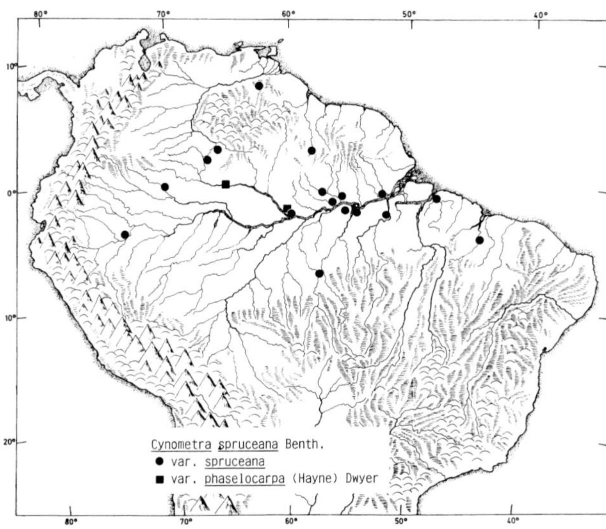

Fig. 7

Fig. 6. *Panopsis rubescens* var. *rubescens*, a tree of sandy river banks wide-spread in Guayana and Amazonia. The fruits are indehiscent, not wettable, and buoyant. (Partly based on SLEUMER 1954 and STEYERMARK 1982). – Fig. 7. *Cynometra spruceana* [= *C. martiana* (HAYNE) BAILL.], a species of sandy and rocky river banks on black water and clear water rivers of Amazonia and western Guayana with buoyant husks. (After DWYER 1958, updated)

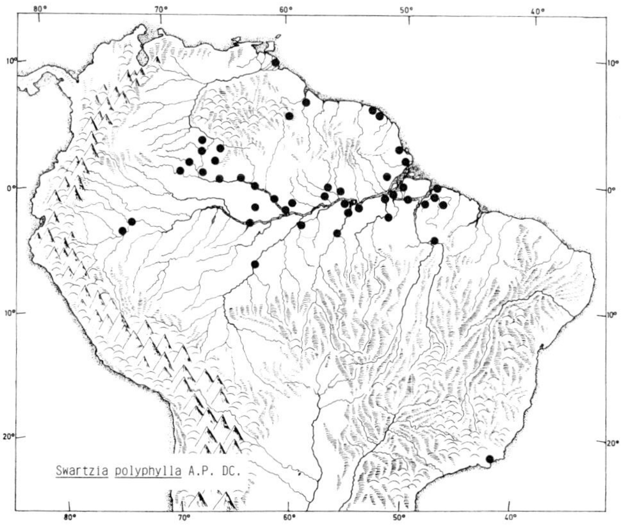

Fig. 8

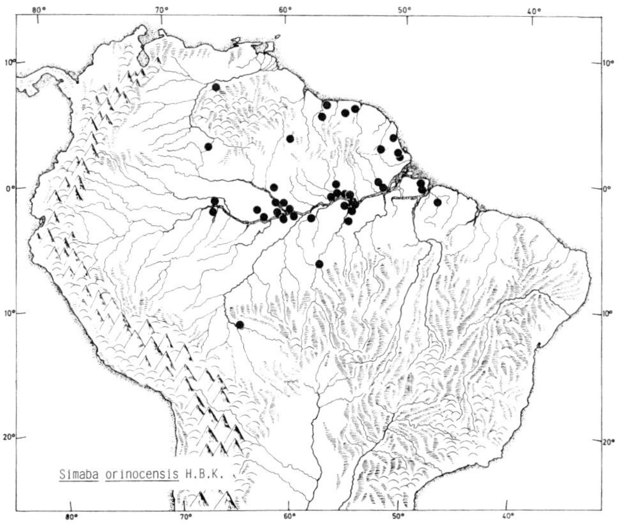

Fig. 9

Fig. 8. *Swartzia polyphylla*, one of the most frequent riverine species of Amazonia and Guayana, occurring preferably in periodically inundated floodplains of clear water rivers, but of black water and white water too. The most closely related species occur partly in flooded forests (*S. schomburgkii* BENTH.), partly in never flooded forests (*S. parvifolia* SCHERY, *S. remiger* AMSH.). The husk is buoyant. (After COWAN 1968, updated). – Fig. 9. *Simaba orinocensis* (incl. *S. multiflora* A. JUSS.) is a typical species of periodically flooded riverine forests of Amazonia and Guayana, preferably in nutrient-poor water. The fruit, which is preferred by fish, has a sour taste, as is often the case with fish-dispersed fruits. (After CAVALCANTE 1983, and THOMAS 1984, updated)

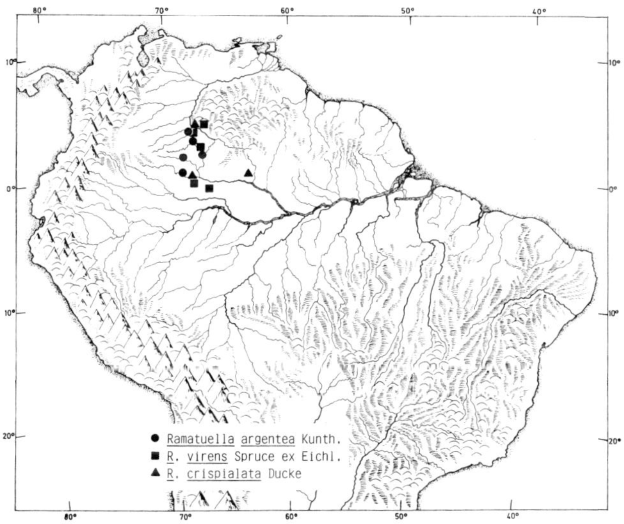

Fig. 10

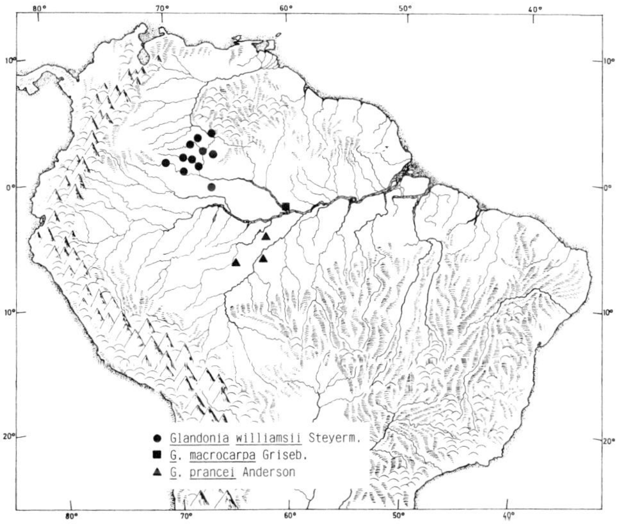

Fig. 11

Fig. 10. *Ramatuella argentea, R. virens* (incl. *R. maguirei* EXELL & STACE and *R. latifolia* MAGUIRE) and *R. crispialata* [incl. *R. obtusa* (MAGUIRE) STACE & EXELL] are restricted to the floodplains and savannahs of the Upper Rio Negro and Upper Orinoco region. (After EXELL & STACE 1963, and pers. comm. by C. A. STACE 1976). – Fig. 11. *Glandonia williamsii, G. macrocarpa*, and *G. prancei* occur on periodically flooded savannahs and ploodplains, the fruit keeps afloat. (After ANDERSON 1981)

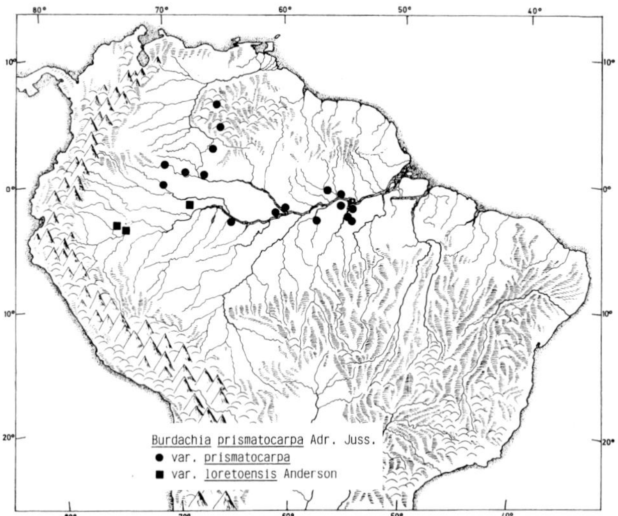

Fig. 12

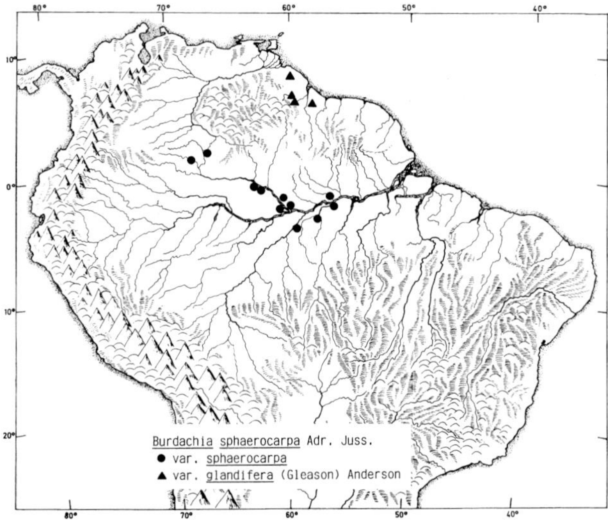

Fig. 13

Fig. 12. *Burdachia prismatocarpa* is restricted to black water and clear water rivers of Amazonia and Guayana. The same is true of the other species of the same genus (Fig. 13) and the three species of the related genus *Glandonia* (Fig. 11). The fruit is buoyant. (After ANDERSON 1981). – Fig. 13. *Burdachia sphaerocarpa*, a species of periodically flooded river banks on nutrient-poor rivers of Amazonia and Guayana. (After ANDERSON 1981, updated)

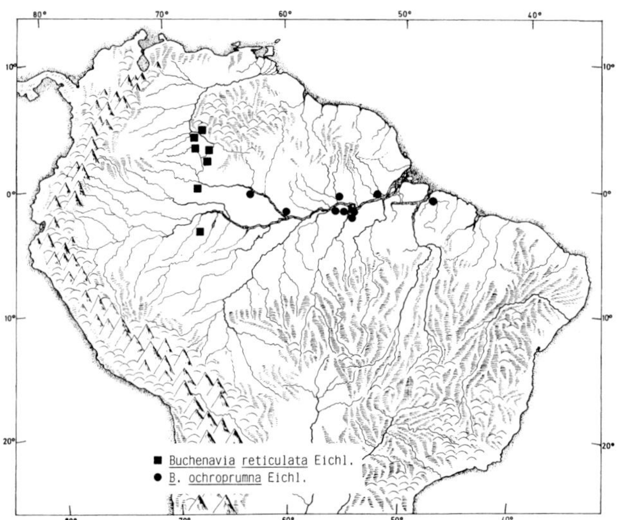

Fig. 14

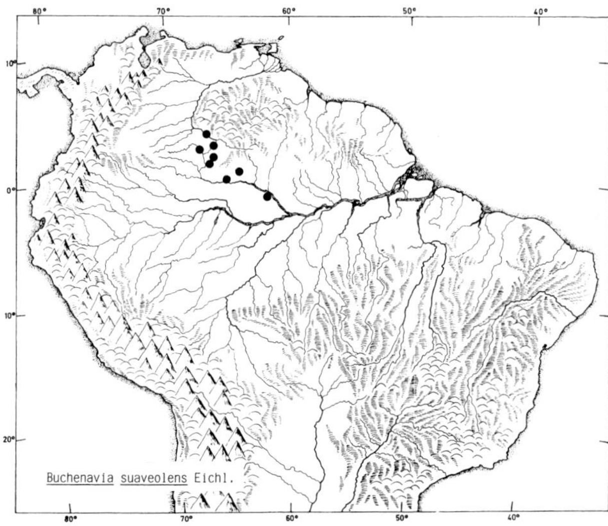

Fig. 15

Fig. 14. *Buchenavia reticulata*, from the Upper Orinoco Basin and western Amazonia, and *B. ochroprumna* from central and eastern Amazonia occur in floodplains of nutrient-poor waters. The differentiation of the whole genus may have occurred in such habitats. (After EXELL & STACE 1963, and pers. comm. by C. A. STACE 1976, updated). – Fig. 15. *Buchenavia suaveolens* (incl. *B. pterocarpa* EXELL & STACE), a floodplain species of the Upper Rio Negro and Upper Orinoco region. (After EXELL & STACE 1963, and pers. comm. by C. A. STACE 1976)

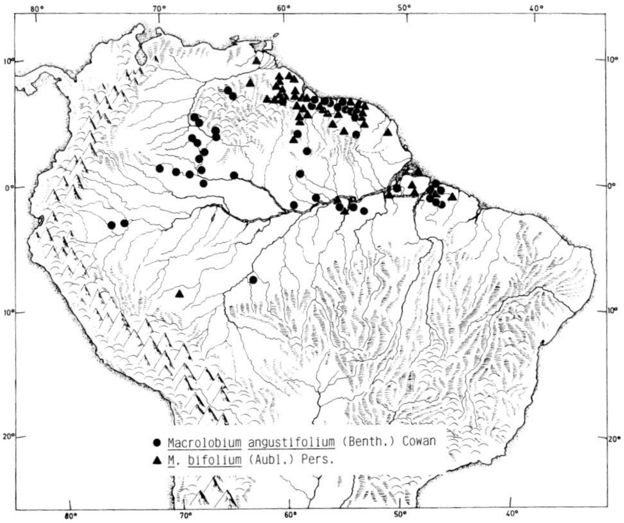

Fig. 16

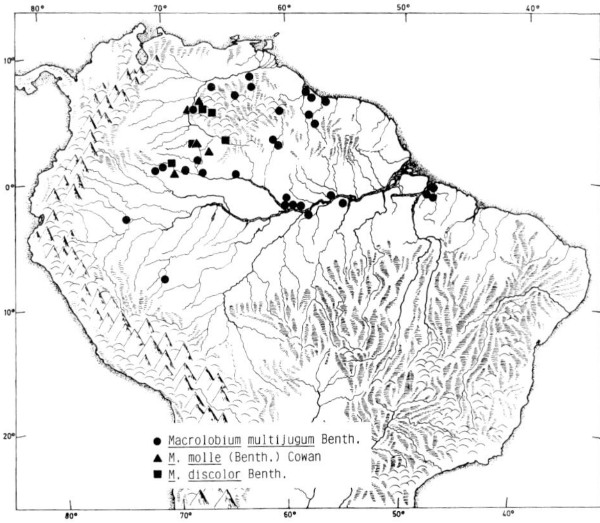

Fig. 17

Fig. 16. *Macrolobium angustifolium*, a frequent tree species of floodplains of white water, clear water and black water rivers in Amazonia and Guayana. Its closest relative, *M. bifolium*, occurs in periodically inundated floodplains and savannahs mainly of the Guayana lowland. (After COWAN 1953, updated). – Fig. 17. *Macrolobium multijugum*, a floodplain tree of nutrient-poor waters of Amazonia and Guayana; the closest relatives, *M. molle* and *M. discolor*, are in the Upper Rio Negro and Upper Orinoco region. (After COWAN 1953, updated)

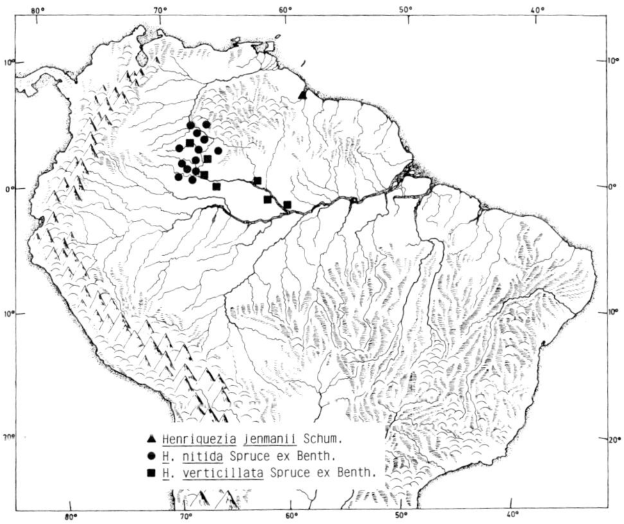

Fig. 18

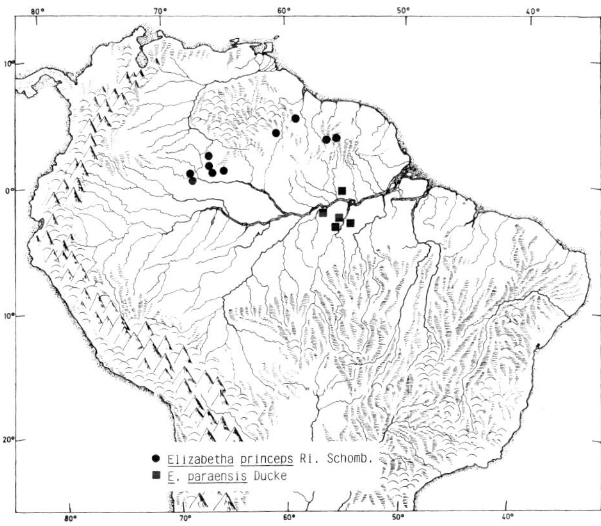

Fig. 19

Fig. 18. *Henriquezia verticillata* is distributed in the Rio Negro Basin, *H. nitida* in the Upper Rio Negro/Rio Orinoco region; all species occur in periodically flooded habitats, *H. jenmanii* also in tidal forests. (After ROGERS 1984). – Fig. 19. Of the 11 known species of *Elizabetha* only *E. princeps* occurs in periodically flooded forests of the Rio Negro Basin and Guayana, while its closest relative, *E. paraensis*, is distributed in eastern Amazonia, where it occurs on terra firme. (After COWAN 1976)

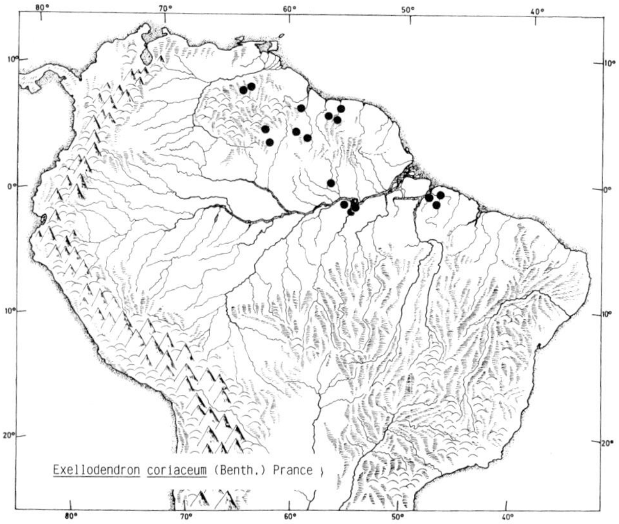

Fig. 20

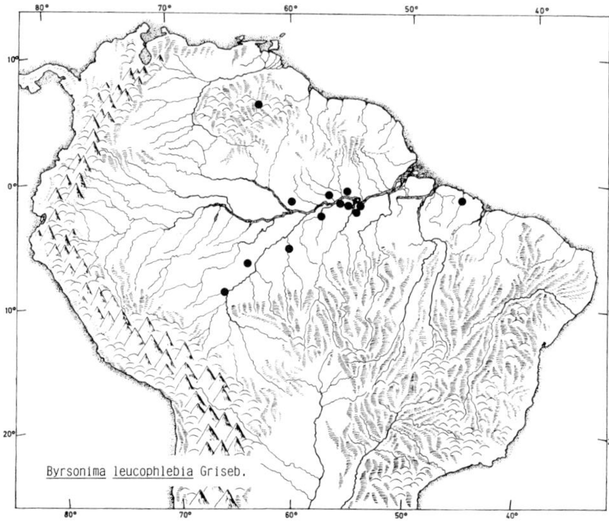

Fig. 21

Fig. 20. *Exellodendron coriaceum* is a species of periodically inundated river banks and savannahs of Guayana and eastern Amazonia. (After PRANCE 1972, updated). – Fig. 21. *Byrsonima leucophlebia* occurs on sandy river banks and savannahs of Amazonia and Guayana. (After ANDERSON 1981, updated)

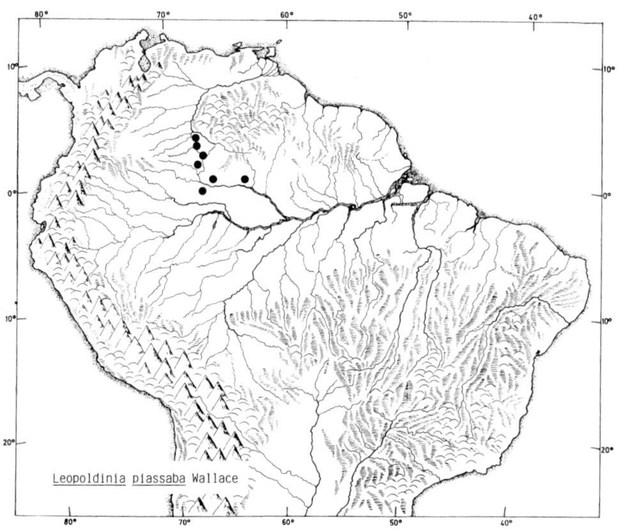

Fig. 22

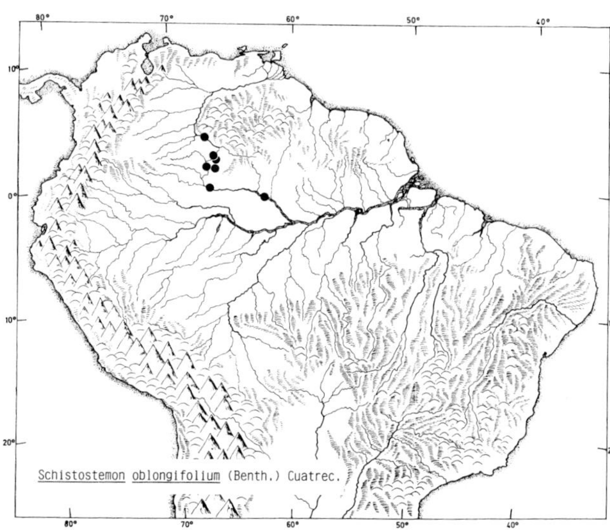

Fig. 23

Fig. 22. *Leopoldinia piassaba* is a species of black water igapó of the Upper and Middle Rio Negro region. (Orig.). – Fig. 23. *Schistostemon oblongifolium* is a tree with buoyant fruits of periodically inundated floodplains of the Upper Orinoco and Rio Negro Basin. (After CUATRECASAS 1961)

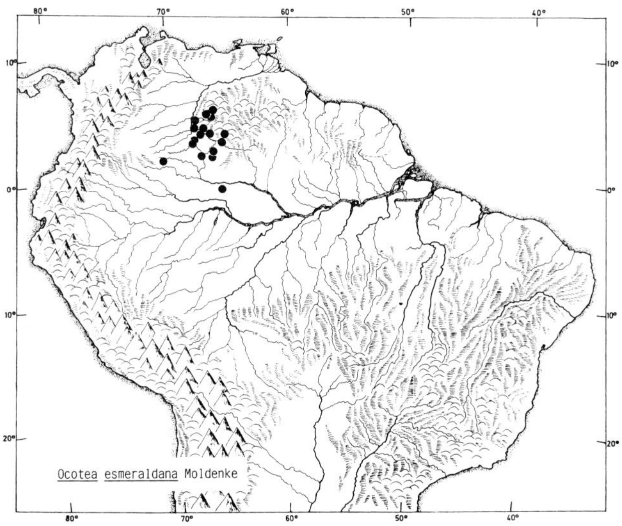

Fig. 24

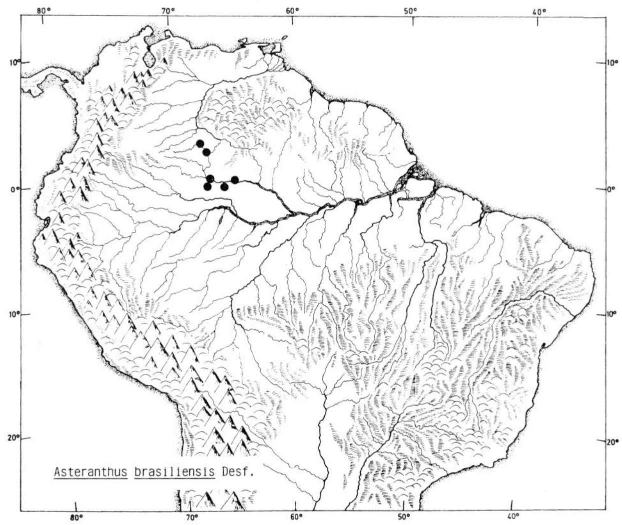

Fig. 25

Fig. 24. *Ocotea esmeraldana* occurs in the Upper Orinoco region in savannahs, in the Upper Rio Negro region in the igapó. (After pers. comm. by O. Huber and J. Rohwer). – Fig. 25. *Asteranthus brasiliensis*, a monotypic of the igapó of the Upper Rio Negro region; the sister group is in tropical Africa. (After Prance & Mori 1980, updated)

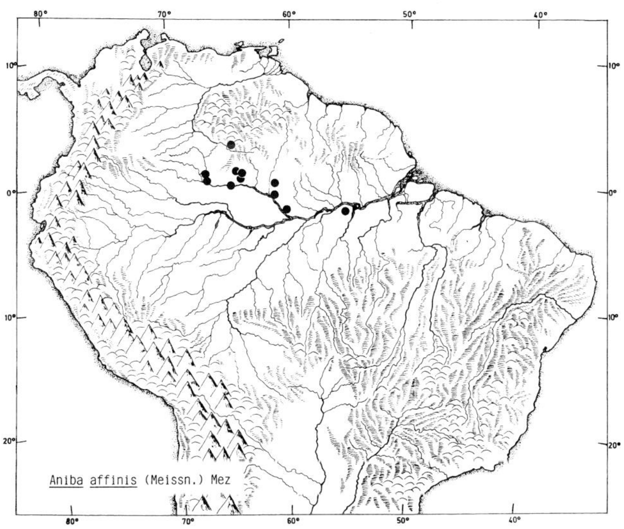

Fig. 26

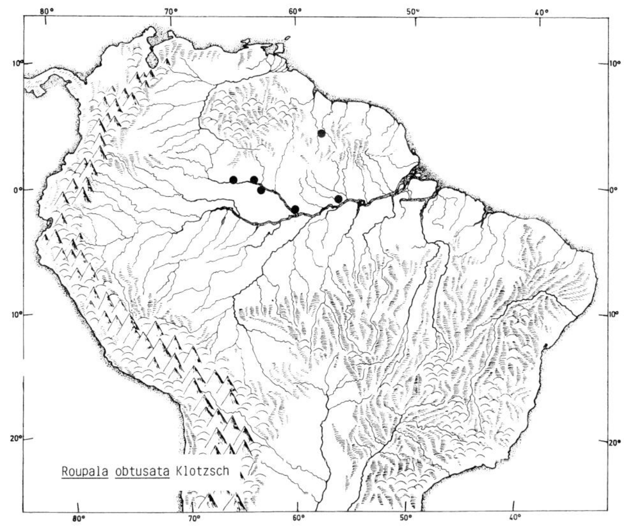

Fig. 27

Fig. 26. *Aniba affinis,* a species of black water igapó of the Rio Negro/Upper Orinoco basins, which occurs also in the clear water Rio Maró (Tapajoz Basin). (Orig.). – Fig. 27. *Roupala obtusata,* a species of floodplains of nutrient-poor rivers, possibly radiating from the Rio Negro Basin. (After Sleumer 1954)

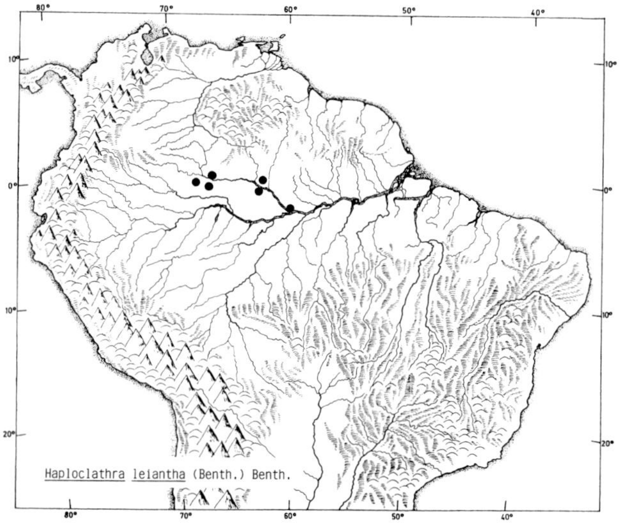

Fig. 28

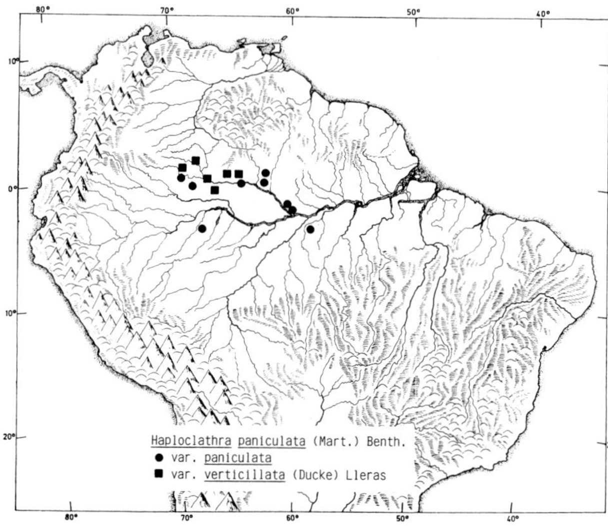

Fig. 29

Fig. 28. *Haploclathra leiantha,* a species of the igapó of the Rio Negro Basin. (Orig.). – Fig. 29. *Haploclathra paniculata,* centred in black water igapó of the Rio Negro Basin. (Orig.)

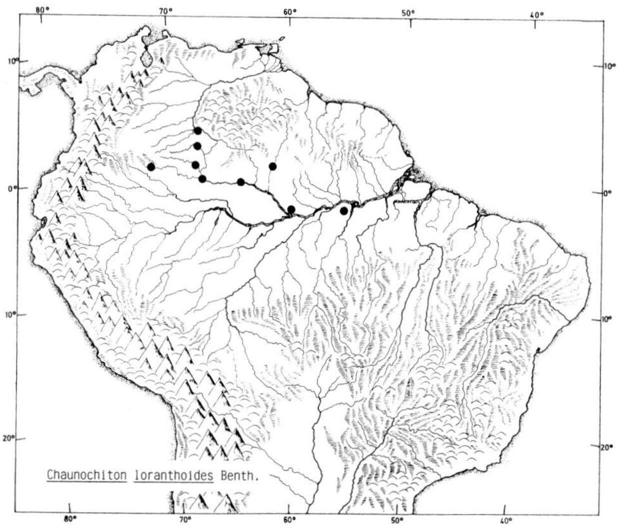

Fig. 30

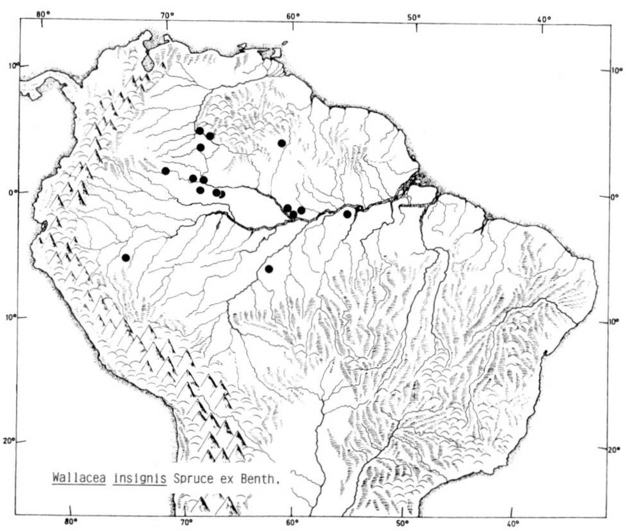

Fig. 31

Fig. 30. *Chaunochiton loranthoides,* a species of upper sandy river banks. (After SLEUMER 1984, updated). – Fig. 31. *Wallacea insignis,* a species of black and clear water igapó of the Rio Negro region and surrounding areas. (Orig.)

forest in the foothills of the Andes is much more fertile than that on the Tertiary sand and clay deposits of Central Amazonia. In this context, the genus *Triplaris* is instructive: Its centre of diversity is in the western Hylaea and its species occur partly on terra firme, partly in the várzea and partly in both habitats (BRANDBYGE 1986).

The flora of the forests bordering nutrient-poor clear and black water rivers has close connexions with the flora of savannahs in the widest sense, especially the oligotrophic woodlands of "campina" and "caatinga" vegetation and savannahs on white sand, but not on more fertile yellow sand. Many species of white sand savannahs of Guayana and Amazonia are able to tolerate short-termed inundations and can occur in the upper parts of sandy river banks of nutrient-poor rivers. One example is *Doliocarpus spraguei* which is frequent on white sand savannahs and is also found on sandy river banks of the Rio Tapajoz and Rio Negro and their affluents. Typical elements of these upper sandy river banks are *Qualea retusa* (Fig. 4), *Panopsis rubescens* (Fig. 6), and representatives of the *Humiriaceae*, such as *Humiria balsamifera* and *Humiriastrum cuspidatum* (Cuatrecasas 1961). *Chaunochiton loranthoides* (Fig. 30) is also restricted to river banks which are inundated only for a short time; its two sister species occur in forests and savannahs (Sleumer 1984). Longer lasting inundations are tolerated by species of the *Myrtaceae, Leguminosae* and *Chrysobalanaceae*; of the latter family, *Exellodendron coriaceum* (Fig. 20), *Licania apetala,* and *Couepia paraensis* are very characteristic. Another example is *Laetia suaveolens*, the sister species of which occur in savannahs, caatingas and similar oligotrophic habitats (Sleumer 1980). In the genus *Elvasia, E. calophyllea* and *E. quinqueloba* prefer upper sandy river banks; they have wind-dispersed diaspores, while *E. hostmannia*, which occurs in the lower part of the floodplain, has buoyant diaspores (Ducke 1949).

There are also genera of which all species are restricted to periodically flooded habitats; in these cases, speciation may have implied geographical separation without ecological diversification; such patterns may be rather recent. Examples include the genera *Ramatuella* (Fig. 10), *Glandonia* (Fig. 11), *Haploclathra* (Figs. 28 and 29), all more or less restricted to the Rio Negro Basin, and the palm genus *Leopoldinia*, ranging from the Rio Negro Basin to Lower Amazonia. The small genus *Burdachia* has a still wider distribution (Figs. 12 and 13); its fruits, like those of the genus *Glandonia* and of *Lophanthera longifolia,* both equally malpighiaceous, keep afloat (Anderson 1981).

Some systematically isolated types are restricted to periodically flooded habitats, such as *Symmeria paniculata,* which apart from Amazonia occurs also in western Africa, *Asteranthus brasiliensis* (Fig. 25), which again has its closest relatives in tropical Africa*, and the two species of the genus *Polygonanthus,* of which one occurs in the Rio Negro basin, the other on clear water rivers of Central Amazonia (Nelson & al. 1987).

The only two species of *Pachira, P. aquatica* and *P. insignis,* occur also only in floodplains. In contrast to their relatives in the genera *Ceiba* and *Pseudobombax* they have a spongy seed coat, but no seed hairs, which are an adaptation for dispersal by wind. Here emerges a problem alread pointed to by Ducke (1930: 60): If single traits that are obviously adaptations for special ecologic conditions

* *Ceiba pentandra* and *Pterocarpus amazonicus* are further examples of species of inundated habitats that occur in eastern S. America and W. Africa. In the genera *Aptandra, Heisteria, Sacoglottis, Heteropterys, Malouetia,* and *Raphia,* there are closely related species in flooded forests on both sides of the Atlantic; others occur in tidal forests; they have been listed by Engler (1905) and Thorne (1973).

are used as a criterion for generic delimitation, the perception of phylogenetic relationships, which is a foremost aim of classification, is obscured.

Chorological relationships

The distribution of many species of periodically inundated forests coincides with the limits of the Amazonian rain forest. Yet more frequent is a distribution pattern which includes lowland Guayana: *Panopsis rubescens* (Fig. 6), *Cynometra spruceana* (Fig. 7), *Swartzia polyphylla* (Fig. 8), *Burdachia prismatocarpa* and *B. sphaerocarpa* (Figs. 12 and 13), *Lophanthera longifolia, Simaba orinocensis* (Fig. 9), and *Piranhea trifoliata* are some examples. More than 250 Amazonian rain forest species are also present in the Atlantic coastal forest of eastern Brazil (Lima 1966), a disjunction probably dating back to the Pliocene. Tree species of periodically flooded forests are absent from the Atlantic forest, since these habitats do not exist there. One remarkable exception is *Caraipa densifolia*, which is a wide-spread element of the Amazonian várzea and igapó, while in the Atlantic coastal forest it occurs on non-flooded ground and is probably ecotypically different from the Amazonian populations (Kubitzki 1978). Perhaps at the time of fractioning of its area, this species was not yet a component of periodically flooded habitats.

Many Amazonian tree genera – those of the flooded forest and of terra firme, too – have geographical links either to the Rio Negro Basin or to lowland Guayana. Both these areas are well-known for their high degree of endemicity. As far as flooded habitats are concerned, many tree genera have species, which are exclusive to either the Upper Rio Negro region, or to Guayana, or to both regions, while their congeners are widely distributed in the Amazon Basin or beyond its limits. In many genera, such as *Macrolobium, Swartzia, Elizabetha, Henriquezia*, in genera of the *Guttiferae, Combretaceae, Euphorbiaceae, Ochnaceae*, etc. examples of this pattern can be found (Figs. 8 and 12 – 19). In other cases species are distributed in one of the two centres and in adjacent parts of Amazonia. *Exellodendron coriaceum* (Fig. 20), *Byrsonima leucophlebia* (Fig. 21), and *Croton maturensis* are species whose distribution includes lowland Guayana and eastern Amazonia. However, the most pronounced distributional relationships of tree species of flooded habitats (at least on nutrient-poor rivers) are with the Rio Negro basin, which is the region of the greatest species diversity and endemicity of Amazonia. Several of its elements have outliers beyond its limits, as is increasingly becoming clear through the progress in the floristic knowledge of Amazonia; *Haploclathra paniculata* (Fig. 29), *Roupala obtusata* (Fig. 27), *Aniba affinis* (Fig. 26), *Chaunochiton loranthoides* (Fig. 30), and *Parkia discolor* (incl. *auriculata*) can be adduced. The four last-mentioned species exemplify the floristic relationships between the sand strands of the Rio Negro area and those of the clear water rivers of eastern Amazonia between the Rio Nhamunda and Rio Tapajoz, to which Ducke (1913) has pointed first.

There is a concentration of endemic species in the Upper Rio Negro Basin including the northern adjacent region south of the Upper Río Orinoco, from the mouth of Rio Uaupes to Río Atabapo, Lower Río Ventuarí, Río Pacimoní and Río Pimichim. This region lies only 100 – 200 m above sea level and is a black water/white sand region with predominating igapó, caatinga and savannah vegetation. It comprises the Rio Negro-Refuge of Steyermark (1982), the Amazonas Savannahs Refuge of Huber (1982), and parts of the Imerí Refuge of Prance

(1973). Tree species of periodically inundated habitats endemic to this area include *Leopoldinia piassaba* (Fig. 22) and *L. maior, Mauritia carana, Mauritiella aculeata, Schistostemon oblongifolium,* the genus *Ramatuella, Ocotea esmeraldana* (Fig. 24), *Glandonia williamsii, Asteranthus brasiliensis, Henriquezia nitida, Vitex calothyrsa,* and several species of *Swartzia* and *Macrolobium*. Of the species mentioned those of *Schistostemon, Glandonia, Vitex,* and several members of the *Leguminosae* have buoyant diaspores.

Causes of endemism

The foregoing leads to the question of the nature and significance of the Rio Negro and Guayana relationship of the flora of inundated riverine forests, a relationship that is also true for the Amazonian forest flora on terra firme. Ducke & Black (1953) tried to explain the floristic richness of the Rio Negro region with its varied ecological conditions. However, this idea is not convincing since the soils of this black water and white sand region vary only as to humidity, while the trophic conditions of waters and soils uniformly lie at the lowermost possible level*.

Another hypothesis tries to relate the centres of biotic diversity of the Amazon Basin, of which the Upper Rio Negro centre is but one, to climatic oscillations of the Pleistocene (for summary, see Prance 1982). According to its proponents, in the humid tropics the glacial periods were arid phases, in which the biota of the rain forests could survive only in reduced areas in which precipitations were sufficient for maintaining them. These refuges were separated by areas where more arid conditions prevailed. It remains unclear whether the refuges acted as simple areas of survival and, after an increase of pluviosity, of re-dispersal, or as centres of speciation. Other hypotheses put emphasis on the actual environmental conditions. Morley (1975), after analyzing the distribution of the species of *Mouriri* (*Melastomataceae*) in Amazonia, found their areas largely in agreement with contemporary rainfall patterns. In a comparative analysis of the species diversity of a wide range of neotropical forest communities, Gentry (1982) demonstrated a close correlation between the amount of precipitation and species diversity.

The two basic attempts to explain species diversity, the historical and the actualistic, relate species diversity ultimately to precipitation, although the nature of this relationship remains dubious. If precipitation were directly responsible for biotic diversity, it would be difficult to understand why this should act both on terra firme species and the elements of periodically flooded habitats. One would expect the latter to depend primarily on edaphic humidity, so that in more arid phases these species could find suitable conditions of survival by simply following

* **Note added in proof.** In a recent study, H. Fölster & O. Huber (1984: Interrelaciones Suelos/vegetación en el Área de Galipero, Territorio Federal Amazonas, Venezuela. Series Informes Técnicos DGSIIA/IT/144) analyzed the interrelationship between soil moisture, nutrients and geomorphological traits in the Galipero region of southern Venezuela. They found the whole gradation from tall forest to open savannah growing upon nearly identical soils. However, slight variations in the depth of the water table combined with almost imperceptible differences of the nutrient reserve in the topsoil were found to determine the distribution of different plant communities. These findings accentuate the paradox that in nutrient-poor habitats in the humid tropics environmental uniformity can be linked with biotic diversity.

the displacement of river beds and their floodplains. Therefore I argue that the relationship between precipitation and biotic diversity must be rather an indirect one, possibly acting via trophic conditions (see also Fittkau 1973, Huston 1979). One could imagine that under poor nutrient supply the advantage of strong competitors would be so much reduced that many weak competitors could coexist with them.

The evolution of periodically flooded forests

Water has been of utmost importance for the shaping of Amazonia always during its geological past. The Amazon Graben system is older than the Atlantic Ocean, and the drainage flow of the "Ur Amazon" was directed into the Pacific Ocean. Only in the Cretaceous, with the break-up of Gondwanaland, the erosion basis of the Atlantic Ocean became available. However, only with the Andean orogenesis, strongly commencing during the Middle Miocene, and culminating in the Plio-/Pleistocene, drainage into the Pacific became impossible, and it was only then that by tapping the sub-Andean catchment the Amazon basin became the world's largest river system (Grabert 1983).

The long-lasting limnic-fluviatil history of the Amazon Basin is documented by the mighty clay and sand deposits known as the Alter do Chão Formation (Upper

Table 1. Species of the igapó (flooded forest on black water and clear water rivers)

Virola carinata Warb.
Acosmium nitens (Vog.) Yakovl.
Aldina latifolia Spruce ex Benth. var. *latifolia*
Cynometra spruceana Benth.
Crudia amazonica Spruce ex Benth.
Heterostemon mimosioides Desf.
Macrolobium multijugum (DC.) Benth.
Ormosia excelsa Spruce ex Benth.
Parkia discolor Spruce ex Benth.
Peltogyne venosa subsp. *densiflora* (Spruce ex Benth.) Freitas da Silva
Pentaclethra macroloba (Willd.) Kuntze
Swartzia argentea Spruce ex Benth.
S. laevicarpa Amsh.
Tachigalia paniculata Aubl.
Couepia paraensis (Mart. & Zucc.) Benth.
Licania apetala (E. May.) Fritsch
L. heteromorpha Benth.
L. macrophylla Benth.
Burdachia prismatocarpa Adr. Juss.
Clonodia racemosa (Adr. Juss.) Niedenzu
Lophanthera longifolia (H. B. K.) Griseb.
Schistostemon macrophyllum (Benth.) Cuatrec.
Panopsis rubescens (Pohl) Pittier var. *rubescens*
Roupala obtusata Klotzsch
Qualea retusa Spruce ex Benth.
Erisma calcaratum (Link) Warming
Mollia lepidota Spruce ex Benth.
Eugenia inundata DC.
Buchenavia suaveolens Eichl.
B. reticulata Eichl.
Eschweilera coriacea (DC.) Berg
Amphirrhox surinamensis Eichl.
Elvasia calophyllea DC.
Blastemanthus gemmiflorus (Mart. & Zucc.) Planch.
Quiina rhytidopus Tul.
Calophyllum brasiliense Camb.
Laetia suaveolens (Poepp.) Benth.
Chaunochiton loranthoides Benth.
Lophostoma ovatum Meissn.
Pera distichophylla (Mart.) Baill.
Micrandra siphonoides Müll.-Arg.
Neoxythece elegans (A. DC.) Aubrév.
Malouetia tamaquarina (Aubl.) A. DC.
Stachyarrhena spicata Hook. f.
Tabebuia barbata (E. Mey.) Sandw.
Astrocaryum jauari Mart.
Leopoldinia pulchra Mart.
L. piassaba Wallace
Mauritia aculeata H. B. K.
M. carana Wallace

Cretaceous?, up to 600 m thick), the Solimões Formation (Eocene to Miocene), and the Barreiras layers (Mio-/Pliocene) (Grabert 1983, Putzer 1984). There is no doubt that conditions for the existence of a riverine and amphibic vegetation had prevailed for a long time. We do not know exactly how long the periodic flooding cycles date back. However, since the Amazon Basin extends over a distance of more than 15° of latitude, and since most of its precipitation falls during "high sun" periods, differences between the moister and drier seasons may have existed all the time. Only during the Pleistocene, repeated changes of precipitation and sea level as a consequence of glacial and interglacial periods, must have had an impact on the hydrographic situation, and the duration and extent of annual floods must have been subject to great fluctuations.

The limnic and fluviatil deposits of the Cretaceous and Tertiary of Amazonia indicate a nutrient-poor environment for the time of their deposition. In contrast, the turbid and nutrient-rich white water, originating from the erosion of the Andes, is geologically a younger phenomenon. This implies a younger age of the floristic differentiation of várzea elements in contradistinction to those of black and clear water igapós. This is in contrast to Grabert's (1984) hypothesis of a holocenic age of the black water region of the Rio Negro. Thus the várzea forests of the Amazon region may date back to the very beginnings of this river system in its present form, i.e., to the Pliocene or even Miocene, while igapó forests must have an older age. The evolution of tolerance for inundation certainly required only a short period, at least in those families that are "programmed" for flooding resistance. The evolution of structural traits, such as those that keep diaspores afloat, may have needed a longer period. This is much more true for the complex mutual dependences that have developed between the fish fauna and the flooded forests.

Table 2. Species of the várzea (flooded forest on white water rivers)

Nectandra amazonum Nees
Virola surinamensis (Rol.) Warb.
Triplaris americana L.
Andira inermis (Sw.) H. B. K.
Cassia grandis L. f.
C. leiandra Benth.
Erythrina glauca Willd.
Lecontea amazonica Ducke
Mora paraensis Ducke
Platymiscium ulei Harms
Pithecellobium niopoides Spruce ex Benth.
Pterocarpus amazonicus J. Hub.
Cecropia spp.
Maquira calophylla (Poepp. & Endl.) C. C. Berg
M. coriacea (Karst.) C. C. Berg
Carapa guianensis Aubl.
Spondias lutea L.
Guazuma ulmifolia Lam.
Sterculia elata Ducke
Ceiba pentandra Gärtn.
Ochroma pyramidale (Cav.) Urb.
Pseudobombax munguba (Mart. & Zucc.) Dugand
Crataeva benthamii Eichl.
Couroupita subsessilis Pilg.
Gustavia augusta L.
Piranhea trifoliata Baill.
Hura crepitans Müll. Arg.
Hevea spruceana Müll. Arg.
Sapium lanceolatum (Müll. Arg.) Hub.
Alchornea castaneifolia (Willd.) Juss.
Bothriospora corymbosa Hook. f.
Calycophyllum spruceanum (Benth.) Hook. f.
Cordia tetrandra Aubl.
Astrocaryum murumuru Mart.
Scheelea martiana Burret.
Montrichardia arborescens Schott

The feeding preferences of fish populations and their migrations into the flooded forests at the time of fruiting contribute to the continuous existence not only of the fish populations but also of the plant populations involved (GOULDING 1980). Hence it appears that these co-evolved systems must be unusually sensitive against the effects of technical civilisation.

The distribution maps of this paper are based on recent monographs, supplemented by data from herbarium material seen or collected by the author. Field work in Amazonia would have been impossible without the support received from the scientists and field assistents of the Instituto Nacional de Pesquisas da Amazônia, Manaus, the Museu Goeldi, Belém, the representation of Ministério de Agricultura at Belém, and the Campus Avançado at Oriximiná of Universidade Federal Fluminense, Niteroi, Brazil. Between 1971 and 1984 the Deutsche Forschungsgemeinschaft has generously provided various travel grants. W. R. ANDERSON, R. COWAN, MARY ENDRESS, G. T. PRANCE, J. ROHWER, C. A. STACE, S. RENNER kindly provided critical plant determinations, and J. M. PIRES, W. A. RODRIGUES, and H.-H. POPPENDIECK gave valuable comments on the manuscript.

References

ADIS, J., 1984: "Seasonal igapó"-forests of Central Amazonian black-water rivers and their terrestrial arthropod fauna. – See SIOLI 1984, pp. 245–268.

ANDERSON, W. R., 1981: *Malpighiaceae*. – Mem. New York Bot. Gard. **32**: 21–305.

BERG, C. C., 1972: *Olmedieae, Brosimeae*. – Flora Neotropica Mongr. **7**. – New York: Hafner.

BRANDBYGE, J., 1986: A revision of the genus *Triplaris* (*Polygonaceae*). – Nordic J. Bot. **6**: 545–570.

CAVALCANTE, P. B., 1983: Revisão taxonômica do gênero *Simaba* AUBL. (*Simaroubaceae*) na America do Sul. – Publ. Avuls. Museu Goeldi, Belem.

COWAN, R. S., 1953: A taxonomic revision of the genus *Macrolobium*. – Mem. New York Bot. Gard. **8/4**: 257–342.

– 1968: Flora Neotropica monogr. **1**: *Swartzia*. – New York: Hafner.

– 1976: A taxonomic revision of *Elizabetha* (*Leg. Caes.*). – Proc. Kon. Nederl. Akad. Wet. **C 79**: 323–346.

CUATRECASAS, J., 1961: A taxonomic revision of the *Humiriaceae*. – Contr. U.S. Herb. **35**: 2.

DUCKE, A., 1913: Explorações scientíficas no Estado do Pará. – Bol. Museu Goeldi **7**: 100–198, 27 t.

– 1930: Plantes nouvelles ou peu connues de la région amazonienne (4e série). – Arch. Jardim Bot. Rio de Janeiro **5**: 101–187.

– 1949: Arvores amazônicas e sua propagação. – Bol. Museu Paraense E. Goeldi **10**: 81–92.

– BLACK, G. A., 1953: Phytogeographical notes on the Brazilian Amazon. – Anais Acad. Brasil. Ciênc. **25**: 1–46.

DWYER, J. D., 1958: The New World species of *Cynometra*. – Ann. Missouri Bot. Gard. **45**: 313–345.

ENGLER, A., 1905: Über floristische Verwandtschaft zwischen dem tropischen Afrika und Amerika, sowie über die Annahme eines versunkenen brasilianisch-äthiopischen Continents. – Sitz. Ber. Kgl. Preuss. Akad. Wiss., Phys.-Math. Cl., VI 1905: 180–231.

EXELL, A. W., STACE, C. A., 1963: A revision of the genera *Buchenavia* and *Ramatuella*. – Bull. Brit. Mus. (Nat. Hist.) London Bot. **3**/1.

FITTKAU, E. J., 1973: Artenmannigfaltigkeit amazonischer Lebensräume aus ökologischer Sicht. – Amazoniana (Kiel) **4**: 321–340.

GENTRY, A. H., 1982: Patterns of neotropical plant species diversity. – Evol. Biol. **15**: 1–83.
GESSNER, F., 1968: Zur ökologischen Problematik der Überschwemmungswälder des Amazonas. – Intern. Revue Ges. Hydrobiol. **53**: 525–547.
GOULDING, M., 1980: The fishes and the forest. – Berkeley: University of California Press.
GOUROU, P., 1950: Observações geográficas na Amazônia. – Rev. Bras. Geogr. **11**: 335–408.
GRABERT, H., 1983: Das Amazonas-Entwässerungssystem in Raum und Zeit. – Geolog. Rundschau **72**: 671–683.
– 1984: Mögliche Wanderwege und Phylogonie der südamerikanischen *Iniidae* (*Cetacea, Mammalia*). – Amazoniana (Kiel) **8**: 365–374.
HUBER, J., 1910: Mattas e madeiras amazonicas. – Bol. Museu Goeldi **6**: 91–225.
HUBER, O., 1982: Significance of savanna vegetation in the Amazon territory of Venezuela. – See PRANCE 1982, pp. 221–244.
HUECK, K., 1966: Die Wälder Südamerikas. – Stuttgart: G. Fischer.
HUSTON, M., 1979: A general hypothesis of species diversity. – Amer. Naturalist **113**: 81–101.
JUNK, W., 1984: Ecology of the várzea, floodplain of Amazonian white-water rivers. – See SIOLI 1984, pp. 215–243.
KEEL, S. H. K., PRANCE, G. T., 1979: Studies of the vegetation of a white-sand blackwater igapó (Rio Negro, Brazil). – Acta Amazonica **9**: 645–655.
KUBITZKI, K., 1978: *Caraipa* and *Mahurea* (*Bonnetiaceae*). Mem. New York Bot. Gard. **29**: 82–138.
LIMA, D. DE A., 1966: Contribuição ao estudo do paralelismo da flora amazônico-nordestina. – Inst. Pesqu. Agron. Pernambuco, Nov. Ser., Bol. tecn. no. **19**. Recife.
MORLEY, T., 1975: The South American distribution of the *Memecyleae* (*Melastomataceae*) in relation to the Guiana area and to the question of forest refuges in Amazonia. – Phytologia **31**: 279–298.
NELSON, B. W., CID FERREIRA, C. A., TODZIA, C. A., ZARUCCHI, J. L., HILL, S. R., 1987: Distributional and ecological notes on *Polygonanthus amazonicus* DUCKE. – Acta Amazonica **15**/1–2 Supl.: 63–69.
PIRES, J. M., 1973: Tipos de vegetação da Amazônia. – Publ. Avulsas Museu Goeldi **20**: 179–202.
PRANCE, G. T., 1972: Flora Neotropica monogr. **7**. *Chrysobalanaceae*. – New York: Hafner.
– 1973: Phytogeographic support to the theory of Pleistocene forest refuges in the Amazon basin. – Acta Amazonica **3**: 5–28.
– 1979: Notes on the vegetation of Amazonia. 3. The terminology of Amazonian forest types subject to inundation. – Brittonia **31**: 26–38.
– (Ed.), 1982: Biological diversification in the tropics. – New York: Columbia University Press.
– MORI, S. A., 1980: *Anthodiscus* (*Caryocaraceae*), um gênero disjunto entre a Amazônia e o leste do Brasil. – Rev. Brasil. Bot. **3**: 63–65.
PUTZER, H., 1984: The geological evolution of the Amazon basin and its mineral resources. – See SIOLI 1984, pp. 15–46.
RODRIGUES, W., 1961: Estudo preliminar de mata de várzea alta de uma ilha do baixo Rio Negro de sólo argiloso e úmido. – Inst. Nac. Pesqu. da Amazônia, Botânica, Publ. no. **10**. Manaus.
ROGERS, G. K., 1984: Flora Neotropica Monogr. **39**. *Gleasonia, Henriquezia*, and *Platycarpum* (*Rubiaceae*). – New York: New York Bot. Gard.
SCHOLANDER, P. F., OLIVEIRA PEREZ, M., 1968: Sap tension in flooded trees and bushes of the Amazon. – Plant Physiol. **43**: 1870–1873.

SINGER, R., 1984: The role of fungi in Amazonian forests and in reforestation. – See SIOLI 1984, pp. 603–614.

SIOLI, H., 1954: Beiträge zur regionalen Limnologie des Amazonasgebietes 2. Der Rio Arapiuns. – Arch. Hydrobiol. **49**: 448–518.

– (Ed.)., 1984: The Amazon. – Monogr. Biol. **56**. – Dordrecht: Dr. W. Junk.

SLEUMER, H., 1954: *Proteaceae* americanae. – Bot. Jahrb. **76**: 139–211.

– 1980: Flora Neotropica Monogr. **22.** *Flacourtiaceae.* – New York: New York Botanical Garden.

– 1984: Flora Neotropica Monogr. **38.** *Olacaceae.* – New York: New York Botanical Garden.

SPRUCE, R., 1908: Notes of a botanist on the Amazon and Andes. 2 vols. – London: Macmillan (reprint New York: Johnson 1970).

STEYERMARK, J. A., 1982: Relationships of some Venezuelan refuges with lowland tropical floras. – See PRANCE 1982, pp. 182–220.

TAKEUCHI, M., 1962: The structure of the Amazonian vegetation 6. Igapó. – J. Fac. Sci. Univ. Tokyo Sect. 3 (Botany) **3**: 297–304.

THOMAS, W. W., 1984: A new species of *Simaba* (*Simaroubaceae*) from Pará, Brazil, with a key to the species north of the Amazon River. – Brittonia **36**: 244–247.

THORNE, R. F., 1973: Floristic relationships between tropical Africa and tropical America. – In MEGGERS, B. J. & al. (Eds.): Tropical forest ecosystems in Africa and South America: a comparative review, pp. 27–47. – Washington, D.C.: Smithsonian Institution Press.

WORBES, M., 1983: Vegetationskundliche Untersuchungen zweier Überschwemmungswälder in Zentralamazonien – vorläufige Ergebnisse. – Amazoniana (Kiel) **8**: 47–65.

Address of the author: Prof. Dr K. KUBITZKI, Institut für Allgemeine Botanik und Botanischer Garten, Ohnhorststrasse 18, D-2000 Hamburg 52, Federal Republic of Germany.

Pl. Syst. Evol. 162, 305–314

Plant Systematics and Evolution

Chorology of the Euxinian and Hyrcanian element in the woody flora of Asia

Kazimierz Browicz

Received January 21, 1987

Key words: Flora of SW. Asia; Euxinian and Hyrcanian Euro-Siberian region, floristic provinces; trees and shrubs.

Abstract: The Euxinian and the Hyrcanian floristic province are analysed in respect to their trees and shrubs. These provinces mark the southern limit of the Euro-Siberian Region in SW. Asia. Mesophyllic forests dominate; they are deciduous, but there is a substantial component of evergreen shrubs in the understorey. Characteristics species frequently have a relic nature. Species lists and some exemplary distribution maps are presented for the Euxinian and the Hyrcanian element, both with endemic and more wide-spread species, including the characteristic Euxine-Hyrcanian group.

The southern limit of the Euro-Siberian region coincides with the southern coastal areas of the Black and Caspian Seas. The mesophyllic vegetation of these areas in the south borders on the xerophytic vegetation of the Irano-Turanian region. From this they are sharply separated by sometimes very high mountain ranges. Only in NW. Anatolia these mountains are much lower and less compact, so that the limit between the Euro-Siberian and the Irano-Turanian regions is more diffuse there.

The western area near the Black Sea is referred to as the Euxinian province, whereas the eastern, near the Caspian Sea, as the Hyrcanian province; they are separated by the Caucasus massif. Sometimes these two provinces are treated jointly as the Euxine-Hyrcanian province, with two subprovinces, the Euxinian and the Hyrcanian. Within Eurasia they represent one of the very characteristic relic refugia for the mesophyllic tree and shrub flora. In spite of considerable similarities, the two provinces (or subprovinces) clearly differ from each other in their floristic composition.

The **Euxinian province** was first described as such by Gajewski (1937). He was using the nomenclature adopted from Braun-Blanquet (1928) calling it a "do maine" and understood it in a very wide sence, including both the Crimea and the Novorossijsk region on the Black Sea. Within it Gajewski (1937) recognized a Hyrcanian sector. Independently, Maleev (1940) limited the Euxinian province to the coastal areas of the Black Sea and considered the vicinity of Varna in Bulgaria as its northern limit in Europe. Both, Gajewski and Maleev, included the province into the Mediterranean region, while Meusel & al. (1965) placed it into the Sub-Mediterranean subregion. As regards the northern limit of the Euxinian province, various opinions were expressed, e.g., by Davis (1971) who includes in it besides the Crimea also the Dobrudzha, a delimitation initially also accepted by Takhtajan (1974). Today, however, following numerous investigations of the area (among others Czeczott 1937, Davis 1965, Zohary 1973, Doluchanov 1980, Browicz 1982–1988) the range of the Euxinian province is sufficiently well established. It covers in Europe the southeastern Black Sea area, in Bulgaria approximately from the Ropotamo valley (or else from the Burgas reservoir), eastern European Turkey (Istranca Daglari), northern Anatolia and the Western Caucasus (Adhsaria, district Kutaisi, Abkhazja, coastal part of Krasnodar district), with its northern limit near Tuapse (Takhtajan 1978). The eastern limit in the Caucasus is not yet exact known.

Davis (1971) divides the Euxinian province (in Turkey only; but I believe this applies to the whole province) into a western and an eastern part (the latter called Colchic sector), placing their boundary in northern Anatolia along Melet valley near Ordu. The basis for this division is the substantial increase in the density of Euxinian species (Fig. 1) and in the total precipitation towards the east.

It is a characteristic feature of the Euxinian province that it is covered by extensive mesophyllic forests, primarily broad-leaved forests. Among them, a dominant role is played by beech woods (*Fagus orientalis*) with an undergrowth of evergreen shrubs, such as *Rhododendron ponticum, Laurocerasus officinalis, Ilex colchica, Daphne pontica, Hedera helix* and *H. colchica*. In the eastern Colchic sector, coniferous forests with *Abies nordmanniana* subsp. *nordmanniana* and *Picea orientalis* are a substantial component.

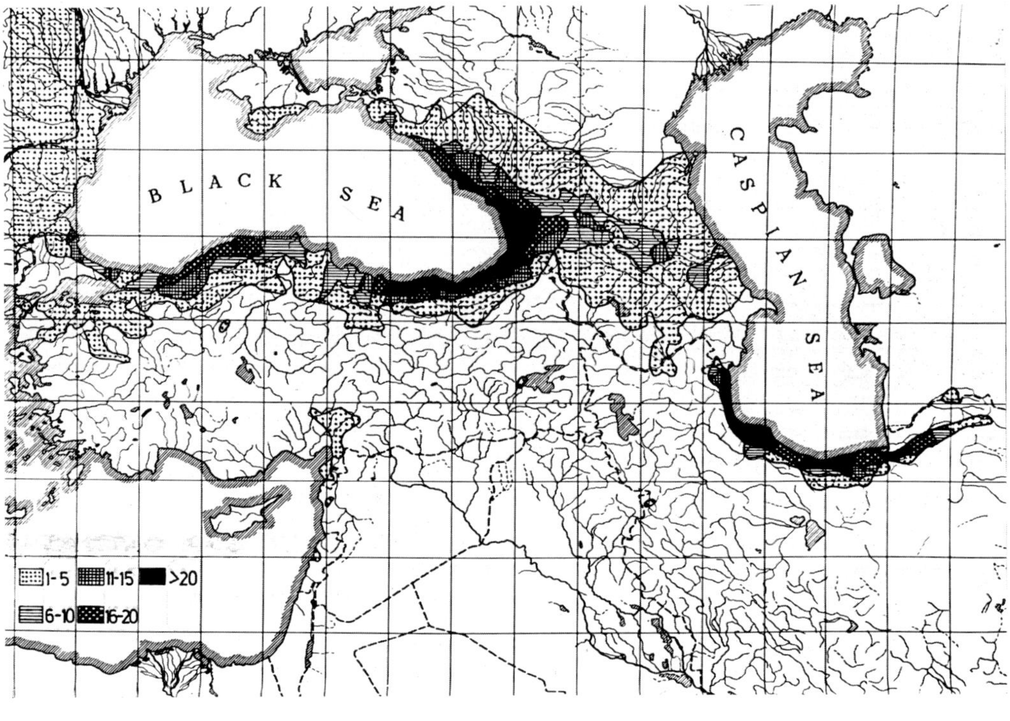

Fig. 1. Density of woody Euxinian and Hyrcanian species

The number of Euxinian species sensu stricto is not fully established yet. To a large extent this is due to the inadequate knowledge of the species ranges or else, to systematic problems. An example is the *Buxus sempervirens* aggregate which is considered to be represented in the Colchic sector by an independent species, *B. colchica*, and in the Hyrcanian province by *B. hyrcana*, species not yet generally recognized. Strictly Euxinian species of trees and shrubs approximately number bout 30, among them a large number (15) of evergreens (marked by *).

Euxinian species: *Abies nordmanniana* (STEV.) SPACH subsp. *nordmanniana**, *Acer trauttveteri* MEDV., *Betula medwedewii* REGEL, *Buxus colchica* POJARK.*, *Daphne glomerata* LAM.*, *Epigaea gaultherioides* (BOISS. & BAL.) TAKHT.*, *Hedera colchica* (K. KOCH) K. KOCH*, *Hypericum calycinum* L.*, *H. xylosteifolium* (SPACH) N. ROBSON, *Ilex colchica* POJARK*, *Osmanthus decorus* (BOISS. & BAL.) KASAPLIGIL*, *Philadelphus caucasicus* KOEHNE, *Picea orientalis* (L.) LINK*, *Pyracantha coccinea* (L.) M. ROEMER*, *Quercus hartwissiana* STEV., *Q. pontica* K. KOCH, *Rhamnus imeretinus* BOOTH, *R. microcarpus* BOISS., *Rhododendron caucasicum* PALLAS*, *R. luteum* SWEET, *R. ponticum* L.*, *R. smirnowii* TRAUTV.*, *R. ungerni* TRAUTV.*, *Rhodothamnus sessilifolius* P. H. DAVIS*, *Staphylea colchica* STEV., *S. pinnata* L., *Sorbus colchica* ZINSERL., and *Viburnum orientale* PALLAS.

The ranges of these Euxinian species coincide with the limits of the province or else (when they also occur in neighbouring regions) are centred there. Relevant and examples of such Euxinian species are, e.g., *Rhododendron luteum*, widely distributed throughout the province, but also in substantial parts of Ukraine and Bialorussia, and extending to Poland and Yugoslavia. Similar is *Staphylea pinnata* whose range mainly occupies the Balkan peninsula, but in the north reaches to southeastern Poland. Finally, *Hypericum androsemum* L. has the major part of its range beyond the Balkan gap, extending into western Europe. Furthermore, there are species not included in the Euxinian list which most probably should be considered as such, e.g., *Castanea sativa* MILLER (very rarely also in the Hyrcanian province) or *Lonicera caucasica* PALLAS.

The Euxinian species can be divided into several groups: First, there are species endemic to the province. Within that group only two species occur almost throughout the province. These are *Quercus hartwissiana* and *Acer trauttveteri*. The others are associated primarily with the Colchic sector: *Abies nordmanniana* subsp. *nordmanniana, Betula medwedewii, Buxus colchica, Daphne glomerata, Hedera colchica, Hypericum xylosteifolium, Philadelphus coronarius, Rhamnus imeretinus, R. microcarpus, Rhododendron caucasicum, Sorbus colchica, Staphylea colchica,* and *Viburnum orientale*.

This Colchic group includes some very localized endemics, such as *Quercus pontica, Epigaea gaulterioides, Rhododendron ungerni* and *Rh. smirnovii, Rhodothamnus sessilifolius* and *Osmanthus decorus* (Fig. 2). In the western part of the Euxinian province there is only one endemic species, i.e., *Hypericum calycinum* (Fig. 3).

Furthermore, there are species extending the second beyond the Euxinian province. Some of these taxa reach far removed areas, e.g., *Rhododendron ponticum*, one of the most important species of the understorey of Euxinian province forests; it has isolated populations in the Lebanon and in the southwestern Iberian Peninsula. *Ilex colchica* (Fig. 4) is also known from southern Turkey (Amanus Mts), and *Pyracantha coccinea* also from Greece, Yugoslavia, and Albania. *Rhododendron luteum* and *Staphylea pinnate* have already been mentioned.

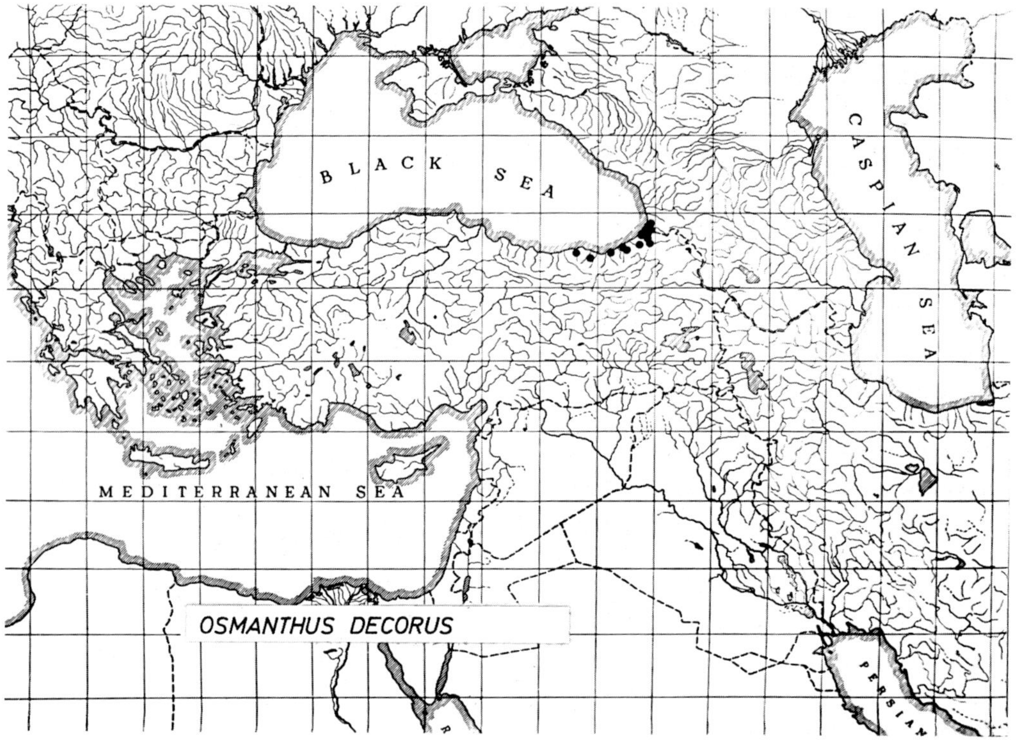

Fig. 2. The range of *Osmanthus decorus*, endemic to the eastern part of the Euxinian province

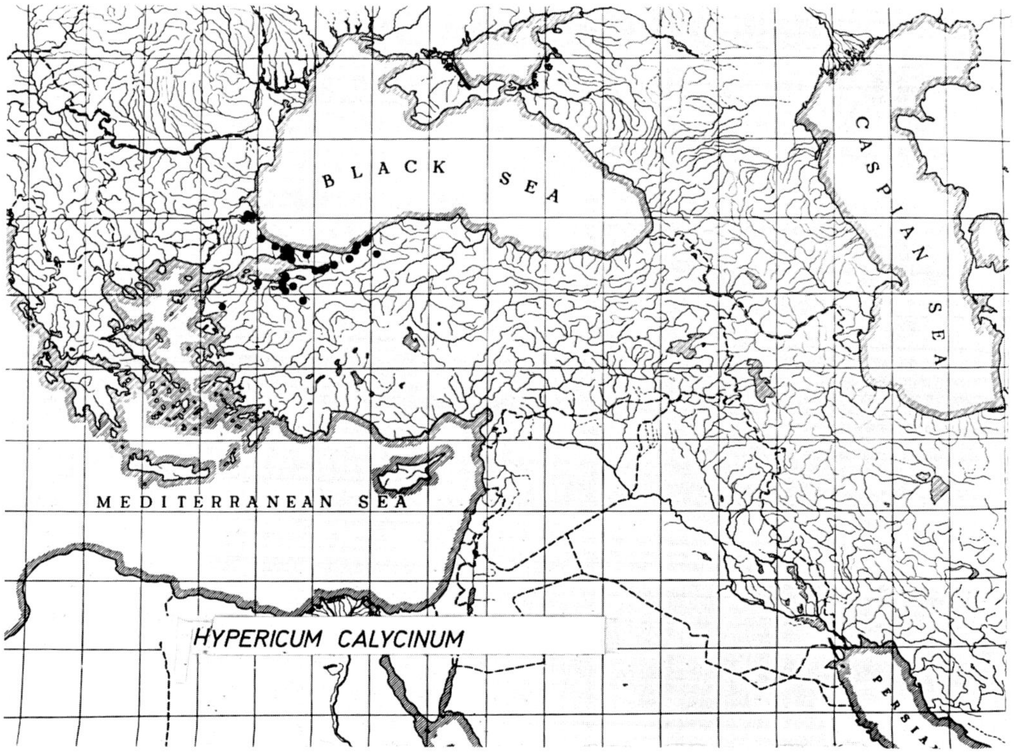

Fig. 3. The range of *Hypericum calycinum*, endemic to the western part of the Euxinian province

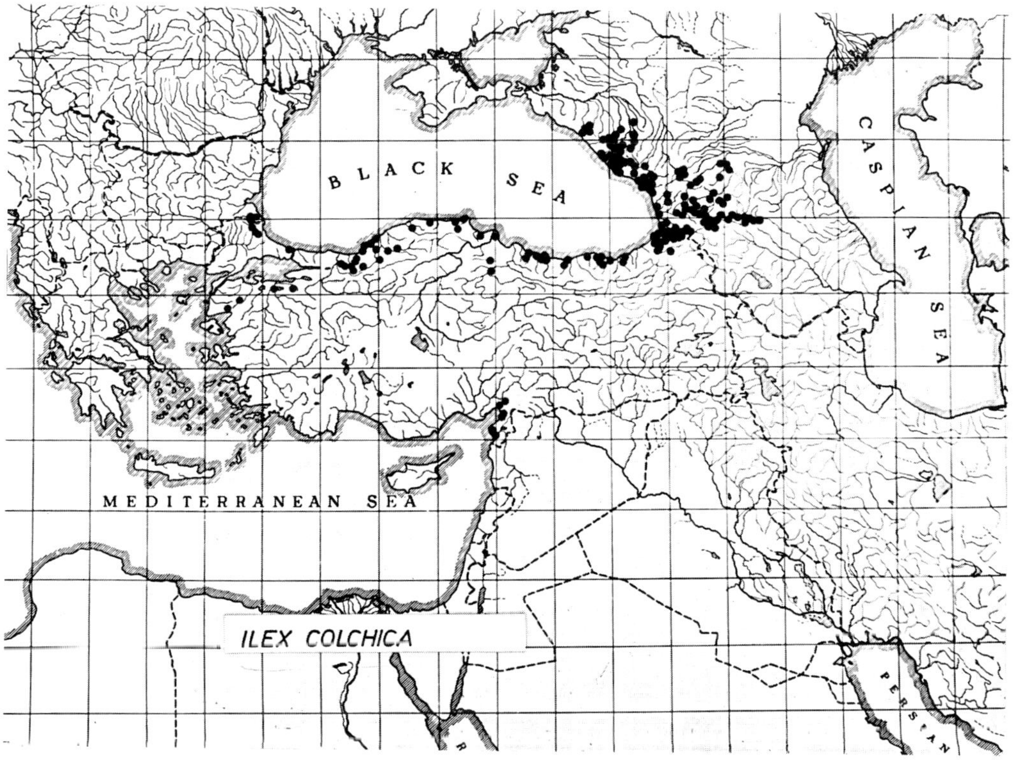

Fig. 4. The range of *Ilex colchica*, an Euxinian species also known from southern Turkey in the Amanus Mts

Species occurring in the Euxinian and the Hyrcanian province usually are being referred to as Euxine-Hyrcanian and number 18, not including *Castanea sativa* and *Lonicera caucasica*.

Euxine-Hyrcanian species: *Acer cappadocicum* GLED., *Alnus barbata* C. MEYER [= *Alnus glutinosa* (L.) GAERTN. subsp. *barbata* YALTIRIK], *Andrachne colchica* FISCH. & C. MEYER (Fig. 5), *Crataegus microphylla* K. KOCH (Fig. 6), *C. pentagyna* WALDST. & KIT., *Daphne pontica* L., *Diospyros lotus* L., *Fagus orientalis* LIPSKY, *Hypericum androsaemum* L., *Jasminum officinale* L., *Laurocerasus officinalis* M. ROEMER, *Mespilus germanica* L., *Pterocarya fraxinifolia* (POIR.) SPACH, *Ribes bibersteinii* BERL. ex DC., *Sorbus subfusca* (LEDEB.) BOISS., *Smilax excelsa* L., *Vaccinium arctostaphylos* L., and *Zelkova carpinifolia* (PALLAS.) K. KOCH.

Some of these are rather rare in the Hyrcanian province, with the majority of their populations in the Euxinian area, e.g., *Andrachne colchica, Daphne pontica, Vaccinium artctostaphylos, Sorbus subfusca,* or *Ribes bibersteinii*. Widely distributed in both provinces as well as in the Caucasus are *Fagus orientalis, Alnus barbata, Crataegus pentagyna, Smilax excelsa* and, to a lesser degree, also *Pterocarya fraxinifolia*.

In the Euxinian province the density of Euxinian and Euxine-Hyrcanian species (Fig. 1) is largest in the western Caucasus and in northeastern Anatolia about 38° E. Long. in the west), where we find from 20 to 35 of these species. West of that line there is a distinct decline, with a minimum in the approximate region between Samsun and Sinop. This region is an obvious enclave of Mediterranean species in

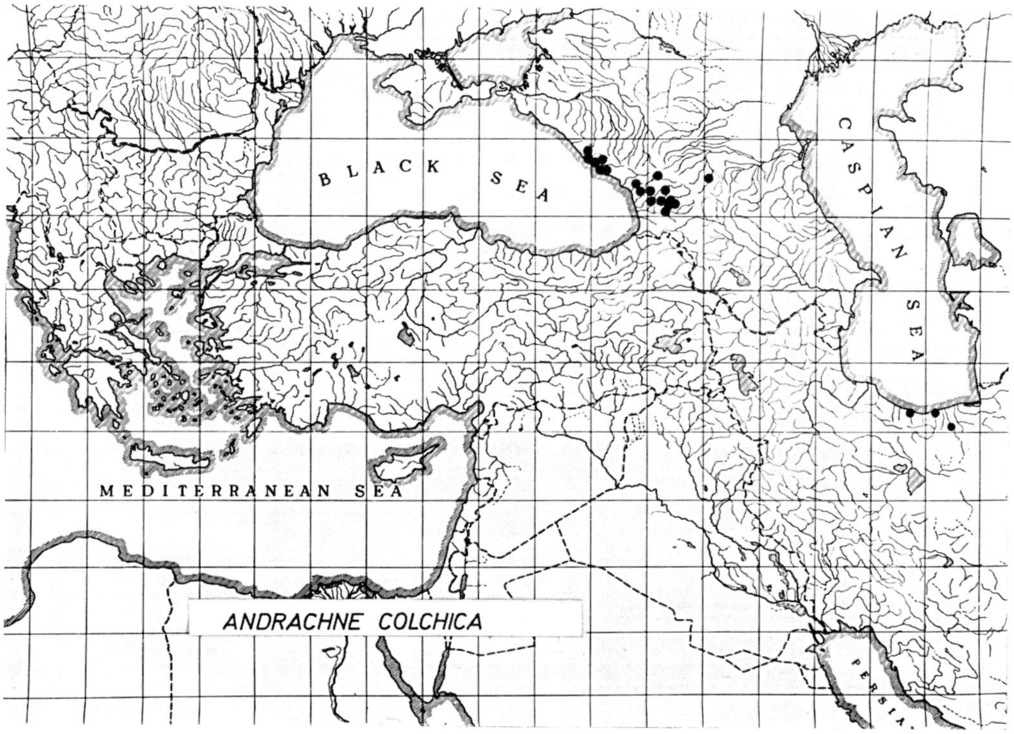

Fig. 5. The range of *Andrachne colchica*, an Euxine-Hyrcanian species, centred in the Euxinian province

northern Anatolia. However, there they do not (or only very locally) form the typical Mediterranean maquis and phyrgana formations. Further westward, the number of Euxinian species increases again but with numbers between 10 and 20 remains substantially lower than in the Colchic sector.

The Hyrcanian province is much smaller than the Euxinian province, but it is more compact and has a much more clearly defined southern boundry. It extends from the southeastern Sovjet Azerbajdzhan (Lenkoran, Talish) as a relatively narrow belt along the shores of the Caspian Sea and including its depressions, through the Iranian provinces of Gilan, Mazandaran and Gorgan to the western tip of the Khurasan province at about 56° E. Long. (where there is a National Park). Some species extend somewhat further eastwards, as far as Bojnurd in Iran and partly reappear in southwestern Turkmeniya (Kopet Dagh Mts), but they do not play a major role in the plant communities there.

Hyracanian species (* evergreen): *Acer velutinum* BOISS., *Albizzia julibrissin* DURAZZ., *Alnus subcordata* C. MEYER, *Buxus hyrcanus* POJARK*, *Danaë racemosa* (L.) MOENCH*, *Euonymus velutinus* FISCH & C. MEYER, *Frangula grandifolia* (FISCH. & C. MEYER) GRUBOV, *Gleditsia caspica* DESF., *Hedera pastuchowii* WORONOW*, *Ilex spinigera* (T. LOESNER) T. LOESNER*, *Parrotia persica* C. MEYER, *Populus caspica* BORNM., *Pyrus boissieriana* BUHSE, *Quercus castaneifolia* C. MEYER, and *Ruscus hyrcanus* WORONOW*.

The **Hyrcanian province**, just as the Euxinian, is characterised by the wide distribution of mesophyllic forests, but they are exlusively broad-leaved here, since

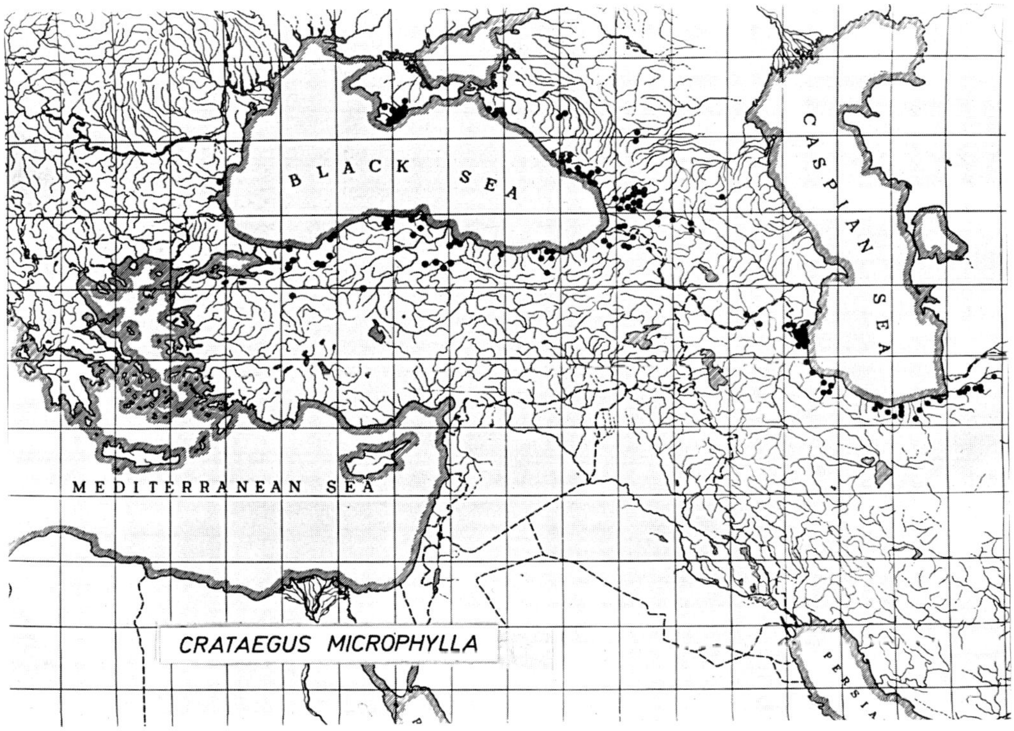

Fig. 6. The range of *Crataegus microphylla*, an Euxine-Hyrcanian species, widely distributed in both provinces

spruce and fir are missing. These forests are composed, besides *Fagus orientalis*, of such species as *Quercus castaneifolia, Alnus barbata* and *A. subcordata, Zelkova carpinifolia, Pterocarya fraxinifolia*, and *Parrotia persica*. Whereas the number of endemic species is clearly smaller in the Hyrcanian as compared with the Euxinian province, the prominence of mesophyllic Hyrcanian forests with their specific composition (including one endemic genus *Parrotia* C. MAYER) does under no circumstances allow its inclusion in the Irano-Turanian region (as proposed by TAKHTAJAN 1978). Local enclaves of Irano-Turanian and even Mediterranean species (e.g., *Cupressus sempervirens* L.), in the Hyrcanian province do not play a prominent role there. One should also remember the presence of characteristic evergreen species, such as *Hedera pastuchovii, Ilex spinigera, Ruscus hyrcanus, Daphne pontica* and *Laurocerasus officinalis*.

As the Euxinian, also the Hyrcanian species can be divided into groups. The first includes species endemic to the Hyrcanian province: *Alnus subcordata, Frangula grandifolia, Gledistia caspica, Ilex spinigera* (Fig. 7), *Populus caspica*, and *Ruscus hyrcanus*. Here, one could perhaps also include *Parrotia persica* and *Quercus castaneifolia* which, however, have also been found at the foothills of the Greater Caucasus in Azerbajdzhan (U.S.S.R.).

The second group includes species which extend only insignificantly beyond the limits of the Hyrcanian province, primarily in a westerly direction, i.e., to the eastern Caucasus.

Here, we can list *Acer velutinum, Buxus hyrcanus, Danaë racemosa, Euonymus*

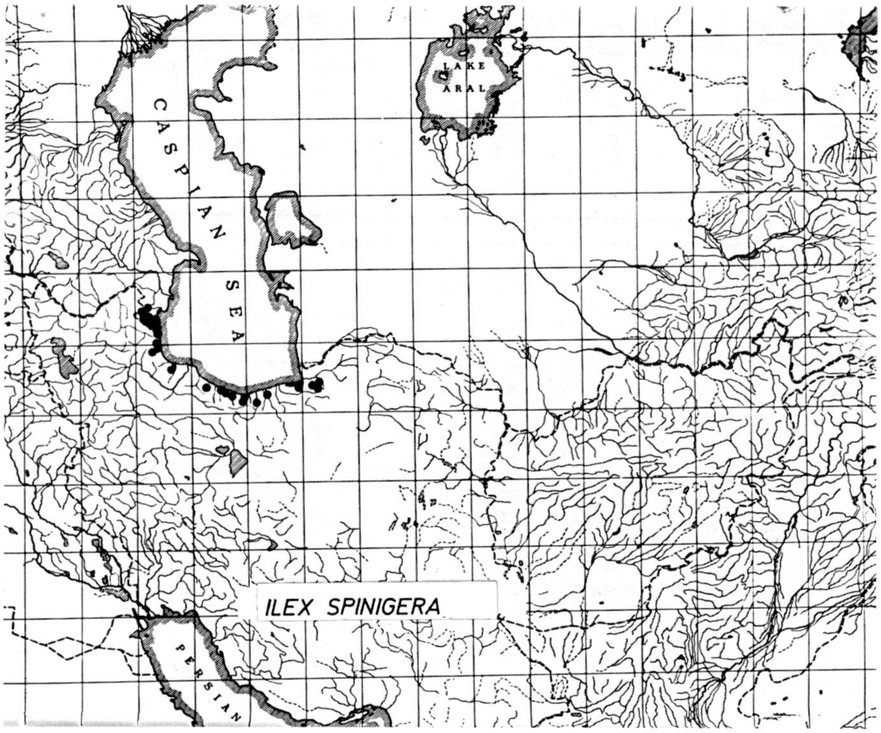

Fig. 7. The range of *Ilex spinigera*, endemic to the Hyrcanian province

velutinus, Hedera pastuchowii, and *Pyrus boissieriana* (grows in the Turkmenian Kopet-Dagh).

Danaë racemosa is characterized by a very unusual disjunction: besides the Hyrcanian province, it also grows in a distance of about 1 000 km and in isolated populations in the Amanus Mts of southern Turkey and northwestern Syria.

Among the more wide-spread species, we have already discussed the Euxine-Hyrcanian groups.

As with *Rhododendron luteum* and *Staphylea pinnata* in the Euxinian group, one could have doubts about *Albizzia julibrissin* being classified as Hyrcanian element. The major part of its range is situated in China, Korea and Japan, at a distance of about 6 000 km from the Caspian Sea.

The **relic nature** of the species of trees and shrubs belonging to the Euxinian and Hyrcanian element is obvious when one compares their systematic affinities and geographic links with other species.

Acer cappadocicum includes varieties (sometimes treated as species) which occur from NE. Afghanistan through the Himalayas to western and central China.

Albizzia julibrissin has the main part of its range in eastern Asia.

Alnus subcordata has closely related species in Europe (Corsica, Italy): *A. cordata* (LOISEL) LOISEL and northeastern Asia: *A. japonica* MIQ.

Betula medwedewii is a member of sect. *Costatae* (RGL.) KOEHNE which has other representatives in eastern Asia (the majority) and in North America.

Danaë racemosa is the only species of the genus *Danaë* MEDIKUS.

Diospyros lotus has the main part of its range in Central and southern China, but also grows in the western Himalayas and in the mountains of Uzbekistan and Tadzhikistan.

Epigaea gaultherioides is one of the three species of the genus *Epigaea* L., occurring in considerable isolation, with *E. asiatica* MAXIM. in Japan and *E. repens* L. in eastern N. America.

Gleditsia caspica has most closely related species in China and Japan (*G. japonica* MIQ. and *G. ferox* DESF.).

Hedera pastuchovii is closely related to the Japanese *H. tobleri* NAKAI.

Jasminum officinalis has the main part of its range from the Himalayas to China.

Laurocerasus officinalis is a species closely related to *L. lusitanica* ROEM. from SE. Europe and NW. Africa. The remaining species of this genus occur in S. Africa and SE. Asia.

Mespilus germanica is the only representative of the genus *Mespilus* L.

Osmanthus decorus is the only representative of the genus *Osmanthus* LOUR. in W. Asia. Other species grow in eastern and southern Asia.

Parrotia persica is an endemic species (and genus) of the Hyrcanian province. It is most closely related to the monotypic genus *Parrotiopsis* C. SCHNEIDER, occurring from eastern Afghanistan to India.

Pterocarya fraxinifolia is the only representative of this genus in western Asia. All remaining species occur in eastern Asia (China and Japan).

Pyracantha coccinea, all remaining species of the genus grow in Asia, from the Himalayas to China and Taiwan.

Pyrus boissieriana, all remaining species of sect. *Pashia* KOEHNE occur in the Himalayas and in E. Asia.

Rhodothamnus sessilifolius belongs to a genus with two species. The other, *Rh. chamaecistus* (L.) REICHB., occurs in the eastern Alps of Europe.

Vaccinium arctostaphylos has its most closely related sister species, *V. padifolium* J. E. SM., on Madeira.

Viburnum orientale is most closely related to *V. sargentii* KOEHNE and *V. trilobum* MARSH. which occur in NE. Asia and E. North America, respectively.

Zelkova carpinifolia belongs to a genus where remaining species grown in E. Asia, and one on Crete [*Z. abelicea* (LAM.) BOISS.].

Finally, one should note that in both provinces, the Euxinian and the Hyrcanian, there occur several Euro-Siberian species of trees and shrubs which are widely distributed and find their optimum conditions in northern and central Europe. Here, near the Black Sea, and to a somewhat lesser degree the Caspian Sea, they attain their southern limit. This group includes, among others: *Acer campestre* L., *Alnus glutinosa* (L.) GAERTN., *Berberis vulgaris* L., *Carpinus betulus* L., *Cerasus avium* (L.) MOENCH, *Cornus sanguinea* L., *Corylus avellana* L., *Daphne mezereum* L., *Euonymus europaeus* L., *Frangula alnus* L., *Fraxinus excelsior* L., *Juniperus communis* L., *Quercus petraea* (MATTUSCHKA) LIEBL., *Rhamnus catharticus* L., *Sorbus aucuparia* L., *Taxus baccata* L., and *Viburnum opulus* L.

Their presence and often substantial participation in Euxinian and Hyrcanian forest communities, gives evidence for the inclusion of these into the Euro-Siberian region.

References

BRAUN-BLANQUET, J., 1928: Pflanzensoziologie. – Berlin: J. Springer.

BROWICZ, K., 1982–1988: Chorology of trees and shrubs in South-West Asia and adjacent regions **1**–**6**. – Poznań: Polish Scientific Publishers.

CZECZOTT, H., 1937: The distribution of some species in Northern Asia Minor and the problem of Pontide. – Izv. Carsk. Prin. Inst. Sofia **10**: 43–68.

DAVIS, P. H., (Ed.)., 1965: Flora of Turkey **1**. – Edinburgh: University Press.

– 1971: Distribution patterns in Anatolia with particular reference to endemism. – In DAVIS, P. H., HARPER, P. C., HEDGE, I. C., (Eds.): Plant life of South-West Asia, pp. 15–27. – The Botanical Society of Edinburgh.

DOLUCHANOV, A. G., 1980: Kolchidskij podlesok. – Tbilisi.

GAJEWSKI, W., 1937: Elementy flory polskiego Podola. – Pl. Polon. **5**: 1–211.

MALEEV, V. P., 1940: Rastitelnos pričernomorskich stran (Evksinskaja provincija Sredizemnomorja) ee proischoždenie i svjazi. – Tr. Inst. Bot. AN S.S.S.R. ser. 3 (Geobotanika) **4**: 135–251.

MEUSEL, H., JÄGER, E., WEINERT, E., 1965: Vergleichende Chorologie der zentraleuropäischen Flora 1 (Text + Karten). – Jena: G. Fischer.

TAKHTAJAN, A. L., 1974: Florističeskoe delenie suši. – Žizn rastenij **1**: Moskva.

– 1978: Florističeskie oblasti zemli. – Leningrad: Nauka.

ZOHARY, M., 1973: Geobotanical foundation of the Middle East, 1–2. – Stuttgart: Gustav Fischer.

Address of the author: KAZIMIERZ BROWICZ, Institute of Dendrology, Polish Academy of Sciences, PL-62-035 Kórnik near Poznań, Poland.

Pl. Syst. Evol. 162, 315–329

—Plant—
Systematics
and
Evolution

Ecogeographical differentiation of the Submediterranean deciduous forest flora

H. Meusel and E. J. Jäger

Received February 18, 1987

Key words: Angiosperms, *Aceraceae, Betulaceae, Celastraceae, Fagaceae, Juglandaceae, Oleaceae, Rosaceae, Staphyleaceae.* – Summer-green woody plants, submediterranean, distribution types, climatic conditions, post-Pleistocene dispersal.

Abstract: The deciduous wood flora of southern Europe is characterized as a special "nemoral-Submediterranean element". The distribution patterns of some representatives (total ranges and local areas in Roumania) are described and explained by recent climatic conditions. The northern limits of Submediterranean and Submediterranean-middle European deciduous forest taxa exhibit a continuous gradation, but only a few species with a wide ecological amplitude extend into the temperate zone.

In the present paper the question is discussed, to what degree the distribution limits and the species richness of European broad-leaved forest plants are affected by recent ecological conditions or by Pleistocene glaciations. For this we rely on new distribution maps and on experience gained in compiling the "Comparative Chorology of the Central European Flora" (see Meusel & al. 1965, 1978, 1989).

The Submediterranean nemoral flora and its distribution types

The flora of the Submediterranean floristic subregion is composed of two very different floristic elements. Most species belong to the typical Mediterranean flora. They are representatives of families (e.g., *Globulariaceae, Cistaceae*) or genera (e.g., *Ophrys, Ononis, Medicago*) whose evolution centres lie in the Macaronesian-Mediterranean region. Their development and distribution depends on the Mediterranean (etesial) type of climate characterized by cool, humid winters and hot and dry summers. In addition to this Mediterranean element, there are in the northern zone of the Mediterranean region (northern parts of the peninsulas, northern Anatolia and the Caucasus) many representatives of the broad-leaved deciduous forests. These nemoral-Submediterranean plants are remnants from a rich forest vegetation of warm and humid periods of the Middle and Late Tertiary. At present, most of these taxa or their close relatives exhibit an abundant development in meridional-temperate East Asia.

Among nemoral-Submediterranean plants we find different characteristic distribution patterns. Only a few extend from the southern coasts of the Caspian Sea

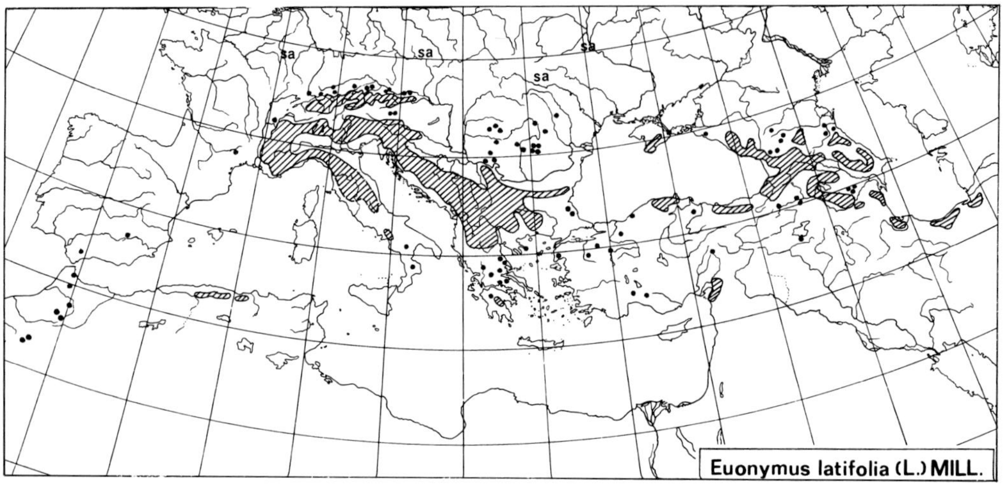

Fig. 1. Range of *Euonymus latifolia* (L.) MILL. (after MEUSEL & al. 1978, with additions)

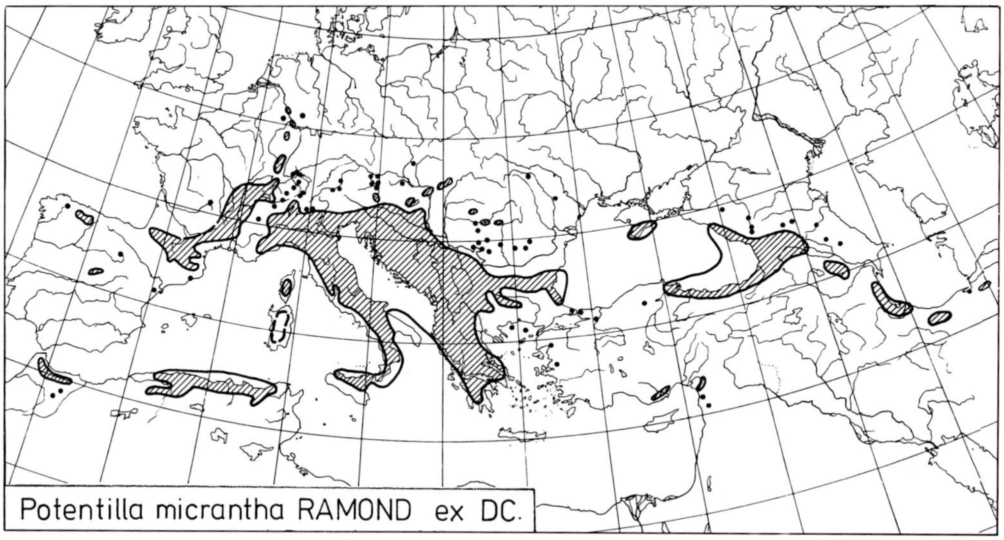

Fig. 2. Range of *Potentilla micrantha* RAM. ex DC. (after MEUSEL & al. 1965, with corrections)

in the East to the Iberian Peninsula and the Rif Mts in the West, where they are confined to relic niches. Examples of this *Euonymus latifolia*-type (Fig. 1) are *Acer* ser. *Opulifolia* (K 277 d), the *Acer monspessulanum*-group (K 277 C), and *Castanea sativa* (K 121 b)*.

Many of the nemoral-Submediterranean plants do not extend over the whole area of the *Euonyms latifolia*-type but are confined to special parts (sectors) of the Submediterranean territory. Of these, only a few occur in the western Submediterranean provinces (e.g., *Acer monspessulanum* subsp. *monspessulanum, A. his-*

* These numbers refer to distribution maps in MEUSEL & al. (1965, 1978, 1989).

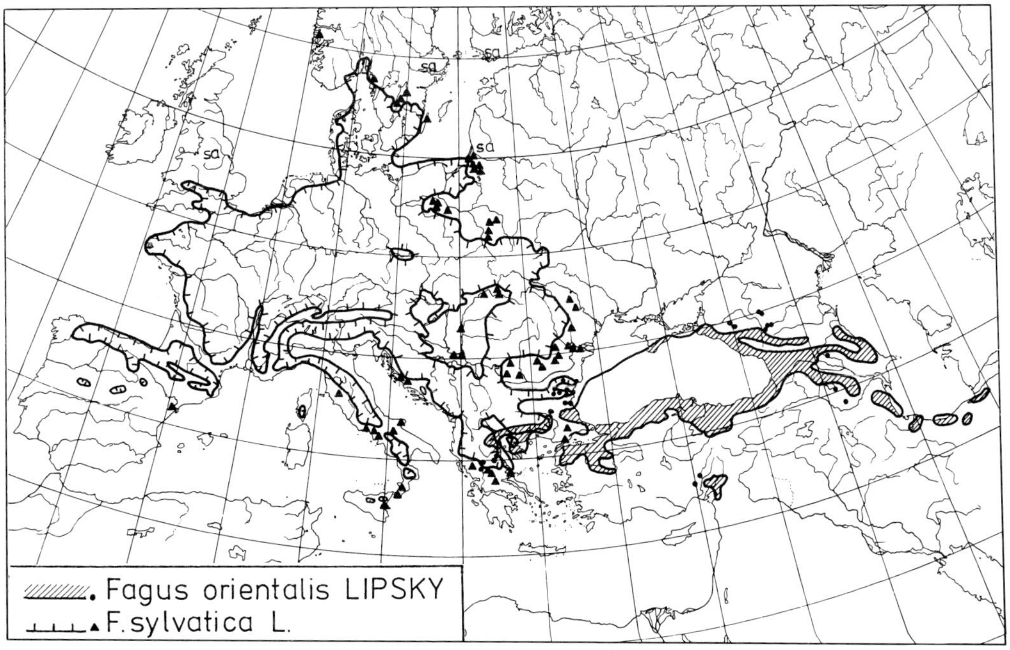

Fig. 3. Range of *Fagus orientalis* Lipsky and *F. sylvatica* L. (after Meusel & al. 1965, Browicz & Zielinski 1982, and Davis 1982)

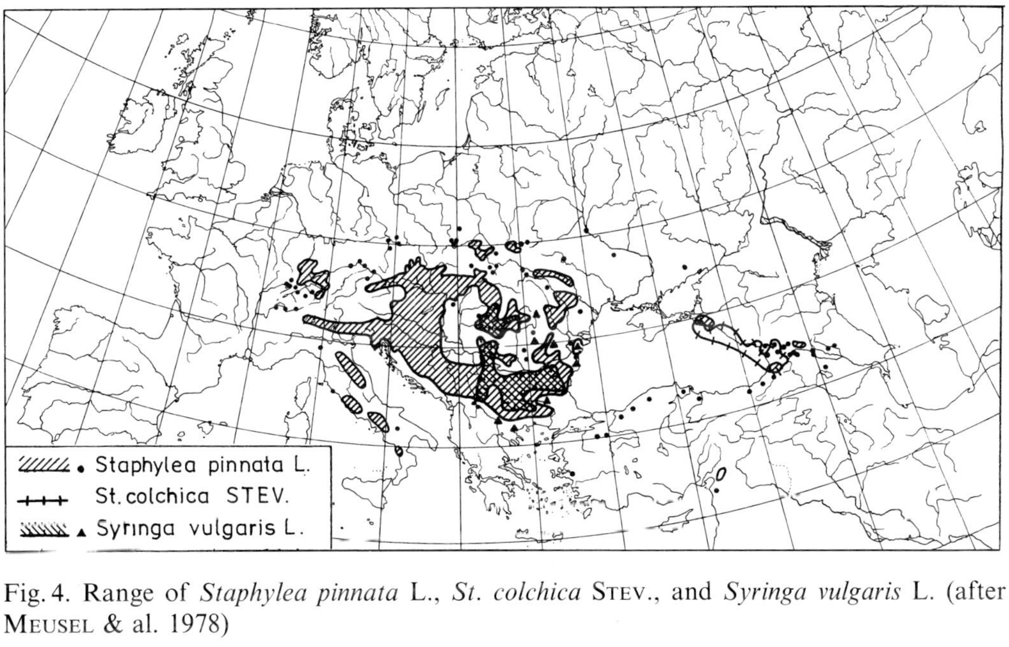

Fig. 4. Range of *Staphylea pinnata* L., *St. colchica* Stev., and *Syringa vulgaris* L. (after Meusel & al. 1978)

panicum, A. opulifolium, K 277 c + d). Evidently, the extremely oceanic climates are not favourable to the growth of summer-green broad-leaved trees. The yearly renewal of the foliage is more effective in suboceanic climates with cold winters (Jäger 1969). On the other hand, the evergreen laurophyllous Submediterranean relics have their best survival chances in the Atlantic West.

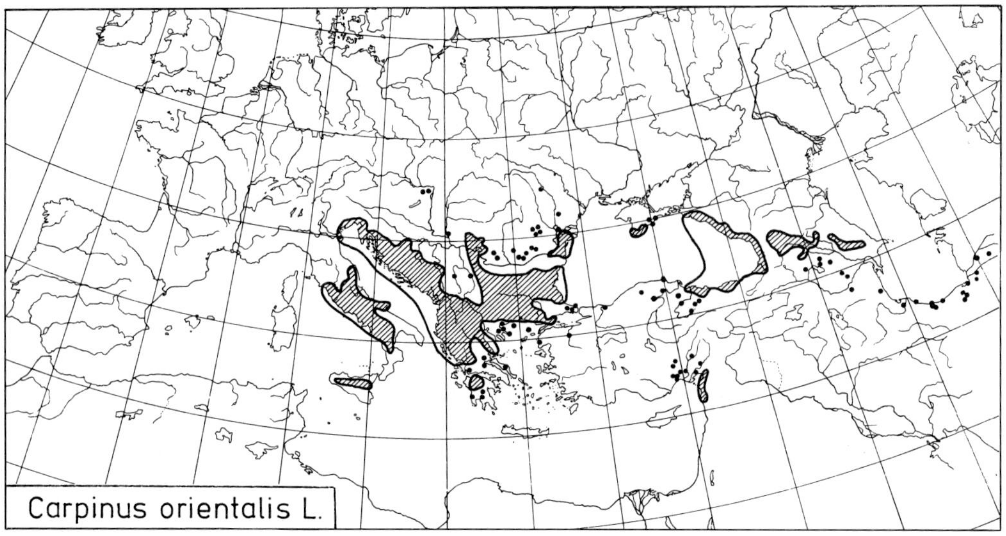

Fig. 5. Range of *Carpinus orientalis* L. (after BROWICZ & ZIELINSKI 1982, JALAS & SUOMINEN 1976, and DAVIS 1982)

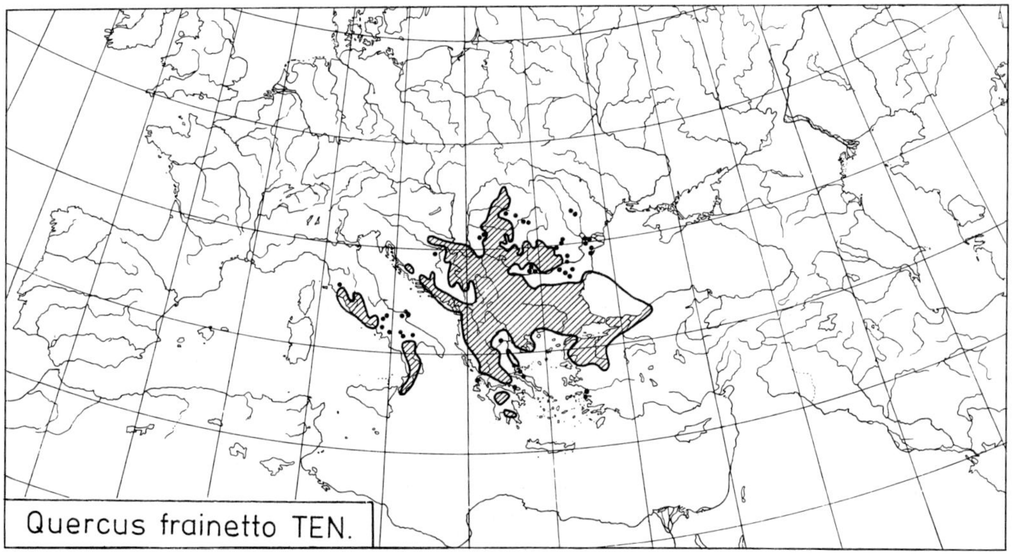

Fig. 6. Range of *Quercus frainetto* TEN. (after BROWICZ & ZIELINSKI 1982, JALAS & SUOMINEN 1976, FUKAREK 1970, ARRIGONI 1974, and DAVIS 1982)

A great part of the deciduous trees and shrubs of the nemoral-Submediterranean element are concentrated in the outermost eastern sector of the *Euonymus latifolia*-area, namely in the Hyrcanian and the Euxinian provinces. To this *Fagus orientalis*-type (Fig. 3) belong endemics like *Pterocarya fraxinifolia* or local endemics like *Staphylea colchica* (Fig. 4 and the contribution by K. BROWICZ in this volume). Further to the west extend the ranges of the Hyrcanian-Euxinian-Illyrian-Appennine species, e.g., *Continus coggygria* (K 273 d) and *Carpinus orientalis* (Fig. 5), and

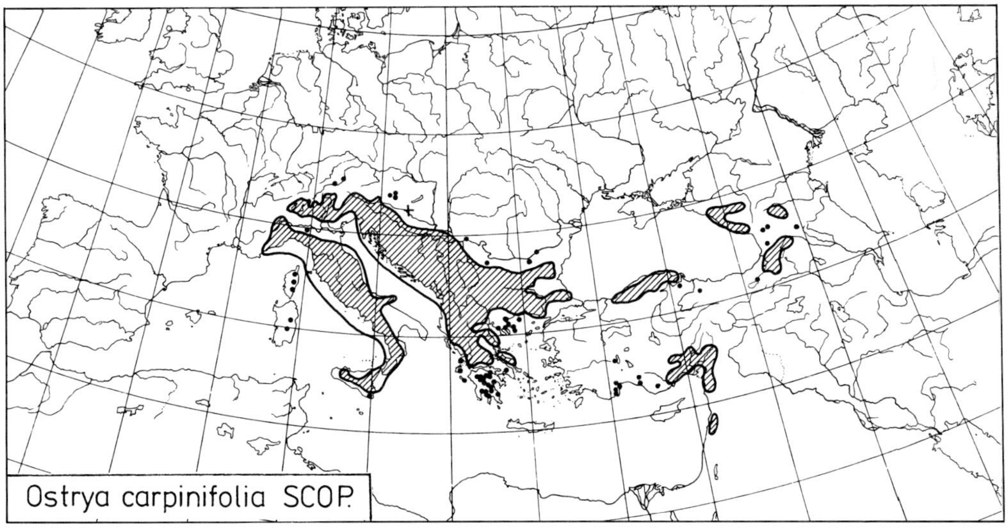

Fig. 7. Range of *Ostrya carpinifolia* SCOP. (after JALAS & SUOMINEN 1976, NIKLFELD 1973, BROWICZ & ZIELINSKI 1982, and DAVIS 1982)

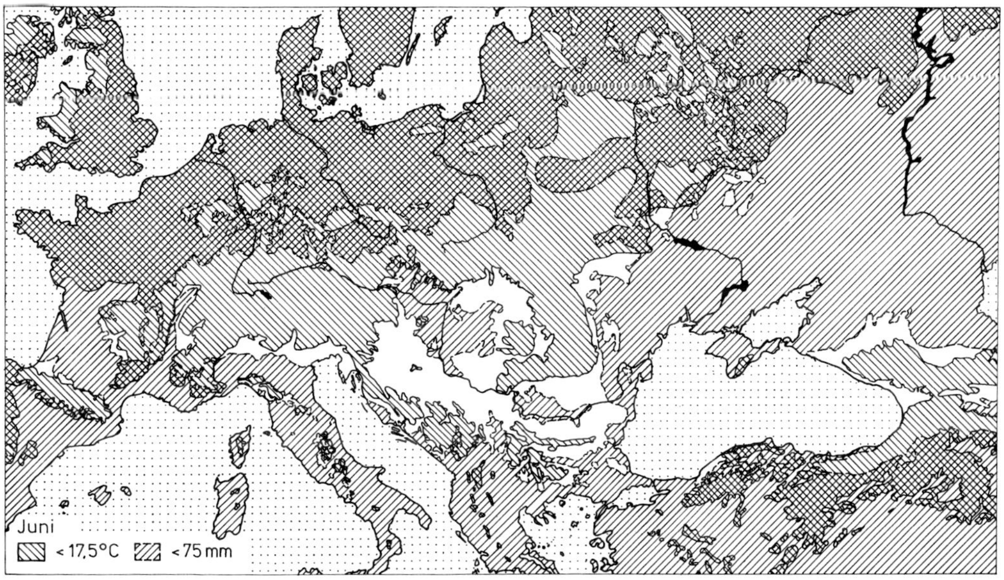

Fig. 8. Climatic map showing the summer-warm and summer-humid region that supports the more sensitive summer-green woody plants in Europe. (Compiled after Climatic atlas of Europe 1970)

some of them reach the southern Alps. All these taxa constitute an eastern group of the nemoral-Submediterranean flora.

Another range group is that of the W. Anatolian-Balkan-Illyrian-Apennine plants (*Staphylea*-type, Fig. 4; compare also *Quercus cerris:* K 123 a, *Qu. frainetto:* Fig. 6; *Ostrya carpinifolia:* Fig. 7, and *Fraxinus ornus:* K 346 c) which grow in the central part of the Submediterranean area. *Tilia tomentosa* (K 281 b) and *Syringa vulgaris* (Fig. 4) are confined to the centre of the *Staphylea*-type area, and *Aesculus hippocastanum* (K 276 a), *Syringa josikaea* (K 347 b) and *Forsythia* grow as endemics within parts of the territory.

Many herbaceous components of the Submediterranean nemoral vegetation have distribution patterns similar to those of the trees and shrubs. Perennials, like *Potentilla micrantha* (Fig. 2), *Viola alba* (K 292 b), *Festuca drymeia* (K 35 a), and *Ruscus hypophyllum* (K 98 B) correspond to the *Euonymus latifolia*-type. Euxinian-Balkan-Illyrian ranges like that of *Carpinus orientalis* are occupied by *Digitalis ferruginea* (K 393 c) and *Calamintha grandiflora* (K 381 d).

Epimedium (K 171 a) grows in three separated relic centres within the territory of the Submediterranean nemoral plants in the Hyrcanian-Euxinian (*E. pinnatum, E. pubigerum*), the northeastern Algerian (*E. perralderianum*), and the south Alpic-NW. Illyrian centre (*E. alpinum*). In the perimontaneous zone of the southern and southeastern Alps and in the NW. Illyrian province there is a striking concentration of herbaceous plants of the broad-leaved summer-green forests (*Epimedium alpinum, Omphalodes verna:* K 366 b, *Hacquetia epipactis:* K 304 b, *Lamium orvala:* K 376 b, *Pseudostellaria europaea:* K 143 a). The centre of diversity of *Galium* ser. *Nemoralia* (Fig. 9) extends from the southern Alps to the Illyrian, Balkan and S. Carpathian mountains, and thus corresponds with the area of the *Staphylea*-type.

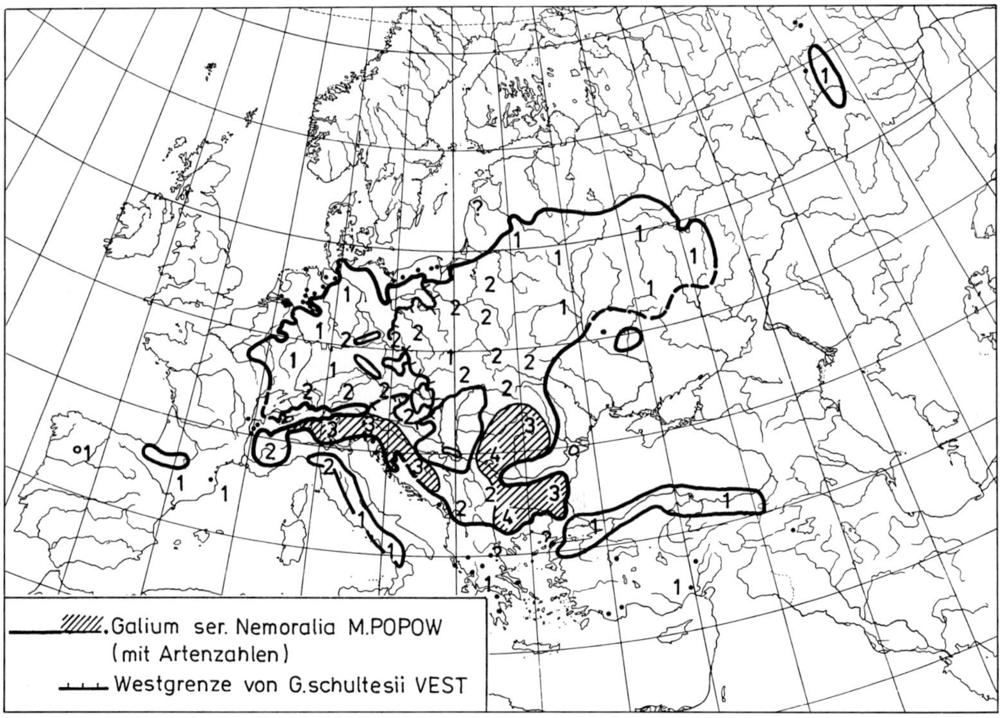

Fig. 9. Range of *Galium* ser. *Nemoralia* M. POPOW. Note the concentration of species in the Submediterranean subregion. In the easternmost part of the temperate range the group is only represented by *G. schultesii* VEST (after MEUSEL & JÄGER mscr.)

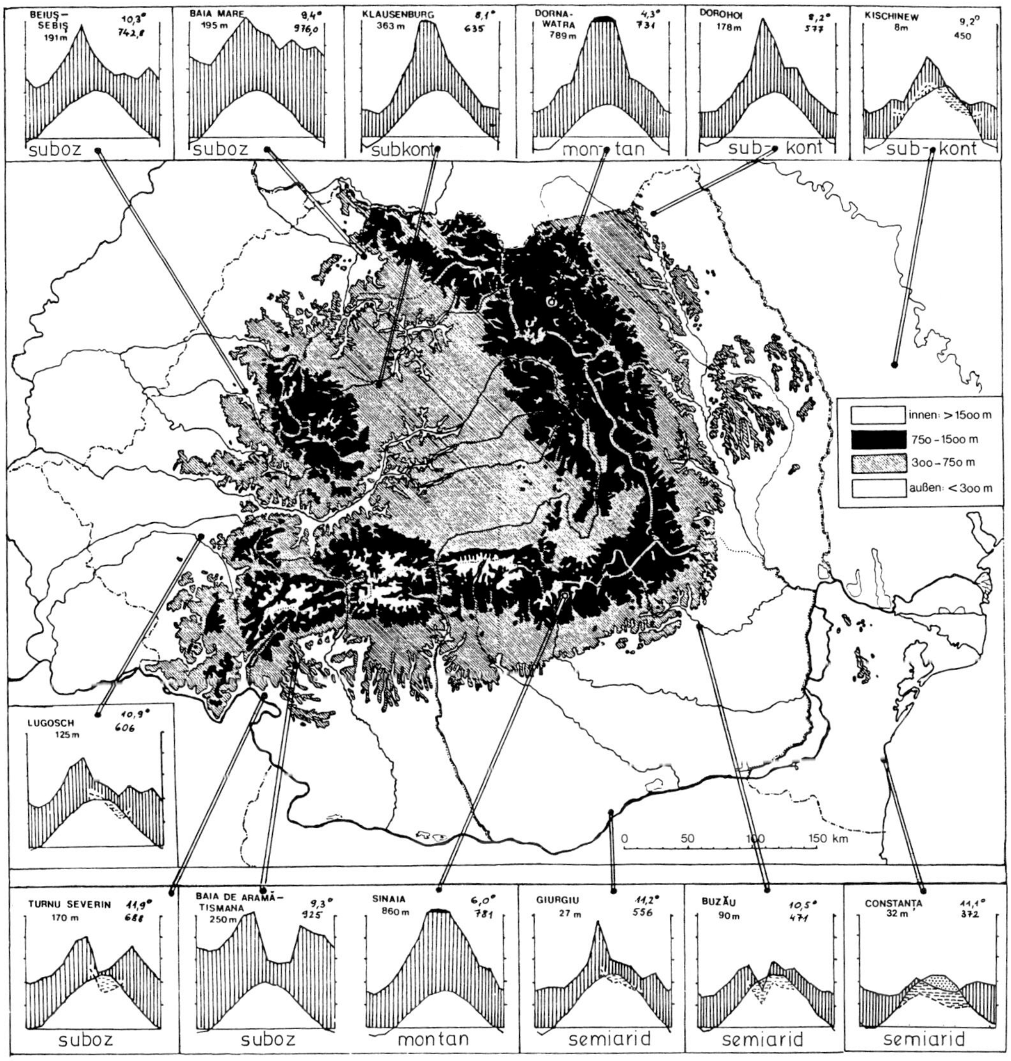

Fig. 10. Climatic diagrams in Roumania (after WALTER & LIETH 1960–67). The deviation from a "normal" diagram of the region is mentioned in the bottom of each diagram

Climatic conditions and distribution patterns

An ecological evaluation of the distribution patterns of the nemoral-Submediterranean element leads to the conclusion that these ranges are basically limited by climatic factors. To the south the distribution of summer-green broad-leaved forest vegetation is restricted by the hot and dry summer climate in the etesian region, where the climax vegetation is formed by sclerophyllous forests and maquis. To the north the temperature sums and/or the length of the growing season are not sufficient for the growth of more sensitive deciduous woody plants and associated herbs. The ranges of many Submediterranean taxa extend into the temperate zone in outpost-niches with favourable summer climate (e.g., *Staphylea pinnata* in the Rhenanian and middle Danubian countries, Fig. 4).

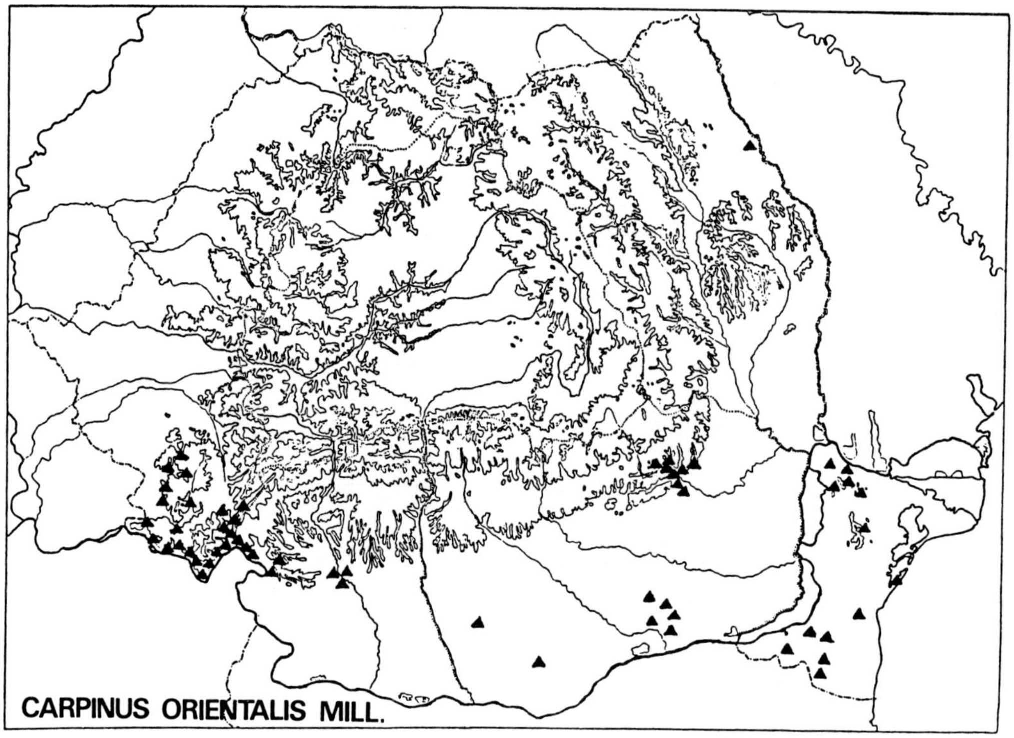

Fig. 11. Distribution of *Carpinus orientalis* MILL. in Roumania (after MEUSEL & NIEDERMAIER 1985)

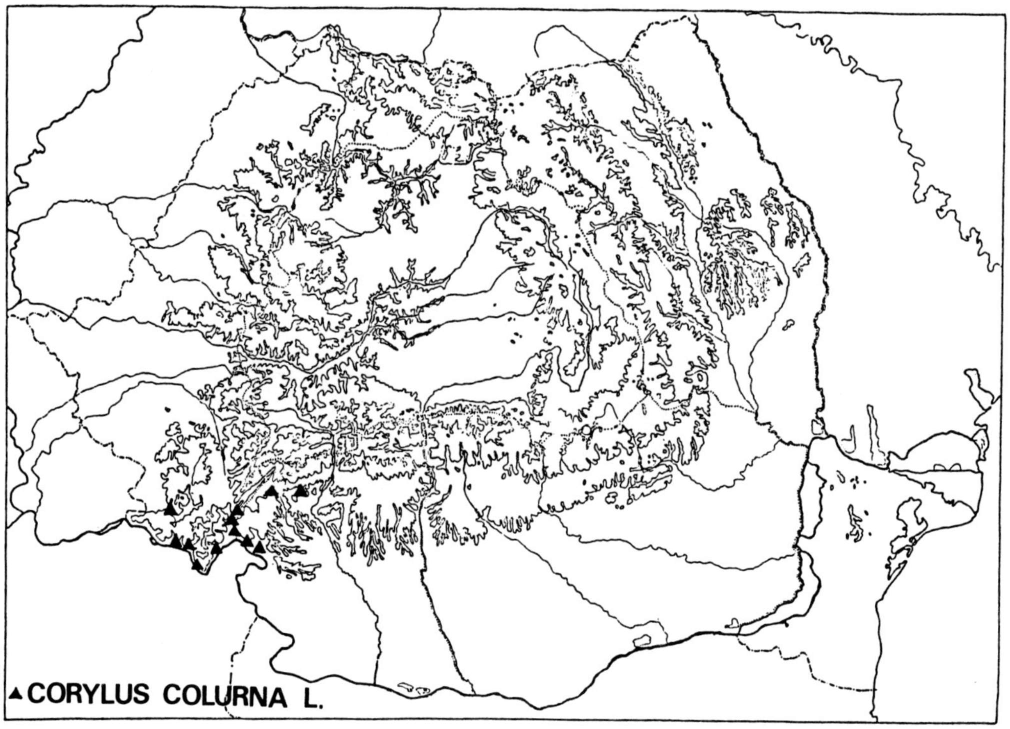

Fig. 12. Distribution of *Corylus colurna* L. in Roumania (after MEUSEL & NIEDERMAIER 1985)

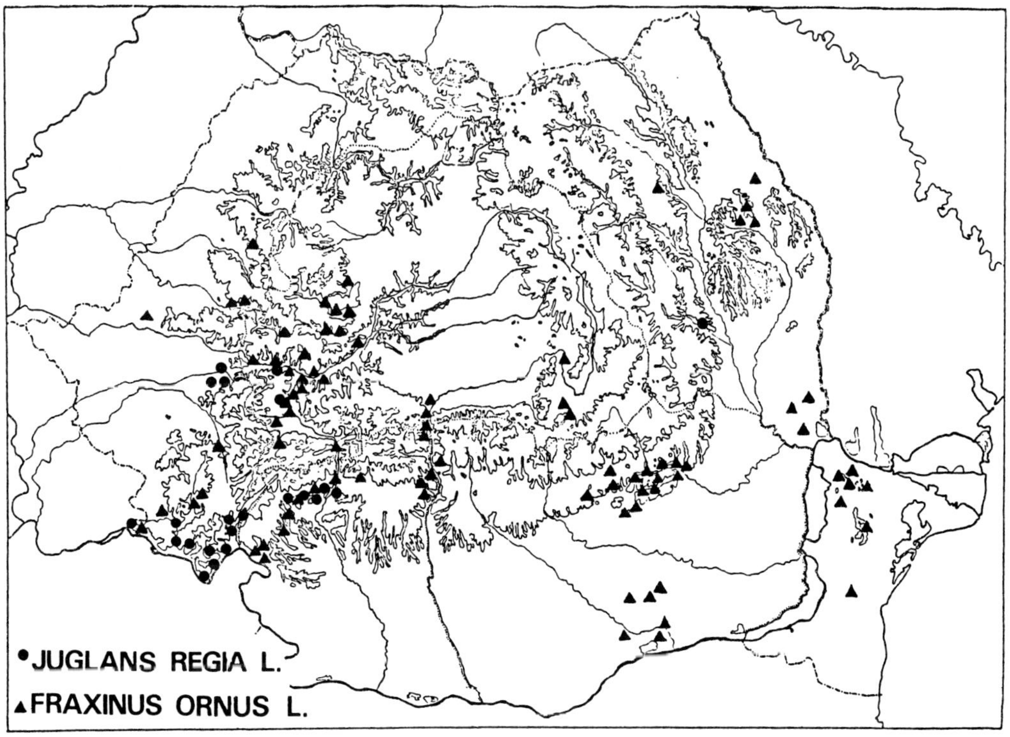

Fig. 13. Distribution of *Fraxinus ornus* L. and *Juglans regia* L. in Roumania (after MEUSEL & NIEDERMAIER 1985)

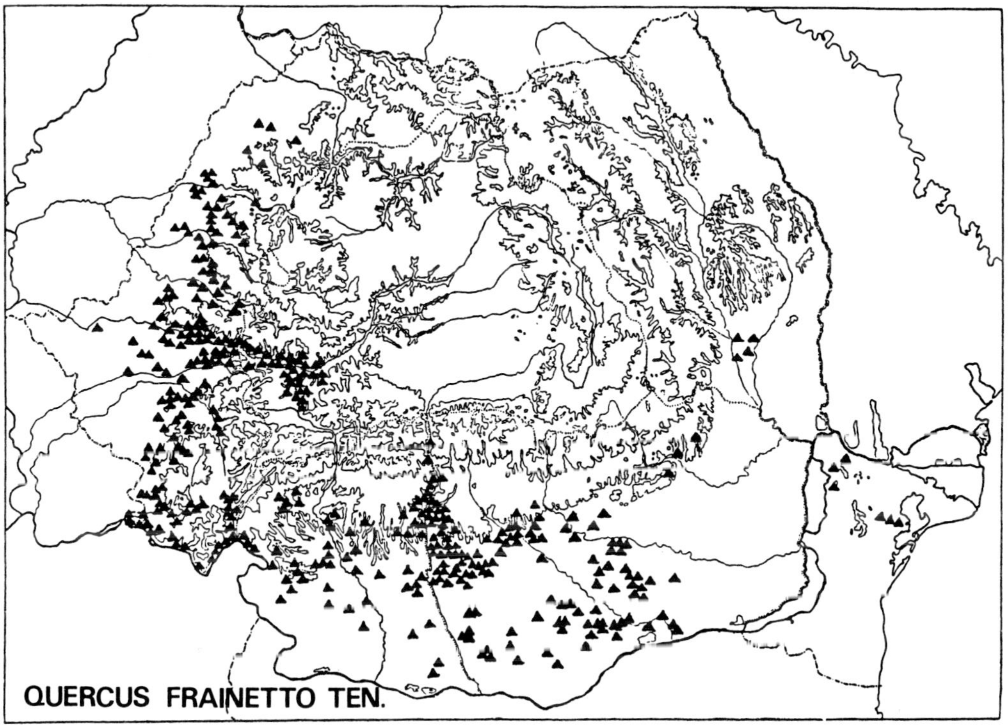

Fig. 14. Distribution of *Quercus frainetto* TEN. in Roumania (after MEUSEL & NIEDERMAIER 1985)

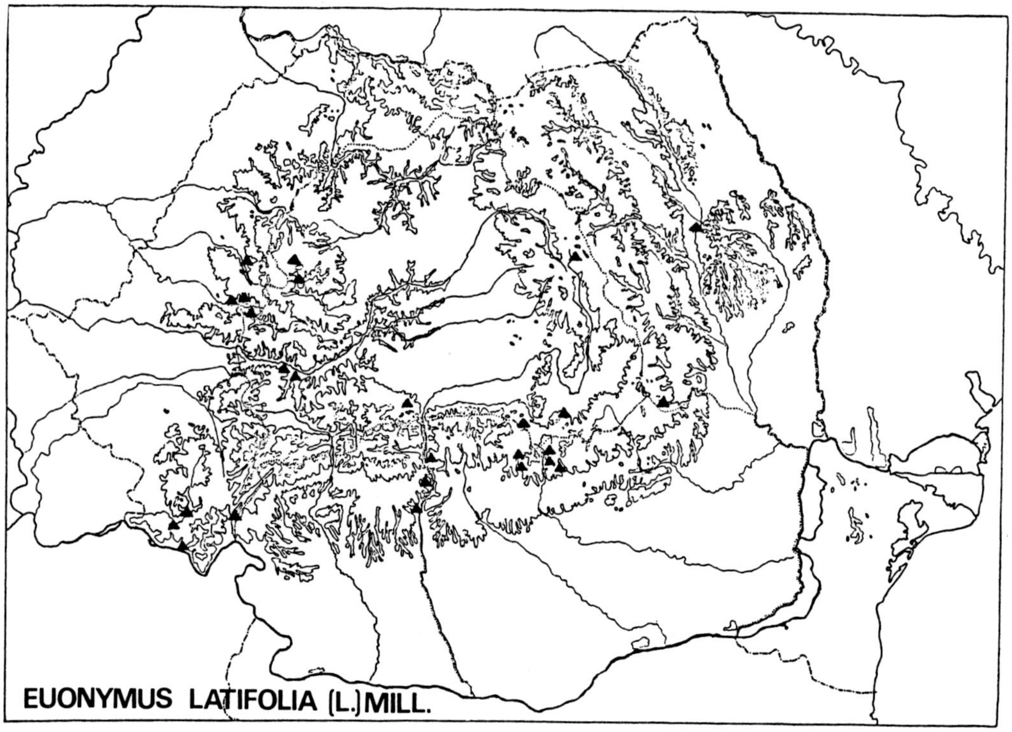

Fig. 15. Distribution of *Euonymus latifolia* (L.) MILL. in Roumania (after MEUSEL & NIEDERMAIER 1985)

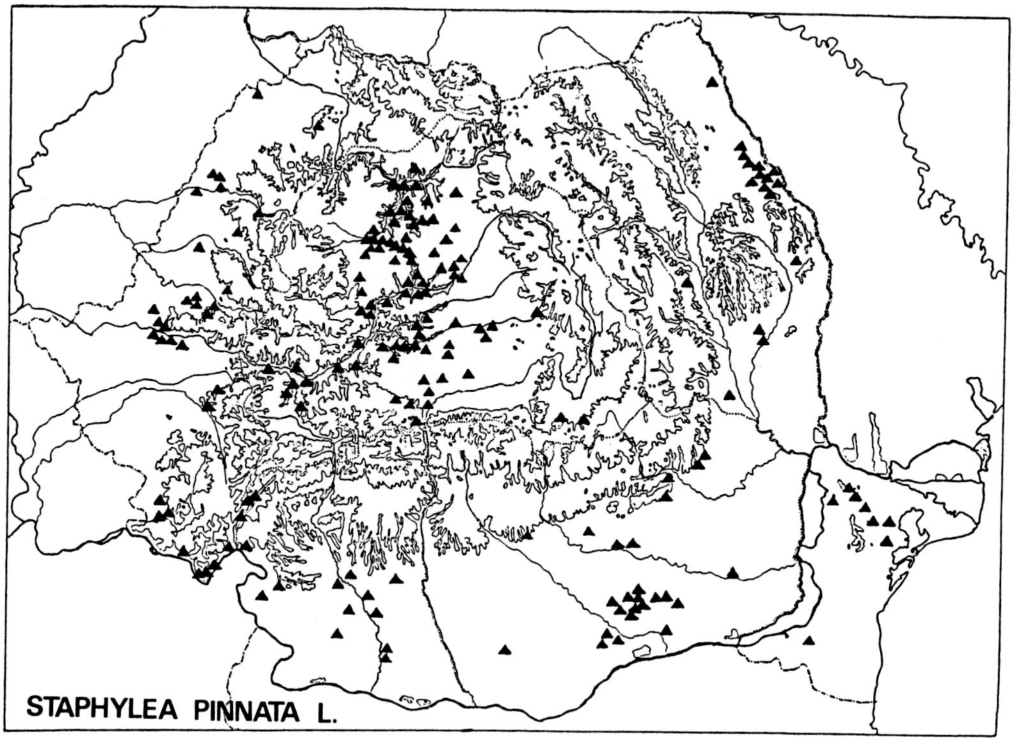

Fig. 16. Distribution of *Staphylea pinnata* L. in Roumania (after MEUSEL & NIEDERMAIER 1985)

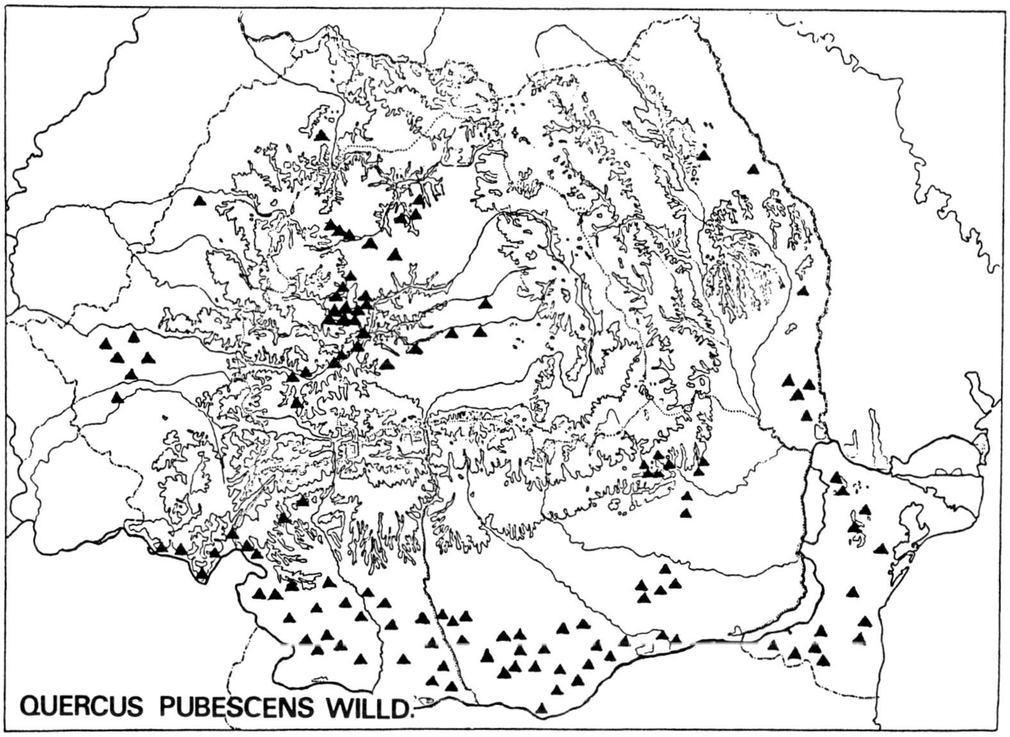

Fig. 17. Distribution of *Quercus pubescens* WILLD. in Roumania (after MEUSEL & NIEDERMAIER 1985)

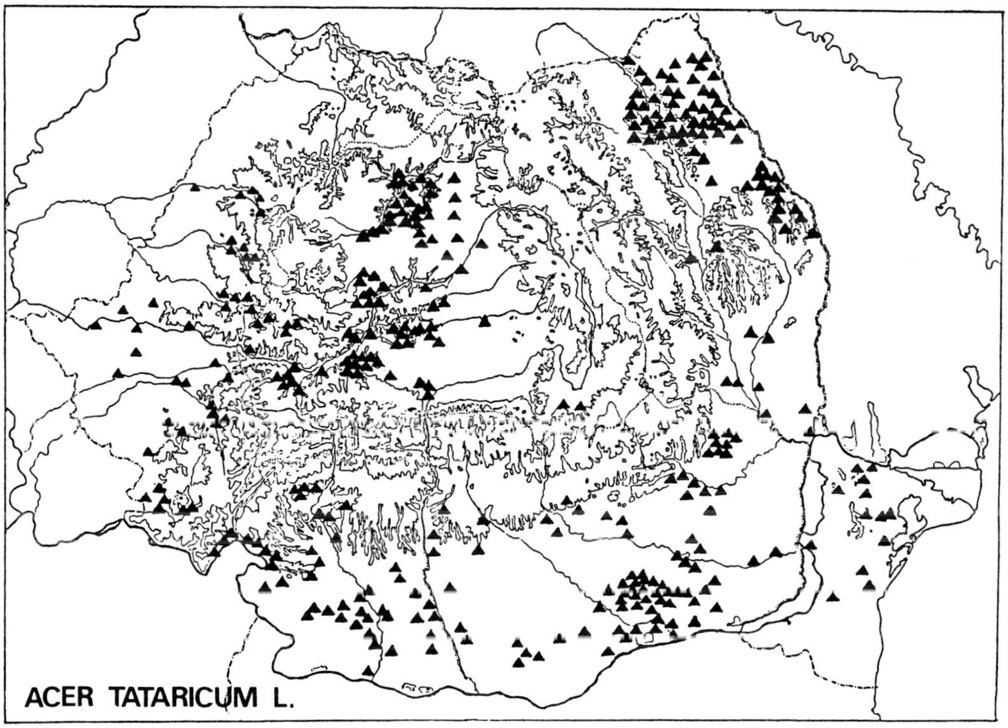

Fig. 18. Distribution of *Acer tataricum* L. in Roumania (after MEUSEL & NIEDERMAIER 1985)

Most territories of the Submediterranean subregion are influenced by the etesian climatic rhythm with semiarid conditions during summer time. Only in some mountain districts and in the Black Sea and the Caspian Sea surroundings the increased humidity allows more sensitive forest taxa to grow. In search of climatic parameters that indicate a situation favourable to deciduous forest plants we have to look for a combination of sufficient humidity and temperature during the vegetation period (and a cold winter!). An attempt to give an idea of the factors limiting the distribution of woody plants as *Staphylea pinnata* (Fig. 5) or *Quercus cerris* is presented in Fig. 8. Here, the isotherme of 17.5 °C average temperature of June combined with the isohyete of 75 mm rainfall for the same month are chosen to define regions with a rather warm and humid summer. This combination is found in Europe only where winters are cold. (Otherwise, one would have to add one of the winter isothermes and thus exclude the winter-mild regions, where the summer-greens are not able to compete). In comparing the distribution map Fig. 5 with Fig. 8 we find that the climatic lines not only run roughly parallel with the range limits, but are correlated with them even in details (e.g., outposts in Bessarabia, Insubria, Appennine Mts).

Clearly, we don't regard such correlation as a direct causal link. These isolines are only symbols of a climatic situation, that influences the competition-equilibrium directly or indirectly in the form of extremes of the meso- and microclimate.

Ecogeographical differentiation of summer-green trees on a local scale (Roumania)

With regard to the climatic diagrams we (MEUSEL & NIEDERMAIER 1985) have tried to study the differentiation of ecological characters of some nemoral-Submediterranean plants in the southern and eastern Carpathians of Roumania. The diagrams (Fig. 10) demonstrate variations within a generally suboceanic warm-temperate climate, favourable to broad-leaved forest vegetation. We recognize differences in the temperature from south to north and in the humidity from the western flank of the Biharian and S. Carpathian mountains to the inner Transsylvanian basin, the eastern and southeastern foothills of the Carpathians and the Bessarabian plains.

From a local aspect we have differentiated the climatic diagrams with regard to the deviation from a "normal" Submediterranean, suboceanic climate in this region, as "subcontinental" (more continental), "subarid" (more Mediterranean-like, dry during summer) and "montane' (with lower temperatures and high precipitations in summer). On the southern slope of the Carpathian mountains *Carpinus orientalis* (Fig. 11) extends from the suboceanic landscapes to the more summer-dry parts further in the east. The distribution of *Corylus colurna* (Fig. 12), *Juglans regia* (Fig. 13), *Fraxinus ornus* (Fig. 13), *Quercus frainetto* (Fig. 14) and *Quercus cerris* is resricted to the suboceanic western parts of the southern Carpathians and the Biharian mountains, but graded from south to north as a series of "equiformal progressive ranges".

Markedly montane is the distribution of *Euonymus latifolia* – both in the general (Fig. 1) and in the local (Fig. 15) distribution.

In contrast to the mentioned suboceanic taxa, *Staphylea pinnata* (Fig. 16), *Quercus pubescens* (Fig. 17) and *Acer tataricum* (Figs. 18, 19) show increasing subcontinental tendencies in their distribution pattern (concentration in the Transsylvanian

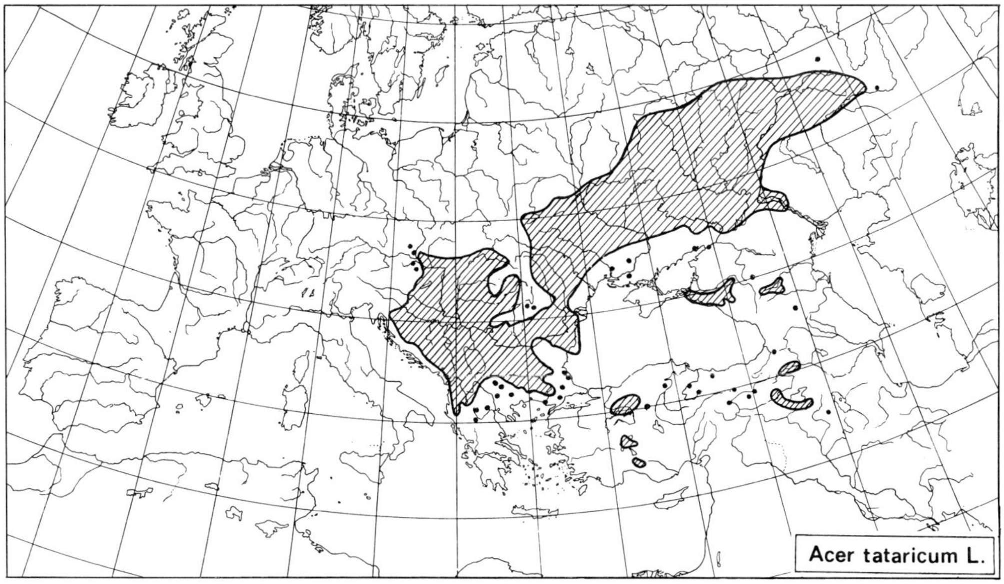

Fig. 19. Range of *Acer tataricum* L. (after MEUSEL & al. 1978, slightly modified)

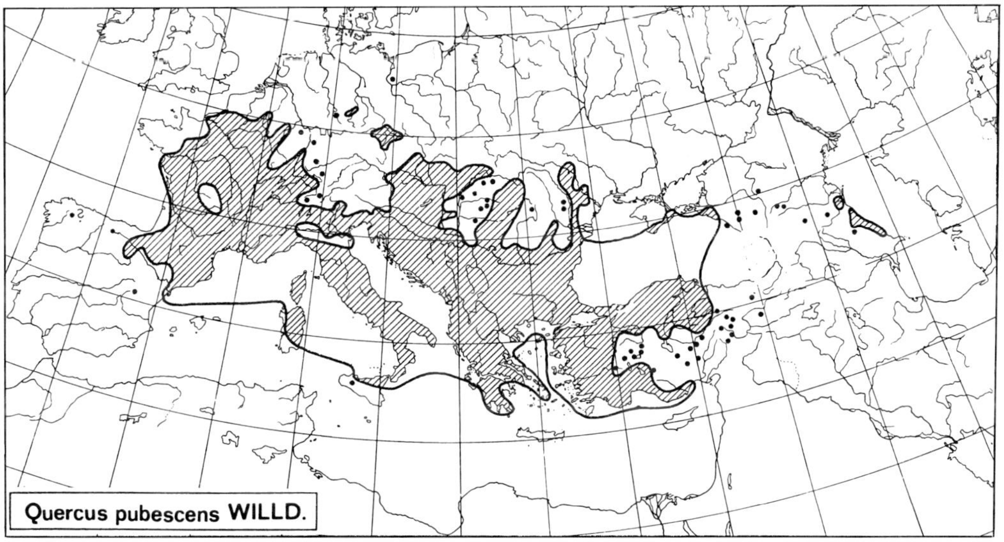

Fig. 20. Range of *Quercus pubescens* WILLD. (after MEUSEL & al. 1965, BROWICZ & ZIELINSKI 1982, and JALAS & SUOMINEN 1976)

basin and in the territories east of the Capathians and in Bessarabia). In this respect, *Acer tataricum* holds an extreme position, both in its local and general distribution.

When the climatic diagrams are compared with the distribution patterns, the strong influence of climatic factors on all these differentiations quickly becomes apparent. Both, in local and general distribution, we suggest that the ranges reflect recent ecological conditions (e.g., in climate), rather than migration routes, dispersal barriers or incomplete migrations.

Some of the nemoral-Submediterranean woody plants, like *Quercus pubescens* (Fig. 20) and *Viburnum lantana,* are not completely restricted to the Submediterranean zone, as they also occur in the southern part of the Middle European region. Here, too, climatic parameters indicate similar conditions but with a graded amplitude of tolerance limits. The northern limits of these taxa run parallel to the isothermes of spring or autumn months (e.g., May and September), or to the number of days with more than a certain temperature middle (5 °C, 10 °C, depending on the species), or to the temperature sums over a certain middle (in the case of *Quercus pubescens,* duration of a period with temperature above 12 °C, or April 8 °C, see JÄGER 1968, 1970). All these values symbolize the length and warmth of the growing season. Southern limits are easily paralleled with the isohyetes (e.g., 50 mm June – August) or humidity values (unfortunately, good isohyete maps for the single seasons do not exist).

Some Middle European taxa of the broad-leaved deciduous forest vegetation may be regarded as intermediate between the nemoral-Submediterranean and the typical European temperate forest element (e.g., *Acer campestre, Ulmus minor, Clematis vitalba*). The ranges of all these taxa include the Submediterranean subregion. The wide extension into the temperate zone, e.g., of *Fagus* (Fig. 3) or the *Galium Nemoralia*-group (Fig. 9), possibly can be attributed to the development of new ecological races during the Late Quaternary (*Fagus sylvatica* as a close relative of *F. orientalis, Galium schultesii* as a polyploid taxon of the *Nemoralia*-group).

Conclusions

In the view of the questions discussed in this symposium we conclude: (1) Chorological studies on extant taxa demonstrate close relations between distribution patterns and climatic conditions. Nemoral-Submediterranean species may be grouped into some characteristic range-types, but each taxon has its own ecogeographical character. (2) The same situation has existed in the past. Therefore, the explanation of fossil plant distribution and dispersal also has to consider ecological causes. (3) A good deal of the representatives of the Submediterranean and Middle European broad-leaved deciduous forest flora is restricted to territories with warm and humid summer conditions in the submeridional zone. The number of species of *Quercus, Acer* (K 276 b) or *Rhamnus* (K 278 d) of submeridional Europe, compared with the forest plants in the corresponding zone in East Asia, is not so strongly reduced as often been maintained. (4) Nearly all deciduous forest plants have their base in the submeridional zone. Their graded northern limits reflect differences in adaptation to climatic gradients. (5) Most of these forest species are sensitive to climatic conditions and strictly Submediterranean; only a few extend into the temperate Middle European region. Their characteristic northern and eastern limits are principally caused by climatic factors during postglacial dispersal, and not (as is often assumed) by migration barriers (e.g., west-east oriented mountain chaines).

References

ARRIGONI, P. V., 1974: Ricerche sulle querce caducifoglie italiane. 3. "*Quercus frainetto*" TEN. in Toscana. – Webbia **29**: 87 – 104.

BROWICZ, K., ZIELINSKI, J., 1982: Chorology of trees and shrubs in South-West Asia and adjacent regions. – Poznan: Polish Scient. Publishers.

Climatic atlas of Europe 1, 1970. (Technical supervisor: F. STEINHAUSER). – Budapest: UNESCO, Cartographia.

DAVIS, P. H., 1982: Flora of Turkey and the East Aegean Islands 7. – Edinburgh: Univ. Press.

FUKAREK, P., 1970: Bemerkungen zu einigen balkanischen und balkano-karpatischen Baum- und Straucharten. – Feddes Repert. **81**: 163–170.

JÄGER, E., 1968: Die pflanzengeographische Ozeanitätsgliederung der Holarktis und die Ozeanitätsbindung der Pflanzenareale. – Feddes Repert. **79**: 157–335.

– 1969: Die klimatischen Bedingungen des Areals der Dunklen Taiga und der sommergrünen Breitlaubwälder. – Ber. Deutsch. Bot. Ges. **81**: 397–408.

– 1970: Charakteristische Typen mediterran-mitteleuropäischer Pflanzenareale. – Feddes Repert. **81**: 67–92.

JALAS, J., SUOMINEN, J., (Eds.), 1976: Atlas Florae Europeae **3.** *Salicaceae* to *Balanophoraceae*. – Helsinki: Committee for Mapping the Flora of Europe.

MEUSEL, H., JÄGER, E., 1989: Vergleichende Chorologie der zentraleuropäischen Flora 3. – Jena: Fischer (in press).

– NIEDERMAIER, K., 1985: Die Breitlaubwälder des Süd- und Ostkarpatenraumes in vergleichend chorologischer Sicht. – Siebenbürgisches Archiv (3. Folge) **20**: 1–42.

– JÄGER, E., WEINERT, E., 1965: Vergleichende Chorologie der zentraleuropäischen Flora. – Jena: Fischer.

– – RAUSCHERT, S., WEINERT, E., 1978: Vergleichende Chorologie der zentraleuropäischen Flora. 2. – Jena: Fischer.

NIKLFELD, H., 1973: Über die Grundzüge der Pflanzenverbreitung in Österreich und einigen Nachbargebieten. – Verh. Zool.-Bot. Ges. Wien **113**: 53–69.

WALTER, H., LIETH, H., 1960–67: Klimadiagramm-Weltatlas. – Jena: Fischer.

Address of the authors: H. MEUSEL and E. J. JÄGER, Sektion Biowissenschaften, Geobotanik und Botanischer Garten der Martin-Luther-Universität, Neuwerk 21, DDR-402 Halle an der Saale, German Democratic Republic.

Verleger: Springer-Verlag KG, Mölkerbastei 5, A-1010 Wien. – Herausgeber: Prof. Dr. Friedrich Ehrendorfer, Vorstand des Instituts für Botanik und Direktor des Botanischen Gartens der Universität Wien, Rennweg 14, A-1030 Wien. – Redaktion: Rennweg 14, A-1030 Wien. – Hersteller: Adolf Holzhausens Nfg., Kandlgasse 19-21, A 1070 Wien. – Verlagsort: Wien. – Herstellungsort: Wien. – Printed in Austria.

Offenlegung gem. §25 Abs. 1 bis 3 Mediengesetz.

Unternehmensgegenstand: Verlag von wissenschaftlichen Büchern und Zeitschriften.

An der Springer-Verlag GmbH & Co. KG ist beteiligt: Dr. Konrad F. Springer, Mölkerbastei 5, A-1010 Wien, als Kommanditist zu 52,38%. Geschäftsführer: Dr. Konrad F. Springer, Prof. Dr. Dietrich Götze, Ing. Wolfram F. Joos, Dipl.-Kfm. Claus Michaletz und Rudolf Siegle, alle Mölkerbastei 5, A-1010 Wien.